W9-BMF-225

Engineering Fundamentals and Problem Solving

ALSO AVAILABLE FROM McGRAW-HILL SCHAUM'S OUTLINE SERIES IN ENGINEERING

Most outlines include basic theory, definitions and hundreds of example problems solved in step-by-step detail, and supplementary problems with answers.

RELATED TITLES ON THE CURRENT LIST INCLUDE:

Acoustics
Computer Graphics
Continuum Mechanics
Descriptive Geometry
Dynamic Structural Analysis
Elementary Statics & Strength of Materials
Engineering Economics
Engineering Mechanics
Fluid Dynamics
Fluid Mechanics & Hydraulics
Heat Transfer
Introductory Surveying
Lagrangian Dynamics
Linear Algebra
Machine Design
Mathematical Handbook of Formulas & Tables
Mechanical Vibrations
Operations Research
Programming with C
Programming with Fortran
Programming with Pascal
Reinforced Concrete Design
Statics & Mechanics of Materials
Strength of Materials
Structural Analysis
Structural Steel Design (LRFD Method)
Theoretical Mechanics
Thermodynamics with Chemical Applications
Vector Analysis

SCHAUM'S SOLVED PROBLEMS BOOKS

Each title in this series is a complete and expert source of solved problems with solutions worked out in step-by-step detail.

Related titles on the current list include:

3000 Solved Problems in Calculus
2500 Solved Problems in Differential Equations
3000 Solved Problems in Electric Circuits
2500 Solved Problems in Fluid Mechanics & Hydraulics
1000 Solved Problems in Heat Transfer
3000 Solved Problems in Linear Algebra
2000 Solved Problems in Mechanical Engineering Thermodynamics
2000 Solved Problems in Numerical Analysis
700 Solved Problems in Vector Mechanics for Engineers: Dynamics
800 Solved Problems in Vector Mechanics for Engineers: Statics

Available at most college bookstores, or for a complete list of titles and prices, write to:

Schaum Division
The McGraw-Hill Companies, Inc.
11 West 19th Street
New York, NY 10011

Third Edition

ENGINEERING FUNDAMENTALS AND PROBLEM SOLVING

Arvid R. Eide

Roland D. Jenison

Lane H. Mashaw

Larry L. Northup

Iowa State University

The McGraw-Hill Companies, Inc.

New York | St. Louis | San Francisco
Auckland | Bogotá | Caracas | Lisbon
London | Madrid | Mexico City | Milan
Montreal | New Delhi | San Juan
Singapore | Sydney | Tokyo | Toronto

McGraw-Hill
A Division of The McGraw-Hill Companies

ENGINEERING FUNDAMENTALS AND PROBLEM SOLVING

This book is printed on acid-free paper.

1 2 3 4 5 6 7 8 9 0 DOC DOC 9 0 9 8 7 6

ISBN 0-07-021306-2

This book was set in Century Schoolbook by York Graphic Services, Inc.
The editors were Eric M. Munson and John M. Morriss;
the production supervisor was Louise Karam.
The cover was designed by John Hite.
New drawings were done by Fineline.
Project supervision was done by York Production Services.
R. R. Donnelley & Sons Company was printer and binder.

Library of Congress Cataloging-in-Publication Data

Engineering fundamentals and problem solving / Arvid R. Eide.—3rd ed.
p. cm.
Includes bibliographical references and index.
ISBN 0-07-021306-2
1. Engineering. I. Eide, Arvid R.
TA147.E52 1997 96-38615
620—dc20

http://www.mhcollege.com

About the Authors

Arvid R. Eide is a native Iowan. He received his baccalaureate degree in mechanical engineering from Iowa State University in 1962. Upon graduation he spent two years in the U.S. Army as a commissioned officer and then returned to Iowa State as a full-time instructor while completing a master's degree in mechanical engineering. Professor Eide has worked for Western Electric, John Deere, and the Trane Company. He received his Ph.D. in 1974 and was appointed professor and chair of Freshman Engineering at Iowa State from 1974 to 1989. Dr. Eide was selected Associate Dean of Academic Affairs from 1989 to 1995 and currently serves as professor of Mechanical Engineering.

Roland D. Jenison is a professor of aerospace engineering and engineering mechanics at Iowa State University. He has 30 years of teaching experience in lower division engineering and engineering technology. He has taught courses in engineering design graphics and engineering problem solving and has published numerous papers in these areas. His scholarly activities include learning-based instruction in design graphics, design education, and design-build projects for lower division engineering students. He is a long-time member of ASEE and served as chair of the Engineering Design Graphics Division in 1986–1987.

Lane H. Mashaw earned his BSCE from the University of Illinois and MSCE from the University of Iowa. He served as a municipal engineer in Champaign, Ill., Rockford, Ill., and Iowa City, Ia., for nine years and then was in private practice in Decatur, Ill. for another nine years. He taught at the University of Iowa from 1964 to 1974 and at Iowa State University from 1974 until his retirement in 1987. He is currently emeritus professor of civil and construction engineering.

About the Author

Larry L. Northup is a professor of civil and construction engineering at Iowa State University. He has more than 30 years of teaching experience, with the past 20 years devoted to lower division engineering courses in problem solving, graphics, and design. He has 2 years of industrial experience and is a registered engineer in Iowa. He has been active in ASEE (Engineering Design Graphics Division), having served as chair of the Freshman Year Committee and Director of Technical and Professional Committees (1981–1984). He also served as chair of the Freshman Programs Constituent Committee (now Division) of ASEE in 1983–1984.

Contents

Preface

TO THE STUDENT

As you begin the study of engineering you are no doubt filled with enthusiasm, curiosity, and a desire to succeed. Your first year will be spent primarily establishing a solid foundation in mathematics, basic sciences, and communications. You may at times question what the benefits of this background material are and when actual engineering experiences will begin. We believe that they begin now. Additionally, we believe that the material presented in this book will provide you a fundamental understanding of how engineers function in today's technological world.

TO THE INSTRUCTOR

Engineering courses for first-year students continue to be in a state of transition. A broad set of course goals, including coverage of prerequisite material, motivation and retention have spawned a variety of first-year activity. The traditional engineering drawing and descriptive geometry courses have been largely replaced with computer graphics and CAD-based courses. Courses in introductory engineering and problem solving are now utilizing spreadsheets and mathematical solvers in addition to teaching the rudiments of a computer language. The World Wide Web (WWW) is rapidly becoming a major instructional tool, providing a wealth of data to supplement your class notes and textbooks. This third edition continues the authors' intent to introduce the profession of engineering and provide students with many of the tools needed to succeed.

Since 1974, students at Iowa State University have taken a computations course that has a major objective of improving their problem-solving skills. Various computational aids have been used from programmable calculators to networked PC's. The third edition of this text has evolved from more than twenty years of experience with teaching engineering problem-solving to thousands of first-year students.

The third edition has the same four broad objectives as the first two editions: (1) to motivate engineering students during their first year when exposure to engineering subject matter is limited; (2) to provide them with experience in solving problems in both SI and customary units while presenting solutions in a logical manner; (3) to introduce students to subject areas common to most engineering disciplines that require the application of fundamental engineering concepts; and (4) to develop their skills in solving open-ended problems.

The material in this book is presented in a manner that allows each instructor to emphasize certain aspects more than others without loss of continuity. The problems that follow most chapters vary in difficulty, so that students can experience success rather quickly and still be challenged as problems become more complex. Most problems in the third edition have been modified and there are more computer-oriented problems.

There is sufficient material in the first thirteen chapters for a three-credit semester course. By omitting selected chapters and/or by varying coverage from term to term, you can present a sound computations course for a two- to four-credit quarter course or a two- to three-credit semester course. Expanded efforts in computer programming and coverage of Chapter 14 on design provide sufficient material for a one-year sequence.

The book can be conveniently divided into six parts. The first, an introduction to engineering, begins with a description and breakdown of the engineering profession. Material concerning most disciplines in engineering are included in the third edition. If a formal orientation course is given separately, Chapter 1 can be simply a reading assignment. Chapters 2 and 3 provide procedures for approaching an engineering problem, determining the necessary data and method of solution, and presenting the results. The authors have found that emphasis in this area will reap benefits when the material and problems become more difficult later on.

Part two, chapters 4, 5 and 7 in the third edition includes engineering estimations, dimensions and units (including both customary and SI units) and statistics. Throughout the book, discussions and example problems tend to emphasize SI metric so that coverage of Chapter 5 is advisable. Other dimension systems do appear in the discussion and many problems contain non-metric units, so that students are exposed to conversions and to units that are commonly used. The inclusion of statistics strengthens the student's ability to deal with data manipulation and interpretation.

The third segment is a preparation for computer solutions. If your program provides separate programming experience, this material can be omitted, although work with flowcharts, independent of a particular computing language, can improve a student's ability to logically think through problem solutions. Because of the variety of computational equipment available to students on engineering campuses, the authors believe that the treatment of the many computer languages would be impractical within this text. This in no way implies that programming is not an important part of the engineering student's course of study. Students should be able to obtain calculated results in the most economical, efficient manner.

A fourth part, Chapter 8, is an introduction to TQM, intended to provide entering students a first look at total quality. It is structured to provide: (1) a history and rationale for its importance; (2) an abbreviated set of definitions and tools; (3) a connection to problem solving, since it is a chapter within a problem solving book; and (4) an introduction to team dynamics. Many new engineering students are placed in teams for collaborative learning, solution of design projects and so forth but are seldom provided any training on what to do or what to expect. This chapter focuses on team interaction and provides you an opportunity to place students in teams, let them do an MBTI exercise and then discuss team activities.

Part 5 allows you a great deal of flexibility. The time available and your personal interests and objectives dictate to what depth any or all of Chapters 9, 10, 11, 12, and 13 can be covered. These chapters have been strengthened by addition of new topics and a greater emphasis on computer solutions. More of the problems are suitable for computer solution if that is appropriate for your course. Material in this part may be covered in any order, since no chapter depends on another for background.

The last part, the design process, is a logical extension of the fundamental problem-solving approach in engineering. A ten-step design process is explained and supplemented with an actual preliminary design performed by a first-year student team. The process as described allows you to supplement the text material with personal examples to bring design experience into the classroom.

Mathematical expertise beyond algebra, trigonometry and analytical geometry is not required for any material in the book. The authors have found, however, that providing additional experience in pre-calculus mathematics is important at this stage of the student's education.

A solution manual is available that contains solutions to most end-of-chapter problems.

ACKNOWLEDGMENTS

The authors are indebted to many who assisted in the development of this edition of the textbook. First we would like to thank the faculty of the former Division of Engineering Fundamentals and Multidisciplinary Design at Iowa State University who have taught the engineering computations courses over the past twenty years. They, with support of engineering faculty from other departments, have made the courses a success by their efforts. Several thousands of students have taken the courses, and we want to thank them for their comments and ideas that have influenced this edition. The many suggestions of faculty and students alike have provided us with much information necessary to improve the previous editions. A special thanks to the reviewers for this edition whose suggestions were extremely valuable and greatly shaped the manuscript. We also express grateful appreciation to Jane Stowe who worked many hours to type the manuscript. Finally we thank our families for their constant support of our efforts.

Arvid R. Eide
Roland D. Jenison
Lane H. Mashaw
Larry L. Northup

Engineering Fundamentals and Problem Solving

1 The Engineering Profession

1.1 Introduction

The rapidly expanding sphere of science and technology may seem overwhelming to the individual seeking a career in a technological field. A technical specialist today may be called either engineer, scientist, technologist, or technician, depending upon education, industrial affiliation, or specific work. For example, nearly 350 colleges and universities offer engineering programs accredited by the Accreditation Board for Engineering and Technology (ABET) or the Canadian Engineering Accreditation Board (CEAB). Included are such traditional specialties as aerospace, agricultural, chemical, civil, electrical, industrial, and mechanical engineering—as well as the expanding areas of computer, energy, environmental, and materials engineering. Programs in construction engineering, engineering science, mining engineering, and petroleum engineering add to a lengthy list of career options in engineering alone. Coupled with thousands of programs in science and technical training offered at hundreds of other schools, the task of choosing the right field no doubt seems formidable.

Since you are reading this book, we assume that you are interested in studying engineering or at least are trying to decide whether or not to do so. Up to this point in your academic life, you have probably had little experience with engineering and have gathered your impressions of engineering from advertising materials, counselors, educators, and perhaps a practicing engineer or two. Now you must investigate as many careers as you can as soon as possible to be sure of making the right choice.

The study of engineering requires a strong background in mathematics and the physical sciences. Section 1.5 discusses typical areas of study within an engineering program that lead to the bachelor's degree. You should also consult with your counselor about specific course requirements. If you are enrolled in an engineering college but have not chosen a specific discipline, consult with an adviser or someone on the engineering faculty about particular course requirements in your areas of interest.

When considering a career in engineering or any closely related fields, you should explore the answers to several questions. What is engineering? What is an engineer? What are the func-

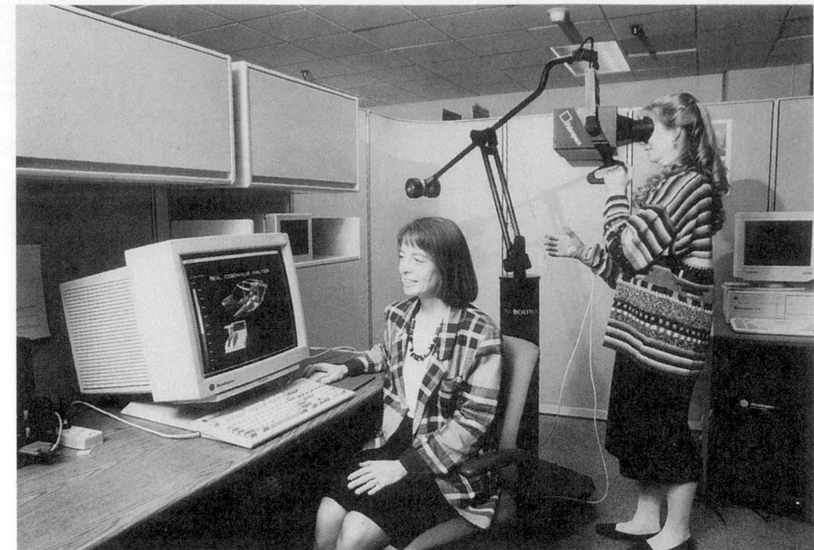

Figure 1.1
An engineering student observes stress formation in an automobile frame with the aid of a virtual reality device. *(Iowa State University.)*

tions of engineering? What are the engineering disciplines? Where does the engineer fit into the technical spectrum? How are engineers educated? What is meant by professionalism and engineering ethics? What have engineers done in the past? What are engineers doing now? What will engineers do in the future? Finding answers to such questions will assist you in assessing your educational goals and obtaining a clearer picture of the technological sphere.

Brief answers to some of these questions are given in this chapter. By no means are they intended to be a complete discussion of engineering and related fields. You can find additional and more detailed technical career information in the reference materials listed in the bibliography at the end of the book and by searching the World Wide Web.

1.2 The Technology Team

In 1876, 15 men led by Thomas Alva Edison gathered in Menlo Park, New Jersey, to work on "inventions." By 1887, the group had secured over 400 patents, including ones for the electric light bulb and the phonograph. Edison's approach typified that used for early engineering developments. Usually one person possessed nearly all the knowledge in one field and directed the research, development, design, and manufacture of new products in this field.

Today, however, technology has become so advanced and sophisticated that one person cannot possibly be aware of all the intricacies of a single device or process. The concept of systems engineering has thus evolved—that is, technological problems are studied and solved by a technology team.

Scientists, engineers, technologists, technicians, and craftspersons form the *technology team.* The functions of the team range across what is often called the *technical spectrum*. At one end of the spectrum are functions which involve work with scientific and engineering principles. At the other end of this technical spectrum are functions which bring designs into reality. Successful technology teams use the unique abilities of all team members to bring about a successful solution to a human need.

Each of the technology team members has a specific function in the technical spectrum, and it is of utmost importance that each specialist understand the role of all team members. It is not difficult to find instances where the education and tasks of team members overlap. For any engineering accomplishment, successful team performance requires cooperation that can be realized only through an understanding of the functions of the technology team. We will now investigate each of the team specialists in more detail. The technology team is one part of a larger team which has the overall responsibility for bringing a device, process, or system into reality. This team, frequently called a project or design team, may include, in addition to the technology team members, managers, sales representatives, field service persons, financial representatives, and purchasing personnel. These project teams meet frequently from the beginning of the project to insure that schedules and design specifications are met, and that potential problems are diagnosed early. This approach, intended to meet or exceed the customer's expectations, is referred to as total quality management (TQM) or continuous improvement (CI). This approach is discussed in Chapter 8.

1.2.1 Scientist

Scientists have as their prime objective increased knowledge of nature (see Fig. 1.2). In the quest for new knowledge, the scientist conducts research in a systematic manner. The research steps referred to as the *scientific method* are often summarized as follows:

1. Formulate a hypothesis to explain a natural phenomenon.
2. Conceive and execute experiments to test the hypothesis.
3. Analyze test results and state conclusions.
4. Generalize the hypothesis into the form of a law or theory if experimental results are in harmony with the hypothesis.
5. Publish the new knowledge.

An open and inquisitive mind is an obvious characteristic of a scientist. Although the scientist's primary objective is that of obtaining an increased knowledge of nature, many scientists are also engaged in the development of their ideas into new and use-

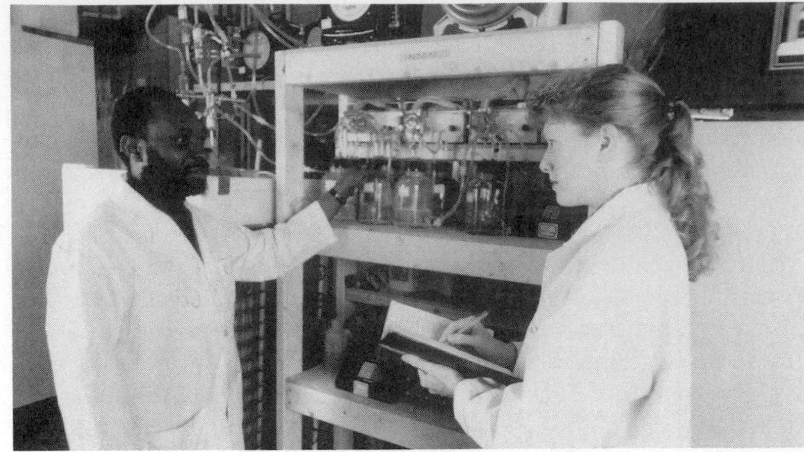

Figure 1.2
Scientific research conducted in an environmental laboratory. *(Iowa State University.)*

ful creations. But to differentiate quite simply between the scientist and engineer, we might say that the true scientist seeks to understand more about natural phenomena, whereas the engineer primarily engages in applying new knowledge.

1.2.2 Engineer

The profession of engineering takes the knowledge of mathematics and natural sciences gained through study, experience, and practice and applies this knowledge with judgment to develop ways to utilize the materials and forces of nature for the benefit of all humans.

An engineer is a person who possesses this knowledge of mathematics and natural sciences, and through the principles of analysis and design, applies this knowledge to the solution of problems and the development of devices, processes, structures, and systems for the benefit of all humans.

Both the engineer and scientist are thoroughly educated in the mathematical and physical sciences, but the scientist primarily uses this knowledge to acquire new knowledge, whereas the engineer applies the knowledge to design and develop usable devices, structures, and processes. In other words, the scientist seeks to know, the engineer aims to do.

You might conclude that the engineer is totally dependent on the scientist for the knowledge to develop ideas for human benefit. Such is not always the case. Scientists learn a great deal from the work of engineers. For example, the science of thermodynamics was developed by a physicist from studies of practical steam engines built by engineers who had no science to guide them. On the other hand, engineers have applied the principles of nuclear fission discovered by scientists to develop nuclear power plants and numerous other devices and systems requiring nuclear reactions for their operation. The scientist's and engineer's functions frequently overlap, leading at times to a somewhat blurred image of the engineer. What distinguishes the engineer from the scientist in broad terms, however, is that the engineer often conducts research, but does so for the purpose of solving a problem.

The end result of an engineering effort—generally referred to as *design*—is a device, structure, system, or process which satisfies a need. A successful design is achieved when a logical procedure is followed to meet a specific need. The procedure, called the *design process,* is similar to the scientific method with respect to a step-by-step routine, but it differs in objectives and end results. The design process encompasses the following activities, all of which must be completed.

1. Identification of a need
2. Problem definition
3. Search
4. Constraints
5. Criteria
6. Alternative solutions
7. Analysis
8. Decision
9. Specification
10. Communication

In the majority of cases, designs are not accomplished by an engineer simply completing the 10 steps shown in the order given. As the designer proceeds through each step, new information may be discovered and new objectives may be specified for the design. If so, the designer must backtrack and repeat steps. For example, if none of the alternatives appear to be economically feasible when the final solution is to be selected, the designer must redefine the problem or possibly relax some of the criteria to admit less expensive alternatives. Thus, because decisions must frequently be made at each step as a result of new developments or unexpected outcomes, the design process becomes iterative.

It is very important that you begin your engineering studies with an appreciation of the thinking process used to arrive at a solution to a problem and ultimately to produce a successful result.

As you progress through your engineering education you will solve problems and learn the design process using the techniques of analysis and synthesis. Analysis is the act of separating a system into its constituent parts, whereas synthesis is the act of combining parts into a useful system. In the design process (Chapter 14), you will observe how analysis and synthesis are utilized to generate a solution to a human need.

Consider the cruise control in an automobile as a system. You can analyze the performance of this system by setting up a test under carefully controlled conditions—that is, you will define and control the operating environment for the system and note the performance of the system. For example, you may determine acceleration or deceleration when a speed change is requested by the driver. You may check to see if the speed returns to the desired level after braking to reduce the speed and using the resume control. During the design of a cruise control system, the system would be modeled on a computer, and performance would be predicted by adjusting the variables and observing the results through various graphical formats on the monitor. You can analyze the physical makeup of the cruise control by actually taking apart the control, identifying the parts according to form and function, and reassembling the control. In general, analysis is the taking of a system, establishing the operating environment, and determining the response (performance) of the system.

If you were attempting to design a new cruise control system, you would consider many methods for sensing speed, ways to adjust engine speed for acceleration and deceleration, ideas for driver interface with the control, and so forth. Many possible solutions will be generated, mostly in the form of conceptual solutions without the details. During the design phase, the computer model may be continually improved by "repeated analysis," that is, finding the best or optimum design by observing the effect of changes in the system variables. This is synthesis and is the inverse of analysis. Synthesis may be said to be the process of defining the desired response (performance) of a system, establishing the operating environment, and, from this, developing the system.

An example will illustrate.

Example problem 1.1 A protective liner exactly 12 m wide is available to line a channel for conveying water from a reservoir to downstream areas. If a trapezoidal-shaped channel (see Fig. 1.3) is constructed so that the liner will cover the surface completely, what is the flow area for $x = 2$ m and $\theta = 45°$? The geometry is defined such that $0 \le x \le 6$ and $0 \le \theta \le 90°$. Flow area multiplied by average flow velocity will yield volume rate of flow, an important parameter in the study of open-channel flows.

Solution The geometry is defined in Fig. 1.3. The flow area is given by the expression for the area of a trapezoid:

$$A = \tfrac{1}{2}(b_1 + b_2)h$$

where $b_1 = 12 - 2x$

$$b_2 = 12 - 2x + 2x\cos\theta$$

$$h = x\sin\theta$$

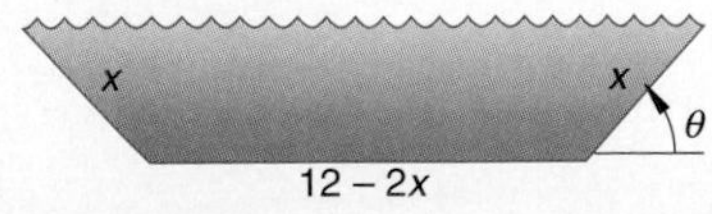

Figure 1.3

Therefore,

$$A = 12x \sin\theta - 2x^2 \sin\theta + x^2 \sin\theta \cos\theta$$

For the situation where $x = 2$ and $\theta = 45°$, the flow area is

$$A = (12)(2)(\sin 45°) - 2(2)^2(\sin 45°) + (2)^2(\sin 45°)(\cos 45°)$$
$$= 13.3 \text{ m}^2$$

Values of A can be quickly found for any combination of x and θ with a spreadsheet. Fig. 1.4a shows areas for $x = 2$ and a series of θ values. You have solved many problems of this nature by analysis; that is, a system is given (the channel as shown in Fig. 1.3), the operating environment is specified (the channel is flowing full), and you must find the system performance (determine the flow area). Analysis usually yields a unique solution.

Example problem 1.2 A protective liner exactly 12 m wide is available to line a channel conveying water from a reservoir to downstream areas. For the trapezoidal cross section shown in Fig. 1.3, what are the values of x and θ for a flow area of 16 m^2?

Solution Based on our work in Example prob. 1.1, we would have

$$16 = 12x \sin\theta - 2x^2 \sin\theta + x^2 \sin\theta \cos\theta$$

The solution procedure is not direct, and the solution is not unique, as it was in Example prob. 1.2. We begin our solution procedure by using a spreadsheet to generate a family of curves that illustrate the behavior of the implicit function of x and θ. Figure 1.4b shows the flow area as a function of θ for five values of x. We quickly observe that for $x = 1$ we cannot generate a flow area of 16 m^2. Also for $x = 4$ and 5, we definitely have two values of θ where a flow of 16 m^2 is possible. The spreadsheet will perform a search for the correct values. Figure 1.4c shows the result of a search between 0 and 90 at $x = 3$ m. In this situation a flow area of 16 m^2 occurs at a θ of 0.698014 radians or 40.0°. Other results are quickly obtainable by simply changing the value of x in the spreadsheet program in Fig. 1.4c.

You probably have not solved many problems of this nature. Example prob. 1.2 is a synthesis problem; that is, the operating environment is specified (channel flowing full), the performance is known (flow area is 16 m^2), and you must determine the system (values for x and θ). Example prob. 1.2 is the inverse problem to Example prob. 1.1. In general, synthesis problems do not have a unique solution, as can be seen from Example prob. 1.2.

Most of us have difficulty synthesizing. We cannot "see" a direct method to find an x and θ that yield a flow area of 16 m^2. Our solution to Example prob. 1.2 involved repeated analysis to "synthesize" the solution. We studied a family of curves (x is constant) of A versus θ, which enabled us to verify the spreadsheet analysis for values of x and θ that yield a specified flow area.

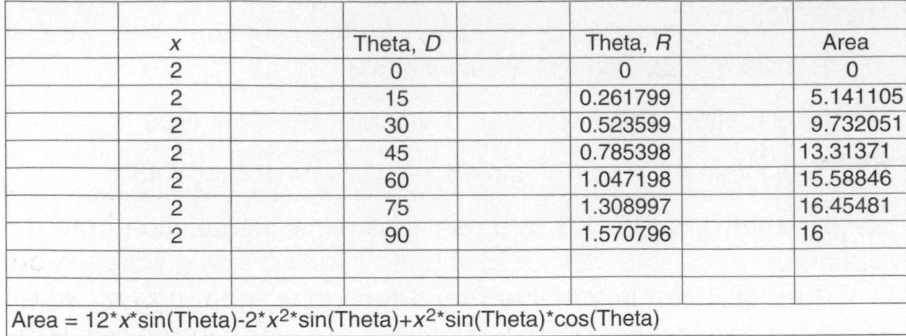

	x		Theta, D		Theta, R		Area
	2		0		0		0
	2		15		0.261799		5.141105
	2		30		0.523599		9.732051
	2		45		0.785398		13.31371
	2		60		1.047198		15.58846
	2		75		1.308997		16.45481
	2		90		1.570796		16
Area = 12*x*sin(Theta)-2*x²*sin(Theta)+x²*sin(Theta)*cos(Theta)							

(a)

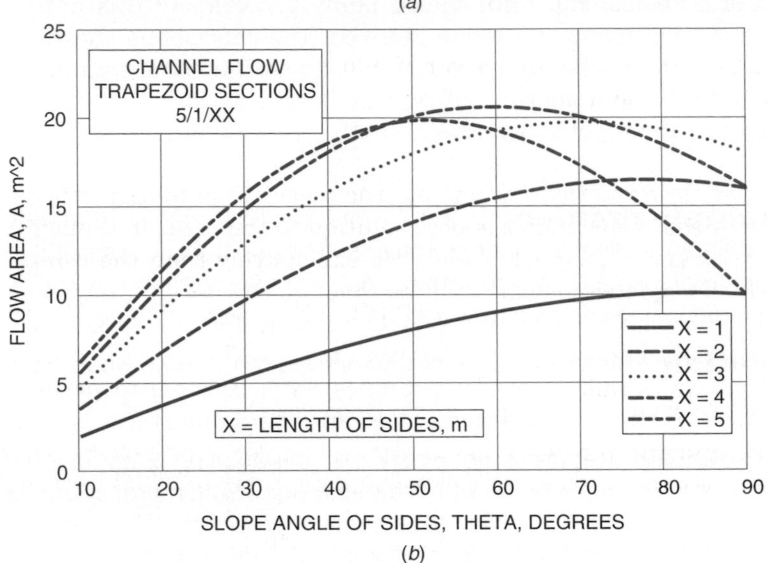

(b)

	x		Theta,D		Theta, R		Area
	3		0		0		0
			90		1.570 796		18
		Area = 16 For Theta =			0.698014 Radians		

(c)

Theta, Degree		Theta, Radians		x, m			
0		0		1			
15		0.261799		2			
35		0.610865		3			
55		0.959931		4			
75		1.308997		5			
90		1.570796		6			
Max flow area		Theta		x			Area
		1.047198		4			20.78461
Area = 12*x*sin(Theta)-2*x²*sin(Theta)+x²*sin(Theta)*cos(Theta)							

(d)

Figure 1.4

Example problem 1.3 For the situation described in Example prob. 1.2, find values of x and θ that yield a maximum flow area.

Solution This is a design problem in which a "best" solution is sought, in this case a maximum flow area for the trapezoidal cross section shown in Fig. 1.3. We can determine the solution from the repeated analysis we did for Example prob. 1.2 The solution is obtained readily from a spreadsheet search over the range of values for x and θ. Figure 1.4d shows the result. You can verify this by checking against Fig. 1.4b.

Analysis and synthesis are very important to the engineering design effort, and a majority of your engineering education will involve techniques of analysis and synthesis in problem solving. We must not, however, forget the engineer's role in the entire design process. In an industrial setting, the objective is to correctly assess a need, determine the best solution to the need, and market the solution more quickly and less expensively than the competition. This demands careful adherence to the design process.

The successful engineer in a technology team will take advantage of computers and computer graphics. Today, with the aid of computers and computer graphics, it is possible to perform analysis, decide among alternatives, and communicate results far more quickly and with more accuracy than ever before. This translates into better engineering and an improved quality of living.

Terms like *computer-aided design* (CAD) and *computer-aided manufacturing* (CAM) label the modern engineering activities that continue to make engineering a challenging profession. Your work with analysis and synthesis techniques will require the use of a computer to a large extent in your education.

In addition to the use of the computer to perform computations and develop models, the information superhighway will provide you with instant access to new technologies, new processes, technical information, current economic conditions, and a multitude of other data that will help you achieve success in the workplace and in all other aspects of life. The Internet, a worldwide collection of computer networks, now has over 3 million servers which can be accessed for information. The World Wide Web (WWW) provides a user-friendly graphics interface to the Internet enabling text, audio, and video to be transmitted. Various search methods exist to help you find the information you are seeking.

Working engineers are now able to communicate with colleagues around the world via electronic mail (e-mail) on common interests and problems. We are able to monitor in real time a field test in a foreign nation of our design as we sit at our desks in the United States. With new company networks now being installed, called Intranets, databases of products, production status, design changes, and field status are at your fingertips.

In your personal life, you are able to join user groups with common interests in sports, music, home maintenance, automobile repair, and the like. The new paradigm of on-line, interactive control of the information we desire is unprecedented. We can get what information we want, when we want it. Not since the Gutenberg press has such a dramatic change occurred in the way we acquire and distribute knowledge.

One of your first tasks as an engineering student should be to locate a computer, find out how you can access the WWW, and learn how to navigate the Internet. You likely will be using this media to conduct research in your courses and to communicate with your instructors and classmates. If you are already capable of "surfing" the Internet through your own computer, you will find this to be most helpful in achieving your educational goals.

1.2.3 Technologist and Technician

Much of the actual work of converting the ideas of scientists and engineers into tangible results is performed by technologists and technicians (see Fig. 1.5). A technologist generally possesses a bachelor's degree and a technician an associate's degree. Technologists

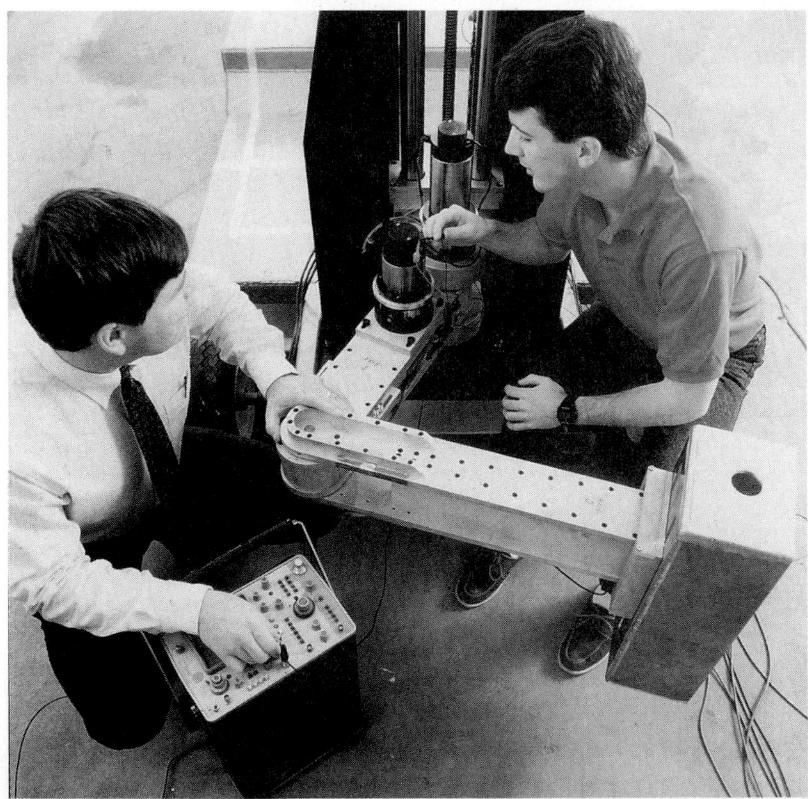

Figure 1.5 Technologists work with engineers on the design and testing of a mobile soil characterization robot arm which, when equipped with a probe, gathers soil samples. *(Ames Laboratory, U.S. Department of Energy.)*

are involved in the direct application of their education and experience to make appropriate modifications in designs as the need arises. Technicians primarily perform computations and experiments and prepare design drawings as requested by engineers and scientists. Thus, technicians (typically) are educated in mathematics and science but not to the depth required of scientists and engineers. Technologists and technicians obtain a basic knowledge of engineering and scientific principles in a specific field and develop certain manual skills that enable them to communicate technically with all members of the technology team. Some tasks commonly performed by technologists and technicians include drafting, estimating, model building, data recording, and reduction, troubleshooting, servicing, and specification. Often they are the vital link between the idea on paper and the idea in practice.

1.2.4 Skilled Trades/Craftspersons

Members of the skilled trades possess the skills necessary to produce parts specified by scientists, engineers, technologists, and technicians. Craftspersons need not have an in-depth knowledge of the principles of science and engineering incorporated in a design (see Fig. 1.6). They are often trained on the job, serving an apprenticeship during which the skills and abilities to build and operate specialized equipment are developed. Some of the specialized jobs of craftspersons include those of welder, machinist, electrician, carpenter, plumber, and mason.

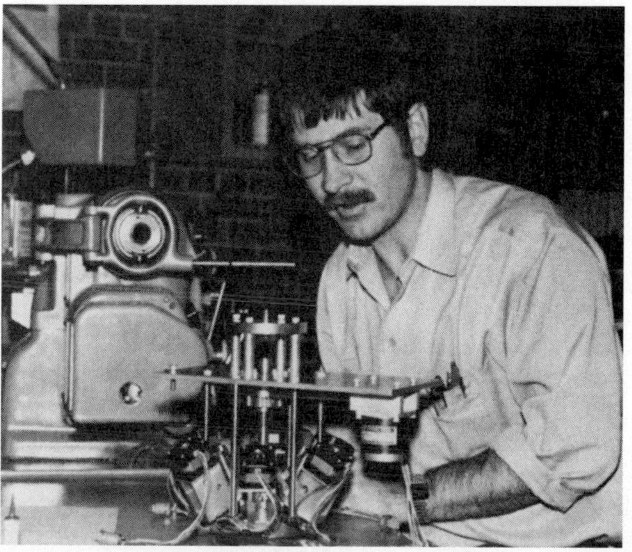

Figure 1.6
A machinist puts the finishing touches to a multiviewing transducer which will assist in locating the characterizing flaws in metals. *(Ames Laboratory, U.S. Department of Energy.)*

1.3 The Functions of the Engineer

As we alluded to in the previous section, engineering feats accomplished from earliest recorded history up to the industrial revolution could best be described as individual accomplishments. The various pyramids of Egypt were usually designed by one individual, who directed tens of thousands of laborers during construction. The person in charge called every move, made every decision, and took the credit if the project was successful or the consequences if the project failed.

With the industrial revolution, there was a rapid increase in scientific findings and technological advances. One-person engineering teams were no longer practical or desirable. We know that today no single aerospace engineer is responsible for the jumbo jets and no one civil engineer completely designs a bridge. Automobile manufacturers assign several thousand engineers to the design of a new model. So we not only have the technology team as described earlier, but we have engineers from many disciplines working together on single projects.

One approach to an explanation of an engineer's role in the technology spectrum is to describe the different types of work that engineers do. For example, civil, electrical, mechanical, and other engineers become involved in design, which is an engineering function. The *engineering functions,* which are discussed briefly in this section, are research, development, design, production, testing, construction, operations, sales, management, consulting, and teaching. Several of the *engineering disciplines* will be discussed in Sec. 1.4.

To avoid confusion between the meaning of the engineering disciplines and engineering functions, let us consider the following. Normally a student selects a curriculum (aerospace, chemical, mechanical, and so forth) either before or soon after admission to an engineering college. When and how the choice is made varies with each school. The point is, the student does not choose a function, but a discipline. To illustrate further, consider a student who has chosen mechanical engineering. This student will, during an undergraduate education, learn how mechanical engineers are involved in the engineering functions of research, development, design, and so on. Some program options allow a student to pursue an interest in a specific subdivision within the curriculum, such as energy conversion in a mechanical engineering program. Most other curricula have similar options.

Upon graduation, when you accept a job with a company, you will be assigned to a functional team performing in a specific area such as research, design, or sales. Within some companies, particularly smaller ones, you may become involved in more than one function—design *and* testing, for example. It is important to realize that regardless of your choice of discipline, you may become involved in one or more of the functions to be discussed in the following paragraphs.

1.3.1 Research

Successful research is one catalyst for starting the activities of a technology team or, in many cases, the activities of an entire industry. The research engineer seeks new findings, as does the scientist; but it must be kept in mind that the research engineer also seeks a way to use the discovery.

Key qualities of a successful research engineer are perceptiveness, patience, and self-confidence. Most students interested in research will pursue the master's and doctor's degrees in order to develop their intellectual abilities and the necessary research skills. An alert and perceptive mind is needed to recognize nature's truths when they are encountered. When attempting to reproduce natural phenomena in the laboratory, cleverness and patience are prime attributes. Research often involves tests, failures, retests, and so on, for long periods of time. Research engineers are therefore often discouraged and frustrated and must strain their abilities and rely on their self-confidence in order to sustain their efforts to a successful conclusion.

Billions of dollars are spent each year on research at colleges and universities, industrial research laboratories, government installations, and independent research institutes. The team approach to research is predominant today primarily because of the need to incorporate a vast amount of technical information into the research effort. Individual research is also carried out but not to the extent it was several years ago. A large share of research monies are channeled into the areas of energy, environment, health, defense, and space exploration. Research funding from fed-

Figure 1.7
Research engineers conducting an electronics experiment in an electrical engineering research laboratory. *(Iowa State University.)*

eral agencies is very sensitive to national and international priorities. During a career as a research engineer, you might expect to work in many diverse, seemingly unrelated areas, but your qualifications will allow you to adapt to many different research efforts.

1.3.2 Development

Using existing knowledge and new discoveries from research, the development engineer attempts to produce a device, structure, or process that is functional. Building and testing scale or pilot models is the primary means by which the development engineer evaluates ideas. A major portion of the development work requires use of well-known devices and processes in conjunction with established theories. Thus reading of available literature and a solid background in the sciences and in engineering principles are necessary for the development engineer's success.

Many people who suffer from heart irregularities are able to function normally today because of the pacemaker, an electronic device that maintains a regular heartbeat. The pacemaker is an excellent example of work of development engineers.

The first model, conceived by medical personnel and developed by engineers at the Electrodyne Company, was an external device that sent pulses of energy through electrodes to the heart. However, the power requirement for stimulus was so great that patients suffered severe burns on their chests. As improvements were being studied, research in surgery and electronics enabled development engineers to devise an external pacemaker with electrodes through the chest attached directly to the heart. Although more efficient from the standpoint of power require-

Figure 1.8
Development engineers use computer-aided design to assist the design and manufacture of prototype microprocessors. *(Iowa State University.)*

ments, the devices were uncomfortable, and patients frequently suffered infection where the wires entered the chest. Finally, two independent teams developed the first internal pacemaker, 8 years after the original pacemaker had been tested. Their experience and research with tiny pulse generators for spacecraft led to this achievement. But the very fine wire used in these early models proved to be inadequate and quite often failed, forcing patients to have the entire pacemaker replaced. A team of engineers at General Electric developed a pacemaker that incorporated a new wire, called a *helicable.* The helicable consisted of 49 strands of wire coiled together and then wound into a spring. The spring diameter was about 46 μm (micrometers), half the diameter of a human hair. Thus, with doctors and development engineers working together, an effective, comfortable device was perfected that has enabled many heart patients to enjoy a more active life. Today pacemakers have been developed that operate at more than one speed, enabling the patient to speed up or slow down heart rate depending on physical activity.

We have discussed the pacemaker in detail to point out that changes in technology can be in part owing to development engineers. Only 13 years to develop an efficient, dependable pacemaker; 5 years to develop the transistor; 25 years to develop the digital computer—it only indicates that modern engineering methods generate and improve products nearly as fast as research generates new knowledge.

Successful development engineers are ingenious and creative. Astute judgment is often required in devising models that can be used to determine whether a project will be successful in performance and economical in production. Obtaining an advanced degree is helpful, but it is not as important as it is for an engineer who will be working in research. Practical experience more than anything else produces the qualities necessary for a career as a development engineer.

Development engineers are often asked to demonstrate that an idea will work. Within certain limits, they do not work out the exact specifications that a final product should possess. Such matters are usually left to the design engineer if the idea is deemed feasible.

1.3.3 Design

The development engineer produces a concept or model that is passed on to the design engineer for converting into a device, process, or structure (see Fig. 1.9). The designer relies on education and experience to evaluate many possible solutions, keeping in mind cost of manufacture, ease of production, availability of materials, and performance requirements. Usually several designs and redesigns will be undertaken before the product is brought before the general public.

Figure 1.9
Student design engineers verify plans for a construction project. *(Iowa State University.)*

To illustrate the role the design engineer plays, we will discuss the development of the over-the-shoulder seat belts for added safety in automobiles, which created something of a design problem. Designers had to decide where and how the anchors for the belts would be fastened to the car body. They had to determine what standard parts could be used and what parts had to be designed from scratch. Consideration was given to passenger comfort, inasmuch as awkward positioning could deter usage. Materials to be used for the anchors and the belt had to be selected. A retraction device had to be designed that would give flawless performance.

From one such part of a car, one can extrapolate the numerous considerations that must be given to the approximately 12,000 other parts that form the modern automobile: optimum placement of engine accessories, comfortable design of seats, maximization of trunk space, and aesthetically pleasing body design all require thousands of engineering hours to be successful in a highly competitive industry.

Like the development engineer, the designer is creative. However, unlike the development engineer, who is usually concerned only with a prototype or model, the designer is restricted by the state of the art in engineering materials, production facilities, and, perhaps most important, economic considerations. An excellent design from the standpoint of performance may be completely impractical when viewed from a monetary point of view. To make the necessary decisions, the designer must have a fundamental knowledge of many engineering specialty subjects as well as an understanding of economics and people.

1.3.4 Production and Testing

When research, development, and design have created a device for use by the public, the production and testing facilities are geared for mass production (see Figs. 1.10 and 1.11). The first

Figure 1.10
A test engineer observes the characteristics of an electronic circuit. *(Iowa State University.)*

step in production is to devise a schedule that will efficiently coordinate materials and personnel. The production engineer is responsible for such tasks as ordering raw materials at the optimum times, setting up the assembly line, and handling and shipping the finished product. The individual who chooses this

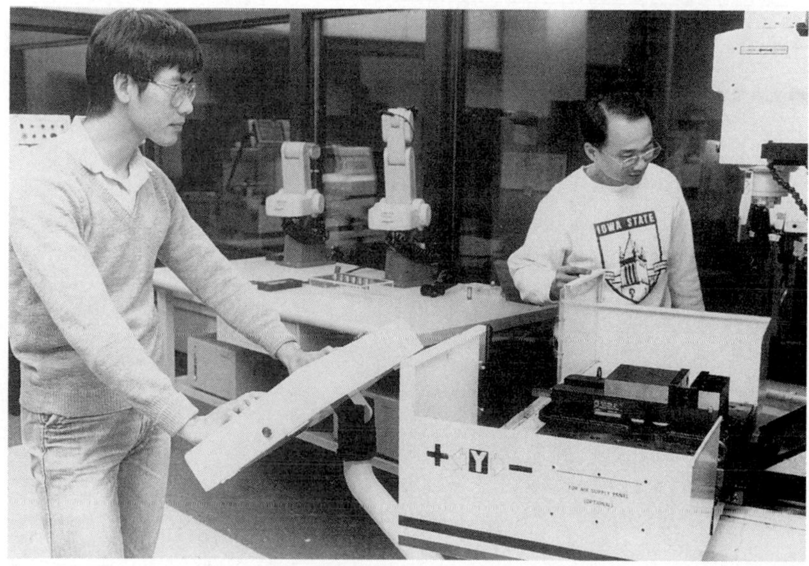

Figure 1.11
Production engineers design a prototype machining center. *(Iowa State University.)*

field must possess the ability to visualize the overall operation of a particular project as well as know each step of the production effort. Knowledge of design, economics, and psychology is of particular importance for production engineers.

Test engineers work with a product from the time it is conceived by the development engineer until such time as it may no longer be manufactured. In the automobile industry, for example, test engineers evaluate new devices and materials that may not appear in automobiles until several years from now. At the same time, they test component parts and completed cars currently coming off the assembly line. They are usually responsible for quality control of the manufacturing process. In addition to the education requirements of the design and production engineers, a fundamental knowledge of statistics is beneficial to the test engineer.

1.3.5 Construction

The counterpart of the production engineer in manufacturing is the construction engineer in the building industry (see Fig. 1.12). When an organization bids on a competitive construction project, the construction engineer begins the process by estimating material, labor, and overhead costs. If the bid is successful, a construction engineer assumes the responsibility of coordinating the project. On large projects, a team of construction engineers may supervise the individual segments of construction such as mechanical (plumbing), electrical (lighting), and civil (building). In

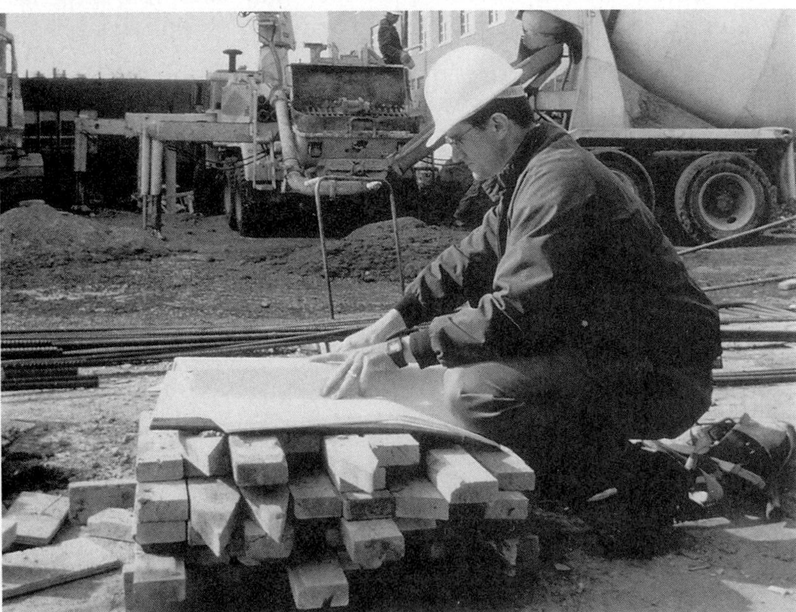

Figure 1.12
Construction engineers quite often work at the job site. *(Iowa State University.)*

addition to a strong background in engineering fundamentals, the construction engineer needs on-the-job experience and an understanding of labor relations.

1.3.6 Operations

Up to this point, discussion has centered around the results of engineering efforts to discover, develop, design, and produce products that are of benefit to human beings. For such work, engineers obviously must have offices, laboratories, and production facilities in which to accomplish it. The major responsibility for supplying such facilities falls on the operations engineer (see Fig. 1.13). Sometimes called a plant engineer, this individual selects sites for facilities, specifies the layout for all facets of the operation, and selects the fixed equipment for climate control, lighting, and communication. Once the facility is in operation, the plant engineer is responsible for maintenance and modifications as requirements demand. Because this phase of engineering comes under the economic category of overhead, the operations

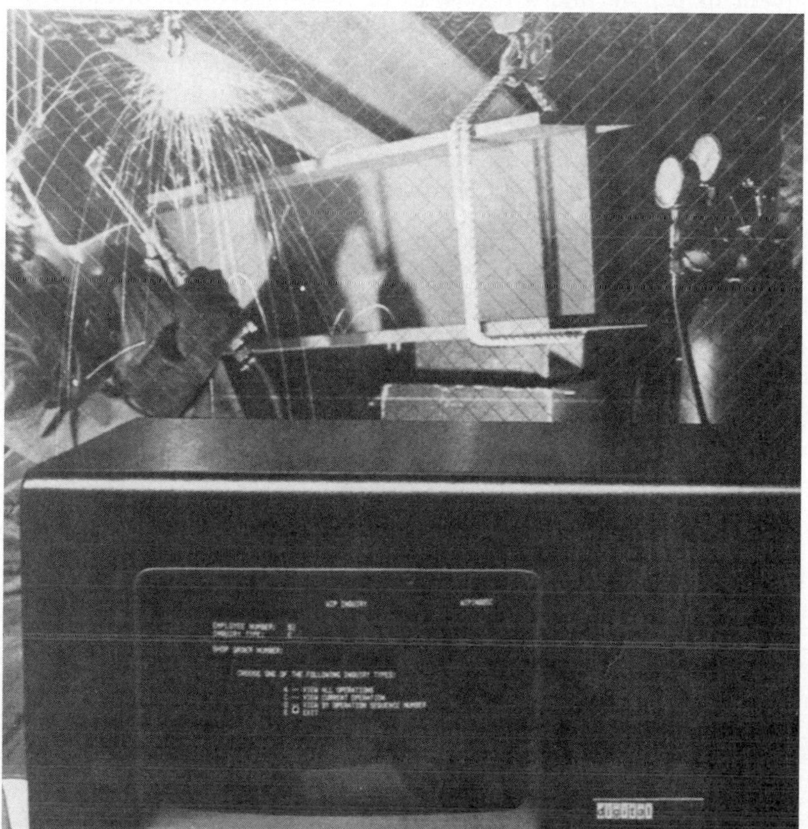

Figure 1.13
An operations engineer is responsible for putting together appropriate facilities such as this computer system used to direct manufacturing operations. *(Digital Equipment Corporation.)*

engineer must be very conscious of cost and keep up with new developments in equipment so that overhead is maintained at the lowest possible level. A knowledge of basic engineering, industrial engineering principles, economics, and law are prime educational requirements of the operations engineer.

1.3.7 Sales

In many respects, all engineers are involved in selling. To the research, development, design, production, construction, and operations engineer, selling means convincing management that money should be allocated for development of particular concepts or expansion of facilities. This is, in essence, selling one's own ideas. Sales engineering, however, means finding or creating a market for a product. The complexity of today's products requires an individual thoroughly familiar with materials in and operational procedures for consumer products to demonstrate to the consumer in layperson's terms how the products can be of benefit. The sales engineer is thus the liaison between the company and the consumer, a very important means of influencing a company's reputation. An engineering background plus a sincere interest in people and a desire to be helpful are the primary attributes of a sales engineer. The sales engineer usually spends a great deal of time in the plant learning about the product to be sold. After a customer purchases a product, the sales engineer is responsible for coordinating service and maintaining customer satisfaction. As important as sales engineering is to a company, it still has not received the interest from new engineering graduates that other engineering functions have. See Fig. 1.14.

Figure 1.14
A team of engineering design students making a sales presentation for their design. *(Iowa State University.)*

Figure 1.15
A management engineer uses an electronic database to continually monitor the performance of the various units in a company. *(Iowa State University.)*

1.3.8 Management

Traditionally, management has consisted of individuals trained in business and groomed to assume positions leading to the top of the corporate ladder. However, with the influx of scientific and technological data being used in business plans and decisions, there has been a need for people in management with knowledge and experience in engineering and science. Recent trends indicate that a growing percentage of management positions are being assumed by engineers and scientists. Inasmuch as one of the principal functions of management is to use company facilities to produce an economically feasible product, and decisions must often be made that may affect thousands of people and involve millions of dollars over periods of several years, a balanced education of engineering or science and business seems to produce the best managerial potential.

1.3.9 Consulting

For someone interested in self-employment, a consulting position may be an attractive one (see Fig. 1.16). Consulting engineers operate alone or in partnership furnishing specialized help to clients who request it. Of course, as in any business, risks must be taken. Moreover, as in all engineering disciplines, a sense of integrity and a knack for correct engineering judgment are primary necessities in consulting.

A consulting engineer must possess a professional engineer's license before beginning practice. Consultants usually spend many years in a specific area before going on their own. A suc-

Figure 1.16 Consulting engineers designed the device being installed on a high-wire tower to measure the "gallop" in the wires caused by winds. *(Iowa State University.)*

cessful consulting engineer maintains a business primarily by being able to solve unique problems for which other companies have neither the time nor capacity. In many cases, large consulting firms maintain a staff of engineers of diverse backgrounds so that a wide range of engineering problems can be contracted.

1.3.10 Teaching

Individuals interested in a career that involves helping others to become engineers will find teaching very rewarding (see Fig. 1.17). The engineering teacher must possess an ability to communicate abstract principles and engineering experiences in a manner that young people can understand and appreciate. By merely following general guidelines, teachers are usually free to develop their own method of teaching and means of evaluating its effectiveness. In addition to teaching, the engineering educator can also become involved in student advising and research.

Engineering teachers today must have a mastery of fundamental engineering and science principles and a knowledge of applications. Customarily, they must obtain an advanced degree in order to improve their understanding of basic principles, to perform research in a specialized area, and perhaps to gain teaching experience on a part-time basis.

If you are interested in a teaching career in engineering or engineering technology, you should observe your teachers carefully as you pursue your degrees. Note how they approach the teaching process, the methodologies they use to stimulate learning, and their evaluation methods. Your initial teaching methods will likely be based on the best methods you observe as a student.

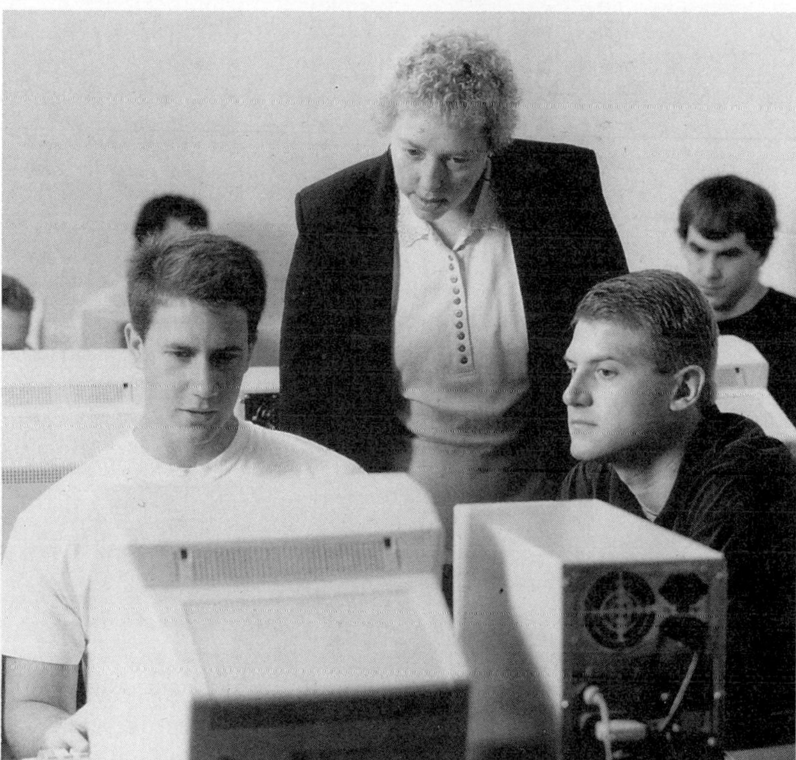

Figure 1.17
Teaching is a rewarding profession with many opportunities to help students realize their educational goals. *(Iowa State University.)*

1.4 The Engineering Disciplines

In the 1994 *Directory of Engineering and Engineering Technology* for undergraduate programs, 24 fields are listed for specialization. The opportunities to work in any of these areas or other specialty areas are numerous. Most engineering colleges offer some combination of the disciplines, primarily as 4-year programs. In some schools, two or more disciplines, such as industrial, management, and manufacturing engineering, are combined within one department which may offer separate degrees, or include one discipline as a specialty within another discipline. In this case a degree in the area of specialty is not offered. Other combinations of engineering disciplines include civil/construction/environmental and electrical/computer.

Figure 1.18 gives a breakdown of the number of engineering degrees in six categories for 1993–1994. Note that each category represents combined disciplines and does not provide information about a specific discipline within that category. The "other" category includes, among others, agricultural, biomedical, ceramic, metallurgical, mining, nuclear, safety, and ocean engineering.

Seven of the individual disciplines will be discussed in the following section. Engineering disciplines which pique your interest may be investigated in more detail by contacting the appropriate department or checking the library at your institution.

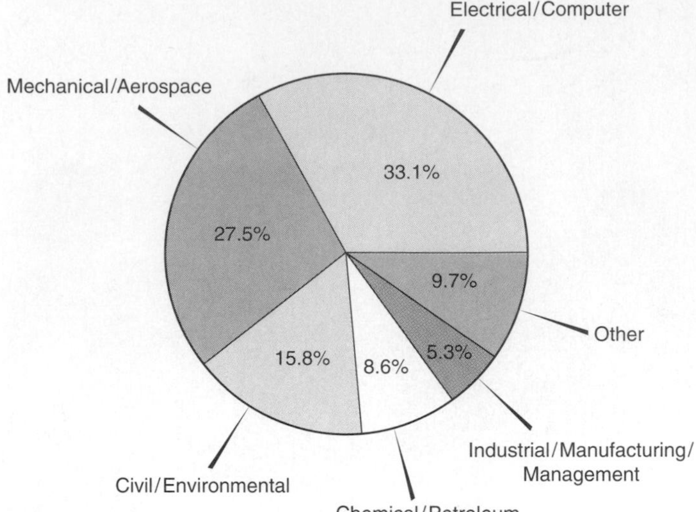

Figure 1.18 Engineering degrees by discipline in 1993–94. Total degrees awarded was 64,946. *(ASEE, PRISM, March 1995.)*

1.4.1 Aerospace Engineering

Aerospace engineers study flight of all types of vehicles in all environments. They design, develop, and test aircraft, missiles, space vehicles, helicopters, hydrofoils, ships, and submerging ocean vehicles. The particular areas of specialty include aerodynamics, propulsion, orbital mechanics, stability and control, structures, design, and testing.

Aerodynamics is the study of the effects of moving a vehicle through the earth's atmosphere. The air produces forces that have both a positive effect on a properly designed vehicle (lift) and a negative effect (drag). In addition, at very high speeds the air generates heat on the vehicle which must be dissipated to protect crews, passengers, and cargo. Aerospace engineering students learn to determine such things as optimum wing and body shapes, vehicle performance, and environmental impact.

The operation and construction of turboprops, turbo and fan jets, rockets, ram and pulse jets, and nuclear and ion propulsion are part of the aerospace engineering student's study of propulsion. Such constraints as efficiency, noise levels, and flight distance enter into the selection of a propulsion system for a flight vehicle.

The aerospace engineer develops plans for interplanetary missions based on a knowledge of orbital mechanics. The problems encountered include determination of trajectories, stabilization, rendezvous with other vehicles, changes in orbit, and interception.

Stability and control involves the study of techniques for maintaining stability and establishing control of vehicles operating in the atmosphere or in space. Automatic control systems for autopilots and unmanned vehicles are part of the study of stability and control.

Figure 1.19
Aerospace engineering students conducting a tensile test on a composite material used for structural support in a wing. *(Iowa State University.)*

The study of structures is primarily involved with thin-shelled, flexible structures that can withstand high stresses and extreme temperature ranges. The structural engineer works closely with the aerodynamics engineer to determine the geometry of wings, fuselages, and control surfaces. The study of structures also involves thick-shelled structures that must withstand extreme pressures at ocean depths and lightweight composite structural materials for high performance vehicles.

The aerospace design engineer combines all the aspects of aerodynamics, propulsion, orbital mechanics, stability and control, and structures into the optimum vehicle. Design engineers work in a team and must learn to compromise in order to determine the best design satisfying all criteria and constraints.

The final proofing of a design involves the physical testing of a prototype. Aerospace test engineers learn to use testing devices such as wind tunnels, lasers, strain gauges, and data-acquisition systems. The testing takes place in structural laboratories, propulsion facilities, and in the flight medium with the actual vehicle.

1.4.2 Chemical Engineering

Chemical engineers deal with the chemical and physical principles that allow us to maintain a suitable environment. They create, design, and operate processes that produce useful materials including fuels, plastics, structural materials, food products, health products, fibers, and fertilizers. As our natural resources grow short, chemical engineers are creating substitutes or finding ways to extend our remaining resources.

The chemical engineer, in the development of new products, in designing processes, and in operating plants, may work in a laboratory, pilot plant, or full-scale plant. In the laboratory, the chemical engineer searches for new products and materials that

Figure 1.20
Process laboratories are a key component in the education of chemical engineers. *(Iowa State University.)*

benefit humankind and the environment. This laboratory work would be classified as research engineering.

In a pilot plant, the chemical engineer is trying to determine the feasibility of carrying on a process on a large scale. There is a great deal of difference between a process working in a test tube in the laboratory and a process working in a production facility. The pilot plant is constructed to develop the necessary unit operations to carry out the process. Unit operations are fundamental chemical and physical processes that are uniquely combined by the chemical engineer to produce the desired product. A unit operation may involve separation of components by mechanical means such as filtering, settling, and floating. Separation may also take place by changing the form of a component—for example, through evaporation, absorption, or crystallization. Unit operations also involve chemical reactions such as oxidation and reduction. Certain chemical processes require the addition or removal of heat or the transfer of mass. The chemical engineer thus works with heat exchanges, furnaces, evaporators, condensers, and refrigeration units in developing large-scale processes.

In a full-scale plant, the chemical engineer will continue to "fine tune" the unit operations to produce the optimum process based on the lowest cost. The day-by-day operations problems in a chemical plant, such as piping, storage, and material handling, are the responsibility of chemical engineers.

1.4.3 Civil Engineering

Civil engineering is the oldest branch of the engineering profession. The term *civil* was used to distinguish this field from military engineers. Military engineers originated in Napoleon's army. The first engineers trained in this country were military engineers at West Point. Civil engineering involves application of the laws, forces, and materials of nature to the design, construction, operation, and maintenance of facilities that serve our needs in an efficient, economical manner. Civil engineers work for consulting firms engaged in private practice, for manufacturing firms, and for federal, state, and local governments. Because of the nature of their work, civil engineers assume a great deal of responsibility, which means that professional registration is an important goal for the civil engineer beginning practice. The specialties within civil engineering include structures, transportation, sanitary and water resources, geotechnical, surveying, and construction.

Structural engineers design bridges, buildings, dams, tunnels, and supporting structures. The designs include consideration of mass, winds, temperature extremes, and other natural phenomena such as earthquakes. Civil engineers with a strong structural background are often found in aerospace and manufacturing firms, playing an integral role in the design of vehicular structures.

Civil engineers in transportation plan, design, construct, operate, and maintain facilities that move people and goods throughout the world. For example, they make the decisions on where a freeway system should be located and describe the economic impact of the system on the affected public. They plan for growth of residential and industrial sectors of the nation. The modern rapid-transit systems are another example of the solution to a public need satisfied by transportation engineers.

Sanitary engineers are concerned with maintaining a healthful environment by proper treatment and distribution of water, treatment of wastewater, and control of all forms of pollution. The water resources engineer specializes in the evaluation of potential sources of new water for increasing or shifting populations, irrigation, and industrial needs.

Before any structure can be erected, a careful study of the soil, rock, and groundwater conditions must be undertaken to ensure stability. In addition to these studies, the geotechnical engineer analyzes building materials such as sand, gravel, and cement to determine proper consistency for concrete and other products.

Surveying engineers develop maps for any type of engineering project. For example, if a road is to be built through a mountain range, the surveyors will determine the exact route and develop the topographical survey which is then used by the transportation engineer to lay out the roadway.

Construction engineering is a significant portion of civil engineering, and many engineering colleges offer a separate degree in this area. Generally, construction engineers will work outside at the actual construction site. They become involved with the initial estimating of construction costs for surveying, excavation,

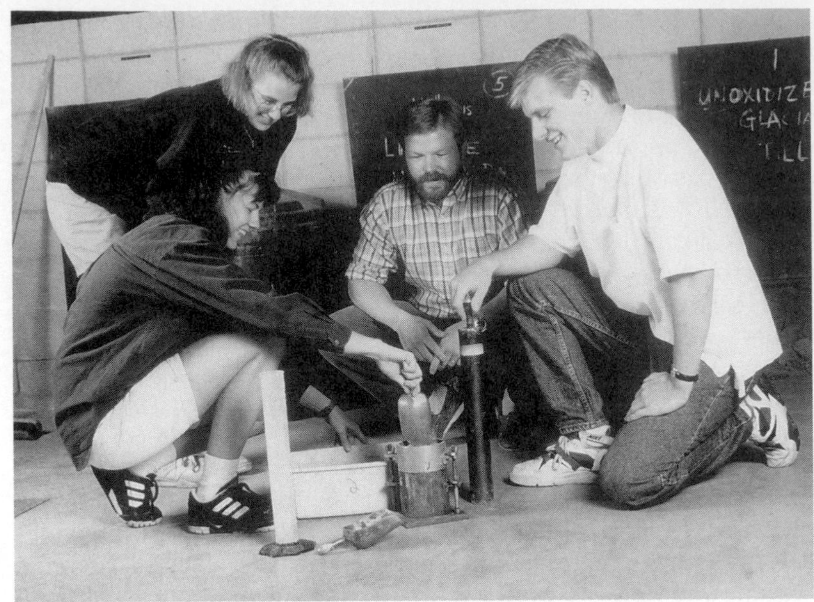

Figure 1.21 Civil engineering students conducting a test on concrete in the materials laboratory. *(Iowa State University.)*

and construction. They will supervise the construction, start-up, and initial operation of the facility until the client is ready to assume operational responsibility. Construction engineers work around the world on many construction projects such as highways, skyscrapers, and power plants.

1.4.4 Electrical/Computer Engineering

Electrical/computer engineering is the largest branch of engineering, representing about 30 percent of the graduates entering the engineering profession. Because of the rapid advances in technology associated with electronics and computers, this branch of engineering is also the fastest growing.

The areas of specialty include communications, power, electronics, measurement, and computers.

We depend almost every minute of our lives on communication equipment developed by electrical engineers. Telephones, television, radio, and radar are common communications devices that we often take for granted. Our national defense system depends heavily on the communications engineer and on the hardware used for our early warning and detection systems.

The power engineer is responsible for producing and distributing the electricity demanded by residential, business, and industrial users throughout the world. The production of electricity requires a generating source such as fossil fuels, nuclear reactions, or hydroelectric dams. The power engineer may be involved with research and development of alternative generation sources such as sun, wind, and fuel cells. Transmission of electricity involves conductors and insulating materials. On the receiving end, appliances are designed by power engineers to be highly efficient in order to reduce both electrical demand and costs.

The area of electronics is the fastest-growing specialty in electrical engineering. The development of solid-state circuits (functional electronic circuits manufactured as one part rather than wired together) has produced high reliability in electronic devices. Microelectronics has revolutionized the computer industry and electronic controls. Circuit components much smaller than 1 micrometer (μm) in width enable reduced costs and higher electronic speeds to be attained in circuitry. The microprocessor, the principal component of a digital computer, is a major result of solid-state circuitry and microelectronics technology. The home computer, automobile control systems, and a multitude of electrical-applications devices conceived, designed, and produced by electronics engineers have greatly improved our standard of living.

Great strides have been made in the control and measurement of phenomena that occur in all types of processes. Physical quantities such as temperature, flow rate, stress, voltage, and acceleration are detected and displayed rapidly and accurately for optimal control of processes. In some cases, the data must be sensed at a remote location and accurately transmitted long distances to receiving stations. The determination of radiation levels is an example of the electrical process called *telemetry.*

The impact of microelectronics on the computer industry has created a multibillion dollar annual business that in turn has enhanced all other industries. The design, construction, and operation of computer systems is the task of computer engineers. This specialty within electrical engineering has in many schools become a separate degree program. Computer engineers deal with both hardware and software problems in the design and application of computer systems. The areas of application include re-

Figure 1.22
Computer engineering students enjoy a light moment with their professor in the software engineering laboratory. *(Iowa State University.)*

search, education, design engineering, scheduling, accounting, control of manufacturing operations, process control, and home computing needs. No single development in history has had as great an impact on our lives in as short a time span as has the computer.

1.4.5 Environmental Engineering

Environmental engineering deals with the appropriate use of our natural resources and the protection of our environment. For the most part, environmental engineering curricula are relatively new and in many instances reside as a specialty within other disciplines such as civil, chemical, and agricultural engineering. In the 1994 *Directory of Engineering and Engineering Technology* for undergraduate programs, 39 environmental engineering programs are listed as compared to 188 civil engineering programs. (Only ABET accredited programs are included in the directory.)

The construction, operation, and maintenance of the facilities we live and work in have a significant impact on the environment. Environmental engineers with a civil engineering background are instrumental in the design of water and wastewater treatment plants, facilities that resist natural disasters such as earthquakes and floods, and facilities that use no hazardous or toxic materials. The design and layout of large cities and urban areas must include protective measures for the disturbed environment.

Environmental engineers with a chemical engineering background are interested in air and water quality which is affected by many by-products of chemical and biological processes. Products which are slow to biodegrade are studied for recycling

Figure 1.23 Environmental engineers become involved in testing for contaminants at waste sites. *(Ames Laboratory, U.S. Department of Energy.)*

possibilities. Other products which may contaminate or be hazardous are being studied to develop better storage or to develop replacement products less dangerous to the environment.

With an agricultural engineering background, environmental engineers study air and water quality which is affected by animal production facilities, chemical run-off from agricultural fertilizers, and weed control chemicals. As we become more environmentally conscious, the demand for designs, processes, and structures which protect the environment will create an increasing demand for environmental engineers. They will provide the leadership for protecting our resources and environment for generations to come.

1.4.6 Industrial Engineering

Industrial engineering covers a broad spectrum of activities in organizations of all sizes. The principal efforts of industrial engineers are directed to the design of production systems for goods and services. As an example, consider the procedures and processes necessary to produce and market a power lawn mower. When the design of the lawn mower is complete, industrial engineers establish the manufacturing sequence from the point of bringing the materials to the manufacturing center to the final step of shipping the assembled lawn mowers to the marketing agencies. Industrial engineers develop a production schedule, oversee the ordering of standard parts (engines, wheels, bolts), develop a plant layout (assembly line) for production of nonstandard parts (frame, height adjustment mechanism), and perform a cost analysis for all phases of production.

As production is ongoing, industrial engineers will perform various studies, called *time and methods studies,* which assist in optimizing the handling of material, the shop processes, and the overall assembly line. In a large organization, industrial engineers will likely specialize in one of the many areas involved in the operation of a plant. In a smaller organization, industrial engineers are likely to be involved in all the plant activities. Because of their general study in many areas of engineering and their knowledge of the overall plant operations, industrial engineers are frequent choices for promotion into management-level positions.

The study of human factors is an important area of industrial engineering. In product design, for example, industrial engineers involved in fashioning automobile interiors study the comfort and fatigue factors of seats and instrumentation. And in the factory they develop training programs for operators and supervisors of new machinery or for new assembly-line operators.

With the rapid development of computer-aided manufacturing (CAM) techniques and of computer-integrated manufacturing

Figure 1.24
Industrial engineering students program a pick-and-place robot in an assembly process. *(Iowa State University.)*

(CIM), the industrial engineer will play a large role in the factories of the future. The industrial engineer of the future will also be involved in retraining the labor force to work in a high-technology environment.

1.4.7 Mechanical Engineering

Mechanical engineering originated when people first began to use levers, ropes, and pulleys to multiply their own strength and to use wind and falling water as a source of energy. Today mechanical engineers are involved with all forms of energy utilization and conversion, machines, manufacturing materials and processes, and engines.

Mechanical engineers utilize energy in many ways for our benefit. Refrigeration systems keep perishable goods fresh for long periods of time, air condition your homes and offices, and aid in various forms of chemical processing. Heating and ventilating systems keep us comfortable when the environment around us changes with the seasons. Ventilating systems help keep the air around us breathable by removing undesirable fumes. Mechanical engineers analyze heat transfer from one object to another and design heat exchangers to effect a desirable heat transfer.

The energy crisis of the 1970s brought to focus a need for new sources of energy as well as new and improved methods of energy conversion. Mechanical engineers are involved in research in solar, geothermal, and wind energy sources, along with research to increase the efficiency of producing electricity from fossil fuel, hydroelectric, and nuclear sources.

Machines and mechanisms used in all forms of manufacturing and transportation are designed and developed by mechanical engineers. Automobiles, airplanes, and trains combine a source of power and a machine to provide transportation. Tractors, combines, and other implements aid the agricultural community. Automated machinery and robotics are rapidly growing areas for mechanical engineers. Lathes, milling machines, grinders, and drills assist in the manufacture of goods. Sorting devices, typewriters, staplers, and mechanical pencils are part of the office environment. Machine design requires a strong mechanical engineering background and a vivid imagination.

In order to drive the machines, a source of power is needed. The mechanical engineer is involved with the generation of electricity by converting chemical energy in fuels to thermal energy in the form of steam, then to mechanical energy through a turbine to drive the electric generator. Internal-combustion devices such as gasoline, turbine, and diesel engines are designed for use in all areas of transportation. The mechanical engineer studies engine cycles, fuel requirements, ignition performance, power output, cooling systems, engine geometry, and lubrication in order to develop high performance, low-energy-consuming engines.

The engines and machines designed by mechanical engineers require many types of materials for construction. The tools that are needed to process the raw material for other machines must be designed. For example, a very strong material is needed for a drill bit that must cut a hole in a steel plate. If the tool is made from steel it must be a higher-quality steel than that found in the plate. Methods of heat treating, tempering, and other metallurgical processes are applied by the mechanical engineer.

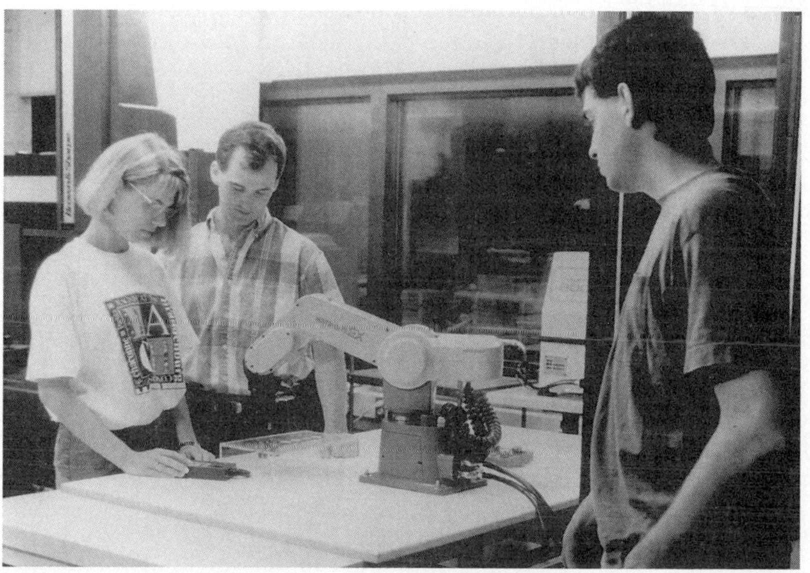

Figure 1.25
Mechanical engineering students test a robot in a manufacturing laboratory. *(Iowa State University.)*

Manufacturing processes such as electric-discharge machining, laser cutting, and modern welding methods are used by mechanical engineers in the development of improved products. Mechanical engineers are also involved in the testing of new materials and products such as composites.

1.5 Education of the Engineer

The amount of information coming from the academic and business world is increasing exponentially, and at the current rate it will double in less than 20 years. More than any other group, engineers are using this knowledge to shape civilization. To keep pace with a changing world, engineers must be educated to solve problems that are as yet unheard of. A large share of the responsibility for this mammoth education task falls on the engineering colleges and universities. But the completion of an engineering program is only the first step toward a lifetime of education. The engineer, with the assistance of the employer and the university, must continue to study. See Fig. 1.26.

Logically, then, an engineering education should provide a broad base in scientific and engineering principles, some study in the humanities and social sciences, and specialized studies in a chosen engineering curriculum. But specific questions concerning engineering education still arise. We will deal here with the questions that are frequently asked by students. What are the desirable characteristics for success in an engineering program? What knowledge and skills should be acquired in college? What is meant by continuing education with respect to an engineering career?

1.5.1 Desirable Characteristics

Years of experience have enabled engineering educators to analyze the performance of students in relation to abilities and de-

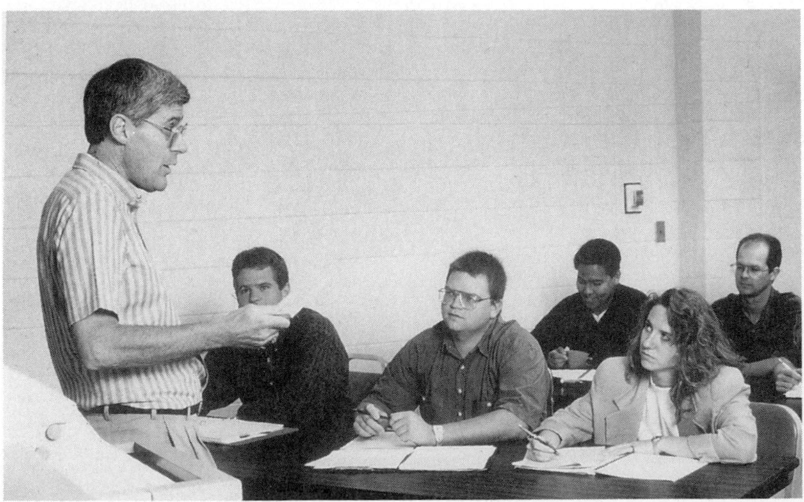

Figure 1.26
Long distance learning via satellite transmission makes it more convenient for working engineers to keep up on new developments in their field. *(Iowa State University.)*

sires they possess entering college. The most important characteristics for an engineering student can be summarized as follows:

1. A strong interest in and ability to work with mathematics and science taken in high school.
2. An ability to think through a problem in a logical manner.
3. A knack for organizing and carrying through to conclusion the solution to a problem.
4. An unusual curiosity about how and why things work.

Although such attributes are desirable, having them is no guarantee of success in an engineering program. Simply a strong desire for the job has made successful engineers of some individuals who did not possess any of these characteristics; and, conversely, many who possessed them did not complete an engineering degree. Moreover, an engineering education is not easy, but it can offer a rewarding career to anyone who accepts the challenge.

1.5.2 Knowledge and Skills Required

As indicated previously, over 340 colleges and universities offer programs in engineering that are accredited by ABET or CEAB. These boards have as their purpose the quality control of engineering and technology programs offered in the United States and Canada. The basis of the boards is the engineering profession, which is represented through the participating professional groups. A listing of the participating bodies is given in Tab. 1.1.

The quality control of engineering programs is effected through the accreditation process. The engineering profession, through ABET and CEAB, has developed standards and criteria for the education of engineers entering the profession. Through visitations, evaluations, and reports, the written criteria and standards are compared with the engineering curricula at a university. Each program, if operating according to the standards and criteria, may receive up to 6 years of accreditation. If some discrepancies appear, accreditations may be granted for a shorter time period or may not be granted at all until appropriate improvements are made.

It is safe to say that for any given engineering discipline, no two schools will have identical offerings. However, close scrutiny will show a framework within which most courses can be placed, with differences occurring only in textbooks used, topics emphasized, and sequences followed. Figure 1.27 depicts this framework and some of the courses that fall within each of the areas. The approximate percentage of time spent on each course grouping is indicated.

The sociohumanistic block is a small portion of most engineering curricula, but it is important because it helps the engi-

Table 1.1 Participating bodies in the accreditation activity. (Compiled from ABET annual report, 1994.)

Organization	Abbreviation
American Academy of Environmental Engineers	AAEE
American Congress on Surveying and Mapping	ACSM
American Institute of Aeronautics and Astronautics, Inc.	AIAA
American Institute of Chemical Engineers	AICHE
American Nuclear Society	ANS
American Society of Agricultural Engineers	ASAE
American Society of Civil Engineers	ASCE
American Society for Engineering Education	ASEE
American Society of Heating, Refrigerating, and Air-Conditioning Engineers, Inc.	ASHRAE
The American Society of Mechanical Engineers	ASME
Institute of Industrial Engineers, Inc.	IIE
The Institute of Electrical and Electronics Engineers, Inc.	IEEE
The Minerals, Metals & Materials Society	TMS
National Council of Examiners for Engineering and Surveying	NCEES
National Institute of Ceramic Engineers	NICE
National Society of Professional Engineers	NSPE
Society of Automotive Engineers	SAE
Society of Manufacturing Engineers	SME
Society for Mining, Metallurgy, and Exploration, Inc.	SME-AIME
Society of Naval Architects and Marine Engineers	SNAME
Society of Petroleum Engineers	SPE

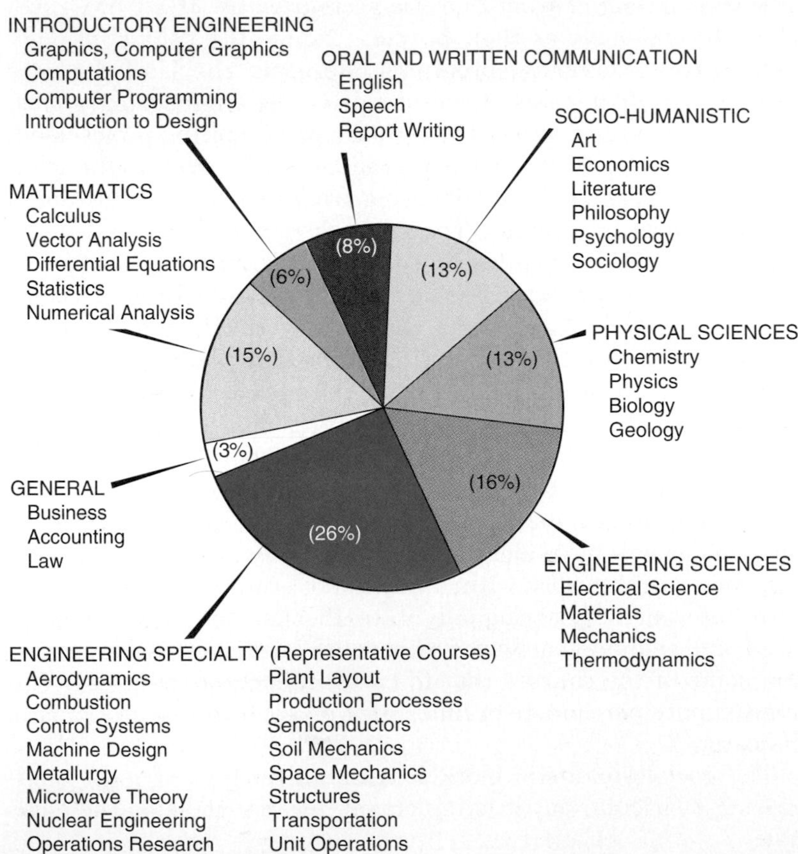

Figure 1.27 Elements of engineering curricula.

neering student understand and develop an appreciation for the potential impact of engineering undertakings on the environment and general society. When the location of a nuclear power plant is being considered, the engineers involved in this decision must respect the concerns and feelings of all individuals who might be affected by the location. Discussions of the interaction between engineers and the general public take place in few engineering courses; sociohumanistic courses are thus needed to furnish engineering students with an insight into the needs and aspirations of society.

Chemistry and physics are almost universally required in engineering. They are fundamental to the study of engineering science. The mathematics normally required for college chemistry and physics is more advanced than that for the corresponding high school courses. Higher-level chemistry and physics may also be required, depending on departmental structure. Finally, other physical science courses may be required in some programs or may be taken as electives.

An engineer cannot be successful without the ability to communicate ideas and the results of work efforts. The research engineer writes reports and orally presents ideas to management. The production engineer must be able to converse with craftspersons in understandable terms. And all engineers have dealings with the public and must be able to communicate on a nontechnical level.

Engineers have been accused of not becoming involved in public affairs. The reason often given for their not becoming involved is that they are not trained sufficiently in oral and written communications. However, the equivalent of one-third of 1 year is spent on formal courses in these these subjects, and additional time is spent in design presentations, written laboratory reports, and the like. A conscious effort by student engineers must be made to improve their abilities in oral and written communication to overcome this nonactivist label.

Mathematics is the most powerful tool the engineer uses to solve problems. The amount of time spent in this area is indicative of its importance. Courses in calculus, vector analysis, and differential equations are common to all degree programs. Statistics, numerical analysis, and other mathematics courses support some engineering specialty areas. Students desiring an advanced degree may want to take mathematics courses beyond the baccalaureate-level requirements.

In the early stages of an engineering education, introductory courses in graphical communication, computational techniques, design, and computer programming are taken. Engineering schools vary somewhat in their emphasis on these areas, but the general intent is to develop skills in the application of theory to practical problem solving and familiarity with engineering terminology. Design is presented from a conceptual point of view to aid the student in creative thinking. Graphics develops the visualization capability and assists the student in transferring mental thoughts into well-defined concepts on paper. The tremendous potential of the computer to assist the engineer has led to the requirement of computer programming in almost all curricula. Use of the computer to perform many tedious calculations has increased the efficiency of the engineer and allows more time for creative thinking. Computer graphics is becoming a part of engineering curricula. Its ability to enhance the visualization of geometry and to depict engineering quantities graphically has increased productivity in the design process.

With a strong background in mathematics and physical sciences, you can begin study of engineering sciences, courses that are fundamental to all engineering specialties. Electrical science includes study of charges, fields, circuits, and electronics. Materials science courses involve study of the properties and chemical compositions of metallic and nonmetallic substances (see Fig. 1.28). Mechanics includes study of statics, dynamics, fluids, and mechanics of materials. Thermodynamics is the science of heat and is the basis for the study of all types of energy and energy transfer. A sound understanding of the engineering sciences is most important for anyone interested in pursuing postgraduate work and research.

Figure 1.27 shows only a few examples of the many specialized engineering courses given. Scanning course descriptions in a college general bulletin or catalog will provide a more detailed insight into the specialized courses required in the various engineering disciplines.

Most curricula allow a student flexibility in selecting a few courses in areas not previously mentioned. For example, a stu-

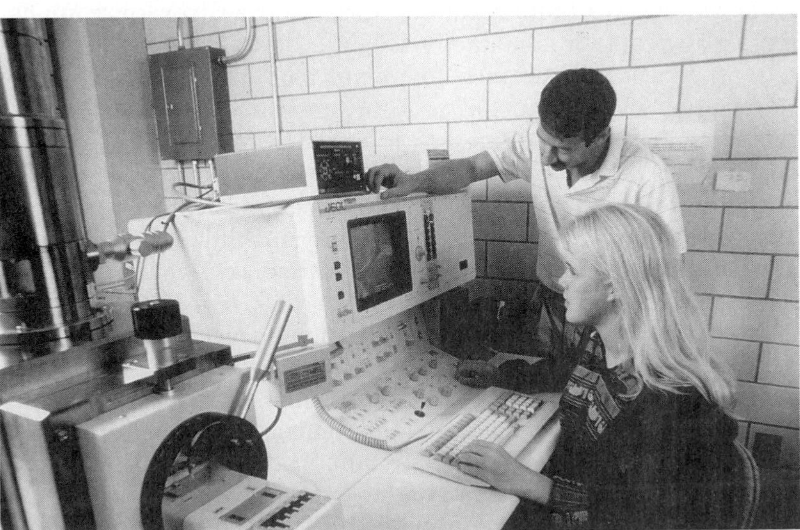

Figure 1.28
Laboratories are instrumental in the learning process for student engineers. *(Iowa State University.)*

dent interested in management may take some courses in business and accounting. Another may desire some background in law or medicine, with the intent of entering a professional school in one of these areas upon graduation from engineering.

1.6 Professionalism and Ethics

Engineering is a learned vocation, demanding an individual with high standards of ethics and sound moral character. When making judgments that may create controversy and affect many people, the engineer must keep foremost in mind a dedication to the betterment of humanity.

The engineering profession has attempted for many years to become unified. However, technical societies that represent individual engineering disciplines have grown strong and tend to keep the various engineering disciplines separated. This contrasts with more unified professions such as law, medicine, and theology.

1.6.1 Professionalism

Professionalism is a way of life. A professional person is one who engages in an activity that requires a specialized and comprehensive education and is motivated by a strong desire to serve humanity. A professional thinks and acts in a manner that brings favor upon the individual and the entire profession. Developing a professional frame of mind begins with your engineering education.

The professional engineer can be said to have the following:

1. Specialized knowledge and skills used for the benefit of humanity.
2. Honesty and impartiality in engineering service.
3. Constant interest in improving the profession.
4. Support of professional and technical societies that represent the professional engineer.

It is clear that these characteristics include not only technical competence but also a positive attitude toward life that is continually reinforced by educational accomplishments and professional service.

A primary reason for the rapid development in science and engineering is the work of the technical societies. The fundamental service provided by a society is the sharing of ideas, which means that technical specialists can publicize their efforts and assist others in promoting excellence in the profession. When information is distributed to other society members, new ideas evolve and duplicated efforts are minimized. The societies conduct meetings on international, national, and local bases. Students of engineering will find a technical society in their specialty that may operate as a branch of the regular society or as a student chapter on campus. The student organization is an important link with the professional workers, providing motivation and the opportunity to make acquaintances that will help students to formulate career objectives.

Table 1.1 is a partial listing of the numerous engineering societies that support the engineering disciplines and functions. These technical societies are linked because of their support of the accreditation process. Over 60 other societies exist for the purpose of supporting the professional status of engineers. Among these are the Society of Women Engineers (SWE), the National Society of Black Engineers (NSBE), the Acoustical Society of America (ASA), the Society of Plastics Engineers (SPE), and the American Society for Quality Control (ASQC).

Many of the technical societies are quite influential in the engineering profession. To unify the profression in the manner of such other professions as law, medicine, and theology would require the cooperation of all the societies. However the individual societies have more than satisfied the professional needs for many engineers, so no pressing desire to unify is apparent. Nonetheless, to preserve the advantages of the technical societies while unifying the entire profession remains a long-range goal of engineers today.

1.6.2 Professional Registration

The power to license engineers rests with each of the 50 states. Since the first registration law in Wyoming in 1907, all states have developed legislation specifying requirements for engineering practice. The purpose of registration laws is to protect the public. Just as one would expect a physician to provide competent medical service, an engineer can be expected to provide competent technical service. However, the laws of registration for engineers are quite different from those for lawyers or physicians. An engineer does not have to be registered to practice engineering. Legally, only the chief engineer of a firm needs to be registered for that firm to perform engineering services. Individuals testifying as expert engineering witnesses in court and individuals offering engineering consulting services need to be registered. In some instances, the practice of engineering is allowed as long as the individual does not advertise as an engineer.

The legal process for becoming a licensed professional engineer consists of four parts, two of which entail examinations. The parts include:

1. An engineering degree from an acceptable institution as defined by the state board for registration. Graduation from an ABET-accredited institution satisfies the degree requirement automatically.

2. Successful completion of the Fundamentals of Engineering Examination (FE) entitles one to the title "engineer-in-training" (EIT). This 8-hour exam may be taken during the last term of an undergraduate program that is ABET-accredited. The first half of the exam covers fundamentals in the areas of mathematics, chemistry, physics, engineering mechanics, electrical science, thermal science, economics, and ethics. The second half is oriented to each discipline, such as mechanical. The passing grade is determined by the state board.
3. Completion of 4 years of engineering practice as an EIT.
4. Successful completion of the Principles and Practice Examination completes the licensing process. This is also an 8-hour examination covering problems normally encountered in the area of specialty, such as mechanical or chemical engineering.

It should be noted that once the license is received, it is permanent, although there is an annual renewal fee. In addition, there is a trend toward specific requirements in continuing education each year in order to maintain the license. Licensed engineers in some states may attend professional meetings in their specialty, take classes, and write professional papers or books to accumulate sufficient professional development activities beyond their job responsibilities to maintain their licenses. This trend is a reflection of the rapidly changing technology and the need for engineers to remain current in their area.

Registration does have many advantages. Most public employment positions, all expert-witness roles in court cases, and some high-level company positions require the professional engineer's license. However, less than one-half of the eligible candidates are currently registered. You should give serious consideration to becoming registered as soon as you qualify. Satisfying the requirements for registration can be started even before graduation from an ABET-accredited curriculum.

1.6.3 Professional Ethics

Ethics is the guide to personal conduct of a professional. Most of the technical societies have a written code of ethics for their members. Because of this, some variations exist; but a general view of ethics for engineers is provided here for two of the technical societies. Figure 1.29 is a code endorsed by the Accreditation Board for Engineering and Technology. Appendix D gives the most widely endorsed code of ethics, that of the National Society of Professional Engineers. As you read both codes, note the many similarities. Figure 1.30 is the "Engineer's Creed" as published by the National Society of Professional Engineers.

*Accreditation Board for Engineering and Technology**

CODE OF ETHICS OF ENGINEERS

THE FUNDAMENTAL PRINCIPLES

Engineers uphold and advance the integrity, honor and dignity of the engineering profession by:

I. using their knowledge and skill for the enhancement of human welfare;

II. being honest and impartial, and serving with fidelity the public, their employers and clients;

III. striving to increase the competence and prestige of the engineering profession; and

IV. supporting the professional and technical societies of their disciplines.

THE FUNDAMENTAL CANONS

1. Engineers shall hold paramount the safety, health and welfare of the public in the performance of their professional duties.

2. Engineers shall perform services only in the areas of their competence.

3. Engineers shall issue public statements only in an objective and truthful manner.

3. Engineers shall act in professional matters for each employer or client as faithful agents or trustees, and shall avoid conflicts of interest.

5. Engineers shall build their professional reputation on the merit of their services and shall not compete unfairly with others.

6. Engineers shall act in such a manner as to uphold and enhance the honor, integrity and dignity of the profession.

7. Engineers shall continue their professional development throughout their careers and shall provide opportunities for the professional development of those engineers under their supervision.

345 East 47th Street New York, NY 10017

*Formely Engineers' Council for Professional Development. (Approved by the ECPD Board of Directors, October 5, 1977)

AB-54 2/85

Figure 1.29
Code of Ethics for Engineers. *(Accreditation Board for Engineering and Technology.)*

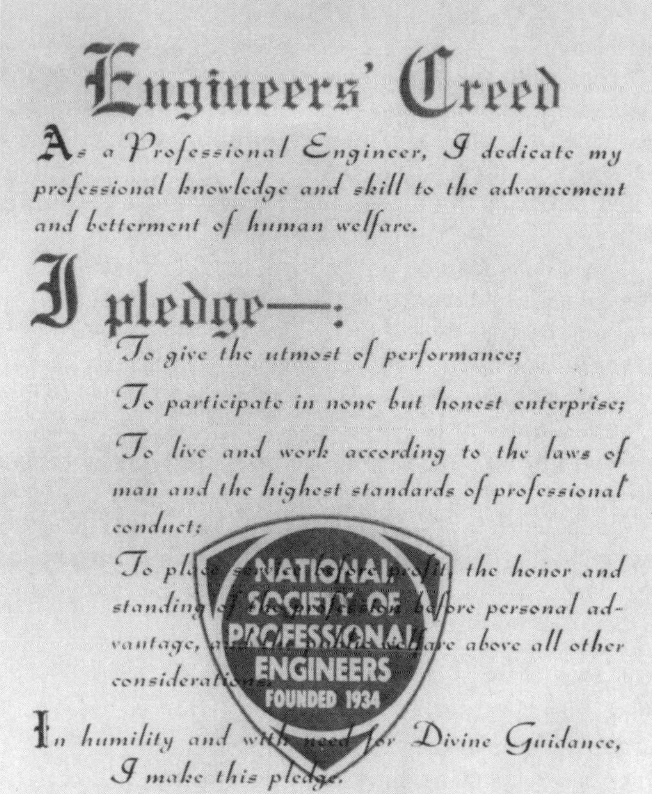

Engineers' Creed

As a Professional Engineer, I dedicate my professional knowledge and skill to the advancement and betterment of human welfare.

I pledge—:

To give the utmost of performance;

To participate in none but honest enterprise;

To live and work according to the laws of man and the highest standards of professional conduct;

To pl[illegible] the honor and standin[illegible]re personal advantage, [illegible]are above all other considera[illegible]

In humility and wit[illegible] Divine Guidance, I make this ple[illegible]

Figure 1.30
Engineer's Creed. *(National Society of Professional Engineers.)*

1.7 Challenges of the Future

The world continues to undergo rapid and sometimes tumultuous change. As a practicing engineer, you will occupy center stage in many of these changes in the near future and will become even more involved in the more distant future. The huge tasks of providing energy, maintaining a supply of water, ensuring a competitive edge in the world marketplace, rebuilding our infrastructure, and preserving our environment will challenge the technical community beyond anyone's imagination.

Engineers of today have nearly instantaneous access to a wealth of information from technical, economic, social, and political sources. A key to the success of engineers in the future will be the ability to study and absorb the appropriate information in the time allotted for producing a design or solution to a problem. A degree in engineering is only the beginning of a lifelong period of study in order to remain informed and competent in the field.

Engineers of tomorrow will have even greater access to information and will use increasingly powerful computer systems to digest this information. They will work with colleagues around the world solving problems and creating new products. They will assume greater roles in making decisions that affect the use of energy, water, and other natural resources.

1.7.1 Energy

In order to develop technologically, nations of the world require vast amounts of energy. With a finite supply of our greatest energy source, fossil fuels, alternate supplies must be developed and existing sources must be controlled with a worldwide usage plan. A key factor in the design of products must be minimum use of energy.

We do not mean to imply that our fossil fuel resources will be gone in a short time. However, as demand increases and supplies become scarcer, the cost of obtaining the energy increases and places additional burdens on already financially strapped regions and individuals. Engineers with great vision are needed to develop alternative sources of energy from the sun, wind, and ocean and to improve the efficiency of existing energy consumption devices.

Along with the production and consumption of energy come the secondary problems of pollution. Such pollutants as smog and acid rain, carbon monoxide, and radiation must receive attention in order to maintain the balance of nature.

Figure 1.31

1.7.2 Water

The basic water cycle—from evaporation to cloud formation, then to rain, runoff, and evaporation again—is taken for granted by most people. However, if the rain or the runoff is polluted, then the cycle is interrupted and our water supply becomes a crucial problem. In addition, some highly populated areas have a limited water supply and must rely on water distribution systems from other areas of the country. Many formerly undeveloped agricultural regions are now productive because of irrigation systems. However, the irrigation systems deplete the underground streams of water that are needed downstream.

These problems must be solved in order for life to continue to exist as we know it. Because of the regional water distribution patterns, the federal government must be a part of the decision-making process for water distribution. One of the concerns that must be eased is the amount of time required to bring a water distribution plan into effect. Government agencies and the private sector are strapped by regulations that cause delays in planning and construction of several years. Greater cooperation and a better informed public are goals that public works engineers must strive to achieve.

Developing nations around the world need an additional water supply because of increasing population. Many of these nations do not have the necessary fresh water and must rely on desalination, a costly process. The continued need for water is a concern for leaders of the world, and engineers will be asked to create additional sources of this life-sustaining resource.

1.7.3 A Competitive Edge in the World Marketplace

We have all purchased or used products that were manufactured outside the country. Many of these products incorporate technology that was developed in the United States. In order to maintain our strong industrial base, we must develop practices and processes that enable us to compete, not just with other U.S. industries, but with international industries.

The goal of any industry is to generate a profit. In today's marketplace this means creating the best product in the shortest time at a lesser price than the competition. A modern design process incorporating sophisticated analysis procedures and supported by high-speed computers with graphical displays increases the capability for developing the "best" product. The concept of integrating the design and manufacturing functions with CAD/CAM and CIM promises to shorten the design-to-market time for new products and for upgraded versions of existing products. The development of the automated factory is an exciting concept that is receiving a great deal of attention from manufacturing engineers today. Remaining competitive by producing at a lesser price requires a national effort involving labor, government, and distribution factors. In any case, engineers are going to have a significant role in the future of our industrial sector.

1.7.4 Infrastructure

All societies depend upon an infrastructure of transportation, waste disposal, and water distribution systems for the benefit of the population. In the United States much of infrastructure is in a state of deterioration without sound plans for upgrading. For example,

1. Commercial jet fleets include aircraft that are 25 to 30 years old. Major programs are now underway to safely extend the service life of these jets. In order to survive economically, airlines must balance new replacement jets with a program to keep older planes flying.

2. One-half of the sewage treatment plants cannot satisfactorily handle the demand.

3. The interstate highway system, nearing 40 years old in many areas, needs major repairs throughout. Local paved roads are deteriorating because of a lack of infrastructure funds.

4. Many bridges are potentially dangerous to the traffic loads on them.

5. Railraods continue to struggle with maintenance of the railbeds and the rolling stock in the face of stiff commercial competition from the air freight and truck transportation industries.

6. Municipal water systems require billions of dollars in repairs and upgrades to meet the demands of the public and the stiffening water quality requirements.

Figure 1.32
As our infrastructure ages, we must earmark the funds necessary to upgrade, expand, and repair these facilities to provide for our needs.

It is estimated that the total value of the public works facilities is over $2 trillion. To protect this investment, innovative thinking and creative funding must be fostered. Some of this is already occurring in road design and repair. For example, a method has been applied successfully that recycles asphalt pavement and actually produces a stronger product. Engineering research is producing extended-life pavement with new additives and structural designs. New, relatively inexpensive methods of strengthening old bridges have been used successfully.

1.7.5 Environment

Our insatiable demand for energy, water, and other national resources creates imbalances in nature which only time and serious conservation efforts can keep under control. The concern for environmental quality is focused in four areas: cleanup, compliance, conservation, and pollution prevention. Partnerships between industry, government, and consumers are working to establish guidelines and regulations in the gathering of raw materials, manufacturing of consumer products, and the disposal of material at the end of its designed use.

The American Plastics Council publishes a document entitled *Designing for the Environment* which describes environmental issues and initiatives affecting product design. All engineers need to be aware of these initiatives and how they apply in their particular industries:

- Design for the environment (DFE)—incorporate environmental considerations into product designs to promote environmental stewardship.
- Environmentally conscious manufacturing (ECM) or green manufacturing—incorporate pollution prevention and toxics use reduction in the making of products to promote environmental stewardship.
- Extended product or producer responsibility—product manufacturers assume responsibility for taking back their products at the end of the products' life and disposing of them according to defined environmental criteria.
- Life-cycle assessment (LCA)—quantified assessment of the environmental impacts associated with all phases of a product's life, often from the extraction of base minerals through the life of the product.
- Pollution prevention—reduce pollution sources in the design phase of products instead of addressing pollution after it is generated.
- Toxic use reduction—reduce the amount, toxicity, and number of toxic chemicals used in manufacturing.

As you can see from these initiatives all engineers, regardless of discipline, must be environmentally conscious in their work. In the next few decades, we will face tough decisions regarding our environment. Engineers will play a major role in making the correct decisions for our small, delicate world.

1.7.6 Conclusion

We have touched only briefly on the possibilities for exciting and rewarding work in all engineering areas. The first step is to obtain the knowledge during your college education that is necessary for your first technical position. After that, you must continue your education, either formally by seeking an advanced degree or degrees, or informally through continuing education courses or appropriate reading to maintain pace with the technology, an absolute necessity for a professional. Many challenges await you. Prepare to meet them well.

Problems

1.1 Compare the definitions of an engineer and a scientist from at least three different sources.

1.2 Compare the definitions of an engineer and a technologist from at least two sources.

1.3 Find three textbooks that introduce the design process. Copy the steps in the process from each textbook. Note similarities and differences and write a paragraph describing your conclusions.

1.4 Find the name of a pioneer engineer in the field of your choice and write a brief paper on the accomplishments of this individual.

1.5 Select a specific branch of engineering and list at least 20 different industrial organizations that utilize engineers from this field.

1.6 Select a branch of engineering, such as mechanical engineering, and an engineering function, such as design. Write a brief paper on some typical activities that are undertaken by the engineer performing the specified function. Sources of information can include books on engineering career opportunities and practicing engineers in the particular branch.

1.7 For a particular branch of engineering, such as electrical engineering, find the program of study for the first 2 years and compare it with the program offered at your school approximately 20 years ago. Comment on the major differences.

1.8 Do Prob. 1.7 for the *last* 2 years of study in a particular branch of engineering.

1.9 List five of your own personal characteristics and compare that list with the list in Sec. 1.5.1.

1.10 Prepare a brief paper on the requirements for professional registration in your state. Include the type and content of the required examinations.

1.11 Prepare a 5-minute talk to present to your class describing one of the technical societies and how it can benefit you as a student.

1.12 Choose one of the following topics (or one suggested by your instructor) and write a paper that discusses technological changes that have occurred in this area in the past 15 years. Include commentary on the social and environmental impact of the changes and on new problems that may have arisen because of the changes.

(*a*) passenger automobiles
(*b*) electric power-generating plants
(*c*) computer graphics
(*d*) heart surgery
(*e*) heating systems (furnaces)
(*f*) microprocessors
(*g*) water treatment
(*h*) road paving (both concrete and asphalt)
(*i*) computer-controlled metal fabrication processes
(*j*) robotics
(*k*) air conditioning

2

Engineering Solutions

2.1 Introduction

The practice of engineering involves the application of accumulated knowledge and experience to a wide variety of technical situations. Two areas, in particular, that are fundamental to all of engineering are design and problem solving. The professional engineer is expected to intelligently and efficiently approach, analyze, and solve a range of technical problems. These problems can vary from single solution, reasonably simple problems to extremely complex, open-ended problems that require a multidisciplinary team of engineers.

Problem solving is a combination of experience, knowledge, process, and art. Most engineers either through training or experience solve many problem by process. The design process, for example, is a series of logical steps that when followed produce an optimal solution given time and resources as two constraints. The total quality (TQ) method is another example of a process. This concept suggests a series of steps leading to desired results while exceeding customer expectations.

This chapter provides a basic guide to problem analysis, organization, and presentation. Early in your education, you must develop an ability to solve and present simple or complex problems in an orderly, logical, and systematic way.

2.2 Problem Analysis

A distinguishing characteristic of a qualified engineer is the ability to solve problems. Mastery of problem solving involves a combination of art and science. By *science* we mean a knowledge of the principles of mathematics, chemistry, physics, mechanics, and other technical subjects that must be learned so that they can be applied correctly. By *art* we mean the proper judgment, experience, common sense, and know-how that must be used to reduce a real-life problem to such a form that science can be applied to its solution. To know when and how rigorously science should be applied and whether the resulting answer reasonably satisfies the original problem is an art.

Much of the science of successful problem solving comes from formal education in school or from continuing education after graduation. But most of the art of problem solving cannot be learned in a formal course; rather, it is a result of experience and common sense. Its application can be more effective, however, if problem solving is approached in a logical and organized method—that is, if it follows a process.

To clarify the distinction, let us suppose that a manufacturing engineer working for an electronics company is given the task of recommending whether the introduction of a new personal computer that will focus on an inexpensive system for the home market can be profitably produced. At the time the engineering task is assigned, the competitive selling price has already been established by the marketing division. Also, the design group has developed working models of the personal computer with specifications of all components, which means that the cost of these components is known. The question of profit thus rests on the cost of assembly. The theory of engineering economy (the science portion of problem solving) is well known by the engineer and is applicable to the cost factors and time frame involved. Once the production methods have been established, the cost of assembly can be computed using standard techniques. Selection of production methods (the art portion of problem solving) depends largely on the experience of the engineer. Knowing what will or will not work in each part of the manufacturing process is a must in the cost estimate, but that data cannot be found in a handbook. It is in the mind of the engineer. It is an art originating from experience, common sense, and good judgment.

Before the solution to any problem is undertaken, whether by a student or by a practicing professional engineer, a number of important ideas must be considered. Consider the following questions: How important is the answer to a given problem? Would a rough, preliminary estimate be satisfactory, or is a high degree of accuracy demanded? How much time do you have and what resources are at your disposal? In an actual situation your answers may depend on the amount of data available or the amount that must be collected, the sophistication of equipment that must be used, the accuracy of the data, the number of people available to assist, and many other factors. Most complex problems require some level of computer support. What about the theory you intend to use? Is it state of the art? Is it valid for this particular application? Do you currently understand the theory, or must time be allocated for review and learning? Can you make assumptions that simplify without sacrificing needed accuracy? Are other assumptions valid and applicable?

The art of problem solving is a skill developed with practice. It is the ability to arrive at a proper balance between the time and resources expended on a problem and the accuracy and validity obtained in the solution. When you can optimize time and resources versus reliability, then problem-solving skills will serve you well.

2.3 The Engineering Method

The engineering method is an example of process. Earlier the engineering design process was mentioned. Although there are different processes that could be listed, a typical process is represented by the following 10 steps:

1. Identify the problem
2. Define the problem
3. Search
4. Constraints
5. Criteria
6. Alternative solutions
7. Analysis
8. Decision
9. Specification
10. Communication

These design steps are simply the overall process that an engineer uses when solving an open-ended problem. One significant portion of this design procedure is step 7—analysis.

Analysis is the use of mathematical and scientific principles to verify the performance of alternative solutions. Analyses conducted by engineers in many design projects normally involve three areas: application of the laws of nature, application of the laws of economics, and application of common sense.

The analysis phase can be used as an example to demonstrate how cyclic the design process is intended to be. Within any given step or phase there are still other processes that can be applied. One very important such process is called the *engineering method.* It consists of six basic steps:

1. Recognize and Understand the Problem

Perhaps the most difficult part of problem solving is developing the ability to recognize and define the problem precisely. This is true at the beginning of the design process and when applying the engineering method to a subpart of the overall problem. Many academic problems that you will be asked to solve have this step completed by the instructor. For example, if your instructor asks you to solve a quadratic algebraic equation but provides you with all the coefficients, the problem has been completely defined before it is given to you and little doubt remains about what the problem is.

If the problem is not well defined, considerable effort must be expended at the beginning in studying the problem, eliminating the things that are unimportant, and focusing on the root problem. Effort at this step pays great dividends by eliminating or reducing false trials, thereby shortening the time taken to complete later steps.

2. Accumulate Data and Verify Accuracy

All pertinent physical facts such as sizes, temperatures, voltages, currents, costs, concentrations, weights, times, and so on must be ascertained. Some problems require that steps 1 and 2 be done simultaneously. In others, step 1 might automatically produce some of the physical facts. Do not mix or confuse these details with data that are suspect or only assumed to be accurate. Deal only with items that can be verified. Sometimes it will pay to actually verify data that you believe to be factual but may actually be in error.

3. Select the Appropriate Theory or Principle

Select appropriate theories or scientific principles that apply to the solution of the problem; understand and identify limitations or constraints that apply to the selected theory.

4. Make Necessary Assumptions

Perfect solutions do not exist to real problems. Simplifications need to be made if they are to be solved. Certain assumptions can be made that do not significantly affect the accuracy of the solution, yet other assumptions may result in a large reduction in accuracy.

Although the selection of a theory or principle is stated in the engineering method as preceding the introduction of simplifying assumptions, there are cases where the order of these two steps should be reversed. For example, if you were solving a material balance problem you often need to assume that the process is steady, uniform, and without chemical reactions, so that the applicable theory can be simplified.

5. Solve the Problem

If steps 3 and 4 have resulted in a mathematical equation (model), it is normally solved by application of mathematical theory, although a trial-and-error solution which employs the use of a computer or perhaps some form of graphical solution may also be applicable. The results will normally be in numerical form with appropriate units.

6. Verify and Check Results

In engineering practice, the work is not finished merely because a solution has been obtained. It must be checked to ensure that it is mathematically correct and that units have been properly specified. Correctness can be verified by reworking the problem using a different technique or by performing the calculations in a different order to be certain that the numbers agree in both trials. The units need to be examined to insure that all equations are dimensionally correct. And finally, the answer must be examined to see if it makes sense. An experienced engineer will generally have a good idea of the order of magnitude to expect.

If the answer doesn't seem reasonable, there is probably an error in the mathematics, in the assumptions, or perhaps in the theory used. Judgment is critical. For example, suppose that you are asked to compute the monthly payment required to repay a car loan of $5,000 over a 3-year period at an annual interest rate of 12 percent. Upon solving this problem, you arrived at an answer of $11,000 per month. Even if you are inexperienced in engineering economy, you know that this answer is not reasonable, so you should reexamine your theory and computations. Examination and evaluation of the reasonableness of an answer is a habit that you should strive to acquire. Your instructor and employer alike will find it unacceptable to be given results which you have indicated to be correct but are obviously incorrect by a significant percentage.

2.4 Problem Presentation

The engineering method of problem solving as presented in the previous section is an adaptation of the well-known *scientific problem-solving method.* It is a time-tested approach to problem solving that should become an everyday part of the engineer's thought process. Engineers should follow this logical approach to the solution of any problem while at the same time learning to translate the information accumulated in to a well-documented problem solution.

The following steps parallel the engineering method and provide reasonable documentation of the solution. If these steps are properly executed during the solution of problems in this text and all other courses, it is our belief that you will gradually develop an ability to solve a wide range of complex problems.

1. Problem Statement

State as concisely as possible the problem to be solved. The statement should be a summary of the given information, but it must contain all essential material. Clearly state what is to be determined. For example, find the temperature (K) and pressure (Pa) at the nozzle exit.

2. Diagram

Prepare a diagram (sketch or computer output) with all pertinent dimensions, flow rates, currents, voltages, weights, and so on. A diagram is a very efficient method of showing given and needed information. It also is an appropriate way of illustrating the physical setup, which may be difficult to describe adequately in words. Data that cannot be placed in a diagram should be listed separately.

3. Theory

The theory used should be presented. In some cases, a properly referenced equation with completely defined variables is sufficient. At other times, an extensive theoretical derivation may be necessary because the appropriate theory has to be derived, developed, or modified.

4. Assumptions

Explicitly list in complete detail any and all pertinent assumptions that have been made to realize your solution to the problem. This step is vitally important for the reader's understanding of the solution and its limitations. Steps 3 and 4 might be reversed in some problems.

5. Solution Steps

Show completely all steps taken in obtaining the solution. This is particularly important in an academic situation because your reader, the instructor, must have the means of judging your understanding of the solution technique. Steps completed but not shown make it difficult for evaluation of your work and, therefore, difficult to provide constructive guidance.

6. Identify Results and Verify Accuracy

Clearly identify (double underline) the final answer. Assign proper units. An answer without units (when it should have units) is meaningless. Remember, this final step of the engineering method requires that the answer be examined to determine if it is realistic, so check solution accuracy and, if possible, verify the results.

2.5 Standards of Problem Presentation

Once the problem has been solved and checked, it is necessary to present the solution according to some standard. The standard will vary from school to school and industry to industry.

On most occasions your solution will be presented to other individuals who are technically trained, but you should remember that many times these individuals do not have an intimate knowledge of the problem. However, on other occasions you will be presenting technical information to persons with nontechnical backgrounds. This may require methods different from those used to communicate with other engineers, so it is always important to understand who will be reviewing the material so that the information can be clearly presented.

One characteristic of engineers is their ability to present information with great clarity in a neat, careful manner. In short, the information must be communicated accurately to the reader. (Discussion of drawings or simple sketches will not be included in this chapter, although they are important in many presentations.)

Employers insist on carefully prepared presentations that completely document all work involved in solving the problems. Thorough documentation may be important in the event of legal considerations, for which the details of the work might be introduced into the court proceedings as evidence. Lack of such documentation may result in the loss of a case that might otherwise have been won. Moreover, internal company use of the work is easier and more efficient if all aspects of the work have been carefully documented and substantiated by data and theory.

Each industrial company, consulting firm, governmental agency, and university has established standards for presenting technical information. These standards vary slightly, but all fall into a basic pattern, which we will discuss. Each organization expects its employees to follow its standards. Details can be easily modified in a particular situation once you are familiar with the general pattern that exists in all of these standards.

It is not possible to specify a single problem layout or format that will accommodate all types of engineering solutions. Such a wide variety of solutions exists that the technique used must be adapted to fit the information to be communicated. In all cases, however, one must lay out a given problem in such a fashion that it can be easily grasped by the reader. No matter what technique is used, it must be logical and understandable.

We have listed guidelines for problem presentation. Acceptable layouts for problems in engineering are also illustrated. The guidelines are not intended as a precise format that must be followed, but rather as a suggestion that should be considered and incorporated whenever applicable.

Two methods of problem presentation are typical in the academic and industrial environments. Presentation formats can be either freehand or computer generated. As hardware technology and software developments continue to provide better tools, the use of the computer as a method of problem presentation will continue to increase.

If a formal report, proposal, or presentation is to be the choice of communication, a computer-generated presentation is the correct approach. The example solutions that are illustrated in Figs. 2.1, 2.2, and 2.3 include both freehand as well as computer output. Check with your instructor to determine which method is appropriate for your assignments.

The following 9 general guidelines should be helpful as you develop the freehand skills needed to provide clear and complete problem documentation. The first two examples, Figs. 2.1 and 2.2, are freehand illustrations, and the third example, Fig. 2.3, is computer generated. These guidelines are most applicable to freehand solutions, but many of the ideas and principles apply to computer generation as well.

1. One common type of paper frequently used is called engineering-problems paper. It is ruled horizontally, and vertically on the reverse side, with only heading and margin rulings on the front. The rulings on the reverse side, which are faintly visible through the paper, help one maintain horizontal lines of lettering and provide guides for

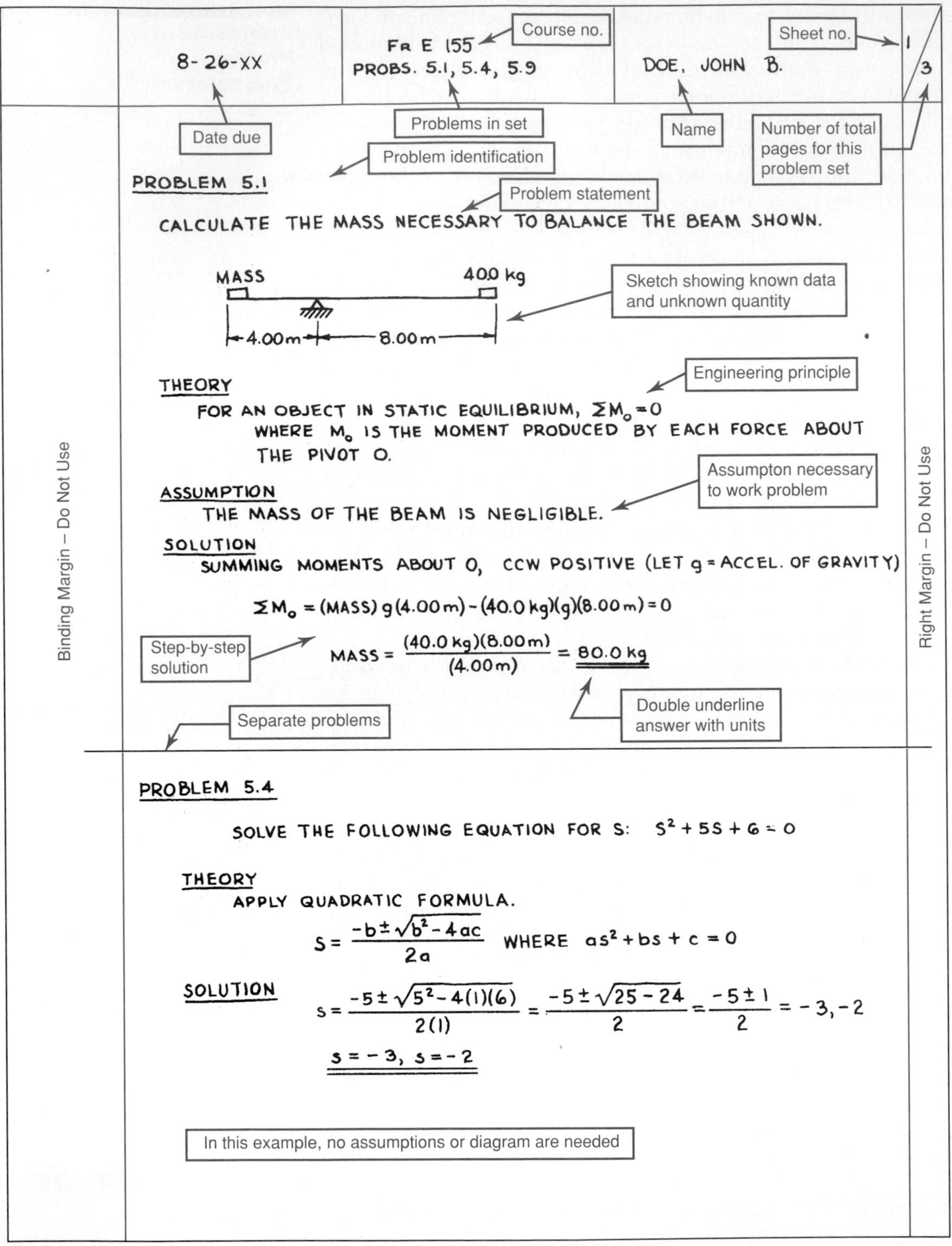

8-26-XX
FR E 155
PROBS. 5.1, 5.4, 5.9
DOE, JOHN B.
1
3
Course no.
Sheet no.
Date due
Problems in set
Name
Number of total pages for this problem set
Problem identification
PROBLEM 5.1
Problem statement
CALCULATE THE MASS NECESSARY TO BALANCE THE BEAM SHOWN.
MASS
40.0 kg
4.00 m
8.00 m
Sketch showing known data and unknown quantity
THEORY
Engineering principle
FOR AN OBJECT IN STATIC EQUILIBRIUM, ΣM_O = 0
WHERE M_O IS THE MOMENT PRODUCED BY EACH FORCE ABOUT THE PIVOT O.
ASSUMPTION
Assumpton necessary to work problem
THE MASS OF THE BEAM IS NEGLIGIBLE.
SOLUTION
SUMMING MOMENTS ABOUT O, CCW POSITIVE (LET g = ACCEL. OF GRAVITY)
ΣM_O = (MASS) g (4.00 m) − (40.0 kg)(g)(8.00 m) = 0
Step-by-step solution
MASS = (40.0 kg)(8.00 m) / (4.00 m) = 80.0 kg
Double underline answer with units
Separate problems
PROBLEM 5.4
SOLVE THE FOLLOWING EQUATION FOR S: S² + 5S + 6 = 0
THEORY
APPLY QUADRATIC FORMULA.
S = (−b ± √(b² − 4ac)) / 2a WHERE as² + bs + c = 0
SOLUTION
s = (−5 ± √(5² − 4(1)(6))) / 2(1) = (−5 ± √(25 − 24)) / 2 = (−5 ± 1) / 2 = −3, −2
s = −3, s = −2
In this example, no assumptions or diagram are needed
Binding Margin – Do Not Use
Right Margin – Do Not Use

Figure 2.1
Elements of a problem layout.

	8-22-XX	FR E 160 PROBLEM 13.1	DOE, JANE A.	1/2

PROBLEM 13.1 SOLVE FOR THE VALUE OF RESISTANCE R IN THE CIRCUIT SHOWN BELOW.

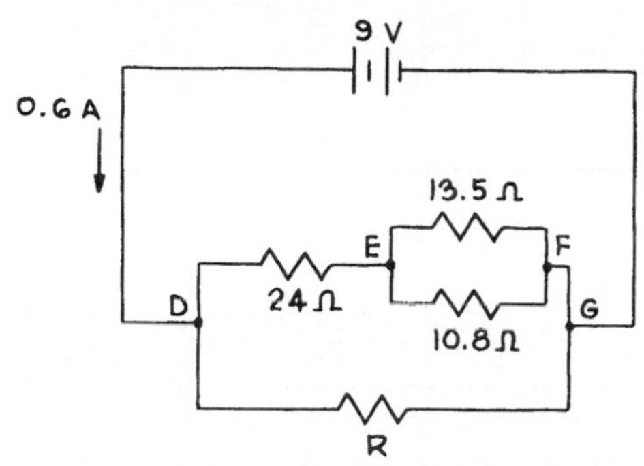

THEORY

- FOR RESISTANCES IN PARALLEL: $\frac{1}{R_{TOTAL}} = \frac{1}{R_1} + \frac{1}{R_2} + \frac{1}{R_3} + \ldots$

  THUS FOR 2 RESISTANCES IN PARALLEL

  $$R_{TOTAL} = \frac{R_1 R_2}{R_1 + R_2}$$

- FOR RESISTANCES IN SERIES: $R_{TOTAL} = R_1 + R_2 + R_3 + \ldots$

- OHM'S LAW: $E = RI$ WHERE E = ELECT. POTENTIAL IN VOLTS
  I = CURRENT IN AMPERES
  R = RESISTANCE IN OHMS

SOLUTION

- CALCULATE EQUIVALENT RESISTANCE BETWEEN POINTS E AND F. RESISTORS ARE IN PARALLEL.

  $$\therefore R_{EF} = \frac{R_1 R_2}{R_1 + R_2} = \frac{(13.5)(10.8)}{13.5 + 10.8} = \frac{145.8}{24.3} = 6.00\ \Omega$$

- CALCULATE EQUIVALENT RESISTANCE OF UPPER LEG BETWEEN D AND G.

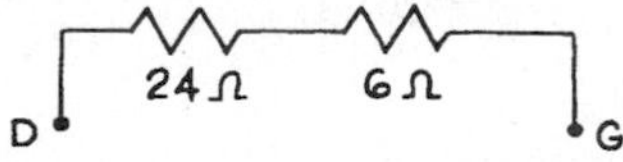

SERIES CIRCUIT

$$\therefore R'_{DG} = R_{24} + R_6 = 24 + 6 = 30\ \Omega$$

In this example, no assumptions were necessary

Figure 2.2
Sample problem presentation.

	8-22-XX	FR E 160 PROBLEM 13.1	DOE, JANE A.	2/2

- CALCULATE EQUIVALENT RESISTANCE BETWEEN D AND G.

PARALLEL RESISTORS, SO

$$R_{DG} = \frac{(R'_{DG})(R)}{R'_{DG} + R}$$

- CALCULATE TOTAL RESISTANCE OF CIRCUIT USING OHM'S LAW.

$$R_{DG} = \frac{E}{I} = \frac{9\text{ V}}{0.6\text{ A}} = 15\ \Omega$$

- CALCULATE VALUE OF R FROM PREVIOUS EQUATIONS.

$$R_{DG} = 15\ \Omega = \frac{(R'_{DG})(R)}{R'_{DG} + R} = \frac{(30)(R)}{30 + R}$$

SOLVING FOR R:

$$(30 + R)(15) = 30R$$

$$30 + R = 2R$$

$$\underline{\underline{R = 30\ \Omega}}$$

Figure 2.2 (cont.)

Date **Engineering** **Name:**__________________

Problem

A tank is to be constructed that will hold 5.00×10^5 L when filled. The shape is to be cylindrical, with a hemispherical top. Costs to construct the cylindrical portion will be $\$300/m^2$, while costs for the hemispherical portion are slightly higher at $\$400/m^2$.

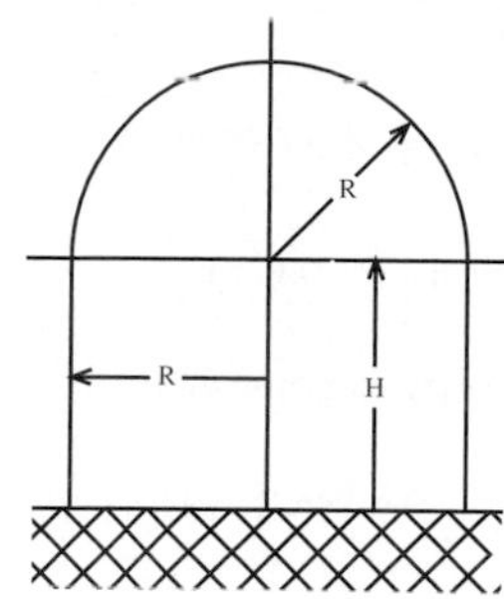

Find

Calculate the tank dimensions that will result in the lowest dollar cost.

Theory

Volume of cylinder is... $V_c = \pi R^2 H$

Volume of hemisphere is... $V_H = \dfrac{2\pi R^3}{3}$

Surface area of cylinder is... $SA_c = 2\pi RH$

Surface area of hemisphere is... $SA_H = 2\pi R^2$

Assumptions

Tank contains no dead air space
Construction costs are independent of size
Concrete slab with hermetic seal is provided for the base.
Cost of the base does not change appreciably with tank dimensions.

Solution

1. Express total volume in meters as a function of height and radius

$$V_{Tank} = f(H, R)$$
$$= V_C + V_H$$
$$500 = \pi R^2 H + \frac{2\pi R^3}{3}$$

Note: $1m^3 = 1000L$

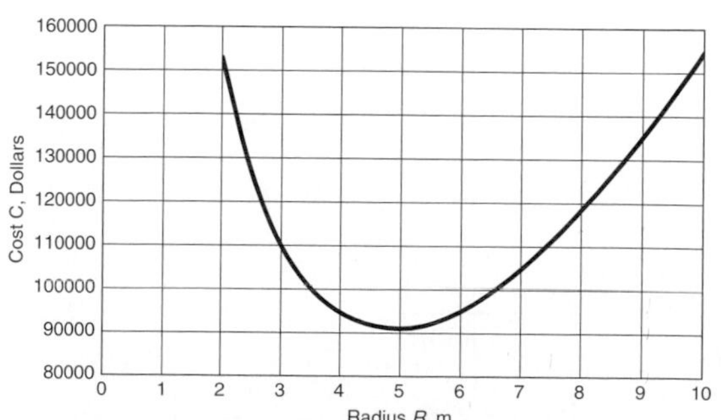

Figure 2.3
Sample problem presentation.

2. Express cost in dollars as a function of height and radius

$$C = C(H, R)$$

$$= 300\,(SA_C) + 400\,(SA_H)$$

$$= 300\,(2\pi RH) + 400\,(2\pi R^2)$$

Note: Cost figures are exact numbers

3. From part #1 solve for H = H (R)

$$H = \frac{500}{\pi R^2} - \frac{2R}{3}$$

4. Solve cost equation, substituting H = H (R)

$$C = 300\left[2\pi R\left(\frac{500}{HR^2} - \frac{2R}{3}\right)\right] + 400\left(2\pi R^2\right)$$

$$C = \frac{300000}{R} + 400\pi R^2$$

5. Develop a table of Cost vs. Radius and plot graph.

Cost vs. Radius

Radius R, m	Cost C, $
1.0	301 257
2.0	155 027
3.0	111 310
4.0	95 106
5.0	91 416
6.0	95 239
7.0	104 432
8.0	117 925
9.0	135 121
10.0	155 664

6. From graph select minimum cost.

R = 5.00m

C = $91 000

7. Calculate H from part 3 above

H = 3.033 m

8. Verification / check of results from the calculus:

$$\frac{dC}{dR} = \frac{d}{dR}\left[\frac{300000}{R} + 400\pi R^2\right]$$

$$= \frac{-300000}{R^2} + 800\pi R = 0$$

$$R^3 = \frac{300000}{800\pi}$$

$$R = 4.92\text{m}$$

Figure 2.3 (cont.)

sketching and simple graph construction. Moreover, the lines on the back of the paper will not be lost as a result of erasures.

2. The completed top heading of the problems paper should include such information as name, date, course number, and sheet number. The upper right-hand block should normally contain a notation such as a/b, where a is the page number of the sheet and b is the total number of sheets in the set.

3. Work should ordinarily be done in pencil using an appropriate lead hardness (F or H) so that the linework is crisp and unsmudged. Erasures should always be complete, with all eraser particles removed. Letters and numbers must be dark enough to insure legibility when photocopies are needed.

4. Either vertical or slant letters may be selected as long as they are not mixed. Care should be taken to produce good, legible lettering but without such care that little work is accomplished.

5. Spelling should be checked for correctness. There is no reasonable excuse for incorrect spelling in a properly done problem solution.

6. Work must be easy to follow and uncrowded. This practice contributes greatly to readability and ease of interpretation.

7. If several problems are included in a set, they must be distinctly separated, usually by a horizontal line drawn completely across the page between problems. Never begin a second problem on the same page if it cannot be completed there. It is usually better to begin each problem on a fresh sheet, except in cases where two or more problems can be completed on one sheet. It is not necessary to use a horizontal separation line if the next problem in a series begins at the top of a new page.

8. Diagrams that are an essential part of a problem presentation should be clear and understandable. Students should strive for neatness, which is a mark of a professional. Often a good sketch is adequate, but using a straightedge can greatly improve the appearance and accuracy of a diagram. A little effort in preparing a sketch to approximate scale can pay great dividends when it is necessary to judge the reasonableness of an answer, particularly if the answer is a physical dimension that can be seen on the sketch.

9. The proper use of symbols is always important, particularly when the International System (SI) of units is used. It involves a strict set of rules that must be followed so that absolutely no confusion of meaning can result. There are also symbols in common and accepted use for engineering quantities that can be found in most engineering handbooks. These symbols should be used whenever possible. It is important that symbols be consistent throughout a solution and that all are defined for the benefit of the reader and for your own reference.

The physical layout of a problem solution logically follows steps similar to those of the engineering method. You should attempt to present the process by which the problem was solved in addition to the solution so that any reader can readily understand all aspects of the solution. Figure 2.1 illustrates the placement of the information.

Figures 2.2 and 2.3 are examples of typical engineering-problem solutions. You may find that they are helpful guides as you prepare your problem presentations.

2.6 Key Terms and Concepts

The following terms are basic to the material in Chapter 2. You should be able to define these terms and to be able to interpret them into various applications.

Process
Analysis
Engineering method
Problem presentation
Solution documentation

Problems

The solution to lengths and angles of oblique triangles can be arrived at by application of fundamental trigonometry. All angles are to be considered precise numbers. Solve the following problems using Fig. 2.4 as a general guide.

Figure 2.4

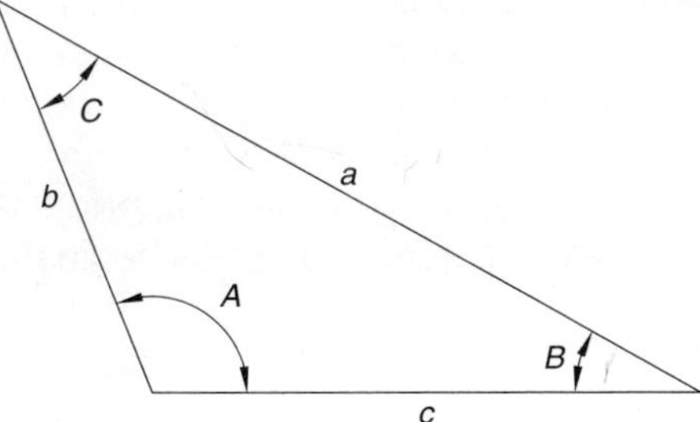

2.1 Given one side and two angles of an oblique triangle.

$C = 75°$ $a = 1\ 255$ m

$A = 30°$

Using laws of sines and sum of angles, determine the angle B and distances b and c.

2.2 Given two sides and the included angle of an oblique triangle.

$C = 40°$ $a = 75$ in

$b = 44$ in

Using the law of cosines and sum of angles, determine the angles A and B and the distance c.

2.3 Given the sides of an oblique triangle.

$a = 440$ ft

$b = 910$ ft

$c = 1\ 285$ ft

Using the law of cosines and sum of angles, determine the angles A, B, and C.

2.4 Given two sides and the included angle of an oblique triangle.

$A = 20°$ $\quad b = 2\ 550$ m

$c = 1\ 825$ m

(*a*) Using the sum of angles, law of sines, and law of tangents, determine the angles B and C and the distance a.

(*b*) Compute the radius (r), of an inscribed circle and the radius (R) of the circumscribed circle.

Vector quantities have both magnitude and direction. Figure 2.5 illustrates convention for Probs. 2.5 and 2.6, with positive angles measured clockwise from the + x-axis.

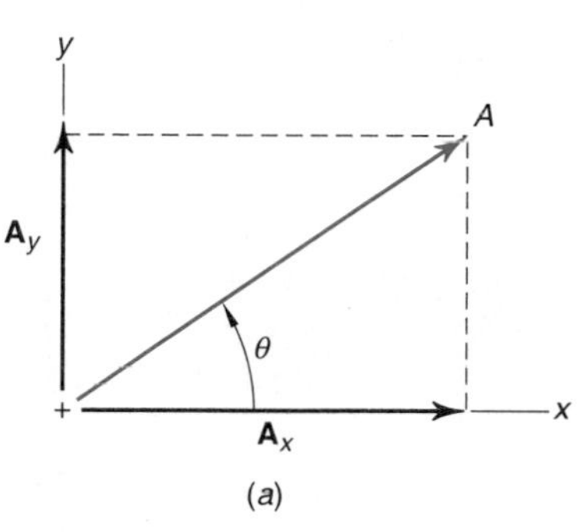

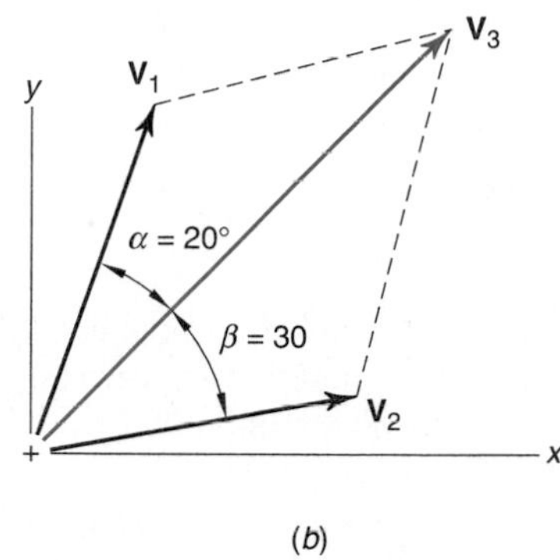

Figure 2.5

2.5 A vector **A** has a magnitude of 325 m at $\theta = -52°$ (ccw) from the + x-axis.

(*a*) Determine the magnitude of $|\mathbf{A}x|$ and $|\mathbf{A}y|$.

(*b*) Determine **A** (magnitude and direction) if $\mathbf{A}x = 85$ m and $\mathbf{A}y = 33$ m.

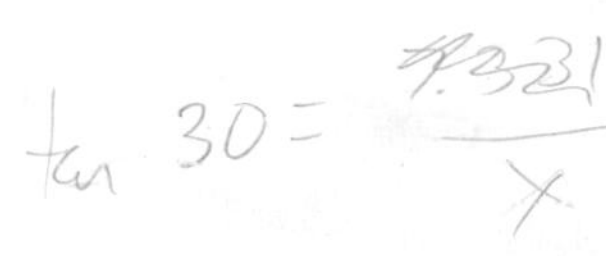

2.6 A wind vector $\mathbf{V}_3$ is the sum of vectors $\mathbf{V}_1$ and $\mathbf{V}_2$. Vector $\mathbf{V}_2$ makes angle of -15 with the + x-axis and has a magnitude of 20.0 mph. Determine the magnitude and direction of vector $\mathbf{V}_3$.

2.7 A survey crew determined the angles and distances given in Fig. 2.6. Calculate the distance across the lake at DE.

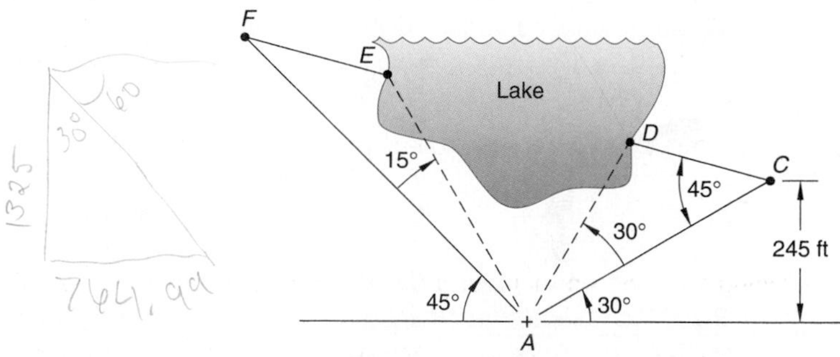

Figure 2.6

2.8 An engineering student in a stationary hot-air balloon is momentarily fixed at 1,325-ft elevation above a level piece of land. The pilot looks down (60° from horizontal) and turns laterally 360°. How many acres are contained within the core generated by his line of sight? How high would the balloon be if, when performing the same procedure, an area 4 times greater is encompassed?

2.9 A small aircraft has a glide ratio of 15:1. (This glide ratio means the plane moves 15 units horizontally for each 1 unit in elevation.) You are exactly in the middle of a 3.0 mile diameter lake at 5.00×10^2 ft when the fuel supply is exhausted. You see a gravel road by the dock (Fig. 2.7) and must decide whether to try a ground or water landing. Show appropriate assumptions and calculations to support your decision.

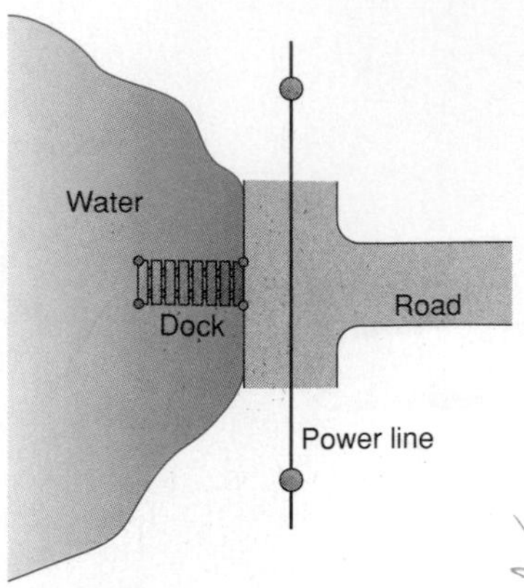

Figure 2.7

2.10 A pilot in an ultralight knows that her aircraft in landing configuration will glide 2.0×10^1 km from a height of 2.0×10^3 m. A TV transmitting tower is located in a direct line with the local runway. If the pilot glides over the tower with 3.0×10^1 m to spare and touches down on the runway at a point 6.5 km from the base of the tower, how high is the tower?

2.11 A simple roof truss design is shown in Fig. 2.8. The lower section *VWXY* is made from three equal-length segments. *UW* and *XZ* are perpendicular to *VT* and *TY*, respectively. If *VWXY* is 24 m, and the height of the truss is 12 m, determine the lengths of *XT* and *XZ*.

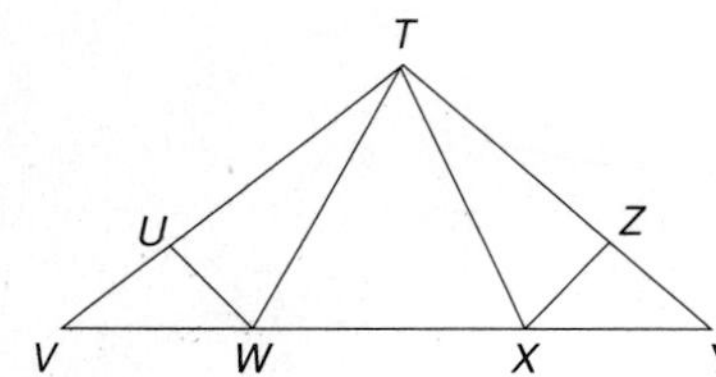

Figure 2.8

2.12 If you are traveling at 65 mph, what is your angular axle speed in RPM. Assume a tire size of P235/75R15, which has an approximate diameter of 30.0 in.

2.13 The wheel of an automobile turns at the rate of 195 RPM. Express this angular speed in (a) revolutions per second and (b) radians per second. If the wheel has a 30.0-in diameter, what is the velocity of the auto in miles per hour?

2.14 A bicycle wheel has a 28.0-in diameter wheel and is rotating 195 RPM. Express this angular speed in radians per second. How far will the bicycle travel (miles) in 45.0 minutes and what will be the velocity in miles per hour?

2.15 Assume the earth's orbit to be circular at 93.0×10^6 mi about the sun. Determine the speed of the earth (in miles per second) around the sun if there are exactly 365 days per year.

2.16 A child swinging on a tree rope 8.00 m long reaches a point 4.00 m above the lowest point. Through what total arc in degrees has the child passed? What distance in meters has the child swung?

2.17 Two engineering students were assigned the job of measuring the height of an inaccessible cliff (Fig. 2.9). The angles and distances shown were measured on a level beach in a vertical plane due south of the cliff. Determine the horizontal and vertical distances from *A* to *B*.

Figure 2.9

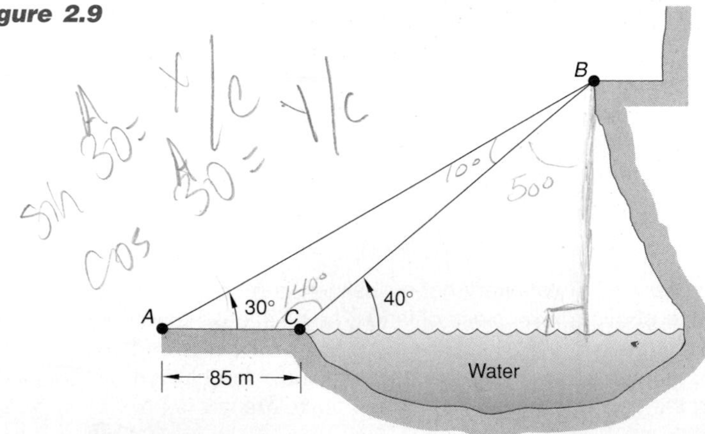

2.18 A survey team with appropriate equipment has been asked to measure a nonrectangular plot of land *ABCD* (Fig. 2.10). The following data were recorded: *CD* = 165.0 m, *DA* = 180.0 m, and *AB* = 115.0 m. Angle *DAB* = 120 and angle *DCB* = 100.

(*a*) Calculate the length of side *BC* and the area of the plot.

(*b*) Estimate the water surface area within the plot and list assumptions.

(*c*) Using answers from part b) what percentage of the total surface area within *ABCD* is land?

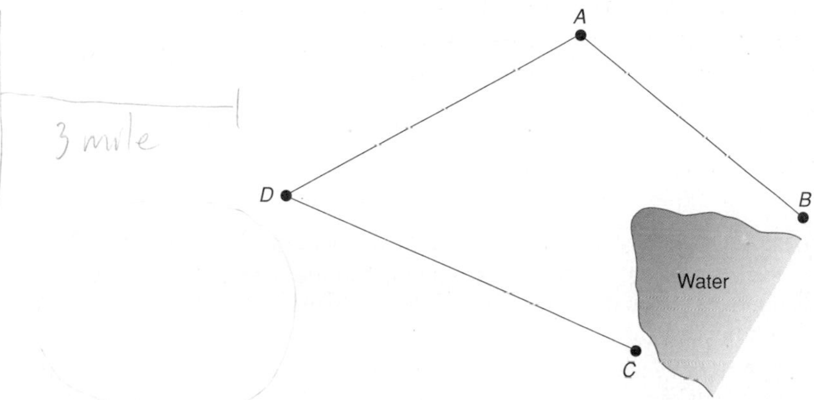

Figure 2.10

2.19 In Fig. 2.11 points *S*, *T*, *U*, and *V* are survey markers in a land development located on level terrain. The distance *ST* measures 355.0 ft. Angles at each marker were recorded as follows: *STU* = 80°, *TUS* = 70°, *VUS* = 50°, and *UVS* = 70°.

(*a*) Calculate the distances *TU*, *UV*, *VS*, and *SU*.

(*b*) Determine the area of *STUV* in acres.

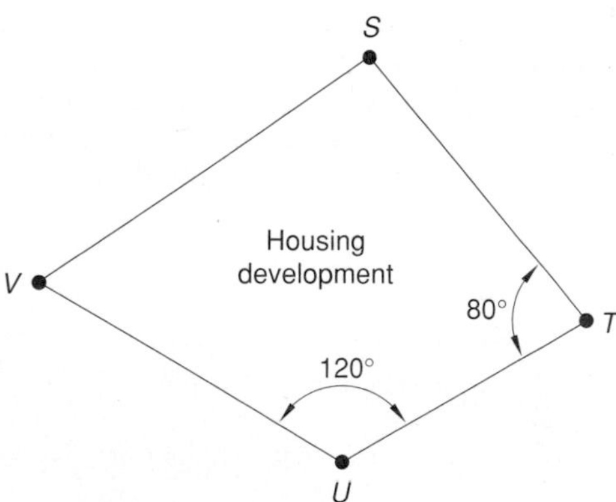

Figure 2.11

2.20 Circular sheets of metal 70.0 cm in diameter are to be used for stamping new highway signs. Calculate the percent of waste when the largest possible inscribed shape is:

(*a*) An equalateral triangle

(*b*) A square

(*c*) An octagon

2.21 Three circles are tangent to each other as in Fig. 2.12. The respective radii are 1.65×10^3, 10.00×10^2, and 7.75×10^2 mm.

(*a*) Find the area of the triangle (in square millimeters) formed by joining the 3 centers.

(*b*) Determine the area within the triangle that is outside of the circles.

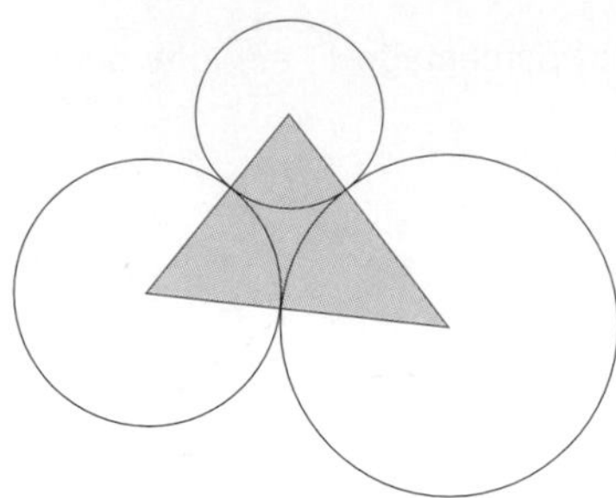

Figure 2.12

2.22 A waterwheel turns a belt on a drive wheel for a flour mill. The pulley on the waterwheel is 2.00 m in diameter, and the drive pulley is 0.500 m in diameter. If the centers of the pulleys are 4.00 m apart, calculate the length of belt needed.

2.23 A narrow, flat belt is used to drive a 50.00 cm (diameter) pulley from a 10.00 cm (diameter) pulley. The centers of the 2 pulleys are 50.00 cm apart. How long must the belt be if the pulleys rotate in the same direction? In opposite directions?

2.24 A homeowner decides to install a family swimming pool. It is 32.0 ft in diameter with a 4.00 ft water level. The cross-section of the pool is illustrated in Fig. 2.13. Consider the following problems.

(*a*) How many cubic feet of soil must be excavated to accommodate the pool liner if the bottom profile is as illustrated in Fig. 2.13?

(*b*) How many gallons of water will be required to fill the pool to 4 ft above ground?

(*c*) If the owner moves the water from a nearby lake, how many tons will be carried? (Density of water is 62.4 lbm/ft^3.)

(*d*) If the owner carries two 5-gal pails per trip and makes 20 trips per evening, how many days will it take to fill the pool?

Figure 2.13

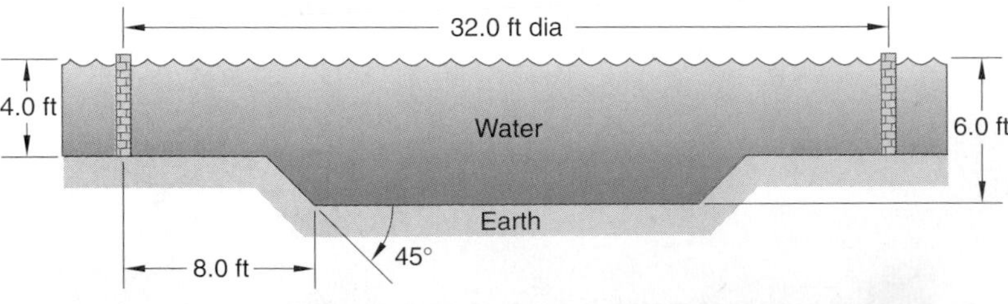

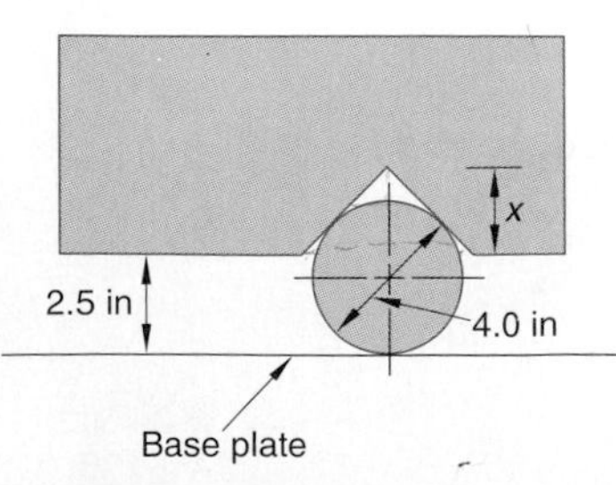

Figure 2.14

2.25 A block of metal has a 90° notch cut from its lower surface. The notched part rests on a circular cylinder 4.0 in in diameter as shown in Fig. 2.14. If the lower surface of the block is 2.5 in above the base plate, how deep is the notch?

2.26 Show that the area of the shaded segment in Fig. 2.15 is given by the expression

$$As = \frac{r^2}{2}(\phi - \sin \phi)$$

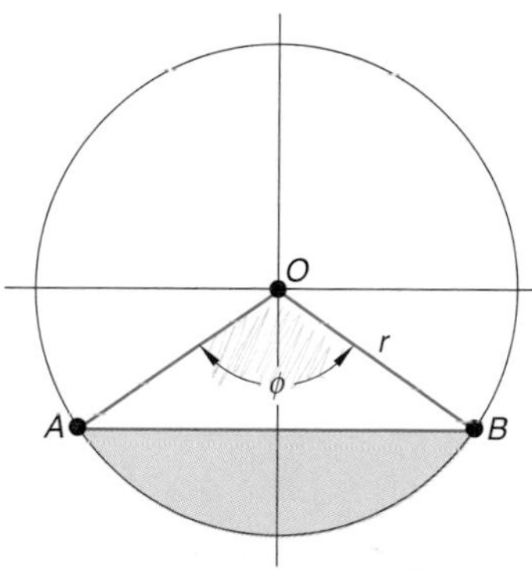

Figure 2.15

2.27 A fuel tank is 10.00 ft in diameter and 20.00 ft long. The tank is buried vertically as shown in Fig. 2.16. Develop an expression for volume (V) in gallons for any depth (h) in feet for this configuration.

2.28 Consider the fuel tank in Prob. 2.27 positioned on its side instead of its base. The tank length is 30.0 ft.

(*a*) Develop an expression for volume (V) in gallons as a function of radius (r) in feet and height (h) in feet.

(*b*) If the empty tank with radius 5.00 ft has a mass of 15.00×10^2 lbm, develop an expression for the mass of tank and fuel (in pounds-mass) as a function of height (h) in feet. Density of the fuel is 815 kg/m^3.

(*c*) How many gallons of fuel are in the tank at a height (h) of 4.50 ft? At that depth, what is the total mass of tank and fuel?

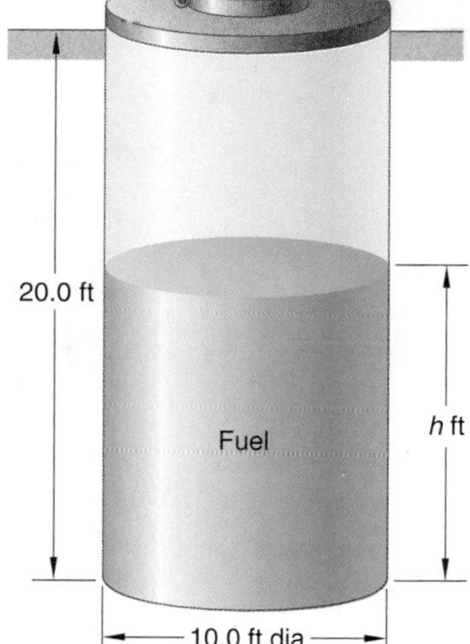

Figure 2.16

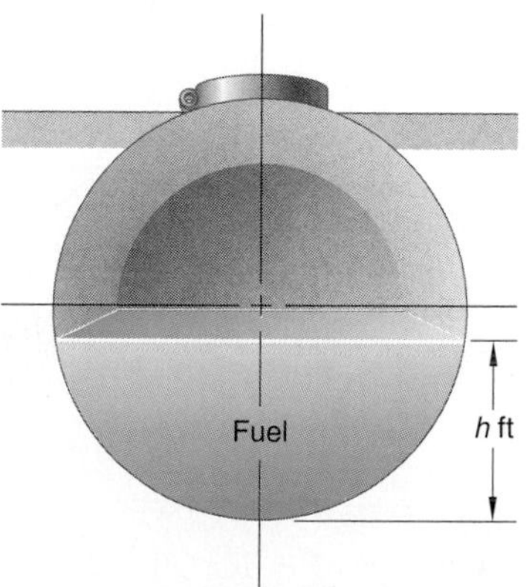

*End view of buried cylindrical tank

Figure 2.17

2.29 Eighteen circular wooden bases are to be cut from a piece of 3/4-in plywood. Each circular base has a diameter of 10.00 in. Assume a negligible saw blade thickness for the following problems.

(*a*) What is the area of triangle *ABC*?
(*b*) Calculate the area of waste material between *X* and *Y* above *BC*.
(*c*) Determine the dimensions of the rectangular piece of plywood shown in Fig. 2.18 to the nearest 0.5 in.
(*d*) What is the area of the largest piece of waste material?
(*e*) What is the percentage of waste given the configuration shown in Fig. 2.18?

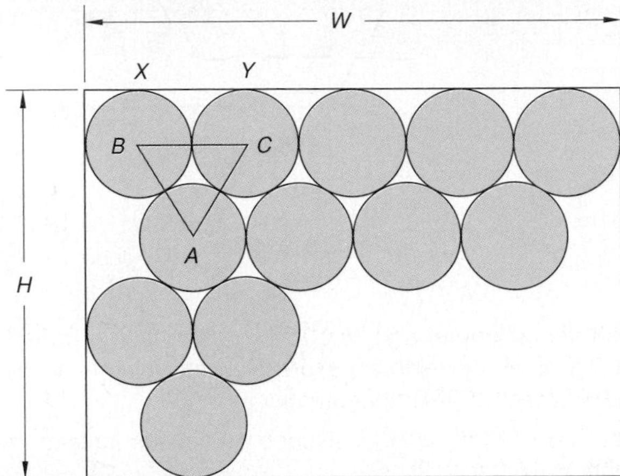

Figure 2.18

3 Representation of Technical Information

3.1 Introduction

This chapter contains examples and guidelines as well as helpful information that will be needed when collecting, recording, plotting, and interpreting technical data. Two areas will be considered in considerable detail: (1) graphical presentation of scientific data and (2) graphical analysis of plotted data.

Graphical presentation of technical data is necessary when calculated or experimental results are recorded in tabular form. Rapid and accurate determination of relationships between numerical values when the information is reported in columns and rows is not an optimal method for understanding. A procedure for graphing the results is needed. This approach provides a visual impression that is a more intuitive method to compare variables, rates of change, or relative magnitudes.

However, complete graphical analysis involves correct and accurate interpretation of data after it has been plotted. At times impressions are not sufficient, so determination of a mathematical model is required.

Computers with their expanding power, versatility, and speed are changing the way we collect, record, display, and analyze data. For example, consider the rapidly rising popularity of the microprocessor. It has changed the way we do many tasks. Microprocessors or small CPU cards can be programmed by a remote host computer, normally a PC. These CPUs are interfaced with some device, perhaps a temperature sensor or a robot. Information can be collected and stored and then based on the feedback of the information to the CPU additional programmed instructions can direct the device to perform corrective actions.

In addition to microprocessors and various other types of data collection instrumentation, there exists a wide variety of commercial software that enables the engineer to reduce the time required for recording, plotting, and analyzing data while increasing the accuracy of the results. Numerous software programs

Table 3-1

Height H, m	Temp T°C	Pressure P, kPa
0	15.0	101.3
300	12.8	97.7
600	11.1	94.2
900	8.9	90.8
1200	6.7	87.5
1500	5.0	84.3
1800	2.8	81.2
2100	1.1	78.2
2400	−1.1	75.3
2700	−2.8	72.4
3000	−5.0	68.7
3300	−7.2	66.9
3600	−8.9	64.4
3900	−11.1	61.9

are available for both technical presentation as well as analysis. These programs provide a wide range of powerful tools.

3.1.1 Software for Recording and Plotting Data

Data are recorded in the field as shown in Tab. 3.1. Many times a quick, freehand plot of the data is produced to provide a visual impression of the results while still in the field (see Fig. 3.1).

Upon returning to the laboratory, however, spreadsheet software, such as EXCEL and LOTUS 123 provide enormous record-

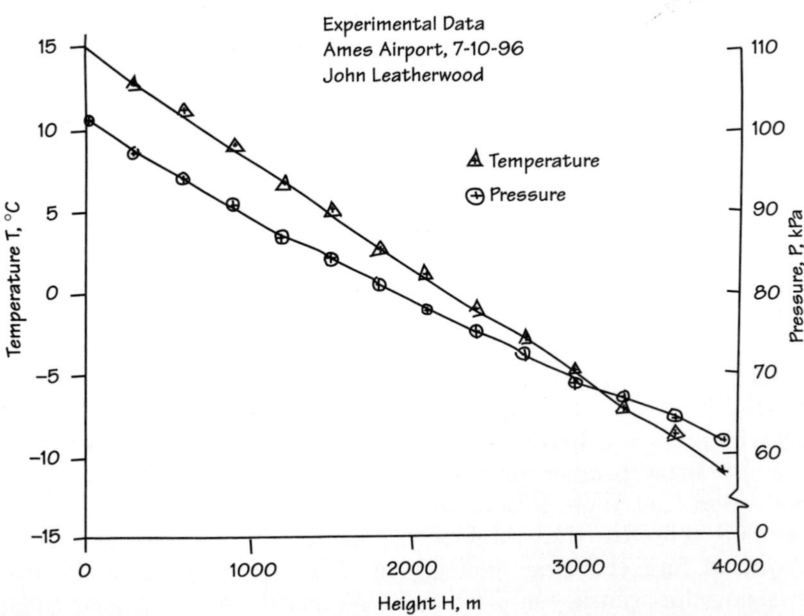

Figure 3.1 Freehand plot of data.

Table 3.2

Height H, m	Temperature T°C	Pressure P, kPa
0	15.0	101.3
300	12.8	97.7
600	11.1	94.2
900	8.9	90.8
1 200	6.7	87.5
1 500	5.0	84.3
1 800	2.8	81.2
2 100	1.1	78.2
2 400	1.1	75.3
2 700	−2.8	72.4
3 000	−5.0	68.7
3 300	−7.2	66.9
3 600	−8.9	64.4
3 900	−11.1	61.9

ing and plotting capability. The data are entered into the computer and by manipulation of software options both the data and a graph of the data can be configured, stored, and printed (Tab. 3.2 and Fig. 3.2 and Fig. 3.3).

Programs, such as Mathematica, Matlab, Mathcad, TK Solver, and many others, provide a range of powerful tools designed to help analyze numerical and symbolic operations as well as present a visual image of the results.

Software is also widely available to provide methods of curve fitting once the data has been collected and recorded. This subject will be treated in a separate chapter.

Even though it is important for the engineer to interpret, analyze, and communicate different types of data, it is not practical to include in this chapter all forms of graphs and charts that may be encountered. For that reason, popular-appeal or adver-

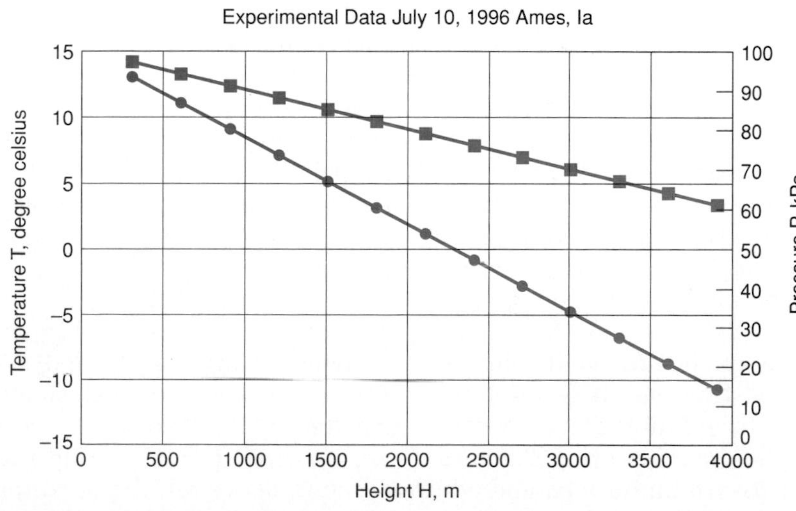

Figure 3.2
Hard copy of computer software analysis.

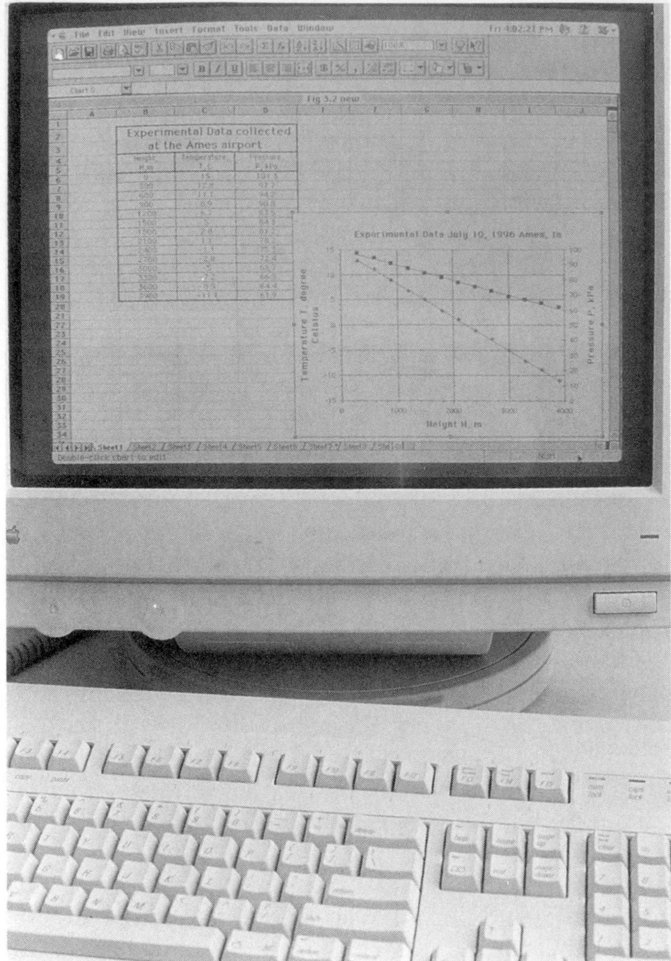

Figure 3.3

tising charts such as bar charts, pie diagrams, and distribution charts, although useful to the engineer, will not be discussed here.

Although commercial software is extremely helpful during the presentation and analysis process, the results are only as good as the original software design and its use by the operator. Some software provides a wide range of tools but only allows limited data applications and minimal flexibility to modify default outputs. Other software provides a high degree of in-depth analysis for a particular subject area with considerable latitude to adjust and modify parameters.

It is inevitable that the computer together with a growing array of software will continue to provide an invaluable analysis tool. However, it is absolutely essential that users be extremely knowledgeable of the software and demonstrate considerable care when manipulating the data. It is important to understand the software limitations and accuracies, but, above all, the operator must know what plotted results are needed and what expectations for appearance and readability are mandated.

It is for this reason that the sections to follow are a combination of manual collection, recording, plotting, and analysis and computer-assisted collection, recording, plotting, and analysis. As we begin the learning process, it is important to know how to mechanically manipulate data so that computer results can be appropriately modified.

3.2 Collecting and Recording Data

3.2.1 Manual

Modern science was founded on scientific measurement. Meticulously designed experiments, carefully analyzed, have produced volumes of scientific data that have been collected, recorded, and documented. For such data to be meaningful, however, certain laboratory procedures must be followed. Formal data sheets, such as those shown in Fig. 3.4, or laboratory notebooks should be used to record all observations. Information about equipment, such as the instruments and experimental apparatus used, should be recorded. Sketches illustrating the physical arrangement of equipment can be very helpful. Under no circumstances should observations be recorded elsewhere or data points erased. The data sheet is the "notebook of original entry." If there is reason for doubting the value of any entry, it may be canceled (that is, not considered) by drawing a line through it. The cancellation should be done in such a manner that the original entry is not obscured, in case you want to refer to it later.

Sometimes a measurement requires minimal precision, so time can be saved by making rough estimates. As a general rule, however, it is advantageous to make all measurements as precisely as time and the economics of the situation will allow. Unfortunately, as different observations are made throughout any experiment, some degree of inconsistency will develop. Errors enter into all experimental work regardless of the amount of care exercised.

It can be seen from what we have just discussed that the analysis of experimental data involves not only measurements and collection of data but also careful documentation and interpretation of results.

Experimental data once collected are normally organized into some tabular form, which is the next step in the process of analysis. Data, such as that shown in Tab. 3.1, should be carefully labeled and neatly lettered so that results are not misinterpreted. This particular collection of data represents atmospheric pressure and temperature measurements recorded at various altitudes by students during a flight in a light aircraft.

Electrical Engineering Laboratory, Iowa State University
of Science and Technology

Title of Test ______________________ No. ________

Test made by ______________________
FOREMAN

Apparatus tested: ______________________ No. ______ Set ______

Rating ______________________

Date __________ Wiring Checked by ________ Data Checked by ________

	1	2	3	4	5	6	7	8	9	10	11	12	13	14
INSTRUMENT NO.														
SHUNT OR C.T. NO.														
MULT. OR P.T. NO.														
RANGE USED														
SCALE READ														
FACTOR														
1														
2														
3														
4														
5														
6														
7														
8														
9														
10														
11														
12														
13														
14														
15														
16														
17														
18														
19														
20														
21														
22														
23														
24														
25														
26														
27														
28														
29														

(*a*)

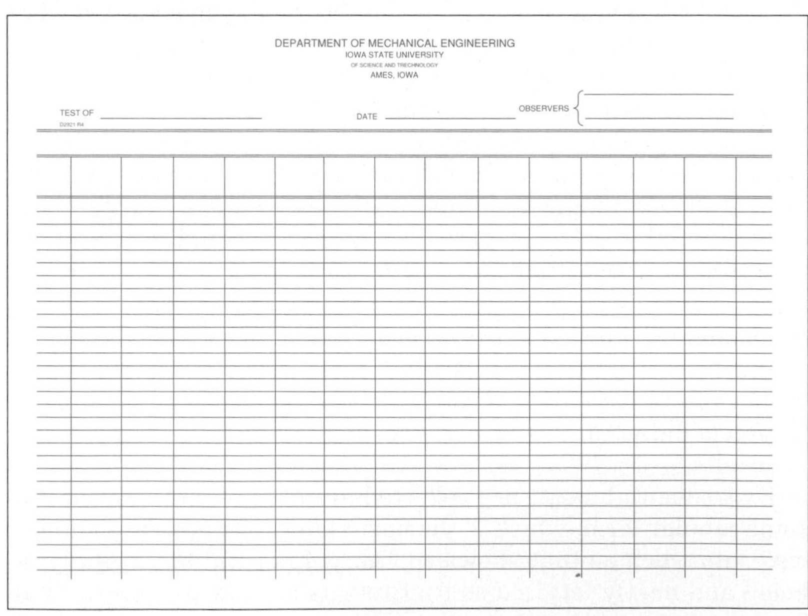

DEPARTMENT OF MECHANICAL ENGINEERING
IOWA STATE UNIVERSITY
OF SCIENCE AND TRECHNOLOGY
AMES, IOWA

TEST OF ______________ DATE ______________ OBSERVERS { ______________ / ______________ }

D2921 R4

(*b*)

Figure 3.4
Sample data sheets used by engineering departments.

Although the manual tabulation of data is frequently a necessary step, you will sometimes find it difficult to visualize a relationship between variables when simply viewing a column of numbers. A most important step in the sequence from collection to analysis is, therefore, the construction of appropriate graphs or charts.

3.2.2 Computer Assisted

In recent years, a variety of equipment has been developed which will automatically sample experimental data for analysis. We expect to see expansion of these techniques along with continuous visual displays that will allow us to interactively control the experiments. As an example the flight data collected onboard the aircraft can be entered directly into a spreadsheet and printed as in Tab. 3.2.

3.3 General Graphing Procedures

Many examples will be used throughout this chapter to illustrate methods of graphical presentation because their effectiveness depends to a large extent on the details of construction.

The proper construction of a graph from tabulated data can be generalized into a series of steps. Each of these steps will be discussed and illustrated in considerable detail in the following subsections.

1. Select the correct type of graph paper and grid spacing.
2. Choose the proper location of the horizontal and vertical axes.
3. Determine the scale units for each axis so that the data can be appropriately displayed.
4. Graduate and calibrate the axes.
5. Identify each axis completely.
6. Plot points and use permissible symbols (that is, ones commonly used and easily understood).
7. Draw the curve or curves.
8. Identify each curve and add the other necessary notes.
9. Darken lines for good reproduction.

3.3.1 Graph Paper

Printed coordinate graph paper is commercially available in various sizes with a variety of grid spacing. Rectilinear ruling can be purchased in a range of lines per inch or lines per centimeter, with an overall paper size of 8.5 × 11 inches considered most typical. Figure 3.5a is an illustration of graph paper having 10 lines per centimeter.

Closely spaced coordinate ruling is generally avoided for results that are to be printed or photoreduced. However, for accu-

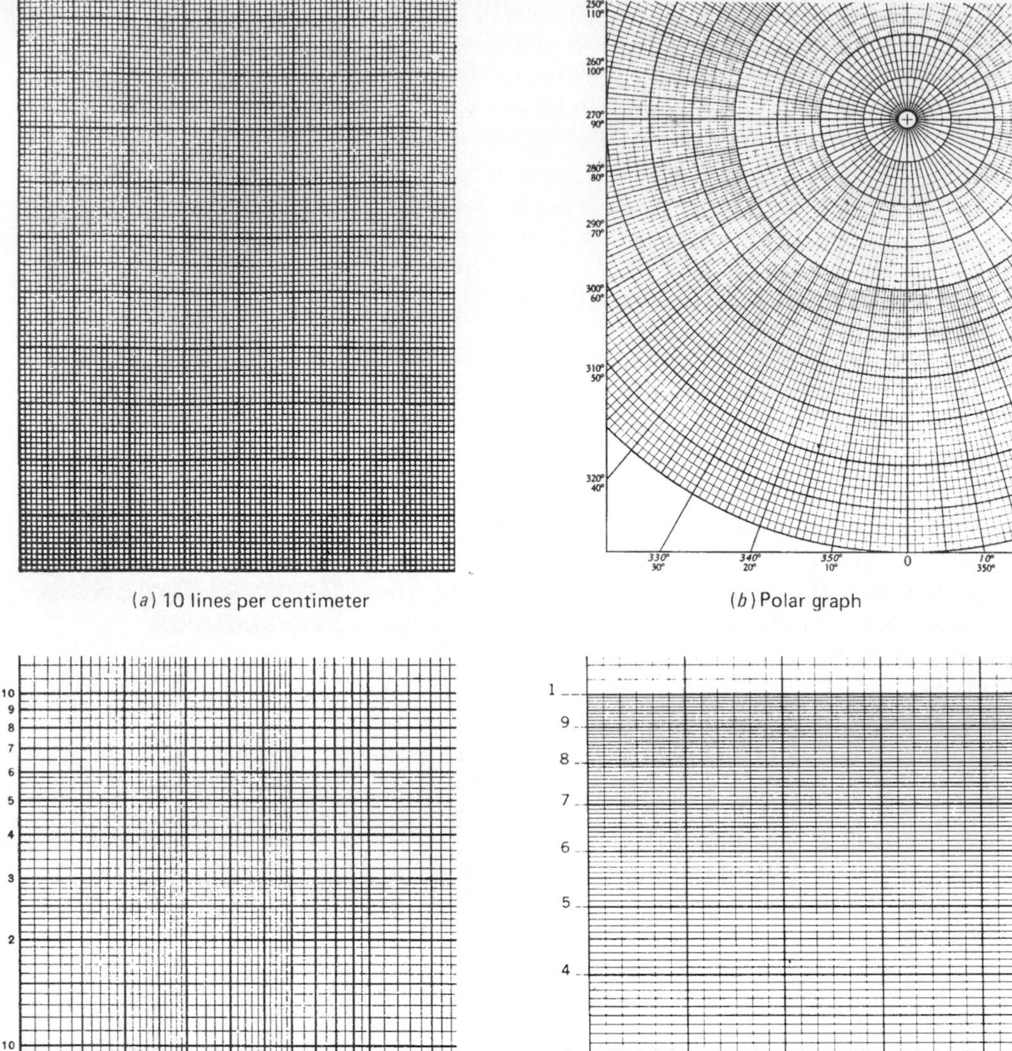

(*a*) 10 lines per centimeter

(*b*) Polar graph

(*c*) Log-log

(*d*) Semilog

Figure 3.5
Commercial graph paper.

rate engineering analyses requiring some amount of interpolation, data are normally plotted on closely spaced, printed coordinate paper. Graph paper is available in a variety of colors, weights, and grades. Translucent paper can be used when the reproduction system requires a material that is not opaque.

If the data require the use of log-log or semilog paper, such paper can also be purchased in different formats, styles, weights, and grades. Both log-log and semilog grids are available in from 1 to 5 cycles per axis. (A later section will discuss different applications of log-log and semilog paper.) Polar-coordinate paper is available in various sizes and graduations. A typical sheet is shown in Fig. 3.5b. Examples of commercially available logarithmic paper are given in Figs. 3.5c and d.

3.3.2 Axes Location and Breaks

The axes of a graph consist of two intersecting straight lines. The horizontal axis, normally called the *x-axis,* is the *abscissa*. The vertical axis, denoted by the *y-axis,* is the *ordinate*. Common practice is to place the independent values along the abscissa and the dependent values along the ordinate, as illustrated in Fig. 3.6.

Many times, mathematical graphs contain both positive and negative values of the variables. This necessitates the division of the coordinate field into 4 quadrants, as shown in Fig. 3.7. Positive values increase toward the right and upward from the origin.

On any graph, a full range of values is desirable, normally beginning at zero and extending slightly beyond the largest value. To avoid crowding, the entire coordinate area should be used as completely as possible. However, certain circumstances require special consideration to avoid wasted space. For example, if val-

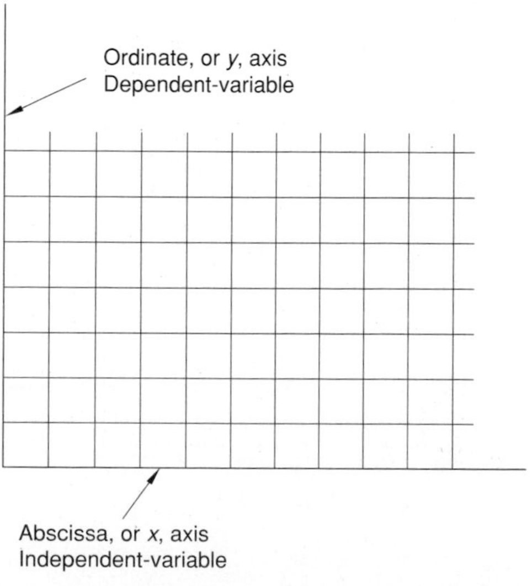

Figure 3.6
Abscissa(x) and ordinate(y) axes.

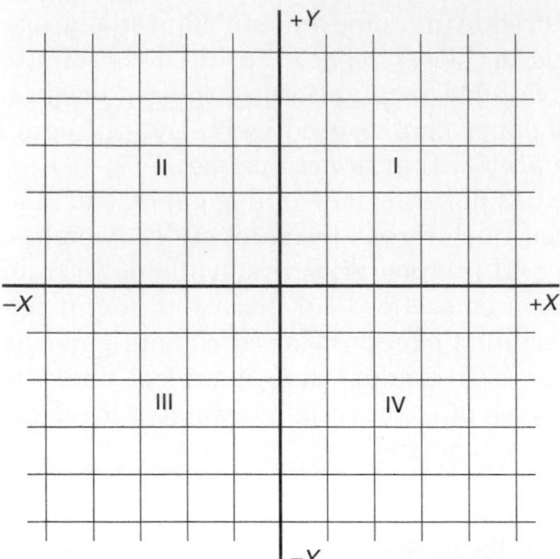

Figure 3.7
Coordinate axes.

ues to be plotted along the axis do not range near zero, a "break" in the grid or the axis may be used, as shown in Figs. 3.8a and b.

When judgments concerning relative amounts of change in a variable are required, the axis or grid should not be broken or the zero line omitted, with the exception of time in years, such as 1970, 1971, and so on, since that designation normally has little relation to zero.

Since most commercially prepared grids do not include sufficient border space for proper labeling, the axes should preferably be placed 20 to 25 mm (approximately 1 in) inside the edge of the printed grid in order to allow ample room for graduations, calibrations, axes labels, reproduction, and binding. The edge of the grid may have to be used on log-log paper, since it is not always feasible to move the axis. However, with careful planning, the vertical and horizontal axes can be repositioned in most cases, depending on the range of the variables.

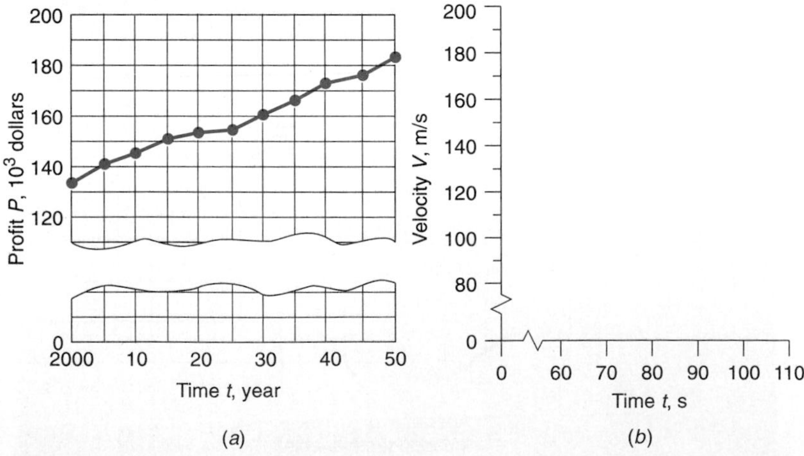

Figure 3.8
Typical axes breaks.

3.3.3 Scale Graduations, Calibrations, and Designations

The scale is a series of marks, called *graduations,* laid down at predetermined distances along the axis. Numerical values assigned to significant graduations are called *calibrations.*

A scale can be uniform, with equal spacing along the stem, as found on the metric, or engineer's, scales. If the scale represents a variable whose exponent is not equal to 1 or a variable that contains trigonometric or logarithmic functions, the scale is called a *nonuniform,* or *functional scale.* Examples of both these scales together with graduations and calibrations are shown in Fig. 3.9. When plotting data, one of the most important considerations is the proper selection of scale graduations. A basic guide to follow is the *1, 2, 5 rule,* which can be stated as follows:

> Scale graduations are to be selected so that the smallest division of the axis is a positive or negative integer power of 10 times 1, 2, or 5.

The justification and logic for this rule are clear. Graduation of an axis by this procedure makes possible interpolation of data between graduations when plotting or reading a graph. Figure 3.10 illustrates both acceptable and nonacceptable examples of scale graduations.

Violations of the 1, 2, 5 rule that are acceptable involve certain units of time as a variable. Days, months, and years can be graduated and calibrated as illustrated in Fig. 3.11.

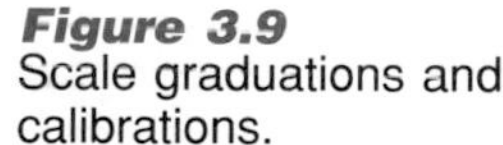
Figure 3.9 Scale graduations and calibrations.

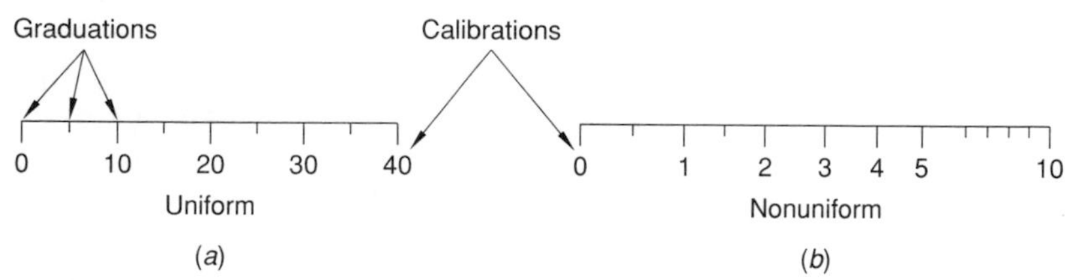

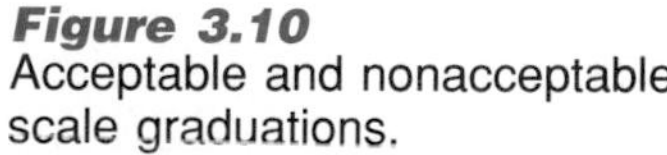
Figure 3.10 Acceptable and nonacceptable scale graduations.

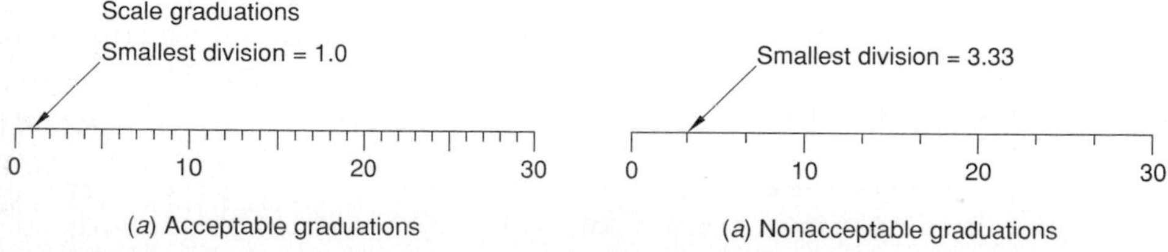

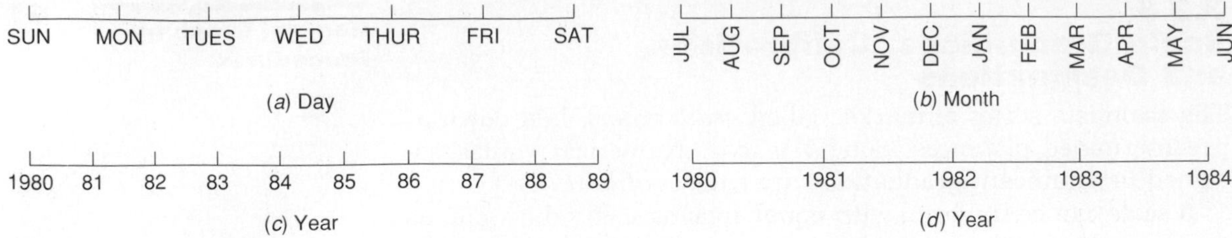

Figure 3.11
Time as a variable.

Scale graduations normally follow a definite rule, but the number of calibrations to be included is primarily a matter of good judgment. Each application requires consideration based on the scale length and range as well as the eventual use. Figure 3.12 demonstrates how calibrations can differ on a scale with the same length and range. Both examples obey the 1, 2, 5 rule, but as you can see, too many closely spaced calibrations make the axis difficult to read.

The selection of a scale deserves attention from another point of view. If the rate of change is to be depicted accurately, then the slope of the curve should represent a true picture of the data. By contracting or expanding the axis or axes, an incorrect impression of the data could be implied. Such a procedure is to be avoided. Figure 3.13 demonstrates how the equation $Y = X$ can be misleading if not properly plotted. Occasionally distortion is desirable, but it should always be carefully labeled and explained to avoid misleading conclusions.

If plotted data consist of very large or small numbers, the SI prefix names (milli-, kilo-, mega-, and so on) may be used to simplify calibrations. As a guide, if the numbers to be plotted and calibrated consist of more than three digits, it is customary to use the appropriate prefix; an example is illustrated in Fig. 3.14.

The length scale calibrations in Fig. 3.14 contain only two digits, but the scale can be read by understanding that the distance between the first and second graduation (0 to 1) is a kilometer; therefore, the calibration at 10 represents 10 km.

Certain quantities, such as temperature in degrees Celsius and altitude in meters, have traditionally been tabulated without the use of prefix multipliers. Figure 3.15 depicts a procedure by which these quantities can be conveniently calibrated. Note in particular that the distance between 0 and 1 on the scale represents 1 000°C. This is another example of how the SI notation is convenient, since the prefix multipliers (micro-, milli-, kilo-, mega-, and so on) allow the calibrations to stay within the three-digit guideline.

Figure 3.12
Acceptable and nonacceptable scale calibrations.

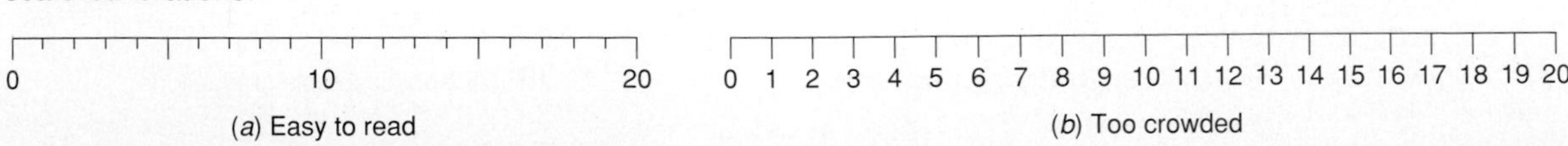

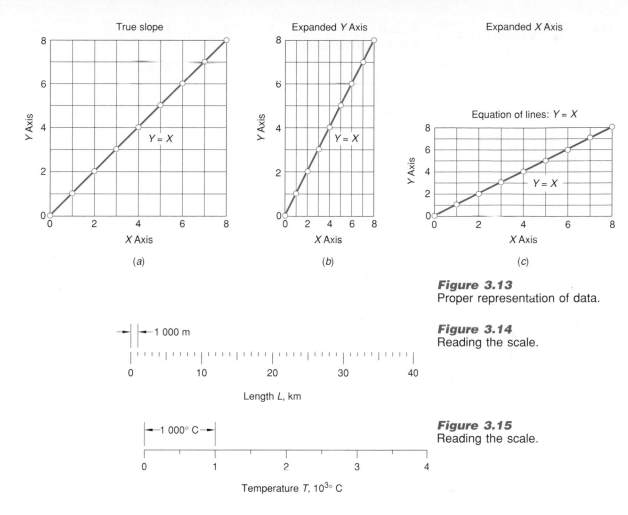

Figure 3.13
Proper representation of data.

Figure 3.14
Reading the scale.

Figure 3.15
Reading the scale.

The calibration of logarithmic scales is illustrated in Fig. 3.16. Since log-cycle designations start and end with powers of 10 (that is, 10^{-1}, 10^0, 10^1, 10^2, and so on) and since commercially purchased paper is normally available with each cycle printed 1 through 10, Figs. 3.16a and b demonstrate two preferred methods of calibration.

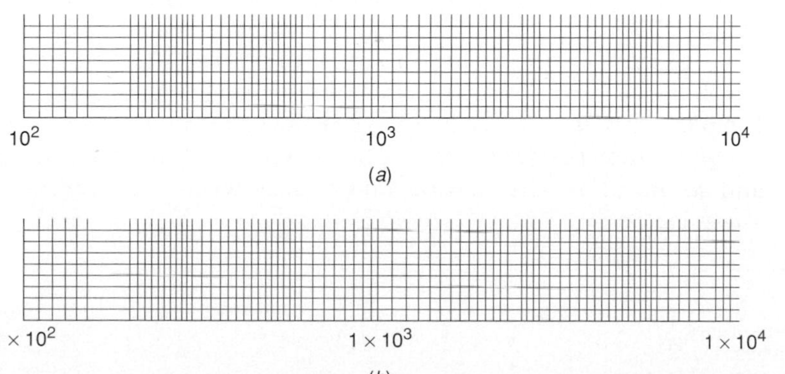

Figure 3.16
Calibration of log scales.

3.3.4 Axis Labeling

Each axis should be clearly identified. At a minimum, the axis label should contain the name of the variable, its symbol, and its units. Since time is frequently the independent variable and is plotted on the x axis, it has been selected as an illustration in Fig. 3.17. Scale designations should preferably be placed outside the axes, where they can be shown clearly. Labels should be lettered parallel to the axis and positioned so that they can be read from the bottom or right side of the page as illustrated in Fig. 3.22.

Figure 3.17
Axis identification.

3.3.5 Point-Plotting Procedure

Data can normally be categorized in one of three general ways: as observed, empirical, or theoretical. Observed and empirical data points are usually located by various symbols, such as a small circle or square around each data point, whereas graphs of theoretical relations (equations) are normally constructed smooth, without use of symbol designation. Figure 3.18 illustrates each type.

Figure 3.18
Plotting data points.

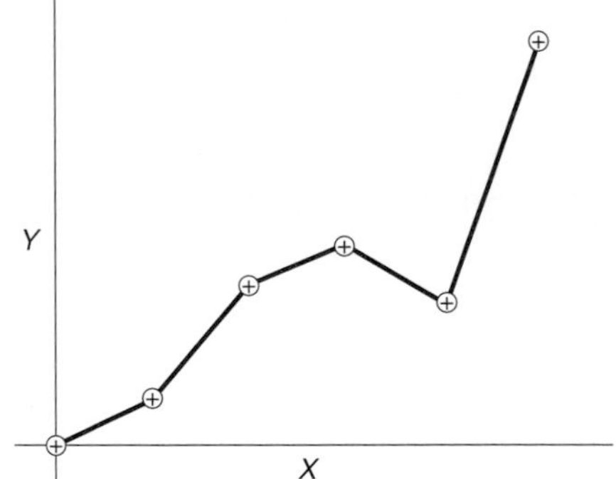

(*a*) Observed: Usually plotted with observed data points connected by straight, irregular line segments. Line does not penetrate the circles.

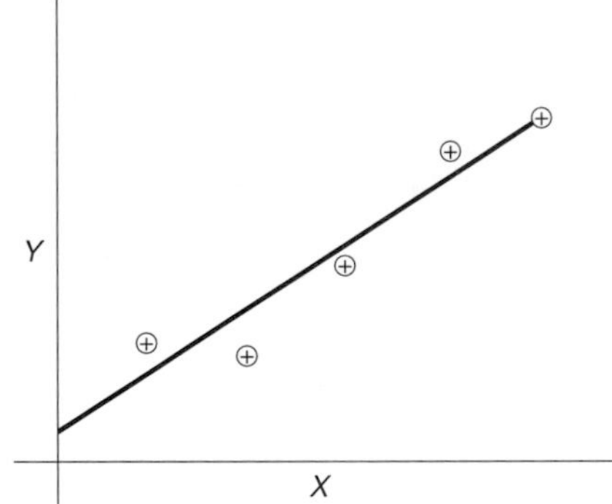

(*b*) Empirical: Reflects the author's interpretation of what occurs between known data points. Normally represented as a smooth curve or straight line fitted to data. Data points may or may not fall on curve.

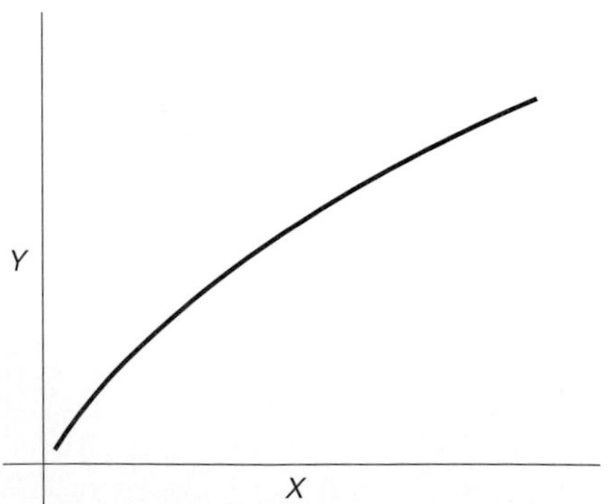

(*c*) Theoretical: Graph of an equation. Curves or lines are smooth and without symbols. Every point on the curve is a data point.

3.3.6 Curves and Symbols

On graphs prepared from observed data resulting from laboratory experiments, points are usually designated by various symbols (see Fig. 3.19). If more than one curve is plotted on the same grid, several of these symbols may be used (one type for each curve). To avoid confusion, however, it is good practice to label each curve. When several curves are plotted on the same grid, another way they can be distinguished from each other is by using different types of lines, as illustrated in Fig. 3.20. Solid lines are normally reserved for single curves, and dashed lines are commonly used for extensions; however, different line representation can be used for each separate curve. The line weight of curves should be heavier than the grid ruling.

A key, or legend, should be placed in an available portion of the grid, preferably enclosed in a border, to define point symbols or line types that are used for curves. Remember that the lines representing each curve should never be drawn through the symbols, so that the precise point is always identifiable. Figure 3.21 demonstrates the use of a key and the practice of breaking the line at each symbol.

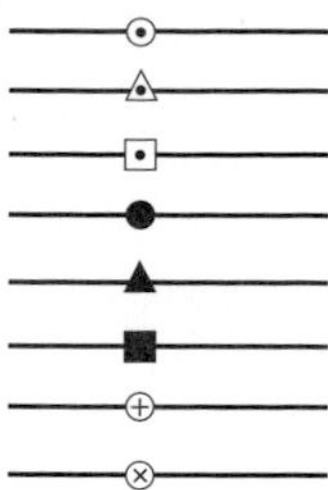

Figure 3.19 Symbols.

Figure 3.20 Line representation.

3.3.7 Titles

Each graph must be identified with a complete title. The title should include a clear, concise statement of the data being represented, along with items such as the name of the author, the data of the experiment, and any and all information concerning

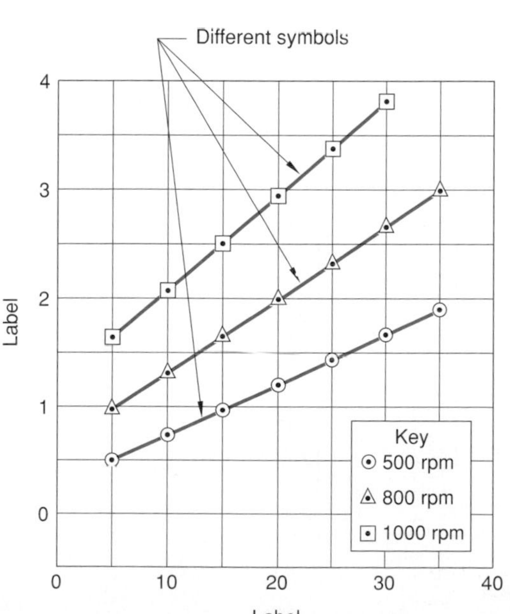

Figure 3.21 Key.

the plot, including the name of the institution or company. Titles are normally enclosed in a border.

All lettering, the axes, and the curves should be sufficiently bold to stand out on the graph paper. Letters should be neat and of standard size. Figure 3.22 is an illustration of plotted experimental data incorporating many of the items discussed in the chapter.

3.3.8 Computer-Assisted Plotting

A number of commercial software packages are available to produce graphs. The quality and accuracy of these computer-generated graphs vary, depending on the sophistication of the software as well as on the plotter or printer employed. Typically, the software will produce an axis scale graduated and calibrated to accommodate the range of data values that will fit the paper. This may or may not produce a readable or interpretable scale. Therefore, it is necessary to apply considerable judgement depending on the results needed. For example, if the default plot does not meet needed scale readability, it may be necessary to specify the scale range to achieve an appropriate scale graduation, since this option allows greater control of the scale drawn.

Figure 3.22
Sample plot.

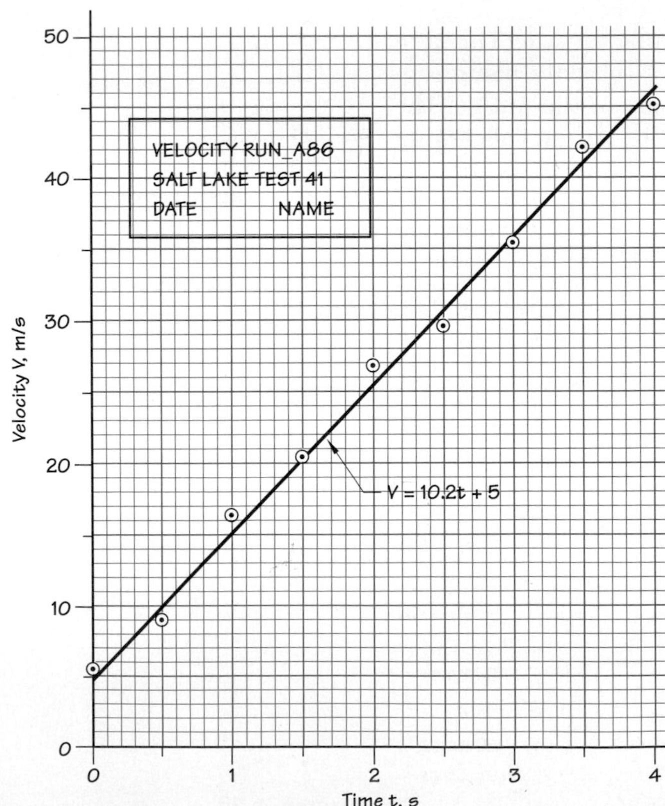

Necessary steps to follow when plotting a graph.

1. Select the type of graph paper (rectilinear, semilog, log-log, etc.) and grid spacing for best representation of the given data.
2. Choose the proper location of the horizontal and vertical axes.
3. Determine the scale units (range) for each axis to display the data appropriately.
4. Graduate and calibrate the axes (1, 2,5 rule).
5. Identify each axis completely.
6. Plot points and use permissible symbols.
7. Draw the curve or curves.
8. Identify each curve and add title and necessary notes.
9. Darken lines for good reproduction.

Computer-produced graphics with uniform scales may not follow the 1, 2, 5 rule, particularly since the software plots the independent variable based on the data collected. If the software has the option of separately specifying the range, that is, plotting the data as an *X-Y* scatter plot, you will be able to achieve scale graduations and calibrations that do follow the 1, 2, 5 rule, making it easier to read values from the graph. The hand-plotted graph that was illustrated in Fig. 3.22 is plotted using EXCEL. The results are shown with the default plot in Fig. 3.23a and using the *X-Y* scatter plot with a linear curve-fit in Fig. 3.23b.

Process

1. Record via keyboard or import data into spreadsheet.
2. Select independent (*x*-axis) and dependent variable(s).
3. Select appropriate graph (style or type) from menu.
4. Produce trial plot with default parameters.
5. Examine (modify as necessary) origin, range, graduation and calibrations: Note, use the 1, 2, 5 rule.
6. Label each axis completely.
7. Select appropriate plotting-point symbols and legend.
8. Create complete title.
9. Examine plot and store the data.
10. Plot or print the data.

3.4 Empirical Functions

Empirical functions are generally described as those based on values obtained by experimentation. Since they are arrived at experimentally, equations normally available from theoretical derivations are not possible. However, mathematical expressions can be modeled to fit experimental functions, and it is possible to classify most empirical results into one of four general categories: (1) linear, (2) exponential, (3) power, or (4) periodic.

A linear function, as the name suggests, will plot as a straight line on uniform rectangular coordinate paper. Likewise, when a curve representing experimental data is a straight line or a close approximation to a straight line, the relationship of the variables can be expressed by a linear equation.

Correspondingly, exponential equations, when plotted on semilog paper, will be linear. The basic form of the equation is $y = be^{mx}$. This can be written in log form and becomes $\log y = mx \log e + \log b$. Alternatively, using natural logarithms, the equation becomes $\ln y = mx + \ln b$ because $\ln e = 1$. The independent variable x is plotted against $\ln y$.

The power equation has the form of $y = bx^m$. Written in log form it becomes $\log y = m \log x + \log b$. This data will plot straight on log-log paper, since the log of the independent variable x is plotted against the log of y.

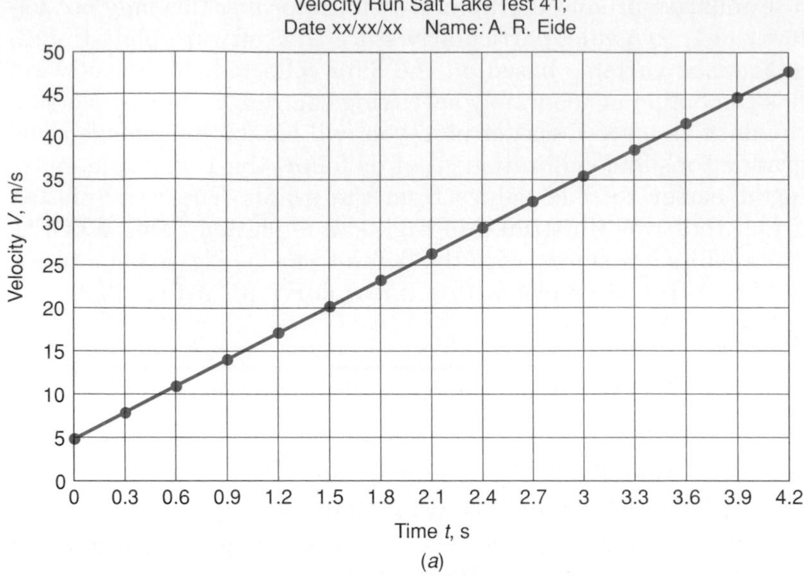

(a)

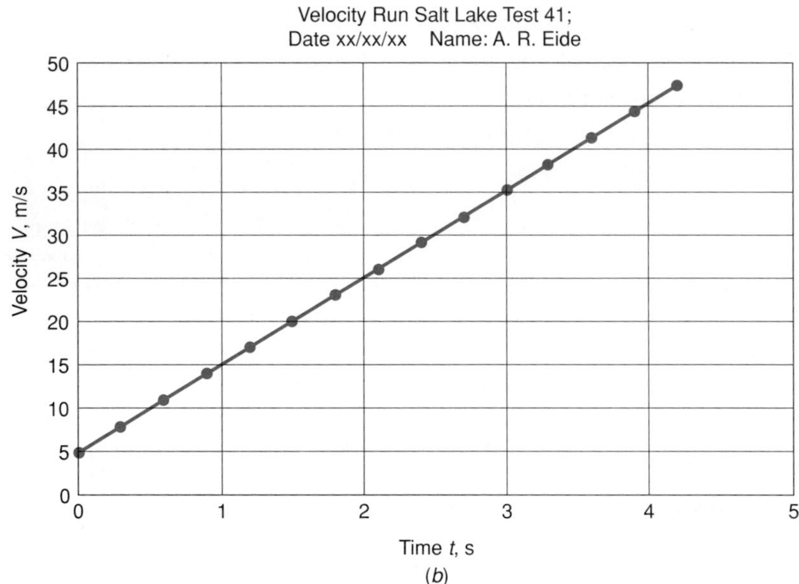

(b)

Figure 3.23

When the data represent experimental results and a series of points are plotted to represent the relationship between the variables, it is improbable that a straight line can be constructed through every point, since some error is inevitable. If all points do not lie on the same line, an approximation scheme or averaging method may be used to arrive at the best possible fit.

3.5 Curve Fitting

Different methods or techniques are available to arrive at the best "straight-line" fit. Three methods commonly employed for finding the best fit are as follows:

1. Method of selected pints
2. Method of averages
3. Method least squares

Each of these techniques is progressively more accurate. The first method will be briefly described in Sec. 3.6. The most accurate method, least squares, is discussed in considerably more detail in the chapter on Statistics. Several examples will be presented in this chapter to demonstrate correct methods for the representation of technical data.

Method 2, the method of averages, is based on the idea that the line location is positioned to make the algebraic sum of the absolute values of the differences between observed and calculated values of the ordinate equal to 0.

In both methods 2 and 3, the procedure involves minimizing what are called *residuals,* or the difference between an observed ordinate and the corresponding computed ordinate. The method of averages will not be applied in this book, but there are any number of reference texts available that adequately cover the concept.

3.6 Method of Selected Points

The method of selected points is a valid method of determining the equation that best fits data that exhibit a linear relationship. Once the data have been plotted and determined to be linear, a line is selected that appears to fit the data best. This is most often accomplished by visually selecting a line that goes through as many data points as possible and has approximately the same number of data points on either side of the line.

Once the line has been constructed, two points, such as A and B, are selected *on the line* and at a reasonable distance apart. The coordinates of both points $A(X_1,Y_1)$ and $B(X_2,Y_2)$ must satisfy the equation of the line, since both are points on the line.

3.7 Empirical Equations—Linear

When experimental data plot as a straight line on rectangular grid paper, the equation of the line belongs to a family of curves whose basic equation is given by

$$y = mx + b \tag{3.1}$$

where m is the slope of the line, a constant, and b is a constant referred to as the *y intercept* (the value of y when $x = 0$).

To demonstrate how the method of selected points works, consider the following example.

Example problem 3.1 The velocity V of an automobile is measured at specified time t intervals. Determine the equation of a straight line constructed through the points recorded in Tab. 3.3. Once an analytic equation has been determined, velocities at intermediate values can be computed.

Table 3.3

Time t, s	0	5	10	15	20	25	30	35	40
Velocity V, m/s	24	33	62	77	105	123	151	170	188

Procedure

1. Plot the data on rectangular paper or do a computer scatter-plot. If the results form a straight line (see Fig. 3.24), the function is linear and the general equation is of the form

$$V = mt + b$$

where m and b are constants.

2. Select two points on the line, $A(t_1, V_1)$ and $B(t_2, V_2)$, and record the value of these points. Points A and B should be widely separated to reduce the effect on m and b of errors in reading values from the graph. Points A and B are identified on Fig. 3.24 for instructional reasons. They should not be shown on a completed graph that is to be displayed.

$A(10, 60)$
$B(35, 165)$

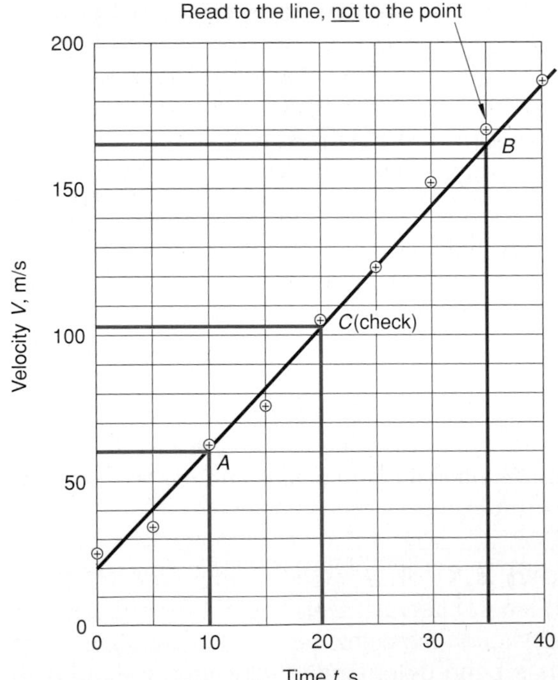

Figure 3.24 Data plot.

3. Substitute the points A and B into $V = mt + b$.

$60 = m(10) + b$

$165 = m(35) + b$

4. The equations are solved simultaneously for the two unknowns.

$m = 4.16$

$b = 18.4$

5. The general equation of the line can be written as

$V = 4.16t + 18.4$

6. Using another point $C(t_3, V_3)$, check for verification:

$C(20, 102)$

$102 = 4.16(20) + 18.4$

$102 = 83.2 + 18.4 = 101.6 \simeq 102$

It is also possible by discreet selection of points to simplify the solution. For example, if point A is selected at (0, 18.4) and this coordinate is substituted into the general equation

$18.4 = m(0) + b$ Note: Approximately 20 on graph

the constant b is immediately known.

3.8 Empirical Equations—Power Curves

When experimentally collected data are plotted on rectangular coordinate graph paper and the points do not form a straight line, you must next determine which family of curves the line most closely approximates. Consider the following familiar example.

Example problem 3.2 Suppose that a solid object is dropped from a tall building, and the values are as recorded in Tab. 3.4.

Table 3.4

Time t, s	Distance s, m/s
0	0
1	4.9
2	19.6
3	44.1
4	78.4
5	122.5
6	176.4

Solution To anyone who has studied fundamental physics, it is apparent that these values should correspond to the general equation for a free-falling body (neglecting air friction):

$s = \frac{1}{2}gt^2$

But assume for a moment that all we have is the table of values.

First it is helpful to make a freehand plot to observe the data visually (see Fig. 3.25). From this quick plot, the data points are more easily recognized as belonging to a family of curves whose general equation can be written

$$y = bx^m \qquad (3.2)$$

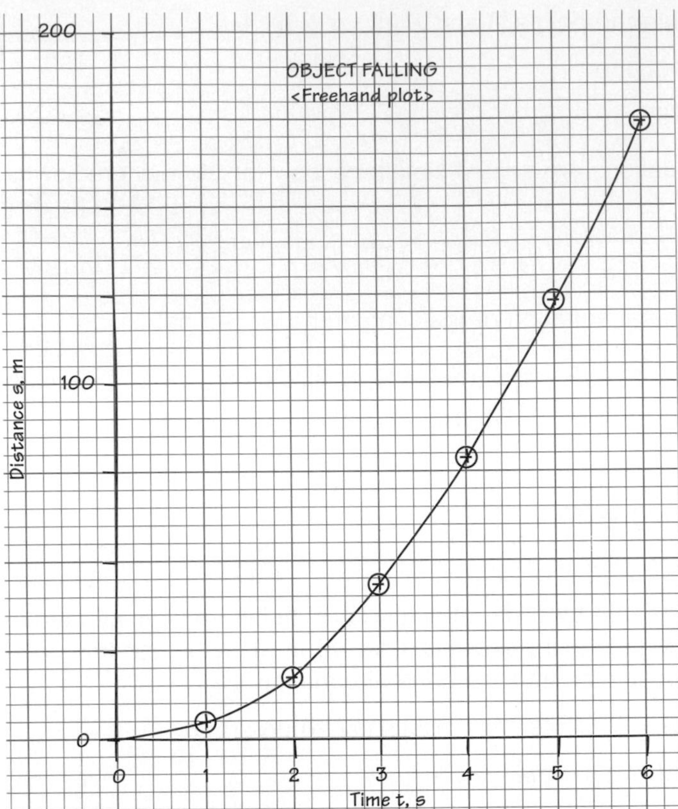

Figure 3.25
Rectilinear paper (freehand).

Remember that before the method of selected points can be applied to determine the equation of the line in this example problem, the plotted line must be straight, because two points on a curved line do not uniquely identify the line. Mathematically, this general equation can be modified by taking the logarithm of both sides,

$$\log y = m \log x + \log b, \text{ or}$$

$$\ln y = m \ln x + \ln b$$

This equation suggests that if the logs of both y and x were determined and the results plotted on rectangular paper, the line would likely be straight.

Realizing that the log of 0 is undefined and plotting the remaining points that are recorded in Tab. 3.5 for Log s versus Log t, the results are shown in Fig. 3.26.

Since the graph of log s versus log t does plot as a straight line, it is now possible to use the general form of the equation

$$\log y = m \log x + \log b$$

and apply the method of selected points.

Table 3.5

Time t, s	Distance s, m	Log t	Log s
0	0		
1	4.9	0.000 0	0.690 2
2	19.6	0.301 0	1.292 3
3	44.1	0.477 1	1.644 4
4	78.4	0.602 1	1.894 3
5	122.5	0.699 0	2.088 1
6	176.4	0.778 2	2.246 5

When reading values for points A and B from the graph, we must remember that the logarithm of each variable has already been determined and the values plotted.

$A(0.2, 1.09)$
$B(0.6, 1.89)$

Points A and B can now be substituted into the general equation $\log s = m \log t + \log b$ and solved simultaneously.

$$1.89 = m(0.6) + \log b$$

$$1.09 = m(0.2) + \log b$$

$$m = 2.0$$

$$\log b = 0.69$$

$$b = 4.9$$

As examination of Fig. 3.26 shows that the value of $\log b$ (0.69) can be read from the graph where $\log t = 0$. This, of course, is where $t = 1$ and is the y intercept for log log plots.

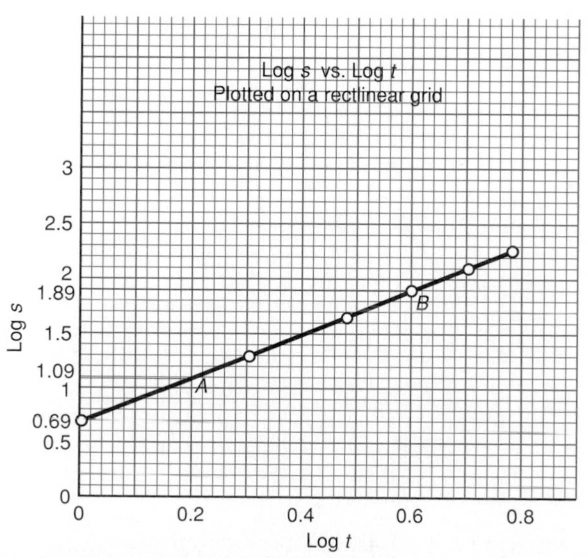

Figure 3.26
Log-log on rectilinear grid paper.

The general equation can then be written as

$$s = 4.9t^{2.0}$$

Or,

$$s = \tfrac{1}{2}gt^2,$$

where $g = 9.8\ m/s^2$.

One obvious inconvenience is the necessity of finding logarithms of each variable and then plotting the logs of these variables.

This step is not necessary since functional paper is commercially available with $\log x$ and $\log y$ scales already constructed. Log-log paper allows the variables themselves to be plotted directly without the need of computing the log of each value.

In the preceding example, once the general form of the equation is determined (Eq. [3.2]), the data can be plotted directly on log-log paper. Since the resulting curve is a straight line, the method of selected points can be used directly (see Fig. 3.27).

The log form of the equation is again used:

$$\log s = m \log t + \log b$$

Select points A and B on the line:

A(1.5, 11)
B(6, 175)

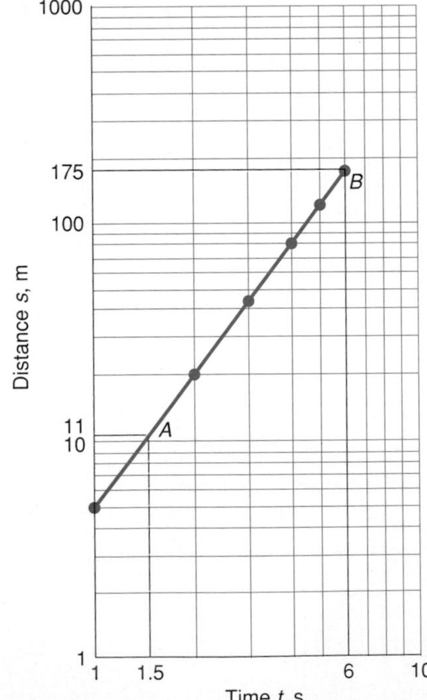

Figure 3.27 Log-log paper.

Substitute the values into the general equation $\log s = m \log t + \log b$, taking careful note that the numbers are the variables and *not* the logs of the variables.

$$\log 175 = m \log 6 + \log b$$

$$\log 11 = m \log 1.5 + \log b$$

Again, solving these two equations simultaneously results in the following approximate values for the constants b and m:

$$b = 4.897\,8 = 4.9$$

$$m = 1.995\,7 = 2.0$$

Identical conclusions can be reached:

$$s = \tfrac{1}{2} g t^2$$

This time, however, one can use functional scales rather than calculate the log of each number.

3.9 Empirical Equations—Exponential Curves

Suppose your data do not plot as a straight line (or nearly straight) on rectangular coordinate paper nor is the line approximately straight on log-log paper. Without experience in analyzing experimental data, you may feel lost as to how to proceed. Normally, when experiments are conducted you have an idea as to how the parameters are related and you are merely trying to quantify that relationship. If you plot your data on semilog graph paper and it produces a reasonably straight line, then it has the general form $y = be^{mx}$.

Example problem 3.3 Assume that an experiment produces the data shown in Tab. 3.6.

Table 3.6

Velocity *V*, m/s	Fuel Consumption *FC*, mm3/s
10	25.2
20	44.6
30	71.7
40	115
50	202
60	367
70	608

Solution The data (Tab. 3.6) when plotted produce the graph shown as Fig. 3.28. To determine the constants in the equation $y = be^{mx}$, write it in linear form, either as

$$\log y = mx \log e + \log b$$

or

$$\ln y = mx + \ln b$$

The method of selected points can now be employed for $\ln Q = mV + \ln b$ (choosing the natural log form). Points $A(15, 33)$ and $B(65, 470)$ are carefully selected on the line, so they must satisfy the equation. Substituting the values of V and Q at points A and B, we get

$$\ln 470 = 65m + \ln b$$

and

$$\ln 33 = 15m + \ln b$$

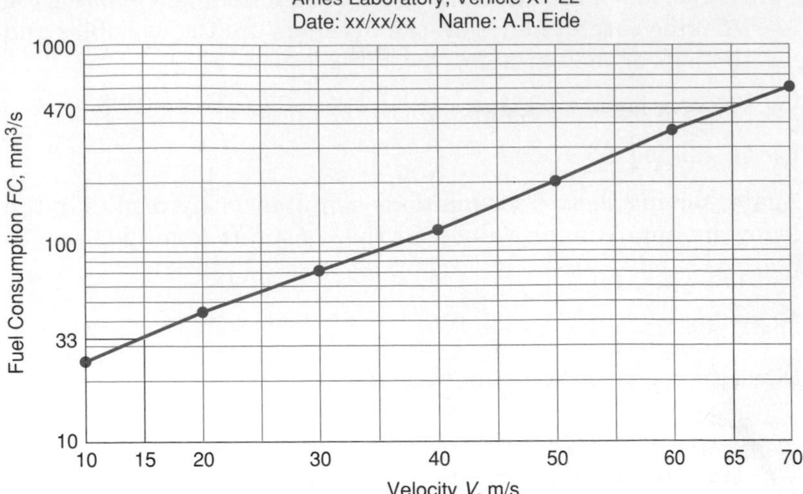

Figure 3.28
Semilog paper.

Solving simultaneously for m and b, we have

$m = 0.053\,1$

and

$b = 14.9$

The desired equation then is determined to be $Q = 15e^{(0.05V)}$. This determination can be checked by choosing a third point, substituting the value for V, and solving for Q.

3.10 Key Terms and Concepts

The following terms are basic to the material in Chapter 3. You should be able to define these terms and be able to interpret them into various applications.

- Scale graduations
- Scale calibrations
- Uniform scales
- Nonuniform scales
- 1, 2, 5 rule
- Plotting procedures
- Graph symbols
- Axis identification
- Line representation
- Titles
- Empirical functions
- Curve fitting
- Method of selected points
- Linear curves
- Power curves
- Exponential curves

3.1 Table 3.7 is data from trial run on the Utah salt flats made by an experimental turbine-powered vehicle.

(*a*) Plot the data on rectilinear paper using time as the independent variable.

(*b*) Determine the equation of the line, using the method of selected points.

(*c*) Interpret the slop of the line.

Table 3.7

Time t, s	Velocity V, m/s
10.0	15.1
20.0	32.2
30.0	63.4
40.0	84.5
50.0	118
60.0	139

3.2 Table 3.8 lists the values of velocity recorded on a ski jump in Colorado this past winter.

(*a*) Plot the data on rectilinear paper using time as the independent variable.

(*b*) Using the method of selected points, determine the equation of the line.

(*c*) Give the average acceleration.

Table 3.8

Time t, s	Velocity V, m/s
1.0	5.3
4.0	18.1
7.0	26.9
10.0	37.0
14.0	55.2

3.3 Table 3.9 is a collection of data for iron-constantan thermocouple. Temperature is in degrees Celsius and the electromotive force (emf) is in millivolts.

(*a*) Plot a graph, using rectilinear paper, showing the relation of temperature to voltage, with voltage as the independent variable.

(*b*) Using the method of selected points, find the equation of the line.

Table 3.9

Temperature t, °C	Voltage (emf), mV
50.0	2.6
100.0	6.7
150.0	8.8
200.0	11.2
300.0	17.0
400.0	22.5
500.0	26.0
600.0	32.5
700.0	37.7
800.0	41.0
900.0	48.0
1 000.0	55.2

3.4 A Sessions pump was tested to determine the power required to produce a range of discharges. The test was performed in Oak Park on July 1, 1984. The results of the test are shown in Tab. 3.10.

(*a*) Plot a graph showing the relation of power required to discharge.

(*b*) Determine the equation of the relationship.

(*c*) Predict the power required to produce a discharge of 37 L/s.

Table 3.10

Discharge Q, L/s	Power P, kW
3.00	28.5
7.00	33.8
10.00	39.1
13.50	43.2
17.00	48.0
20.00	51.8
25.00	60.0

Table 3.11

Deflection D, mm	Load L, N
2.25	35.0
12.0	80.0
20.0	120.0
28.0	160.0
35.0	200.0
45.0	250.0
55.0	300.0

3.5 A spring was tested at AMAC Manufacturing. The test of spring ZX-15 produced the data shown in Tab. 3.11.

(*a*) Plot the data on rectangular graph paper and determine the equation that expresses the deflection to be expected under a given load.
(*b*) Predict the load required to produce a deflection of 30 mm.
(*c*) What deflection would be expected to produce a load of 50 N?

3.6 A Johnson furnace was tested in northern Minnesota to determine the heat generated, expressed in thousands of British thermal units per cubic foot of furnace volume, at varying temperatures. The results are shown in Tab. 3.12.

(*a*) Plot data on rectilinear paper, with temperature as the independent variable.
(*b*) Plot the data on log-log graph paper, with temperature as the independent variable.
(*c*) Using the method of selected points, determine the equation that best fits the data.
(*d*) What would be the heat released for a temperature of 1000°F?

Table 3.12

Heat released H, 103 Btu/ft3	Temperature T, °F
0.200	172
0.600	241
2.00	392
4.00	483
8.00	608
20.00	812
40.00	959
80.00	1 305

3.7 The capacity of a 20-cm screw conveyor that is moving dry ground corn is expressed in liters per second and the conveyor speed in revolutions per minute. A test was conducted in Cleveland on conveyor model JD172 last week. The results of the test are given in Tab. 3.13.

(*a*) Plot the data on rectilinear graph paper.
(*b*) Plot the data on semilog paper.
(*c*) Plot the data on log-log paper.
(*d*) Determine the equation that expresses velocity in terms of capacity utilizing the proper plot.

Table 3.13

Capacity C, L/s	Angular velocity V, r/min
3.01	10.0
6.07	21.0
15.0	58.2
30.0	140.6
50.0	245
80.0	410
110.0	521

3.8 The resistance of a class and shape of electrical conductor was tested over a wide range of sizes at constant temperature. The test was performed in Madison, Wisconsin, on April 4, 1984, at the Acme Electrical Labs. The test results are shown in Tab. 3.14. The resistance is expressed in milliohms per meter of conductor length.

(*a*) Plot the data on log-log paper.

(*b*) Using the method of selected points, find the equation that expresses resistance as a function of the area of the conductors.

Table 3.14

Area *A*, mm2	Resistance *R*, mΩ/m
0.021	505
0.062	182
0.202	55.3
0.523	22.2
1.008	11.3
3.32	4.17
7.29	1.75

3.9 The area of a circle can be expressed by the formula $A = \pi R^2$. If the radius varies from 0.5 to 5 cm, perform the following:

(*a*) Construct a table of radius versus area mathematically. Use radius increments of 0.5 cm.

(*b*) Construct a second table of log *R* versus log *A*.

(*c*) Plot the values from (a) on log-log paper and determine the equation of the line.

(*d*) Plot the values from (b) on rectilinear paper and determine the equation of the line.

3.10 The volume of a sphere is $V = \frac{4}{3}\pi R^3$.

(*a*) Prepare a table of volume versus radius, allowing the radius to vary from 2.0 to 10.0 m in 1-m increments.

(*b*) Plot a graph on log-log paper showing the relation of volume to radius using the values from the table in (a).

(*c*) Verify the equation, $V = \frac{4}{3}\pi R^3$, by the method of selected points.

3.11 A 90° triangular weir is commonly used to measure flow rate in a stream. Data on the discharge through the weir were collected and recorded as shown in Tab. 3.15.

(*a*) Plot the data on log-log paper, with height as the independent variable.

(*b*) Determine the equation of the line using the method of selected points.

Table 3.15

Discharge, *Q*, m3/s	1.5	8	22	45	78	124	182	254
Height, *h*, *m*	1	2	3	4	5	6	7	8

3.12 A pitot tube is a device for measuring the velocity of flow of a fluid (see Fig. 3.29). A stagnation point occurs at point 2; by recording the height differential *h*, the velocity at point 1 can be calculated. Assume for this problem that the velocity at point 1 is known corresponding to the height differential *h*. Table 3.16 records these values.

(*a*) Plot the data on log-log paper using height as the independent variable.

(*b*) Determine the equation of the line using the method of selected points.

Table 3.16

Velocity *V*, m/s	1.4	2.0	2.8	3.4	4.0	4.4
Height *h*, m	0.1	0.2	0.4	0.6	0.8	1.0

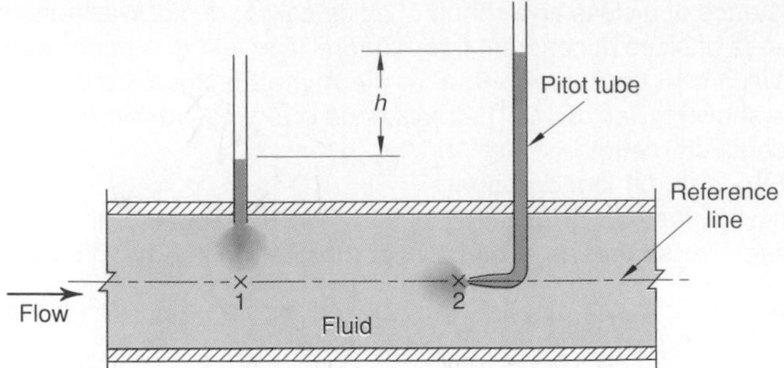

Figure 3.29

3.13 A new production facility manufactured 29 parts the first month, but then increased production, as shown in Tab. 3.17.

(*a*) Plot the data on semilog paper.

(*b*) Using the time variable that defines January to be 1, February to be 2, and so on, determine the equation.

Table 3.17

Month	Jan	Feb	Mar	Apr	May	Jun	Jul	Aug
Number	29	40	48	58	85	115	124	180

Table 3.18

Time t, s	Voltage V, V
6	98
10	62
17	23
25	9.5
32	3.5
38	1.9
42	1.33

3.14 The voltage across a capacitor during discharge was recorded as a function of time (see Tab. 3.18).

(*a*) Plot the data on semilog paper, with time as the independent variable.

(*b*) Determine the equation of the line using the method of selected points.

3.15 When a capacitor is to be discharged, the current flows until the voltage across the capacitor is 0. This current flow when measured as a function of time resulted in the data given in Tab. 3.19.

(*a*) Plot the data points on semilog paper, with time as the independent variable.

(*b*) Determine the equation of the line using the method of selected points.

Table 3.19

Current I, A	1.81	1.64	1.48	1.34	1.21	0.73
Time t, s	0.1	0.2	0.3	0.4	0.5	1.0

3.16 When fluid is flowing in the line, it is relatively easy to begin closing a valve that is wide open. But as the valve approaches a more nearly closed position, it becomes considerably more difficult to force movement. Visualize a circular pipe with a simple flap hinged at one edge being closed over the end of the pipe. The fully open position is $\theta = 0$, and the fully closed condition is $\theta = 90°$ (see Fig. 3.30).

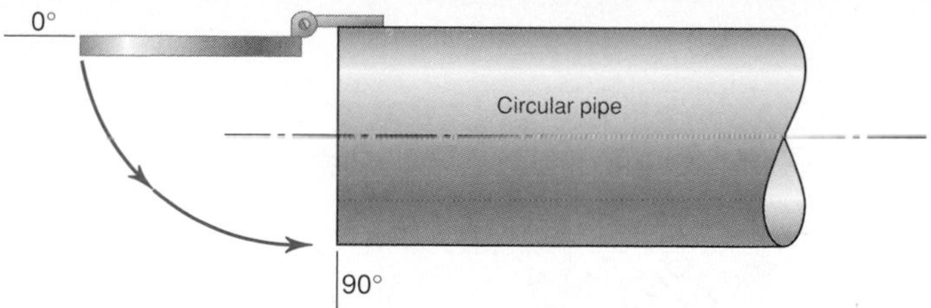

Figure 3.30

A test was conducted on such a valve by applying a constant torque at the hinge position and measuring the angular movement of the valve. The test data are shown in Tab. 3.20.

(*a*) Plot the data, with angle as the independent variable.
(*b*) By the method of selected points, find the equation relating torque to angular movement.

Table 3.20

Torque T, N · m	Movement θ, degrees
3.0	5.2
6.0	29.3
10.0	40.9
20.0	56.3
35.0	71.0
50.0	84.8

3.17 All materials are elastic to some extent. It is desirable that certain parts of some designs compress when a load is applied to assist in making the part airtight or watertight (such as a jar lid). The test results shown in Tab. 3.21 resulted from a test made at the Herndon Test Labs in Houston on a material known as SILON Q-177.

(*a*) Plot the data on semilog graph paper.
(*b*) Using the method of selected points, find the equation of the relationship.
(*c*) What pressure would cause a 10 percent compression?

Table 3.21

Pressure P, Mpa	Relative compression R, %
1.12	27.3
3.08	37.6
5.25	46.0
8.75	50.6
12.3	56.1
16.1	59.2
30.2	65.0

3.18 The rate of absorption of radiation by metal plates varies with the plate thickness and the nature of the source of radiation. A test was made at Ames Labs on October 11, 1982, using a Geiger counter and a constant source of radiation; the results are shown in Tab. 3.22.

(*a*) Plot the data on semilog graph paper.
(*b*) Find the equation of the relationship between the parameters.
(*c*) What level of radiation would you estimate to pass a 2-in-thick plate of the metal used in the test described above?

Table 3.22

Plate thickness W, mm	Geiger counter C, counts per second
0.20	5 500
5.00	3 720
10.00	2 550
20.00	1 320
27.5	720
32.5	480

4 Engineering Estimations and Approximations

4.1 Introduction

In Chapter 1, we discussed a myriad of engineering disciplines and functions. Within each discipline there are specialty areas which give the appearance that engineering is a diverse profession with little commonality in the tasks performed. However, engineers are problem solvers, creating new designs which satisfy a need and improve the living standard. During the design process, engineers of all specialists will need to acquire physical measurements pertaining to the product or system being designed, the environment in which the design will operate, or both.

The nineteenth century physicist Lord Kelvin stated that man's knowledge and understanding are not of high quality unless the information can be expressed in numbers. We all have made or heard statements such as "the water is too hot." This statement may or may not give us an indication of the temperature of the water. At a given temperature water may be too hot for taking a bath but not hot enough for making instant coffee or tea.

The truth is that pronouncements such as "hot," "too hot," "not very hot," and so on, are relative to a standard selected by the speaker and have meaning only to those who know what that standard is.

Engineers make measurements of a vast array of physical quantities that control the design solution. Skill in making and interpreting measurements is an essential element in our practice of engineering.

4.2 Significant Digits

Any physical measurement cannot be assumed to be exact. Errors are likely to be present regardless of the precautions used when making the measurement. Quantities determined by analytical means are not always exact either. Often assumptions are made to arrive at an analytical expression which is then used to calculate a numerical value.

Figure 4.1 Measurements from a wind tunnel experiment are acquired electronically and immediately analyzed with software. The test can be quickly modified as needed. *(Iowa State University.)*

It is clear that a method of expressing results and measurements is needed that will convey how "good" these numbers are. The use of significant digits gives us this capability without resorting to the more rigorous approach of computing an estimated percentage error to be specified with each numerical result or measurement.

A *significant digit,* or *figure,* is defined as any digit used in writing a number, *except* those zeros that are used only for location of the decimal point or those zeros that do not have any nonzero digit on their left. When you read the number 0.001 5, only the digits 1 and 5 are significant, since the three zeros have no nonzero digit to their left. We would say then that this number has two significant figures. If the number is written 0.001 50, it contains three significant figures; the rightmost zero is significant.

Numbers 10 or larger that are not written in scientific notation and that are not counts (exact values) can cause difficulties in interpretation when zeros are present. For example, 2 000 could contain one, two, three, or four significant digits; it is not clear which. If you write the number in scientific notation as 2.000×10^3, then clearly four significant digits are intended. If you want to show only two significant digits, you would write 2.0×10^3. It is our recommendation that, if uncertainty results from using standard decimal notation, you switch to scientific notation so your reader can clearly understand your intent. Table 4.1 shows the number of significant figures for several quantities.

Table 4.1

Quantity	Number of significant figures
4 784	4
36	2
60	1 or 2
600	1, 2, or 3
6.00×10^2	3
31.72	4
30.02	4
46.0	3
0.02	1
0.020	2
600.00	5

You may find yourself as the user of values where the writer was not careful to properly show significant figures. What then? Assuming that the number is not a count or a known exact value, about all you can do is establish a reasonable number of significant figures based on the context of the value and on your experience. Once you have decided on a reasonable number of significant digits, you can then use the number in any calculations that are required.

When reading instruments, such as an engineer's scale, thermometer, or fuel gauge, the last digit will normally be an estimate. That is, the instrument is read by estimating between the smallest graduations on the scale to get the final digit. In Fig. 4.2a, you may estimate the reading from the engineer's scale to be 1.27, with the 7 being a doubtful digit of the three significant figures. It is standard practice to count the doubtful digit as significant, thus the 1.27 reading has three significant figures. Similarly, the thermometer in Fig. 4.2b may be read as 52.8° with the 8 being doubtful.

In Fig. 4.2c, the graduations create a more difficult task for reading a fuel level. Each graduation is one-sixth of a full tank. The reading appears to be about three-fourths of the distance between one-sixth and two-sixths making the reading seven twenty-fourths of a tank or 0.292. How many significant figures are there? In this case one significant figure is all that can be obtained, so the answer should be rounded to 0.3. The difficulty in this example is not the significant figures, but the scale of the fuel gauge. It is meant to convey a general impression of the fuel level and not a numerically significant value. Furthermore, the automobile manufacturer did not deem that the cost of a more accurate and precise fuel-measurement system was justified. Therefore, the selection of the instrument is an important factor in physical measurements.

Calculators and computers commonly work with numbers having as few as 7 digits or as many as 16 or 17 digits. This is true no matter how many significant digits an input value or calculated value should have. Therefore, you will need to exercise care in reporting values from a calculator display or from a computer output. Most high-level computer languages allow you to control the number of digits that are to be displayed or printed. If a computer output is to be a part of your final solution presentation, you will need to carefully control the output form. If the output is only an intermediate step, you can round the results to a reasonable number of significant figures in your presentation.

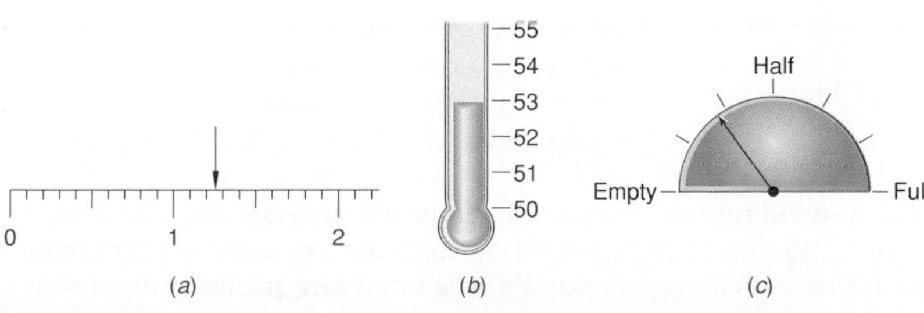

Figure 4.2
Reading graduations on instruments will include a doubtful, or estimated, value.

As you perform arithmetic operations, it is important that you not lose the significance of your measurements or, conversely, imply precision that does not exist. Rules for determining the number of significant figures that should be reported following computations have been developed by engineering associations. The following rules customarily apply.

1. Multiplication and Division

The product or quotient should contain the same number of significant digits as are contained in the number with the fewest significant digits.

Examples

a. (2.43)(17.675) = 42.950 25
If each number in the product is exact, the answer should be reported as 42.950 25. If the numbers are not exact, as is normally the case, 2.43 has three significant figures and 17.675 has five. Applying the rule, the answer should contain three significant figures and be reported as 43.0 or 4.30×10^1.

b. (2.479 h)(60 min/h) = 148.74 min
In this case, the conversion factor is exact (a definition) and could be thought of as having an infinite number of significant figures. Thus, 2.479, which has four significant figures, controls the precision, and the answer is 148.7 min, or 1.487×10^2 min.

c. (4.00×10^2 kg)(2.204 6 lbm/kg) = 881.84 lbm
Here, the conversion factor is not exact, but you should not let the conversion factor dictate the precision of the answer if it can be avoided. You should attempt to maintain the precision of the value being converted; you cannot improve its precision. Therefore, you should use a conversion factor that has one or two more significant figures than will be reported in the answer. In this situation, three significant figures should be reported, yielding 882 lbm.

d. 589.62/1.246 = 473.210 27
The answer, to four significant figures, is 473.2.

2. Addition and Subtraction

The answer should show significant digits only as far to the right as is seen in the least precise number in the calculation.

Example

a.
```
1 725.463
  189.2
   16.73
---------
1 931.393
```
The least precise number in this group is 189.2 so, according to the rule, the answer should be reported as 1 931.4. Using alternative reasoning, suppose these numbers are instrument read-

ings, which means the last reported number in each is a doubtful digit. A column addition that contains a doubtful digit will result in a doubtful digit in the sum. So all three digits to the right of the decimal in the answer are doubtful. Normally we report only one; thus the answer is 1 931.4 after rounding.

b. 897.0
− 0.092 2
896.907 8
Application of the rule results in an answer of 896.9.

3. Combined Operations

If products or quotients are to be added or subtracted, perform the multiplication or division first, establish the correct number of significant figures in the subanswer, perform the addition or subtraction, and round to proper significant figures. Note, however, that in calculator or computer applications it is not practical to perform intermediate rounding. It is normal practice to perform the entire calculation and then report a reasonable number of significant figures.

If results from additions or subtractions are to be multiplied or divided, an intermediate determination of significant figures can be made when the calculations are performed manually. Use the suggestion already mentioned for calculator or computer answers.

Subtractions that occur in the denominator of a quotient can be a particular problem when the numbers to be subtracted are very nearly the same. For example, 39.7/(772.3 − 772.26) gives 992.5 if intermediate roundoff is not done. If, however, the subtraction in the denominator is reported with one digit to the right of the decimal, the denominator becomes zero and the result becomes undefined. Commonsense application of the rules is necessary to avoid problems.

4. Rounding

In rounding a value to the proper number of significant figures, *increase the last digit retained by 1 if the first figure dropped is 5 or greater.* This is the rule normally built into a calculator display control or a control language.

Examples

a. 827.48 rounds to 827.5 or 827 for four and three significant digits, respectively.
b. 23.650 rounds to 23.7 for three significant figures.
c. 0.014 3 rounds to 0.014 for two significant figures.

4.3 Accuracy and Precision

In measurements, accuracy and precision have different meanings and cannot be used interchangeably. *Accuracy* is a measure of the nearness of a value to the correct or true value. *Precision* refers to the repeatability of a measurement, that is how close successive measurements are to each other. Figure 4.3 illustrates accuracy and precision of the results of four dart throwers. Thrower a is both inaccurate and imprecise because the results are away from the bullseye (accuracy) and widely scattered (precision). Thrower b is accurate because the throws are evenly distributed about the desired result but imprecise because of the wide scatter. Thrower c is precise with the tight cluster of throws but inaccurate because the results are away from the desired bullseye. Finally, thrower d demonstrates accuracy and precision with tight cluster of throws around the center of the target. Throwers a, b, and c can improve their performance by analyzing the causes for the errors. Body position, arm motion, and release point could cause deviation from the desired result.

Engineers making physical measurements encounter two types of errors, systematic and random. These will be discussed in the next section.

Measurements can be reported as a value plus or minus (±) a number, for example, 32.3 ± 0.2. This indicates a range of values which are equally representative of the indicated value (32.3). Thus 32.3, 32.1, and 32.5 are among the "acceptable" values for this measurement. A range of permissible error can also be specified as a percentage of the indicated value. For example, a ther-

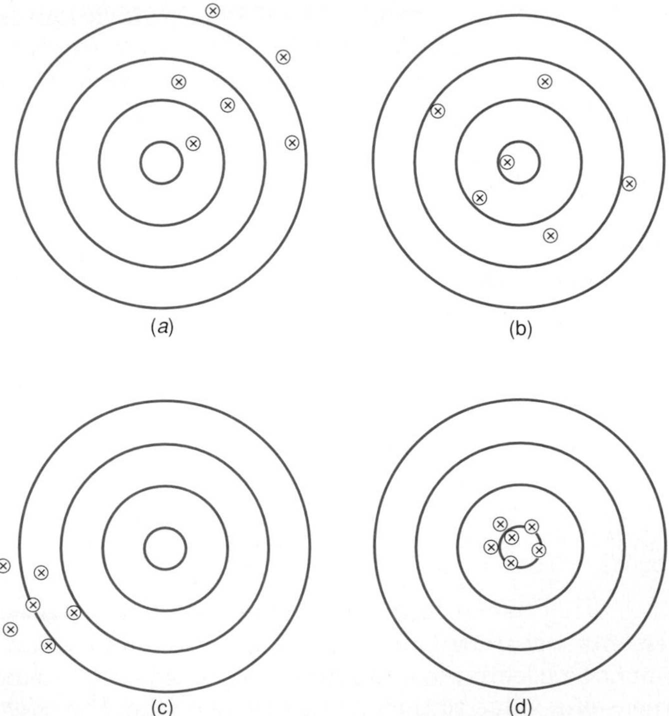

Figure 4.3 Illustration of the difference between accuracy and precision in physical measurements.

mometer's accuracy may be specified as ±1.0 percent of full scale reading. Thus if the full scale reading is 220°F, readings should be within ±2.2 of the true value (220 × 0.01 = 2.2).

4.4 Errors

To measure is to err! Any time a measurement is taken, the result is being compared to a true value, which itself may not be known exactly. If we measure the dimensions of a room, why doesn't a repeat of the measurements yield the same results? Did the same person make all measurements? Was the same measuring instrument used? Were the readings all made from exactly the same eye position? Was the measuring instrument correctly graduated? It is obvious that errors will occur in each measurement. We must try to identify the errors if we can and correct them in our results. If we can't identify the error, we must provide some conclusions as to the resulting accuracy and precision of our measurements.

Identifiable and correctable errors are classified as systematic; accidental or other nonidentifiable errors are classified as random.

4.4.1 Systematic Errors

Our task is to measure the distance between two fixed points. Assume that the distance is in the range of 1 200 m and that we are experienced and competent and have equipment of high quality to do the measurement. Some of the errors that occur will always have the same sign (+ or −) and are said to be systematic. Assume that a 25 m steel tape is to be used, one that has been compared with the standard at the U.S. Bureau of Standards in Washington, D.C. If the tape is not exactly 25.000 m long, then there will be a systematic error each of the 48 times that we use the tape to measure out the 1 200 m.

However, the error can be removed by applying a correction. A second source of error can stem from a difference between the temperature at the time of use and at the time when the tape was compared with the standard. Such an error can be removed if we measure the temperature of the tape and apply a mathematical correction. The coefficient of thermal expansion for steel is 11.7×1.0^{-6} per kelvin). The accuracy of such a correction depends on the accuracy of the thermometer and on our ability to measure the temperature of the tape instead of the temperature of the surrounding air. Another source of systematic error can be found in the difference in the tension applied to the tape while in use and the tension employed during standardization. Again, scales can be used but, as before, their accuracy will be suspect. In all probability, the tape was standardized by laying it on a smooth surface and supporting it throughout. But such surfaces are seldom available in the field. The tape is suspended at times, at least partially. But, knowing the weight of the tape, the tension that is applied, and the length of the suspended tape, we can calculate a correction and apply it.

The sources of systematic error just discussed are not all the possible sources, but they illustrate an important problem even encountered in taking comparatively simple measurements. Similar problems occur in all types of measurements: mechanical quantities, electrical quantities, mass, sound, odors, and so forth. We must be aware of the presence of systematic errors, eliminate those that we can, and quantify and correct for those remaining.

4.4.2 Random Errors

In reading Sec. 4.4.1 you may have realized that even if it had been possible to eliminate all the systematic errors, the measurement is still not exact. To elaborate on this point, we will continue with the example of the task of measuring the 1 200 m distance. Several random errors can creep in, as follows. When reading the thermometer, we must estimate the reading when the indicator falls between graduations. Moreover, it may appear that the reading is exactly on a graduation when it is actually slightly above or below the graduation. Furthermore, the thermometer may not be accurately measuring the tape temperature but may be influenced instead by the temperature of the ambient air. These errors can thus produce measurements that are either too large or too small. Regarding sign and magnitude, the error is therefore random.

Errors can also result from our correcting for the sag in a suspended tape. In such a correction, it is necessary to determine the weight of the tape, its cross-sectional area, its modulus of elasticity, and the applied tension. In all such cases, the construction of the instruments used for acquiring these quantities can be a source of both systematic and random errors.

The major difficulty we encounter with respect to random errors is that, although their presence is obvious by the scatter in the data, it is impossible to predict the magnitude and sign of the accidental error that is present in any one measurement. Repeating measurements and averaging the results will reduce the random error in the average. However, repeating measurements will not reduce the systematic error in the average result.

Refinement of the apparatus and care in its use can reduce the magnitude of the error; indeed, many engineers have devoted their careers to this task.

Likewise, awareness of the problem, knowledge about the degree of precision of the equipment, skill with measurement procedures, and proficiency in the use of statistics allow us to determine the approximate magnitude of the error remaining in

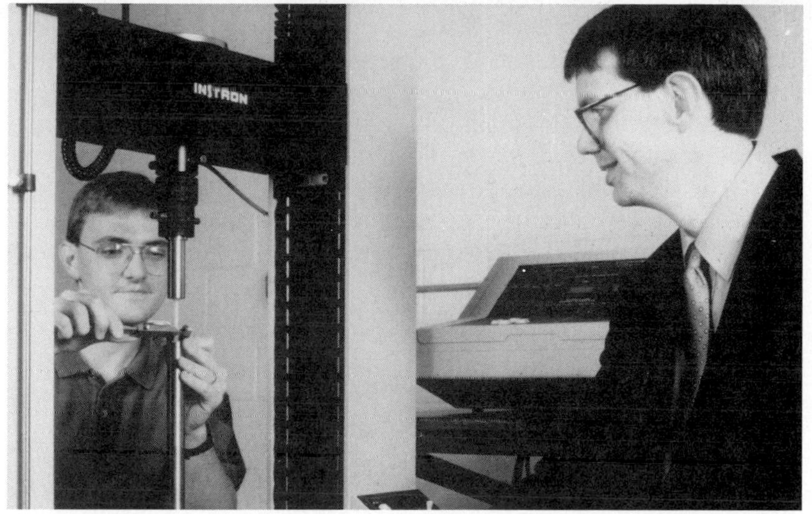

Figure 4.4
Measurements being taken on a test specimen in a materials laboratory. (Iowa State University.)

measurements. This knowledge, in turn, allows us to accept the error or develop different apparatus and/or methods in our work. It is beyond the scope of this text to discuss quantifying accidental errors. However, Chapter 7 includes a brief discussion of central tendency and standard deviation, which are part of the analysis of random errors.

4.5 Approximations

Engineers strive for a high level of precision in their work. However, it is also important to be aware of the expected precision and the time and cost of attaining it. There are many instances where an engineer is expected to make an approximation to an answer, that is estimate the result with reasonable accuracy but under tight time and cost constraints. To do this engineers rely on their basic understanding of the problem under discussion coupled with their previous experience. This knowledge and experience is what distinguishes an "approximation" from a "guess." If greater accuracy is needed, the initial approximation can be refined when time and funds are available and the necessary data for refining the result are available.

In the area of our highest competency, we are expected to be able to make rough estimates to provide figures that can be used for tentative decisions. These estimates may be in error by perhaps 10 to 20 percent or even more. The accuracy of these estimates depends strongly on what reference materials we have available, how much time is allotted for the estimate, and, of course, how experienced we are with similar problems. The first example we present will attempt to illustrate what a professional engineer might be called upon to do in a few minutes with no references. It is not the type of problem you, as a beginning student, would be expected to do because your have not yet gained the necessary experience.

Example problem 4.1 A civil engineer is asked to meet with a city council committee to discuss their needs with respect to the disposal of solid wastes (garbage or refuse). The community, a city of 12 000 persons, must begin supplying refuse collection and disposal for its citizens for the first time. In reviewing various alternatives for disposal, a sanitary landfill is suggested. One of the council members is concerned about how much land is going to be needed, so he asks the engineer how many acres will be required within the next 10 years.

Discussion The engineer quickly estimates as follows: The national average solid-waste production is 2.75 kg/(capita)(day). We can determine that each citizen will thus produce 1 000 kg of refuse per year by the following calculation:

$$(2.75 \text{ kg/day})(365 \text{ days/year}) \cong 1\,000 \text{ kg/year}$$

Experience indicates that refuse will probably be compacted to a density of 400 to 600 kg/m^3. On this basis, the per capita landfill volume will be 2 m^3 each year; and 1 acre filled 1 m deep will contain the collected refuse of 2 000 people for a year (1 acre = 4 047 m^2). Therefore, the requirement for 12 000 people will be 1 acre filled 6 m deep. However, knowledge of the geology of the particular area indicates that bedrock occurs at approximately 6 m below the ground surface. The completed landfill should therefore have an average depth of 4 m; consequently, 1.5 acres a year, or 15 acres in 10 years, will be required. The patterns of the recent past indicate that some growth in population and solid-waste generation should be expected. It is finally suggested that the city should plan to use about 20 acres in the next 10 years.

This calculation took only minutes and required no computational device other than pencil and paper. The engineer's experience, rapid calculations, sound basic assumptions, and sensible rounding of figures were the main requirements. And a usable estimate, designed to neither mislead nor to sell a point of view, was provided. If this project proceeds to the actual development of a sanitary landfill, the civil engineer will then gather actual data, refine the calculations, and prepare estimates upon which one would risk a professional reputation.

Example prob. 4.2 is an illustration of a problem you might be assigned. Here you have the necessary experience to perform the estimation. Not counting the final written presentation, you should be able to do a similar problem in one-half to 1 hour.

Example problem 4.2 Suppose that your instructor assigns the following problem: Determine the number of pieces of lumber 5 cm × 10 cm × 2.40 m that can be sawn from the tree nearest to the southeast corner of the building in which you are now meeting. How would you proceed? See Fig. 4.5 for one student's response.

	9-24-XX	ENGR 160 SPECIAL PROBLEM	JAMES JOHNSON	2

PROBLEM

ESTIMATE THE NUMBER OF 5cm × 10cm × 2.40m BOARDS THAT CAN BE SAWN FROM THE FIR TREE NEAR THE S.E. CORNER OF THE ENGINEERING BUILDING.

ASSUMPTIONS

1. THE TREE TRUNK IS CONICAL.
2. THE LIMBS CANNOT BE USED - TOO SMALL.
3. ALL PIECES THAT ARE TOO SMALL WILL BE DISCARDED - NO PARTIALS OR PARTICLE BOARDS.
4. THE TREE WILL BE CUT 0.3m ABOVE THE GROUND.

COLLECTED DATA

1. I AM 180cm TALL AND MY SHADOW WAS 135cm.
2. THE TREES SHADOW WAS 14m.
3. THE BASE OF THE TREE HAS A CIRCUMFERENCE OF 120cm.

SOLUTION

HEIGHT OF TREE $\left(\frac{180}{135}\right)$ 14 m = 18.67m – APPROX ~ 19m

DIAMETER OF TREE AT GROUND $\frac{120}{\pi}$ = 38.2cm

APPROX ~ 38cm

DIAMETER REDUCTION = $\frac{38\text{cm}}{19\text{m}} = 2.0\,\frac{\text{cm}}{\text{m}}$

FIRST SECTION (0.3m – 2.7m)

EFFECTIVE CROSS SECTION @ 2.7m MEASURING FROM GROUND = 38cm – (2.7 × 2.0) = 32.6 cm

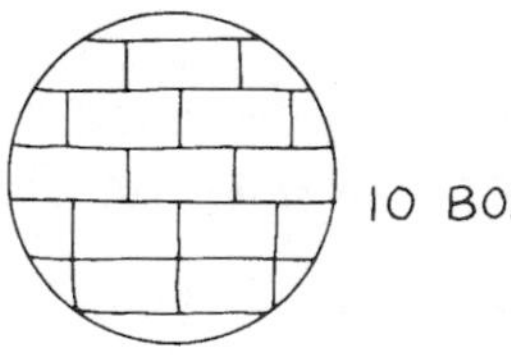

10 BOARDS

SECOND SECTION (2.7m – 5.1m)

38 – (5.1)(2.0) = 27.8 cm

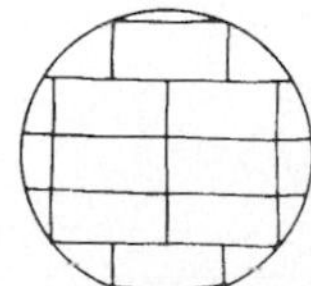

8 BOARDS

Figure 4.5 Student presentation for Example prob. 4.2.

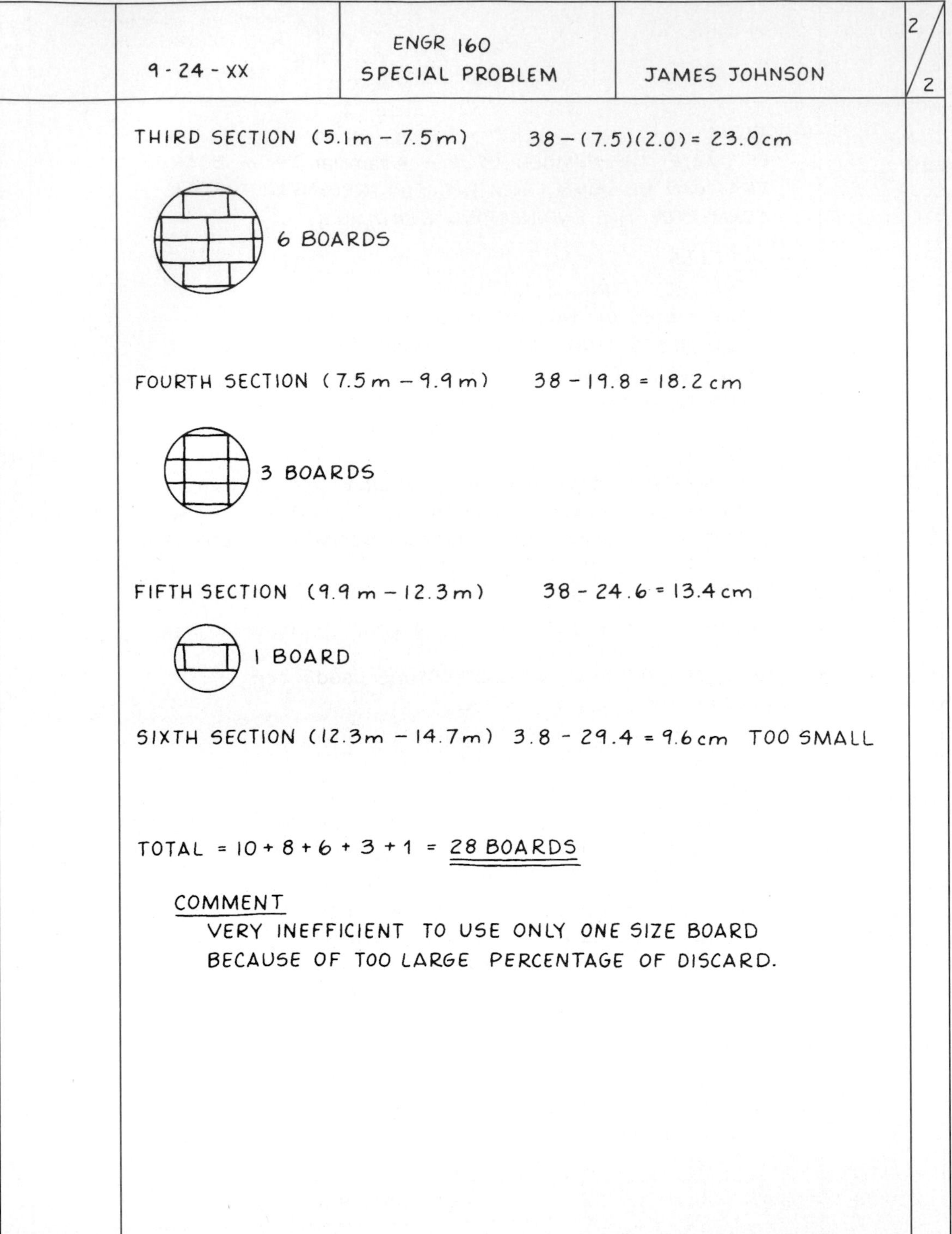

	9 - 24 - XX	ENGR 160 SPECIAL PROBLEM	JAMES JOHNSON	2/2

THIRD SECTION (5.1m – 7.5m)   38 – (7.5)(2.0) = 23.0cm

6 BOARDS

FOURTH SECTION (7.5m – 9.9m)   38 – 19.8 = 18.2cm

3 BOARDS

FIFTH SECTION (9.9m – 12.3m)   38 - 24.6 = 13.4cm

1 BOARD

SIXTH SECTION (12.3m – 14.7m) 3.8 - 29.4 = 9.6cm TOO SMALL

TOTAL = 10 + 8 + 6 + 3 + 1 = 28 BOARDS

COMMENT

VERY INEFFICIENT TO USE ONLY ONE SIZE BOARD BECAUSE OF TOO LARGE PERCENTAGE OF DISCARD.

Figure 4.5 *(cont.)*

Discussion The assumptions that Jim made seem to be reasonable. Although he did not allow for the width of the saw cut nor for the thickness of the bark, the omissions may very well be consistent with the degree of accuracy involved in his calculations. Most Boy and Girl Scouts have learned to measure heights by the method he used; and all freshman engineering students should be familiar with similar triangles. After determining the height and diameter of the tree, Jim applied a graphical technique for determining how many boards could be cut from each 2.4 m section of the tree. He correctly used the upper (smaller) diameter of the section. His task was then reduced to a simple counting of the boards.

We have mentioned previously that both time allotted and physical circumstances (reference material at hand, access to a phone, and so on are major factors that influence the type of estimate you can produce. Example probs. 4.3 and 4.4 will show reasonable estimates of the same quantity (paper used) under two different sets of circumstances. In the first case, the student had about 10 minutes to do the estimate under exam conditions; that is, the student could not leave the classroom seat and had no reference material. The second case resulted from a homework assignment where 1 to 2 hours were available and the student could obtain data needed for the estimate.

Example problem 4.3 Estimate the amount of paper used by the students of this college for homework, quizzes, and examinations during this academic year. Express your answer in kilograms. You may use no reference materials.

Discussion The problem requires that data be known or assumed. The calculation is simple and straightforward once the data are established (see Fig. 4.6). Because much data must be assumed without verification, the estimate cannot be expected to be very precise, so only one, or at most two, significant figures should be reported. Two significant figures represent a maximum error range of about 10 percent. For example, 4 200 represents numbers between 4 150 and 4 250, an error of 100 (4 200 $\pm$ 50) in 4 200 or about 2.4 percent.

Example problem 4.4 Estimate the amount of paper used by the students of this college for homework, quizzes, and examinations during this academic year. Express your answer in kilograms. Plan to use 1 to 2 hours as a homework assignment.

Discussion The results of this estimation are found in Fig. 4.7. Many of the assumptions that had to be made in Example prob. 4.3 are no longer necessary because time is available to obtain information. The small sample in the survey still produces some uncertainty. In each case, your reader should be reasonably convinced by your techniques; for example, anyone would be willing to accept student population obtained from the registrar.

3-16-XX	ENGR 160 PROBLEM 4.11	SARA KENILWORTH	1/1

PROBLEM 4.11

ESTIMATE THE MASS (kg) OF PAPER USED BY STUDENTS IN THIS COLLEGE DURING THIS ACADEMIC YEAR FOR HOMEWORK, QUIZZES, AND EXAMS.

ASSUMPTIONS

1. COLLEGE HAS 4200 STUDENTS.
2. 2 SHEETS ARE USED PER CREDIT HOUR PER WEEK.
3. 32 WEEKS PER ACADEMIC YEAR.
4. TABLET OF 50 SHEETS OF PAPER HAS A MASS OF ABOUT ½ lbm.
5. STUDENTS AVERAGE 15 CREDITS PER TERM.

CALCULATIONS

$$\text{MASS OF SHEET} = \left(\frac{\frac{1}{2}\ \text{lbm}}{50\ \text{SHEETS}}\right)\left(\frac{454\ \text{g}}{\text{lbm}}\right) = 4.54\ \text{g/SHEET}$$

$$\text{MASS OF PAPER} = (4200\ \text{STUDENTS})\left(\frac{15\ \text{CREDITS}}{\text{STUDENT}}\right)\left(\frac{2\ \text{SHEETS}}{(\text{CREDIT})(\text{WEEK})}\right)\left(\frac{32\ \text{WEEKS}}{\text{YEAR}}\right)\left(\frac{4.54\ \text{g}}{\text{SHEET}}\right)\left(\frac{1\ \text{Kg}}{1000\ \text{g}}\right)$$

$$= 18305.28\ \text{kg/YEAR}$$

$$\cong 1.8 \times 10^{4}\ \text{kg/YEAR}$$

Figure 4.6
Student presentation for Example prob. 4.3.

	3-16-XX	ENGR 160 PROBLEM 4.11	SARA KENILWORTH	1/2

PROBLEM 4.11

ESTIMATE THE MASS (kg) OF PAPER USED BY STUDENTS IN THIS COLLEGE DURING THIS ACADEMIC YEAR FOR HOMEWORK, QUIZZES, AND EXAMS.

ASSUMPTIONS

1. THIS COLLEGE REFERS TO THE COLLEGE OF ENGINEERING, STATE UNIVERSITY
2. ONLY UNDERGRADUATE STUDENTS WILL BE CONSIDERED.
3. THE COMPUTATION IS FOR THE 1983-84 ACADEMIC YEAR.
4. ENGINEERING-PROBLEMS PAPER IS THE STANDARD FOR THE COLLEGE.
5. THE RESULTS OF THE SURVEY BELOW REASONABLY REPRESENT THE ENGINEERING STUDENTS AS A WHOLE.

COLLECTED DATA

1. ACCORDING TO THE UNIVERSITY REGISTRAR, 4 256 ENGINEERING UNDERGRADUATES WERE REGISTERED FOR FALL SEMESTER AND 4 028 FOR SPRING SEMESTER.
2. THE GENERAL CATALOG SHOWS 15 WEEKS OF CLASSES EACH SEMESTER PLUS FINAL EXAM WEEK EACH TERM.
3. A SURVEY OF 8 ENGINEERING STUDENTS, 2 FROM EACH CLASS, FRESHMAN THROUGH SENIOR, PROVIDED THE FOLLOWING DATA:

STUDENTS	SHEETS/WEEK	SHEETS/EXAM WEEK	COMMENTS
1	35	15	
2	18	8	PART TIME
3	45	12	
4	52	20	20 CREDIT HOURS
5	28	10	
6	25	14	
7	38	15	
8	42	18	
AVE	35.4	14	

4. A REAM (500 SHEETS) OF ENGINEERING-PROBLEMS PAPER HAS A MASS OF 3.75 lbm AS DETERMINED WITH A POSTAL SCALE.

THEORY

MASS PER SHEET = (MASS PER REAM)/(500 SHEETS PER REAM)

MASS OF PAPER = MASS FOR FALL SEMESTER + MASS FOR SPRING SEMESTER

WHERE

MASS FOR FALL = (NUMBER OF STUDENTS)[(SHEETS/WEEK)(CLASS WEEKS) + (SHEETS/EXAM WEEK)](MASS/SHEET)

MASS FOR SPRING = (NUMBER OF STUDENTS)[(SHEETS/WEEK)(CLASS WEEKS) + (SHEETS/EXAM WEEK)](MASS/SHEET)

Figure 4.7 Student presentation for Example prob. 4.4.

	3-16-XX	ENGR 160 PROBLEM 4.11	SARA KENILWORTH	2/2

CALCULATIONS

MASS PER SHEET = [(3.75 lbm PER REAM)/(500 SHEETS)](0.453 59 kg/lbm)
= 0.003 402 kg/SHEET

MASS FOR FALL = (4 256 STUDENTS)[(35.4 SHEETS/WEEK)(15 WEEKS) +
14 SHEETS](0.003 402 kg/SHEET)
= 7 891 kg

MASS FOR SPRING = (4 028 STUDENTS)[(35.4 SHEETS/WEEK)(15 WEEKS) +
14 SHEETS](0.00:XX 02 kg/SHEET)
= 7 468 kg

TOTAL MASS OF PAPER = 7 891 kg + 7 468 kg
= 15 359 kg

ESTIMATE OF TOTAL MASS = $\underline{\underline{1.5 \times 10^4 \text{ kg}}}$

Figure 4.7 *(cont.)*

Another type of estimate that engineers are called upon to make is one where a choice is involved. Estimates are prepared of the various alternatives available, and a decision as to which alternative to follow is then made; perhaps the one that is least expensive is chosen. Example prob. 4.5 is designed to show estimating for decision purposes. This problem requires experience that a student might have.

Example problem 4.5 Approximate the amount of gasoline that will be used by the students of Iowa State University during the next Christmas–New Year recess for the purpose of visiting their homes and returning. Provide the answer in gallons. What will be the total cost of this gasoline?

Discussion Figure 4.8 is the result of the approximation. A number of assumptions were made in order to obtain a solution (without writeup) in less than 30 minutes. Note that the writeup was prepared on a word processor.

PROBLEM 4.5

Estimate the amount of gasoline that will be used by Iowa State University in traveling home and returning during the upcoming Christmas-New Year's holiday break. Also estimate the cost of the gasoline.

ASSUMPTIONS

1. Only automobile usage is considered.
2. Ten percent of the students fly home. Only the trip to the airport will be considered.
3. Ten percent of the students remain on campus.
4. An average of two students travel together in the same automobile.
5. The average automobile used for the trip gets 20 miles to the gallon.
6. Travel distance will be expressed in terms of median round trip distances of 100, 300, 500, 700, 900 miles. For longer distances, students are assumed to fly.

COLLECTED DATA

1. Iowa State has 24 500 students (FROM STUDENT PROFILES BROCHURE)
   Iowa residents 65%
   Other U.S. 25%
   International 10%
2. Cost of gasoline is $1.08/gal
3. Distance to airport is 45 miles

Figure 4.8
Student presentation, produced on a word processor, for Example prob. 4.5.

Travel distance estimate:

round trip distance	percent of students	number of students
0-200	35	8 575
200-400	30	7 350
400-600	5	1 225
600-800	5	1 225
800-1000	5	1 225
Flying	10	2 450
Not traveling	10	2 450
	100	24 500

Mileage total:

0-200 mi	[(8 575 students)/(2 students/car)][100 mi/car]	= 428 750 mi
200-400 mi	[7 350 /2][300]	= 1 102 500 mi
400-600 mi	[1 225/2][500]	= 306 250 mi
600-800 mi	[1 225/2][700]	= 428 750 mi
800-1000 mi	[1 225/2][900]	= 551 250 mi
Flying	[2 450/2][90]	= 110 250 mi
		2 927 750 mi

gasoline amount:

(2 927 750 mi)/(20 mi/gal) = 150 000 gal

gasoline cost:

(150 000 gal)($1.08/gal) = $160 000

4.6 Key Terms and Concepts

- Physical measurement
- Significant digits
- Scientific notation
- Accuracy
- Precision
- Systematic error (in a measurement)
- Random error (in a measurement)
- Engineering estimates
- Engineering approximations

Problems

4.1 How many significant digits are contained in each of the following quantities?

(*a*) 4 930
(*b*) 4.930
(*c*) 0.049 3
(*d*) 206.0
(*e*) 200
(*f*) 6.220×10^2
(*g*) 9.009
(*h*) 0.000 3
(*i*) 0.320 0
(*j*) 40 620

4.2 How many significant digits are contained in each of the following quantities?

(*a*) 322
(*b*) 3.22
(*c*) 0.032 2
(*d*) 0.032 20
(*e*) $3.220\,3 \times 10^5$
(*f*) 0.0030
(*g*) 3 600 s/h
(*h*) 5 280 ft/mi
(*i*) 2.006×10^4
(*j*) 0.090 50

4.3 Perform the following computations and report the answers with the proper number of significant digits.

(*a*) (2.05)(360)
(*b*) 26.35/14
(*c*) [(4.91)(32.2)]/12.03
(*d*) $[(14.2 \times 10^3)(7.3 \times 10^{-1})(1\,021)]/[(16)(21.0 \times 10^{-3})]$
(*e*) $(4.597\,6 \times 10^2) + (4.721 \times 10^1)$
(*f*) 0.009 024 + 0.065 320
(*g*) (0.127 60 − 0.073 2)/14.06
(*h*) 163/(0.021 68 − 0.021 66)

4.4 Perform the following computations. Record results with the proper number of significant digits.

(*a*) 648.2/14.0
(*b*) 436.2 mi/1.06h
(*c*) $(4.010 \times 10^3$ s)/(3 600 s/h)
(*d*) ($95.99)(0.15)
(*e*) $[(4.83 \times 10^3) - (14.63 \times 10^2)]/136.2$

4.5 Express the results of the following computations with the proper number of significant digits.

(*a*) $Q = 4.0P^2 - 35\,P - 23$ for $P = 10.4$
(*b*) $(46.13)^2/4$
(*c*) $Y - 4.33 \sin x - 12.0 \sin x \cos x$ for $x = 1.2$ radians
(*d*) ($36 ticket)(23 850 tickets)
(*e*) $[(3.0004)/(8.660\,0 \times 10^{-3})] - 345.6$

4.6 (*a*) The accuracy of a multimeter is given as ± 0.8 percent of full scale. If the full scale reading is 130 volts, what is the range of values associated with a full scale reading?

(*b*) A thermometer is found to yield readings that are consistently 2.5 percent high. If a reading of 82.4°F is taken, what is the correct temperature?

4.7 (*a*) Compute the percent error for the following cases.

	Case A	Case B
Measured	1 203.7	17.4
True	1 204.8	16.3

(*b*) From the results in a, discuss the precision of the measurements in the two cases.

4.8 Estimate the total volume in cubic meters and mass in kilograms of the concrete and asphalt paving in the largest parking lot on your campus.

4.9 Estimate the volume (capacity) in cubic meters of a water tower located on your campus or nearby (as specified by your instructor).

4.10 How much paint (primer plus two coats) is needed to paint a water tower on your campus or nearby (as specified by your instructor)? Consider only the tank portion, not the supporting structure, if any. Express result in gallons.

4.11 Compute the surface area (in square meters) of external glass in your classroom building or any other structure designated by your instructor. What is the mass of this glass in kilograms?

4.12 Suppose your school's basketball team plans to play in a 3-day tournament in Kansas City. Your job is to estimate the total cost of transporting the team, food and housing for the team, and so on. Consider the essential personnel (team, coaches, and managers) only.

4.13 Your school has plans to require all entering engineering students to purchase their own microcomputers. Estimate the total cost to students for the next fall term based on a system specification worked out by the class or provided by the instructor.

4.14 Based on your local electric energy rates, estimate the cost of lighting for your classroom for this academic year.

4.15 By your personal observation, estimate the total number of computer terminals available for general student use on your campus. Determine the maximum number of students that could be served for 2 hours each week, assuming the terminals are available 24 hours per day, 7 days per week. How does this compare with the number of students on campus?

4.16 Estimate the number of audio tapes and CDs owned by a typical engineering student at your school. Include a well-documented survey in your solutions.

4.17 Each year rain and/or snow falls on your campus. For a specified parking lot or rooftop, estimate the total volume (in cubic meters) and mass (in kilograms) of water that must be carried away during the entire year.

4.18 By utilizing a reasonable survey technique, compute the amount of money annually spent by all engineering students enrolled at your school for long-distance telephone calls.

4.19 If your campus has a body of water (lake, pond, or fountain), determine the volume of water contained in it (in cubic feet) and the surface area (in square feet). Do the estimate without getting wet or getting in trouble with campus security.

4.20 Estimate the mass of all textbooks you will purchase to complete your first engineering degree. If you have not selected a major, use one that is a likely candidate.

4.21 What is the total floor area (in square feet) and volume (in cubic feet) of the classrooms in the building where this course meets?

4.22 Assuming that the interior of the classroom where this class meets is painted, estimate the amount of paint necessary to refurbish it and make a significant change in wall color. (Alternative: Do the estimate for another interior space designated by your instructor.)

4.23 Following an accident where a student was cut by broken glass in a door, your school has decided to replace all door glass with plastic (Plexiglas or some other brand) in the building where this class meets. Considering interior and exterior doors, estimate the amount (in square feet) of material needed and its approximate cost. Exclude labor costs in this estimate.

4.24 For a building (or portion of a building) on your campus that is carpeted (the student union building, perhaps), estimate the amount (in square yards) of material reqiured to recarpet the space. What is the expected cost of the materials?

Problems 4.25 through 4.28 require a significant amount of data to be collected prior to performing the estimates. You should allow at least 2 hours (excluding writeup) to complete each problem.

4.25 Estimate the amount of each of the ingredients required to make the concrete used in all the designated interstate highways in your state.

4.26 Estimate the amount of water used in a 1-year period by a family of four who own their own house. Determine the cost from local utility rates.

4.27 Estimate the number of family dwellings that could be supplied by the electricity generated by all the nuclear power plants in the United States.

4.28 Estimate the distance (in kilometers) that you walk (and run) during a typical week while attending classes at your school. Include all activities each day.

5 Dimensions, Units, and Conversions

5.1 Introduction

Years ago, when countries were more isolated from one another, individual governments tended to develop and use their own set of measures. As the rapid increase in global communication and travel brought countries closer together and the world advanced in technology, the need for a universal system of measurement became abundantly clear. There was such a growth of information but diversity of reporting among nations that a standard set of dimensions, units, and measurements was vital if this wealth of knowledge was to be of benefit at all. This chapter deals with the difference between dimensions and units and at the same time explains how there can be an orderly transition from many systems of units to one system—that is, an international standard.

The standard currently accepted in most industrial nations (it is optional in the United States) is the international metric system, or Systeme International d'Unites, abbreviated SI. The SI units are a modification and refinement of an earlier version of the metric system (MKS) that designated the meter, kilogram, and second as fundamental units.

France was the first country, in 1840, to officially legislate adoption of the metric system and decree that its use be mandatory.

The United States almost adopted the metric system 150 years ago. In fact, the metric system was made legal in the United States in 1866, but its use was not made compulsory. In spite of many attempts since that time, full conversion to the metric system has not yet been realized in the United States, but significant steps in that direction are continuously underway.

5.2 Physical Quantities

Engineers are constantly concerned with the measurements of fundamental physical quantities such as length, time, temperature, force, and so on. In order to specify a physical quantity fully, it is not sufficient to indicate merely a numerical value. The magnitude of physical quantities can be understood only when they are compared with predetermined reference amounts, called *units*. Any measurement is, in effect, a comparison of how many

Look for the **km/h** tab below the maximum speed limit sign, indicating that this is the new speed in metric.

100 km/h This speed limit will likely be the most common on freeways. On most rural two-lane roadways, **80 km/h** will be typical.

50 km/h A **50 km/h** speed limit will apply in most cities. Actual speed limits will be established in accordance with local regulations.

HIGHWAY Speed km/h

Metric Commission Canada — Commission du système métrique Canada

Surveillez l'indication de l'unité de vitesse **km/h**; ce symbole signifie que va vitesse est mesurée selon le système métrique.

100 km/h Sur les autoroutes, la vitesse maximale la plus courante sera de **100 km/h** tandis que sur les routes à grande circulation, elle sera de **80 km/h**.

50 km/h Dans la plupart des grands centres, la vitesse maximale sera de **50 km/h**.

Les vitesses maximales en vigueur dans votre société seront établies selon les règlements municipaux.

Commission du système métrique Canada — Metric Commission Canada

Figure 5.1
Highway signs in Canada
(Metric Commission of Canada)

(the numerical value V) units U are contained within the physical quantity Q. If we consider length the physical quantity L, and 20.0 the numerical value V, with meters as the designated units U, then $L = 20.0$ m can be represented in a general fashion by the expression

$$Q = VU$$

For this relationship to be valid, the exact reproduction of a unit must be theoretically possible at any time. Therefore standards must be established. These standards are a set of fundamental unit quantities kept under normalized conditions in order to preserve their values as accurately as possible. We shall speak more about standards and their importance later.

5.3 Dimensions

Dimensions are used to describe physical quantities. An important element to remember is that dimensions are independent of units. As outlined above, the physical quantity length can be represented by the dimension L, for which there are a large number of possibilities available when selecting a unit. For example, in ancient Egypt, the cubit was related to the length of the arm from the tip of the middle finger to the elbow. Measurements were thus a function of physical stature, with variation from one individual to another. Much later, in Britain, the inch was specified as the distance covered by three barley corns, round and dry, laid end to end.

Today we require more precision. For example, the meter is defined in terms of the distance traveled by light in a vacuum in a specified amount of time. We can draw two important points from this discussion: (1) Physical quantities can be accurately measured, and (2) each of these units (cubit, inch, and meter), although distinctly different, has in common the quality of being a length and not an area or a volume.

A technique used to distinguish between units and dimensions is to call all quantities of length simply L. In this way, each new physical quantity gives rise to a new dimension, such as T for time, F for force, M for mass, and so on. (Note that there are as many dimensions as there are kinds of physical quantities.)

Moreover, dimensions can be divided into two areas—fundamental and derived. A fundamental dimension is a dimension that can be conveniently and usefully manipulated when expressing all physical quantities of a particular field. Derived dimensions are a combination of fundamental dimensions. Velocity, for example, could be defined as fundamental dimension V, but it is more customary as well as more convenient to consider velocity as a combination of fundamental dimensions, so that it becomes a derived dimension, $V = (L)(T)^{-1}$. L and T are fundamental dimensions, and V is a derived dimension because it is made up of two fundamental dimensions (L,T).

It is advantageous to use as few fundamental dimensions as possible, but the selection of what is to be fundamental and what is to be derived is not fixed. In actuality, any dimension can be selected as a fundamental dimension in a particular field of engineering or science; and for reasons of convenience, it may be a derived dimension in another field.

A *dimensional system* can be defined as the smallest number of fundamental dimensions which will form a consistent and complete set for a field of science. For example, three fundamental dimensions are necessary to form a complete mechanical dimensional system. Depending on the discipline these dimensions may be specified as either length (L), time (T), and mass (M) or length (L), time (T), and force (F). If temperature is important to the application, a fourth dimension must be added.

The *absolute system* (so called because dimensions used are not affected by gravity) has as its fundamental dimensions L, T, and M. An advantage of this system is that comparisons of masses at various locations can be made with an ordinary balance, because the local acceleration of gravity has no influence upon the results.

Table 5.1 Two basic dimensional systems.

Quantity	Absolute	Gravitational
Length	L	L
Time	T	T^{-}
Mass	M	$FL^{-1}T^{2}$
Force	MLT^{-2}	F
Velocity	LT^{-1}	LT^{-1}
Pressure	$ML^{-1}T^{-2}$	FL^{-2}
Momentum	MLT^{-1}	FT
Energy	$ML^{2}T^{-2}$	FL
Power	$ML^{2}T^{-3}$	FLT^{-1}
Torque	$ML^{2}T^{-2}$	FL

The *gravitational system* has as its fundamental dimensions L, T, and F. It is widely used in many engineering branches because it simplifies computations when weight is a fundamental quantity in the computations. Table 5.1 illustrates two of the more basic dimensional systems; however, a number of other dimensional systems are commonly used for heat, electromagnetism, electrical dimensions, and so forth.

5.4 Units

Once a consistent dimensional system has been selected, one must select a unit system by choosing a specific unit for each fundamental dimension. The problem one encounters when working with units is that there can be a large number of unit systems to choose from for each complete dimension system, as we have already suggested. It is obviously desirable to limit the number of systems and combinations of systems. The SI previously alluded to is intended to serve as an international standard that will provide worldwide consistency.

There are two fundamental systems of units commonly used in mechanics today. One system used in almost every industrial country of the world is called the *metric system*. It is a decimal-absolute system based on the meter, kilogram, and second (MKS) as the units of length, mass, and time, respectively. The United States has used the other system, normally referred to as the British gravitational system. It is based on the foot, pound-force, and second.

Numerous international conferences on weights and measures over the past 40 years have gradually modified the MKS system to the point that all countries previously using various forms of the metric system are beginning to standardize. The Systeme International d'Unites (SI) is now considered the new international system of units. The United States has adopted the system, but full use will be preceded by a long and expensive period of change. During this transition period, engineers will have to not only be familiar with SI but also other systems and the necessary conversion process between or among systems.

5.5 SI Units and Symbols

The International System of Units (SI), developed and maintained by the General Conference on Weights and Measures (Conference Generale des Poids et Mesures, CGPM), is intended as a basis for worldwide standardization of measurements. The name and abbreviation were set forth in 1960. SI at the present time is a complete system that is being universally adopted.

This new international system is divided into three classes of units:

1. Base units
2. Supplementary units
3. Derived units

There are seven base units in the SI. The units (except the kilogram) are defined in such a way that they can be reproduced anywhere in the world.

Table 5.2 lists each base unit along with its name and proper symbol.

In the following list, each of the base units is defined as established at the international CGPM:

1. Length: The meter is a length equal to the distance traveled by light in a vacuum during 1/299 792 458 s. The meter was defined by the CGPM that met in 1983.
2. Time: The second is the duration of 9 192 631 770 periods of radiation corresponding to the transition between the two hyperfine levels of the ground state of the cesium-133 atom. The second was adopted by the thirteenth CGPM in 1967.
3. Mass: The standard for the unit of mass, the kilogram, is a cylinder of platinum-iridium alloy kept by the International Bureau of Weights and Measures in France. A duplicate copy is maintained in

Table 5.2 Base units

Quantity	Name	Symbol
Length	meter	m
Mass	kilogram	kg
Time	second	s
Electric current	ampere	A
Thermodynamic temp.	kelvin	K
Amount of substance	mole	mol
Luminous intensity	candela	cd

the United States. The unit of mass was adopted by the First and Third CGPMs in 1889 and 1901. It is the only base unit nonreproducible in a properly equipped lab.

4. Electric current: The ampere is a constant current which, if maintained in two straight parallel conductors of infinite length and of negligible circular cross sections and placed one meter apart in volume, would produce between these conductors a force equal to 2×10^{-7} newton per meter of length. The ampere was adopted by the Ninth CGPM in 1948.

5. Temperature: The kelvin, a unit of thermodynamic temperature, is the fraction 1/273.16 of the thermodynamic temperature of the triple point of water. The kelvin was adopted by the Thirteenth CGPM in 1967.

6. Amount of substance: The mole is the amount of substance of a system that contains as many elementary entities as there are atoms in 0.012 kilogram of carbon-12. The mole was defined by the Fourteenth CGPM in 1971.

7. Luminous intensity: The base unit candela is the luminous intensity in a given direction of a source that emits monochromatic radiation of frequency 540×10^{12} hertz and has a radiant intensity in that direction of 1/683 watts per steradian.

The units listed in Tab. 5.3 are called *supplementary units* and may be regarded as either base units or as derived units.

The unit for a plane angle is the radian, a unit that is used frequently in engineering. The steradian is not as commonly used. These units can be defined in the following way:

1. Plane angle: The radian is the plane angle between two radii of a circle that cut off on the circumference of an arc equal in length to the radius.

2. Solid angle: The steradian is the solid angle which, having its vertex in the center of a sphere, cuts off an area of the sphere equal to that of a square with sides of length equal to the radius of the sphere.

As indicated earlier, derived units are formed by combining base, supplementary, or other derived units. Symbols for them are carefully selected to avoid confusion. Those which have special names and symbols, as interpreted for the United States by the National Bureau of Standards, are listed in Tab. 5.4 together with their definitions in terms of base units.

Table 5.3 Supplementary units

Quantity	Name	Symbol
Plane angle	radian	rad
Solid angle	steradian	sr

Table 5.4 Derived units

Quantity	SI unit symbol	Name	Base units
Frequency	Hz	hertz	s^{-1}
Force	N	newton	$kg \cdot m \cdot s^{-2}$
Pressure stress	Pa	pascal	$kg \cdot m^{-1} \cdot s^{-2}$
Energy or work	J	joule	$kg \cdot m^2 \cdot s^{-2}$
Quantity of heat	J	joule	$kg \cdot m^2 \cdot s^{-2}$
Power radiant flux	W	watt	$kg \cdot m^2 \cdot s^{-3}$
Electric charge	C	coulomb	$A \cdot s$
Electric potential	V	volt	$kg \cdot m^2 \cdot s^{-3} \cdot A^{-1}$
Potential difference	V	volt	$kg \cdot m^2 \cdot s^{-3} \cdot A^{-1}$
Electromotive force	V	volt	$kg \cdot m^2 \cdot s^{-3} \cdot A^{-1}$
Capacitance	F	farad	$A^2 \cdot s^4 \cdot kg^{-1} \cdot m^{-2}$
Electric resistance	Ω	ohm	$kg \cdot m^2 \cdot s^{-3} \cdot A^{-2}$
Conductance	S	siemens	$kg^{-1} \cdot m^{-2} \cdot s^3 \cdot A^2$
Magnetic flux	Wb	weber	$kg^{-1} \cdot m \cdot s^{-2} \cdot A^{-1}$
Magnetic flux density	T	tesla	$kg \cdot s^2 \cdot A^{-1}$
Inductance	H	henry	$kg \cdot m^2 \cdot s^{-2} \cdot A^{-2}$
Luminous flux	lm	lumen	cd·sn
Illuminance	lx	lux	$cd \cdot sn \cdot m^{-2}$
Celsius temperature*	°C	degree Celsius	K
Activity (radionuclides)	Bq	becqueret	s^{-1}
Absorbed dose	Gy	gray	$m^2 \cdot s^{-2}$
Dose equivalent	S	sievert	$m^2 \cdot s^{-2}$

*The thermodynamic temperature (T_k) expressed in kelvins is related to Celsius temperature ($t_{°C}$) expressed in degrees Celsius by the equation $t_{°C} = T_k - 273.15$.

At first glance, Fig. 5.2 may appear complex, even confusing; however, a considerable amount of information is presented in this concise flowchart. To get the point of it quickly, be aware that the solid lines denote multiplication and the broken lines indicate division. The arrows pointing toward the units (circled) are significant and arrows going away have no meaning for that particular unit. Consider the pascal, as an example: Two arrows point toward the circle—one solid and one broken. This means that the unit pascal is formed from the newton and meter squared, or N/m^2.

Other derived units, such as those included in Tab. 5.5, have no special names but are combinations of base units and units with special names.

Being a decimal system, the SI is convenient to use because by simply affixing a prefix to the base, a quantity can be increased or decreased by factors of 10 and the numerical quantity can be kept within manageable limits. Tab. 5.6 lists the multiplication factors with their prefix names and symbols.

The proper selection of prefixes will also help eliminate nonsignificant zeros and leading zeros in decimal fractions. One rule to follow is that the numerical value of any measurement should be recorded as a number between 0.1 and 1 000. This rule is suggested because it is easier to make realistic judgments when working with numbers between 0.1 and 1 000. For example, suppose that you are asked the distance to a nearby town. It would

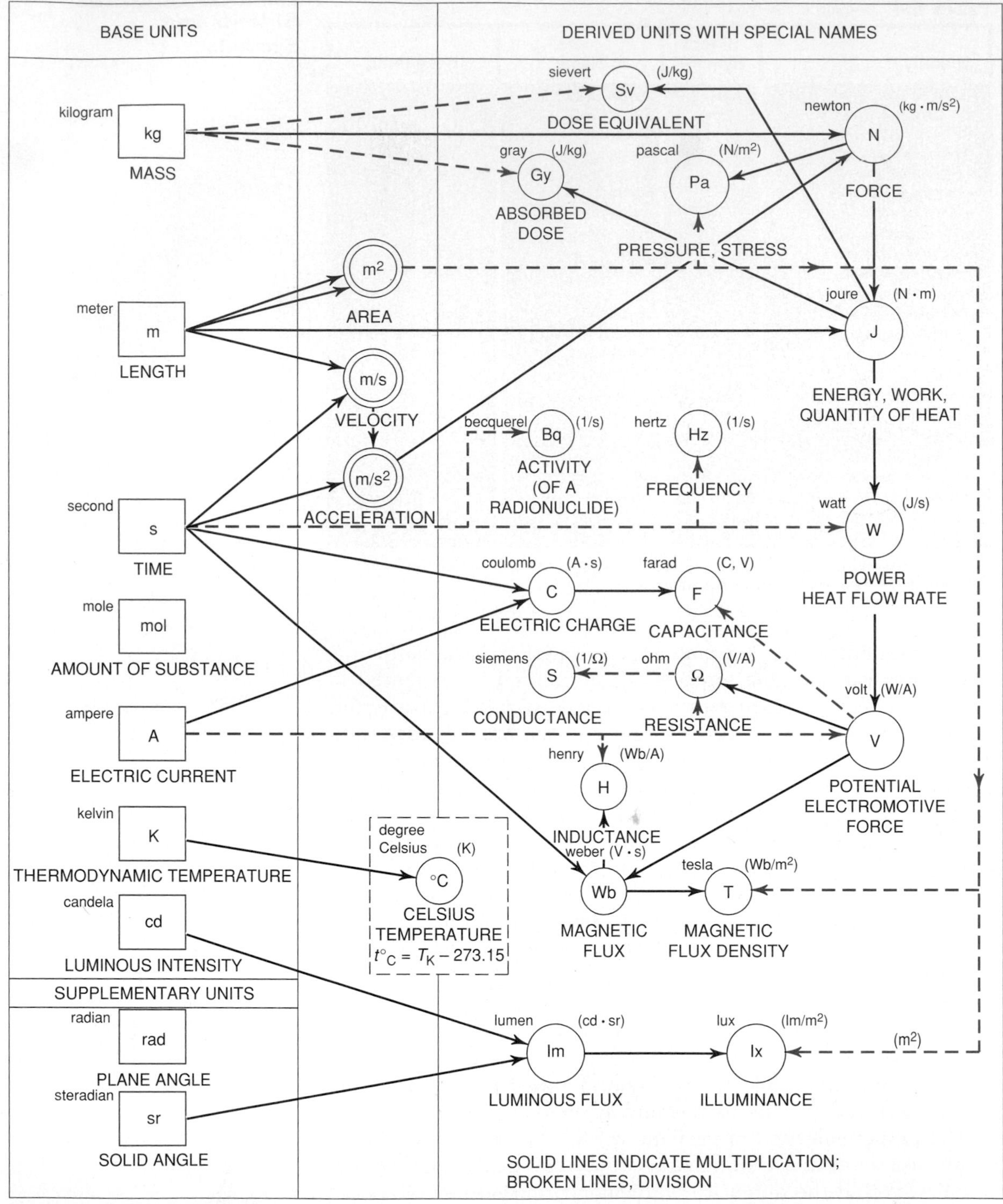

Figure 5.2
Graphical illustration of how certain SI units are derived in a coherent fashion from base and supplementary units. *(National Bureau of Standards.)*

Table 5.5 Common derived units

Quantity	Units	Quantity	Units
Acceleration	$m \cdot s^{-2}$	Molar entropy	$J \cdot mol^{-1} \cdot K^{-1}$
Angular acceleration	$rad \cdot s^{-2}$	Molar heat capacity	$J \cdot mol^{-1} \cdot K^{-1}$
Angular velocity	$rad \cdot s^{-1}$	Moment of force	$N \cdot m$
Area	m^2	Permeability	$H \cdot m^{-1}$
Concentration	$mol \cdot m^{-3}$	Permittivity	$F \cdot m^{-1}$
Current density	$A \cdot m^{-2}$	Radiance	$W \cdot m^{-2} \cdot sr^{-1}$
Density, mass	$kg \cdot m^{-3}$	Radiant intensity	$W \cdot sr^{-1}$
Electric charge density	$C \cdot m^{-3}$	Specific heat capacity	$J \cdot kg^{-1} \cdot K^{-1}$
Electric field strength	$V \cdot m^{-1}$	Specific energy	$J \cdot kg^{-1}$
Electric flux density	$C \cdot m^{-2}$	Specific entropy	$J \cdot kg^{-1} \cdot K^{-1}$
Energy density	$J \cdot m^{-3}$	Specific volume	$m^3 \cdot kg^{-1}$
Entropy	$J \cdot K^{-1}$	Surface tension	$N \cdot m^{-1}$
Heat capacity	$J \cdot K^{-1}$	Thermal conductivity	$W \cdot m^{-1} \cdot K^{-1}$
Heat flux density	$W \cdot m^{-2}$	Velocity	$m \cdot s^{-1}$
Irradiance	$W \cdot m^{-2}$	Viscosity, dynamic	$Pa \cdot s$
Luminance	$cd \cdot m^{-2}$	Viscosily, kinematic	$m^2 \cdot s^{-1}$
Magnetic field strength	$A \cdot m^{-1}$	Volume	m^3
Molar energy	$J \cdot mol^{-1}$	Wavelength	m

be more understandable to respond in kilometers than meters. That is, it is easier to visualize 10 km than 10 000 m.

Moreover, the use of certain prefixes is preferred over that of others. Those representing powers of 1 000, such as kilo-, mega-, milli-, and micro-, will reduce the number you must remember. These preferred prefixes should be used, with the following three exceptions which are still common because of convention:

1. When expressing area and volume, the prefixes hecto-, deka-, deci-, and centi- may be used, for example, cubic centimeter.

2. When discussing different values of the same quantity or expressing them in a table, calculations are simpler to perform when you use the same unit multiple throughout.

Table 5.6 Decimal multiples

Multiplier	Prefix name	Symbol
10^{18}	exa	E
10^{15}	peta	P
10^{12}	tera	T
10^{9}	giga	G
10^{6}	*mega	M
10^{3}	*kilo	k
10^{2}	hecto	h
10^{1}	deka	da
10^{-1}	deci	d
10^{-2}	centi	c
10^{-3}	*milli	m
10^{-6}	*micro	μ
10^{-9}	nano	n
10^{-12}	plco	p
10^{-15}	femto	f
10^{-18}	atto	a

*Most often used.

3. Sometimes a particular multiple is recommended as a consistent unit even though its use violates the 0.1 to 1 000 rule. For example, many companies use the millimeter for linear dimensions even when the values lie far outside this suggested range. The cubic decimeter (commonly called liter) is also used.

Recalling the discussion of significant figures, we see that SI prefix notations can be used to a definite advantage.

Consider the previous example of 10 km. When giving an estimate of distance to the nearest town, there is certainly an implied approximation in the use of a round number. Suppose that we were talking about a 10 000 m Olympic track and field event. The accuracy of such a distance must certainly be greater than something between 5 000 and 15 000 m. This example is intended to illustrate the significance of the four zeros. If all four zeros are in fact significant, then the race is accurate within 1 m (9 999.5 to 10 000.5). If only three zeros are significant, then the race is accurate to within 10 m

(9 995 to 10 005).

There are two logical and acceptable methods available of eliminating confusion concerning zeros:

1. Use proper prefixes to denote intended significance.

Distance	Precision
10.000 km	9 999.5 to 10 000.5 m
10.00 km	9 995 to 10 005 m
10.0 km	9 950 to 10 050 m
10 km	5 000 to 15 000 m

2. Use scientific notation to indicate significance.

Distance	Precision
10.000×10^3 m	9 999.5 to 10 000.5 m
10.00×10^3 m	9 995 to 10 005 m
10.0×10^3 m	9 950 to 10 050 m
10×10^3 m	5 000 to 15 000 m

Selection of a proper prefix is customarily the logical way to handle problems of significant figures; however, there are conventions that do not lend themselves to the prefix notation. An example would be temperature in degrees Celsius; that is, $4.00(10^3)$°C is the conventional way to handle it, not 4.00 k°C.

5.6 Rules for Using SI Units

Along with the adoption of SI comes the responsibility to thoroughly understand and properly apply the new system. Obsolete practices involving both English and metric units are widespread. This section provides rules that should be followed when working with SI units.

5.6.1 Unit Symbols and Names

1. Periods are never used after symbols unless the symbol is at the end of a sentence (that is, SI unit symbols are not abbreviations).

2. Unit symbols are written in lowercase letters unless the symbol derives from a proper name, in which case the first letter is capitalized.

Lowercase	Uppercase
m, kg, s, mol, cd	A, K, Hz, Pa, C

3. Symbols rather than self-styled abbreviations should always be used to represent units.

Correct	Not correct
A	amp
s	sec

4. An s is never added to the symbol to denote plural.

5. A space is always left between the numerical value and the unit symbol.

Correct	Not correct
43.7 km	43.7km
0.25 Pa	0.25Pa

Exception: No space should be left between numerical values and the symbols for degree, minute, and second of angles and for degree Celsius.

6. There should be no space between the prefix and the unit symbols.

Correct	Not correct
mm, MΩ	k m, μ F

7. When writing unit names, all letters are lowercase except at the beginning of a sentence, even if the unit is derived from a proper name.

8. Plurals are used as required when writing unit names. For example, henries is plural for henry. The following exceptions are noted:

Singular	Plural
lux	lux
hertz	hertz
siemens	siemens

With these exceptions, unit names form their plurals in the usual manner.

9. No hyphen or space should be left between a prefix and the unit name. In three cases the final vowel in the prefix is omitted: megohm, kilohm, and hectare.

10. The symbol should be used in preference to the unit name because unit symbols are standardized. An exception to this is made when a number is written in words preceding the unit, for example, we would write *ten meters,* not *ten m.* The same is true the other way, for example, 10 m, not 10 meters.

5.6.2 Multiplication and Division

1. When writing unit names as a product, always use a space (preferred) or a hyphen.

Correct usage
newton meter or newton-meter

2. When expressing a quotient using unit names, always use the word *per* and not a solidus (/). The solidus, or slash mark, is reserved for use with symbols.

Correct	Not correct
meter per second	meter/second

3. When writing a unit name that requires a power, use a modifier, such as squared or cubed, after the unit name. For area or volume, the modifier can be placed before the unit name.

Correct	Not correct
millimeter squared	square millimeter

4. When expressing products using unit symbols, the center dot is preferred.

Correct
N·m for newton meter

5. When denoting a quotient by unit symbols, any of the following methods are accepted form:

Correct
m/s or $m \cdot s^{-1}$ or $\frac{m}{s}$

In more complicated cases, negative powers or parentheses should be considered. Use m/s^2 or $m \cdot s^{-2}$ but not m/s/s for acceleration; use $kg \cdot m^2/(s^3 \cdot A)$ or $kg \cdot m^2 \cdot s^{-3} \cdot A^{-1}$ but not $kg \cdot m^2/s^3/A$ for electric potential.

5.6.3 Numbers

1. To denote a decimal point, use a period on the line. When expressing numbers less than 1, a zero should be written before the decimal marker.

Example

15.6
0.93

2. Since a comma is used in many countries to denote a decimal point, its use is to be avoided in grouping data. When it is desired to avoid this confusion, recommended practice calls for separating the digits into groups of three, counting from the decimal to the left or right, and using a small space to separate the groups.

Correct and recommended procedure

6.513 824   76 851   7 434   0.187 62

5.6.4 Calculating with SI Units

Before we look at some suggested procedures that will simplify calculations in SI, the following positive characteristics of the system should be reviewed.

Only one unit is used to represent each physical quantity, such as the meter for length, the second for time, and so on. The SI metric units are *coherent;* that is, each new derived unit is a product or quotient of the fundamental and supplementary units without any numerical factors. Since coherency is a strength of the SI system, it would be worthwhile to demonstrate this characteristic by using two examples. Consider the use of the newton as the unit of force instead of pound-force (lbf). It is defined by Newton's second law, $F = ma$. It is the force that imparts an acceleration of one meter per second squared to a mass of 1 kilogram. Thus,

$$1 \text{ N} = (1 \text{ kg})(1 \text{ m/s}^2)$$

Consider also the Joule, a unit that replaces the British Thermal Unit, calorie, foot-pound-force, electronvolt, and horsepower-hour to stand for any form of energy. It is defined as the amount of work done when an applied force of one newton acts through a distance of one meter in the direction of the force. Thus,

$$1 \text{ J} = (1 \text{ N})(1 \text{ m})$$

To maintain the coherency of units, however, time must be expressed in seconds rather than minutes or hours, since the second is the base unit. Once coherency is violated, then a conversion factor must be included and the advantage of the system is diminished.

But there are certain units *outside* SI that are accepted for use in the United States, even though they diminish the system's coherence. These exceptions are listed in Tab. 5.7.

Calculations using SI can be simplified if you

1. Remember that all fundamental relationships such as the following still apply, since they are independent of units.

$$F = ma \quad KE = \frac{1}{2}mv^2 \quad E = RI$$

2. Recognize how to manipulate units and gain a proficiency in doing so. Since Watt = J/s = N·m/s, you should realize that N·m/s = $(N/m^2)(m^3/s)$ = (pressure)(volume flow rate).

3. Understand the advantage of occasionally adjusting all variables to base units. Replacing N with $kg \cdot m/s^2$, Pa with $kg \cdot m^{-1} \cdot s^{-2}$, and so on.

4. Develop a proficiency with exponential notation of numbers to be used in conjunction with unit prefixes.

$$1\ mm^3 = (10^{-3}\ m)^3 = 10^{-9}\ m^3$$

$$1\ ns^{-1} = (10^{-9}\ s)^{-1} = 10^9\ s^{-1}$$

5.7 Special Characteristics

A term that should be avoided when using SI is "weight." Frequently we hear statements such as "The man weighs 100 kg." A better statement would be "The man has a mass of 100 kg." To clarify any confusion, let's look at some basic definitions.

Table 5.7 Non-SI units accepted for use in the United States

Quantity	Name	Symbol	SI equivalent
Time	minute	min	60 s
	hour	h	3 600 s
	day	d	86 400 s
Plane angle	degree	°	π/180 rad
	minute	′	π/10 800 rad
	second	″	π/648 000 rad
Volume	liter	L*	10^{-3} m^3
Mass	metric ton	t	10^3 kg
	unified atomic mass unit	u	1.660 57 × 10^{-27} kg (approx)
Land area	hectare	ha	10^4 m^2
Energy	electronvolt	eV	1.602 × 10^{-19} J (approx)

*Both "L" and "l" are acceptable international symbols for liter. The uppercase letter is recommended for use in the United States because the lowercase "l" can be confused with the numeral 1.

First, the term *mass* should be used to indicate only a quantity of matter. Mass, as we know, is measured in kilograms against an international standard.

Force, as defined previously, is measured in newtons. It denotes an acceleration of one meter per second squared to a mass of one kilogram.

The acceleration of gravity varies at different points on the surface of the earth as well as distance from the earth's surface. The accepted standard value of gravitational acceleration is 9.806 650 m/s^2.

Gravity is instrumental in measuring mass with a balance or scale. If you use a beam balance to compare an unknown quantity against a standard mass, the effect of gravity on the two masses cancels out. If you use a spring scale, mass is measured indirectly, since the instrument responds to the local force of gravity. Such a scale can be calibrated in mass units and be reasonably accurate when used where the variation in the acceleration of gravity is not significant.

In the English gravitational system, the unit "pound" is sometimes used to denote both mass and force. We will use the convention that pound-mass (lbm) is a unit of mass and pound-force (lbf) is a unit of force. Thus, pound-mass can be directly converted to the SI unit kilogram, and pound-force units convert to newtons. Another English unit describing mass is the slug.

A word of caution when using English gravitational units in Newton's second law ($F = ma$). The combination of units, lbf, lbm, and ft/s^2, is not a coherent (consistent) set that is, 1 lbf imparts an acceleration of 32.147 ft/s^2 to 1 lbm rather than 1 ft/s^2 required for coherency. A coherent set of English units is lbf, slug, and ft/s^2 because 1 lbf = 1 slug $\times$ 1ft/s^2. You can convert mass quantities from lbm to slugs before substituting into $F = ma$. The slug is 32.174 times the size of the pound-mass (1 slug = 32.174 lbm).

The following example problem clarifies the confusion that exists in the use of the term *weight* to mean either force or mass. In everyday use, the term *weight* nearly always means mass; thus, when a person's weight is discussed, the quantity referred to is mass.

Example problem 5.1 A "weight" of 100.kg (the unit itself indicates mass) is suspended by a rope (see Fig. 5.3). Calculate the tension in the rope in newtons when the mass is lifted vertically at constant velocity and the local gravitational acceleration is (a) 9.807 m/s^2 and (b) 1.63 m/s^2 (approximate value for the surface of the moon).

Theory Tension in the rope when the mass is at rest or moving at constant velocity is

$$F = mg$$

where g is the local acceleration of gravity and m is the mass of object.

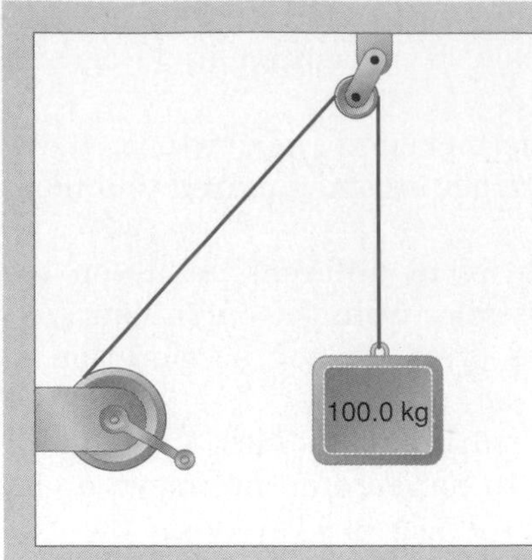

Figure 5.3

Assumption Neglect the mass of the rope.

Solution

(a) For $g = 9.807\ \text{m/s}^2$ (given to four significant figures)

$$F = (100.0\ \text{kg})(9.807\ \text{m/s}^2)$$
$$= \underline{\underline{0.980\,7\ \text{kN}}}$$

(b) For $g = 1.63\ \text{m/s}^2$

$$F = (100.0\ \text{kg})(1.63\ \text{m/s}^2)$$
$$= \underline{\underline{0.163\ \text{kN}}}$$

5.8 Conversion of Units

Although the SI system is the international standard, there are many other systems in use today. It would be fair to say that most of the current work force of graduate engineers has been schooled using terminology such as slugs, pound-mass, pound-force, and so on, and a very high percentage of the total United States population is more familiar with degrees Fahrenheit than degrees Celsius.

For this reason and because it will be some time before the SI total system becomes the single standard in this country, you must be able to convert between unit systems.

Four typical systems of mechanical units presently being used in the United States are listed in Table 5.8. The table does not provide a complete list of all possible quantities; it is presented to demonstrate the different terminology that is associated with each unique system. If a physical quantity is expressed in any system, it is a simple matter to convert the units from that system to another. To do this, the basic unit conversion must be known and a logical unit analysis must be followed.

Table 5.8 Mechanical units

Quantity	Absolute system		Gravitational system	
	MKS	CGS	Type I	Type II
Length	m	cm	ft	ft
Mass	kg	g	slug	lbm
Time	s	s	s	s
Force	N	dyne	lbf	lbf
Velocity	$m \cdot s^{-1}$	$cm \cdot s^{-1}$	$ft \cdot s^{-1}$	$ft \cdot s^{-1}$
Acceleration	$m \cdot s^{-2}$	$cm \cdot s^{-2}$	$ft \cdot s^{-2}$	$ft \cdot s^{-2}$
Torque	N·m	dyne·cm	lbf·ft	lbf·ft
Moment of inertia	$kg \cdot m^2$	$g \cdot cm^2$	$slug \cdot ft^2$	$lbm \cdot ft^2$
Pressure	$N \cdot m^{-2}$	$dyne \cdot cm^{-2}$	$lbf \cdot ft^{-2}$	$lb \cdot ft^{-2}$
Energy	J	erg	ft·lbf	ft·lbf
Power	W	$erg \cdot s^{-1}$	$ft \cdot lbf \cdot s^{-1}$	$ft \cdot lbf \cdot s^{-1}$
Momentum	$kg \cdot m \cdot s^{-1}$	$g \cdot cm \cdot s^{-1}$	$slug \cdot ft \cdot s^{-1}$	$lbm \cdot ft \cdot s^{-1}$
Impulse	N·s	dyne·s	lbf·s	lbf·s

Mistakes can be minimized if you remember that a conversion factor simply relates the same physical quantity in two different unit systems. For example, 1.0 in and 25.4 mm each describe the same length quantity. Thus, when using the conversion factor 25.4 mm / in to convert a quantity in inches to millimeters, you are multiplying a factor that is not numerically 1 but is physically one. This fact allows you to readily avoid the most common error, that of using the reciprocal of a conversion. Just imagine that the value in the numerator of the conversion must describe the same physical quantity as that in the denominator. When so doing, you will never use the incorrect factor 0.304 8 ft/m, since 0.304 8 ft is clearly not the same length as 1 m.

Example prob. 5.2 demonstrates a systematic procedure to use when performing a unit conversion. The construction of a series of horizontal and vertical lines separating the individual quantities will aid the thought process and help ensure a correct unit

Figure 5.4
Military systems are designed in accordance with SI standards. *(General Dynamics, Pomona Division)*

analysis. In other words, the units to be eliminated will cancel out, leaving the desired results. The final answer should be checked to make sure it is reasonable. For example, the results of converting from inches to millimeters should be approximately 25 times larger than the original number.

Example problem 5.2 Convert 6.7 in to millimeters.

Solution Write the identity

$$6.7 \text{ in} = \frac{6.7 \text{ in}}{1}$$

and multiply by conversion factor.

$$\frac{6.7 \text{ in}}{1} \left| \frac{25.4 \text{ mm}}{1 \text{ in}} = 1.7 \times 10^2 \text{ mm} \right.$$

Example problem 5.3 Convert 85.0 lbm/ft^3 to kilograms per cubic meter.

Solution

$$85.0 \text{ lbm/ft}^3 = \frac{850 \text{ lbm}}{1 \text{ ft}^3} \left| \frac{1^3 \text{ ft}^3}{(6.304\,8)^3 \text{ m}^3} \right| \frac{0.453\,6 \text{ kg}}{1 \text{ lbm}}$$

$$= 1.36 \times 10^3 \text{ kg/m}^3$$

Example problem 5.4 Determine the gravitational force (in newtons) on an auto with a mass of 3 645 lbm. The acceleration of gravity is known to be 32.2 ft/s^2.

Solution A Force, mass, and acceleration of gravity are related by $F = mg$.

$$m = \frac{3645 \text{ lbm}}{1} \left| \frac{1 \text{ kg}}{2.204\,6 \text{ lbm}} = 1\,653.36 \text{ kg} \right.$$

$$g = \frac{32.2 \text{ ft}}{1 \text{ s}^2} \left| \frac{0.304\,8 \text{ m}}{1 \text{ ft}} = 9.814\,6 \text{ m/s}^2 \right.$$

$$F = mg = (1\,653.36 \text{ kg})(9.814\,6 \text{ m/s}^2) = 16\,227 \text{ N} \cong 16.2 \text{ kN}$$

Note: Intermediate values were not rounded to final precision, and we have used either exact or conversion factors with at least one more significant figure than contained in the final answer.

Solution B

$$F = mg = \frac{3\,645 \text{ lbm}}{1} \left| \frac{32.2 \text{ ft}}{1 \text{ s}^2} \right| \frac{1 \text{ kg}}{2.204\,6 \text{ lbm}} \left| \frac{0.304\,8 \text{ m}}{1 \text{ ft}} \right.$$

$$= 16\,227 \text{ N} \cong 16.2 \text{ kN}$$

Note: It is often convenient to include conversions with the appropriate engineering relationship in a single calculation.

Example problem 5.5 Convert a mass flow rate of 195 kg/s (typical of the airflow through a turbofan engine) to slugs per minute.

Solution

$$195 \text{ kg/s} = \frac{195 \text{ kg}}{1 \text{ s}} \left| \frac{1 \text{ slug}}{14.954 \text{ kg}} \right| \frac{60 \text{ s}}{1 \text{ min}} = 782 \text{ slug/min}$$

Example problem 5.6 Compute the power output of a 225-hp engine in (a) British thermal units per minute and (b) kilowatts.

Solution

$$\text{(a) } 225 \text{ hp} = \frac{225 \text{ hp}}{1} \left| \frac{2.546\,1 \times 10^3 \text{ Btu}}{1 \text{ hp·h}} \right| \frac{1 \text{ h}}{60 \text{ min}}$$

$$= 9.55 \times 10^3 \text{ Btu/min}$$

$$\text{(b) } 225 \text{ hp} = \frac{225 \text{ hp}}{1} \left| \frac{0.745\,70 \text{ kW}}{1 \text{ hp}} \right. = 168 \text{ kW}$$

The problem of unit conversion becomes more complex if an equation has a constant with hidden dimensions. It is necessary to work through the equation converting the constant K_1 to a new constant K_2 consistent with the equation units.

Consider the following example problem given with English units.

Example problem 5.7 The velocity of sound in air (c) can be expressed as a function of temperature (T):

$$c = 49.02\sqrt{T}$$

where c is in feet per second and T is in degrees Rankine.

Find an equivalent relationship when c is in meters per second and T is in kelvins.

Procedure

1. First, the given equation must have consistent units; that is, it must have the same units on both sides. Squaring both sides we see that

$$c^2\text{ft}^2/\text{s}^2 = 49.02^2 T°R$$

It is obvious that the constant $(49.02)^2$ must have units in order to maintain unit consistency. (The constant must have the same units as c^2/T.)

Solving for the constant,

$$(49.02)^2 = c^2 \frac{\text{ft}^2}{\text{s}^2}\left[\frac{1}{}\right]$$

$$= \frac{c^2}{T}\left[\frac{\text{ft}^2}{\text{s}^2 °R}\right]$$

2. The next step is to convert the constant $49.02^2\ ft^2/(s^2R)$ to a new constant that will allow us to calculate c in meters per second given T in kelvins. We recognize that the new constant must have units of square meters per second squared-kelvin.

$$\frac{(49.02)^2\ ft^2}{1\ s^2{}^\circ R} = \frac{(49.02)^2\ ft^2}{1\ s^2{}^\circ R}\left|\frac{(0.304\ 8)^2\ m^2}{1\ ft^2}\right|\frac{9^\circ R}{} = \frac{401.84\ m^2}{1\ s^2K}$$

3. Substitute this new constant 401.84 back into the original equation

$$c^2 = 401.84T$$

$$c = 20.05\sqrt{T}$$

where c is in meters per second and T is in kelvins.

5.9 Key Terms

Physical quantities	Fundamental dimensions
Units	Derived dimensions
Base units	Absolute system
Supplemental units	Gravitational system
Derived units	Metric system
Dimensions	Symbols

Problems

5.1 Using the correct number of significant figures, convert the following physical quantities to the proper SI units.

(*a*) 645 lbm
(*b*) 98.2 °F
(*c*) 4.75×10^2 acres
(*d*) 55×10^2 gal
(*e*) 110.0×10^3 gal/h
(*f*) 88 ft/s
(*g*) 285 hp
(*h*) 2025 in
(*i*) $1.255 \times 10^2\ ft^3/min$

5.2 Convert the following to SI units. Use correct significant figures.

(*a*) 750.5 Btu/min
(*b*) 65.2 hp·h
(*c*) 4.500×10^3 mi
(*d*) 1.00×10^2 mi/h
(*e*) 225 lbf
(*f*) $8.255 \times 10^4\ lbm/ft^3$
(*g*) 1.955 atm
(*h*) 212°F
(*i*) 5280.0 ft

5.3 Convert as indicated giving answer with proper significant figures.

(*a*) 85.5 in to centimeters
(*b*) 505 L to cubic feet
(*c*) 78.8°C to degrees Fahrenheit
(*d*) 10 750 bushels to cubic centimeters
(*e*) 65.5×10^5 Btu/h to kilowatts

5.4 Convert as indicated giving answer with proper sign figures.

(*a*) 7.550×10^3 km to feet
(*b*) 285 K to degrees Fahrenheit
(*c*) 6.85×10^4 ft lbf to joules
(*d*) $14.7\ lbf/in^2$ to pascals
(*e*) $77.7\ slug/ft^3$ to grams per cubic centimeter

5.5 Using the rules for expressing SI units, express each of the following in correct form if given incorrectly.

(*a*) 11.5 cm's
(*b*) 475 N
(*c*) 9.5 m/s/s
(*d*) 5000 K
(*e*) 8,000 pa
(*f*) 25 amp
(*g*) 100.1 m m
(*h*) 62.5 j
(*i*) 300 degrees Kelvin

5.6 Using the rules for expressing SI units, express each of the following in correct form if given incorrectly.

(*a*) 53 m per sec	(*d*) 25 farads	(*g*) .5 mm
(*b*) 101C	(*e*) 1 000 N	(*h*) 40 nM
(*c*) 75 A's	(*f*) 48 Kg	(*i*) 8050 N/m/m

5.7 What force in newtons would be required to lift with uniform velocity a 275 lbm vise under the following conditions?

(*a*) Acceleration of gravity is 32.2 ft/s^2
(*b*) Acceleration of gravity is 9.80 m/s^2
(*c*) Acceleration of gravity is 25 ft/s^2

5.8 Determine the acceleration of gravity (meters per second squared) if the force required to lift at uniform velocity a 4.000×10^2 kg object is

(*a*) 1295 lbf
(*b*) 1055 lbf
(*c*) 585 N

5.9 What work is done to lift a 25.0×10^2 lbm object 4.50×10^1 ft vertically if the acceleration of gravity is 9.807 m/s^2? Express answer in joules. *Note:* Work = force $\times$ distance traveled in force direction

5.10 Determine the engine power (kilowatts) and horsepower required to move an automobile on level ground if the resistance of the tires and air resistance is 35.0 lbf. The auto is traveling at 88.0 ft/s. *Note:* Power = (force)(velocity).

5.11 The density of water at 70°F is about 1.936 slug/ft^3. First determine the volume of a 50.00 ft diameter spherical tank. Then compute the mass of water contained if the tank is full. Express volume in cubic meters and mass in kilograms.

5.12 A cylindrical tank 8.0 ft long and 6.0 ft in diameter with hemispherical ends is placed so its longitudinal axis is horizontal. How many gallons of gasoline are in the tank if the fluid level is 4.5 ft above the bottom of the tank? What is the mass of the gasoline in kilograms if its specific gravity is 0.67?

5.13 Determine the total force in the ground exerted by the air above a sports field 55.0 by 100 yd if the air pressure measured at field level is 2088 lbf/ft^2. Express your answer in newtons.

5.14 Compute the mass (in kilograms) of gravel stored in a rectangular feeder bin 17.5 by 25.0 ft. The depth of the gravel is 15.0 ft and its density is 97 lbm/ft^3. If the feeder bin is elevated on supports 20.0 ft above the ground, what vertical load in newtons does the gravel place on the supports? Assume standard gravitational attraction.

5.15 Shelled corn is often piled on the ground because of insufficient storage facilities. Compute the diameter of a pile in feet, number of bushels in the pile, volume of the pile in cubic meters, and the mass of the pile in kilograms if the height of the pile is 17.5 ft. *Hint:* The angle of repose for dry shelled corn is about 12 degrees and its density is 56 lbm/bushel.

5.16 Use a spreadsheet to do the computation in Prob. 5.15 for a range of heights from 10.0 to 30 ft in increments of 0.50 ft. Prepare a printed copy so that a grain dealer can estimate the amount of corn in a pile by measuring its height.

5.17 Do Prob. 5.16 but modify the spreadsheet so that the dealer can simply measure the circumference of the base of the pile instead of the height. Use a range of circumferences from 35.0 to 275 ft in increments of 5.00 ft.

5.18 The Darcy-Weisback friction formula for pipes allows us to compute the frictional energy loss per unit mass (ft lbf/lbm or J/km) of a fluid flowing through a pipe by

$$h_L = (f)\left(\frac{L}{D}\right)\left(\frac{v^2}{2g}\right)$$

where h_L = energy loss per unit mass

f = friction factor, dimensionless
L = length of pipe
D = diameter of pipe
v = average velocity of fluid
g = acceleration of gravity

Calculate h_L in J/km and in ft·lbf/lbm for water flowing at 25.0 ft/s through a 1.00-in ID cast-iron pipe, 275 ft long. Assume $f = 0.040$ for this situation.

5.19 A weir is used to measure flow-rates in open channels. For a rectangular weir, the expression can be written

$$Q = 5.35\ LH^{3/2}$$

where

Q = discharge rate, ft^3/s

L = length of weir, ft

H = height of fluid above crest, ft

(*a*) Determine a new constant so the expression can be applied with Q in gallons per hour, and L and H in inches.
(*b*) Write a computer program or prepare a spreadsheet to produce a table of values of Q in both ft^3/s and gallons per hour, with L (inches) and H (inches). Use values of L that range from 1 to 15 ft in 1 inch increments. Let H range from 6 to 36 in increments of 2 inches.

5.20 For certain conditions, the law of conservation of energy can be written

$$V = 4.429\sqrt{h}$$

where

V = velocity, m/s

h = distance, m

Determine a new constant so that the equation is valid for h in ft and V in miles per hour.

5.21 The specific fuel consumption of an engine may be written as

$$\text{sfc} = \frac{2545}{\eta Q}$$

where

sfc = specific fuel consumption, lbm/hp·h

η = thermal efficiency, dimensionless

Q = heat of combustion per unit mass, Btu/lbm

Determine a new constant for Q in J/kg and sfc in kg/kW·h.

5.22 The universal law of gravity for the force of attraction between two masses may be written as

$$F = (6.673 \times 10^{11}) \frac{m_1 m_2}{r^2}$$

where F is the force of attraction, N

m_1, m_2 are masses, kg
r is separation distance, m

Determine a new constant for F in lbf, m_1 and m_2 in lbm, and r in mi.

5.23 A hollow aluminum sphere, 260 mm in diameter (outside) with a wall thickness of 3.10 mm.

(*a*) Compute the outside surface area (square millimeters).
(*b*) Compute the mass of the ball (kilograms).
(*c*) Determine whether the ball will float in water (density of 62.4 lbm/ft^3).

5.24 A piece of oak (density = 47 lbm/ft^3) has been cut into the shape of a right rectangular pyramid with a base of 4.00 in by 3.00 in and a height of 8.00 in. Determine

(*a*) The outside surface area (square inches).
(*b*) The mass in lbm.
(*c*) The height of the base above the surface of water (density of 62.4 lbm/ft^3), assuming the object will float point down.

5.25 An object in the shape of a right cone with a base diameter of 290 mm and a height of 320 mm is floated point down in alcohol (density of 49 lbm/ft^3). The base of the cone extends 75 mm above the surface of the alcohol.

(*a*) What is the density of the cone in kg/m^3?
(*b*) What is the mass of the cone in kg?
(*c*) What is the surface area in square millimeters of the cone that is below the surface of the alcohol (wetted area)?

6 Preparation for Computer Solutions

6.1 Introduction

An engineering method of problem solving is presented here for convenience.

1. Recognize and understand the problem.
2. Accumulate facts.
3. Select the appropriate theory or principle.
4. Make necessary assumptions.
5. Solve the problem.
6. Verify and check results.

The mechanics of computation is a significant part of step 5. At this point you must examine the complexity of the problem and decide how to do the computation. A simple problem may require only brief hand calculations or the use of an inexpensive engineering calculator. If the complexity of the solutions is high or the amount of data or the number of cases to be processed is large, other methods should be considered. Perhaps a spreadsheet is a good choice. How about software tools such as Mathematica® or MathCAD®? Should you write a program in FORTRAN or C to do the calculations?

If your decision is to use a software tool or to create your own program, you must prepare a well-thought-out procedure or a series of steps that solve the problem called an *algorithm.* Once the algorithm is developed, you must create the detailed steps needed by the software, programming language, or computing platform that you have chosen. The detailed step or code production is highly dependent on your software, coding language, and computing system and will not be covered in this text. For further information consult manuals and texts specifically prepared for your unique combination of software and equipment. We will devote the remainder of the chapter to a discussion of the algorithm preparation.

6.2 Flowcharting

An algorithm can be described in terms of textlike statements (called pseudocode) or graphically in a form known as a *flowchart.* A flowchart provides a picture of the logic and steps involved in solving a problem.

As you develop a flowchart, it is advantageous to think in terms of the big picture before focusing on details. For example, when designing a house, an architect must first plan where the kitchen, bathrooms, bedrooms, and other rooms are to be located before specifying where electric, water, and sewer lines should be placed. Similarly, a flowchart is designed by working out large blocks to assure that global logic is satisfied before deciding what detailed procedure should be used within each large block.

A set of graphical symbols is used to describe each step of the flowchart (see Fig. 6.1). Although there are many *flowchart symbols* in general use, we will only define a small subset that is generic in nature; that is, each symbol does not denote any particular device or method for performing the operation. For example, a general *input/output* symbol is used that does not suggest a document, tape disk storage or display. The symbol only means that communication with the software should occur by using whatever device or method is available and appropriate. However, when coding the input or output process, the programmer will have to be specific about the method or device.

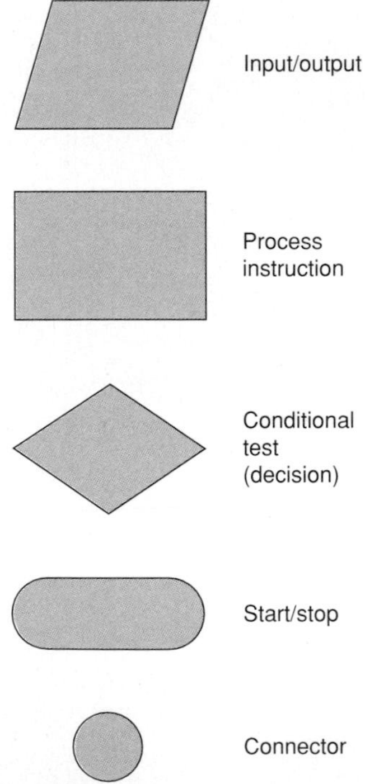

Figure 6.1 Flowchart symbols.

Three general structures are used in algorithm, or flowchart development: the *sequential structure,* the *selection structure,* and the *repetition structure.* The latter two have some variations that will be illustrated in the following paragraphs.

A *sequential structure* defines a series of steps that are performed in order, beginning at the start position and proceeding sequentially from operation to operation until the stop symbol is reached. No decisions are made and no step or series of steps is repeated. A sequential structure is shown in Fig. 6.2. Dashed lines in this flowchart and others described later simply mean that repeated symbols have been omitted.

The fundamental *selection structure* is illustrated in Fig. 6.3. It contains a conditional test symbol that asks a question with a yes/no (true/false) answer or states a condition with two possible outcomes. Thus, based on the outcome of the decision, one of the two sets of operations will be performed. Each branch of the selection structure may contain as many operations as are necessary. There may be one operation or several in a branch, or even no operations in one of the branches. With this structure, flow proceeds through one or the other of the branches and continues on into a later section of the flowchart. Therefore, during a single pass through the structure one of the branches is not used.

A somewhat generalized version of the selection structure is shown in Fig. 6.4. Here the decision or condition at the top of the structure has more than two outcomes, called cases. Each outcome or case can have a unique set of steps to be performed. This structure is useful, for example, in a sorting procedure where one is treating data differently with each of several defined ranges of values.

The *repetition structure* (looping structure) consists of a step or series of steps that are performed repeatedly until some condition (perhaps a specified number of repetitions) is satisfied, at which time the next step after the loop is executed. Two common repetition structures are presented, one where the conditional text is performed as the last step of the structure and one where the test is the first step of the structure.

The repetition structure where the conditional test is the last step is illustrated in Fig. 6.5. One or more processes are placed in the forward section of the structure. The conditional test can be reversed; that is, the true and false flow lines can be interchanged depending on the nature of the condition to be tested. There may be no need for the process block in the reverse loop depending on the action in the process blocks in the forward section. You may or may not include the reverse-process block as your logic dictates. Frequently, the reverse-loop block performs

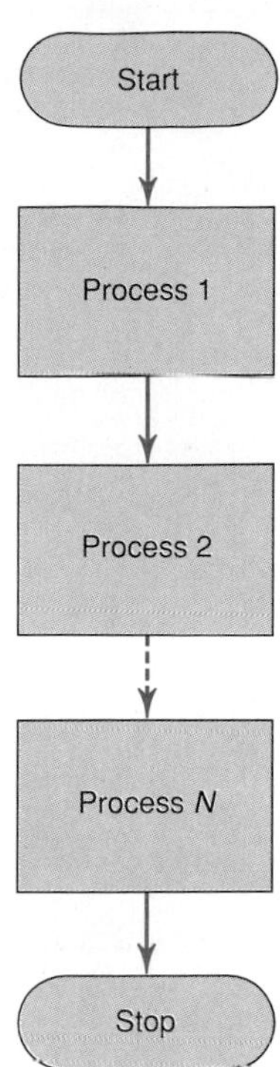

Figure 6.2
Simple sequence.

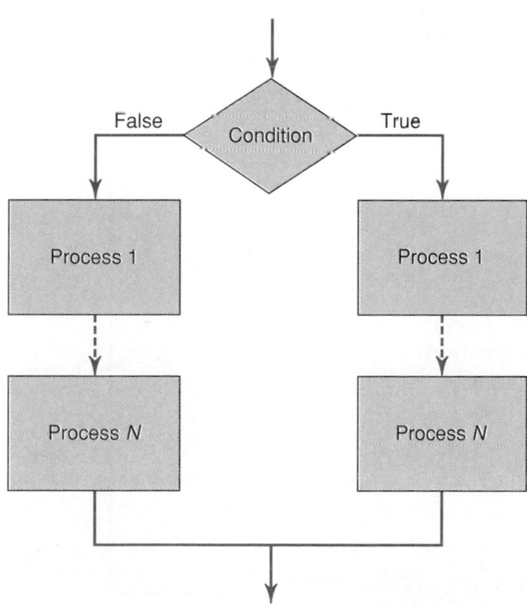

Figure 6.3
Selection structure.

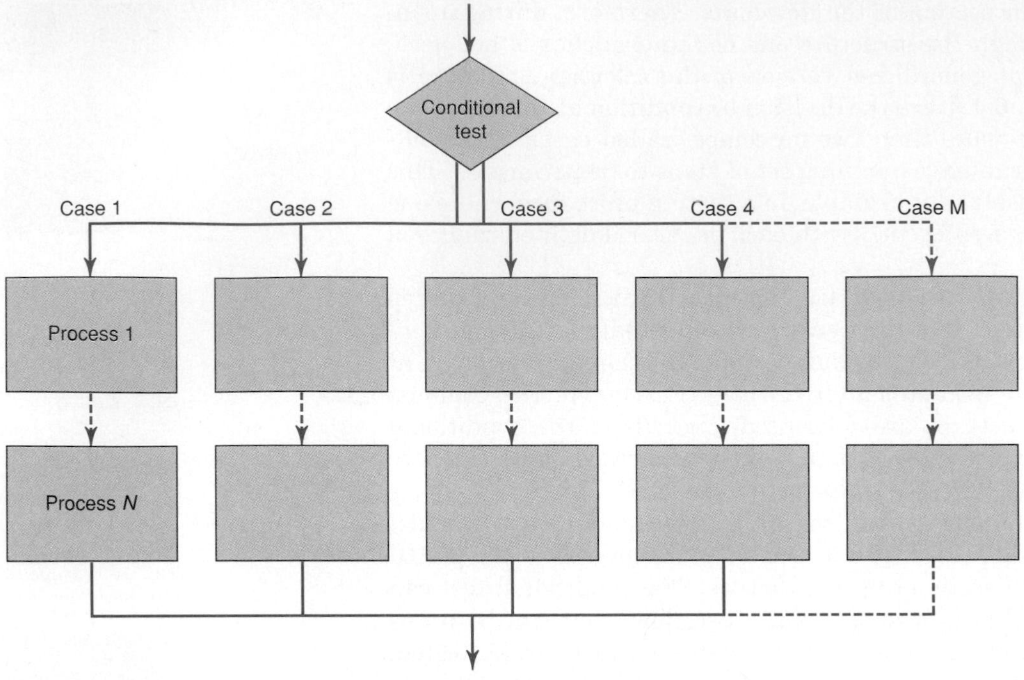

Figure 6.4
Generalized selection structure (case structure).

the action of a counter. Operations such as $X = X + 1$ or $Z = Z + 5$ might appear here. These are not algebraic equations since they are clearly not mathematically correct. They are instructions to replace the current value of X by a new value 1 greater or to replace Z by $Z + 5$. Therefore, they can count the number of times through the loop, as does X, or can increment a variable by a con-

Figure 6.5
Repetition structure with conditional test last.

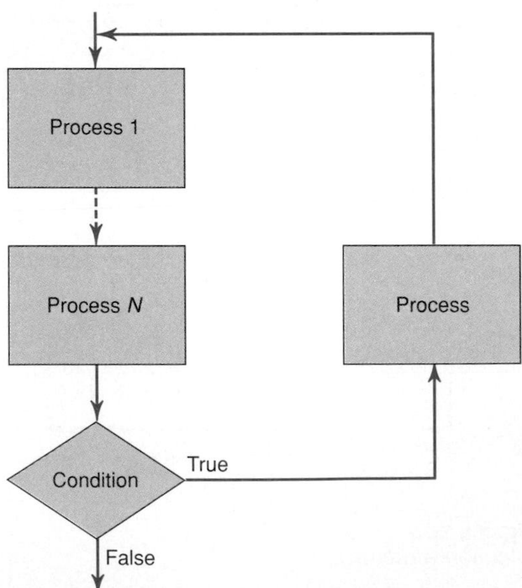

stant, as in the case of Z. Negative increments are also possible so that you can count backward or decrement a variable.

Figure 6.6 shows the repetition structure where the conditional test is performed first. Again, the true and false branches can be reserved to match the chosen conditional test. As many process blocks as desired may be used. One of them could be a counter or incrementing block.

Several selection structures or repetition structures can be combined by a method called *nesting.* In this way one or more loops can be contained within a loop. Similarly, a selection structure can be placed within another selection structure. Figure 6.7 shows an example of how a nested loop might appear. The inner loop is performed until condition 2 is satisfied; then flow returns to the decision block for condition 1. In each pass through the outer loop the inner loop will be repeated until condition 2 is satisfied. Eventually, condition 1 will be satisfied and flow will pass to the next part of the flowchart.

Examples of various combinations of the structures just discussed can be seen in the problems that follow.

Example problem 6.1 Construct a flowchart for calculating the sum of the squares of the even integers from $N1$ to $N2$.

Procedure For purposes of this example, $N1$ and $N2$ will be restricted to even integers only and $N2 > N1$. We could use the sequential structure shown in Fig. 6.8. This flowchart will result in a variable called SUM as the desired value; SUM is then output. Because of the repetitive nature of the steps, it is far more convenient to use a repetition structure as seen in Fig. 6.9.

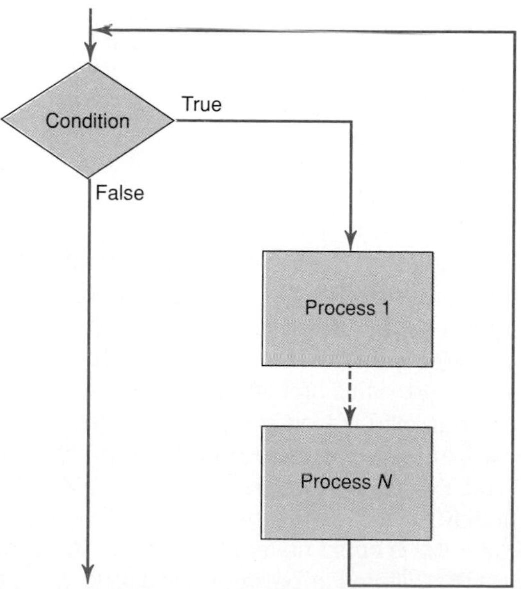

Figure 6.6
Repetition structure with conditional test first.

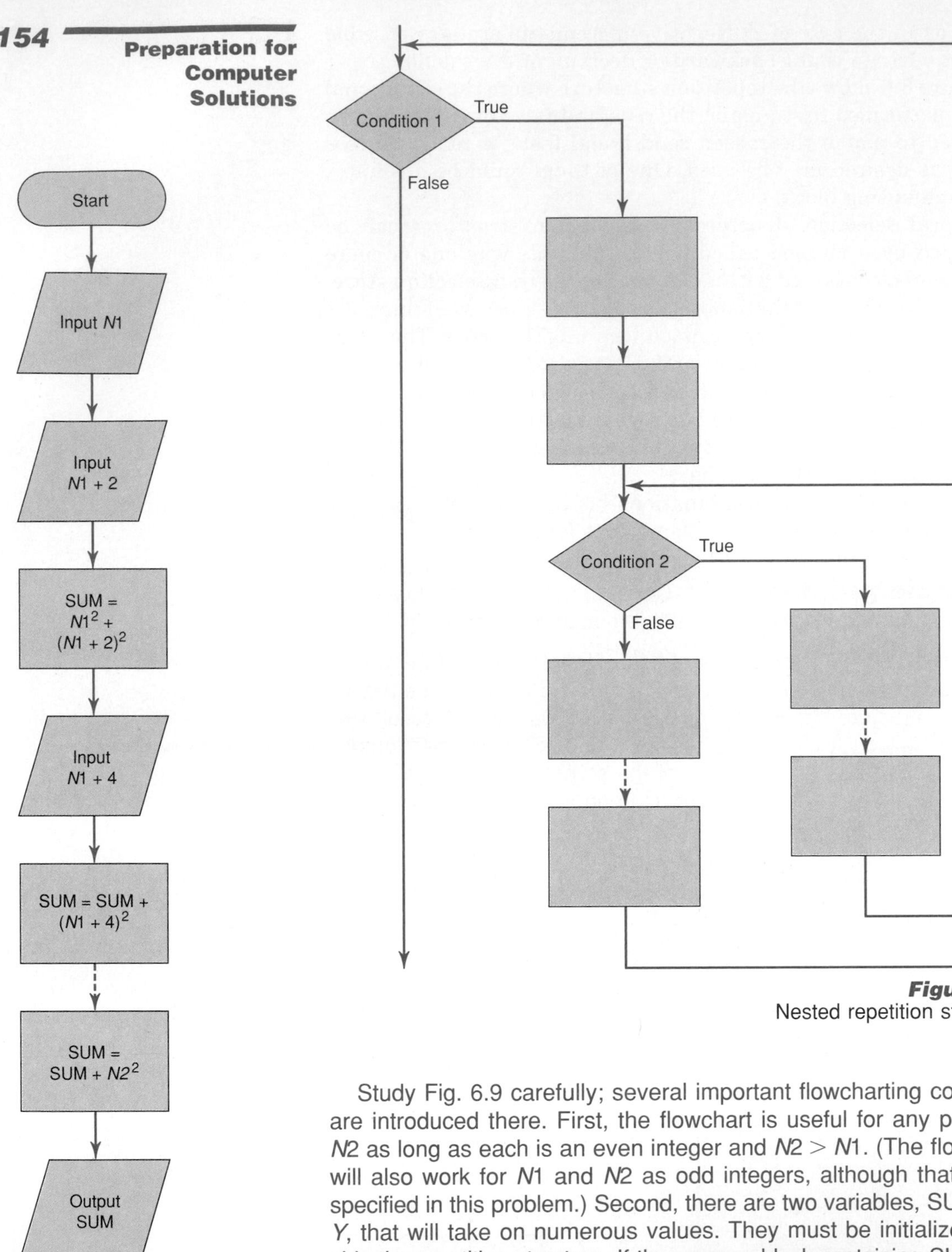

Figure 6.7
Nested repetition structure (loop).

Figure 6.8

Study Fig. 6.9 carefully; several important flowcharting concepts are introduced there. First, the flowchart is useful for any pair $N1$, $N2$ as long as each is an even integer and $N2 > N1$. (The flowchart will also work for $N1$ and $N2$ as odd integers, although that is not specified in this problem.) Second, there are two variables, SUM and Y, that will take on numerous values. They must be initialized outside the repetition structure. If the process block containing SUM $= 0$ and $Y = N1$ was inside the loop, SUM and Y would be reset to their initial values each time through the loop and the decision block ($Y = N2$?) could never be satisfied, thereby creating an infinite loop.

The difference between initializing a variable and inputting a variable is important. As a general rule, variables that must have initial val-

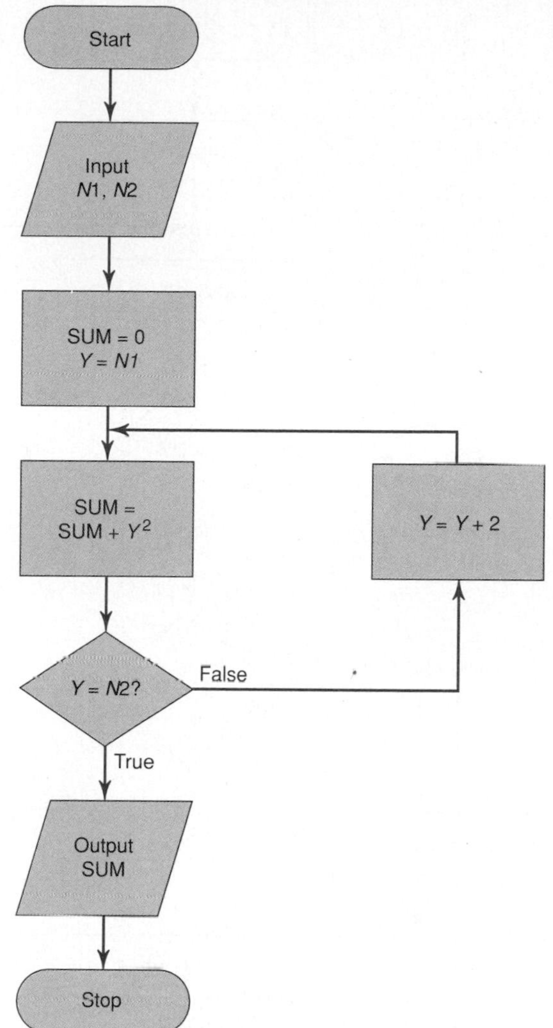

Figure 6.9

ues but whose values will not change from one use of the flowchart to another (one run of the resulting program to another) should be placed in a process block. Variables that you wish to change from one run to another should be placed in an input block. This provides you with the necessary flexibility of reusing the flowchart (or program) without having to input the variables that do not change from run to run.

Example problem 6.2 Draw a flowchart that will calculate the future sum of a principal (an amount of money) for a given interest rate and number of interest periods. Allow the user to decide if simple or compound interest is to be used and to compute as many future sums as desired.

Procedure For simple interest $S = P(1 + ni)$, and for compound interest $S = P(1 + i)^n$, where S = future sum, P = principal

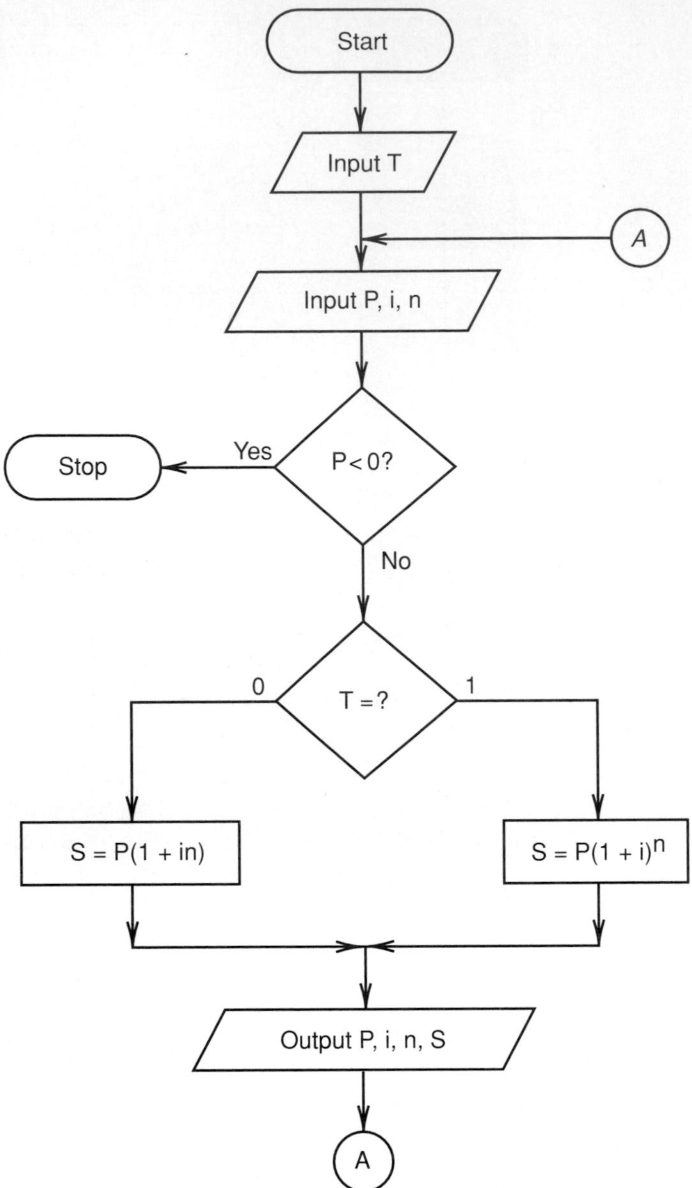

Figure 6.10

amount, i = interest rate per period, and n = number of interest periods. One possible flowchart is given in Fig. 6.10. The user is asked to input $T = 0$ for simple interest and $T = 1$ for compound interest. A decision block checks on P before performing any further calculation. If $P < 0$ (a value not expected to be used), the process terminates. Thus, using a unique value for one of the input variables is one method of terminating the processing. A logical alternative would be to construct a counter and to check to see if a specified number of variable sets has been reached.

The connector symbol has been used to avoid drawing a long flow line in this example. A letter or number (the letter A was used in this case) is placed in the symbols to define uniquely the pair of symbols that should be connected. Connector symbols are also used when a flowchart occupies more than one page and flow lines cannot physically connect portions of the flowchart.

The flowchart in Fig. 6.10 was prepared by using one of several commercially available programs that make flowchart preparation and modification convenient.

Example problem 6.3 The sine of an angle can be approximately calculated from the following series expansion.

$$\sin x \cong \sum_{i=1}^{N}(-1)^{i+1}\left[\frac{x^{2i-1}}{(2i-1)!}\right]$$

$$= x - \frac{x^3}{3!} + \frac{x^5}{5!} - \frac{x^7}{7!} + \cdots + (-1)^{N+1}\left[\frac{x^{2N-1}}{(2N-1)!}\right]$$

where x is the angle in radians. The degree of accuracy is determined by the number of terms in the series that are summed for a given value of x. Prepare a flowchart to calculate the sine of an angle of P degrees and cease the summation when the last term in the series calculated has a magnitude less than 10^{-7}. Of course, an exact answer for the sine of the angle would require the summation of an infinite number of terms. Include the procedure for evaluating a factorial in the flowchart.

Procedure One possible flowchart is shown in Fig. 6.11. We will discuss several features of this flowchart, after which you should track on paper the first three or four terms of the series expansion to make sure you understand that the flowchart is correctly handling the problem. Note that the general term has been used in the loop and that specific values of each variable are calculated in order to produce the required term each time through the loop:

1. The magnitude of the quantity controlling the number of terms summed is called ERR. It is input so that a magnitude of other than 10^{-7} can be used in a future run.

2. i denotes the summation variable and is initialized as 1.

3. M represents $(2i-1)!$ and is initialized as 1 (its value in the first term of the series).

4. SUM is the accumulated value of the series as each term is added to the previous total. It is initialized as 0.

5. The angle is input in degrees and then immediately converted to radians by multiplying by $\pi/180$.

6. TERM is the value of each term beginning with

$$(-1)^{1+1}\{x^{2(1)-1}/[2(1)-1]!\}$$

or simply x. The factor $(-1)^{i+1}$ causes TERM to alternate signs.

7. SUM is equal to $0 + x$ the first time through the procedure.

8. The absolute value of TERM is now checked against the control value ERR to see if computations should cease. Note that the

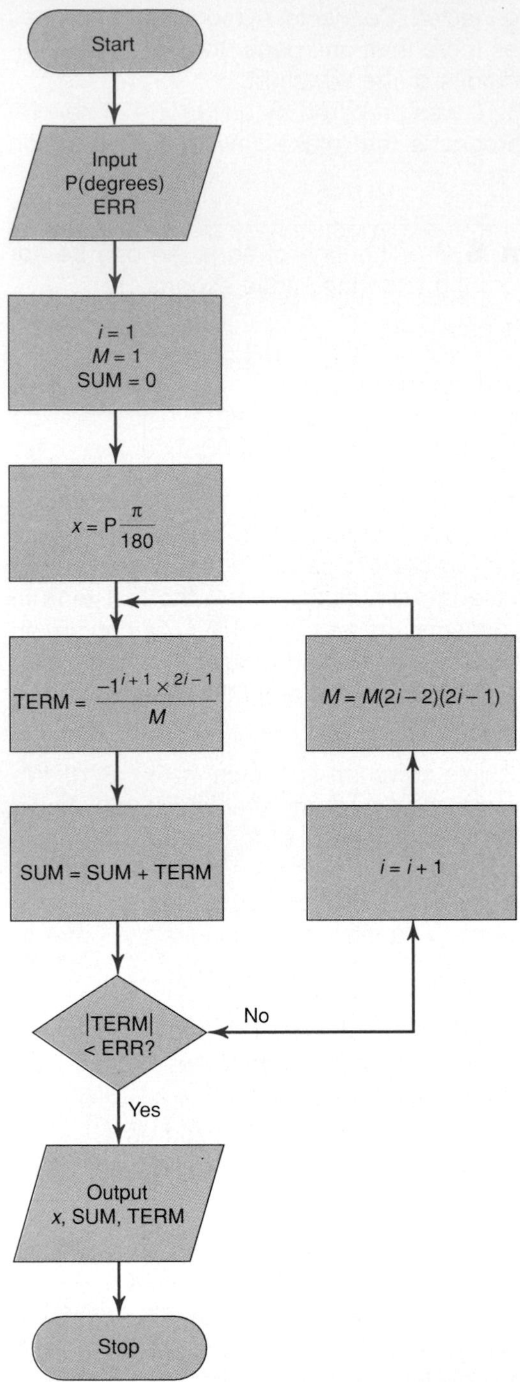

Figure 6.11

absolute value must be used because of the alternating signs. If a "no" answer is received to this conditional test, the appropriate incrementing of the variable must be undertaken.

9. *i* is increased by 1. This time through *i* becomes 2 since it was initialized as 1.

10. M becomes $(1)(2 \times 2 - 2)(2 \times 2 - 1) = 1 \times 2 \times 3 = 3!$

11. Following the flowchart directions, we now return to the evaluation of TERM with the new value of i.

$$\text{TERM} = (-1)^{2+1}x^{2(2)-1}/3! = -x^3/3!$$

12. SUM then becomes $x - x^3/3!$. TERM is again tested against ERR.

13. The process repeats until the magnitude of the last term is less than ERR (10^{-7} in this case). Then the value of x, sin x, and the value of the last term are reported to make certain that the standard of accuracy has been attained.

You should check several terms for a given value of x and then repeat the process for different angles. You will note that as the size of the angle varies, the number of terms required to achieve the standard of accuracy also varies. For example, when using a value of $P = 5°$, the third term of the series is about $4(10^{-8})$, less than the 10^{-7} requirement. But for $P = 80°$ the third term is approximately $4.4(10^{-2})$. It requires seven terms of the series before the magnitude of the last term becomes less than 10^{-7}.

The preceding examples should give you some insight into the construction of a flowchart as a prelude to writing a computer program or to using other software tools. The mechanisms for calculating, testing, incrementing, looping, and so on vary with the computational device and the programming language or software tool. A flowchart, however, should be valid for all computer systems and software tools because it graphically portrays the steps that must be completed to solve the problem.

6.3 Key Terms

The following are some of the terms, procedures, and processes that you should recognize and understand.

- Algorithm
- Flowchart
- Flowchart symbols
  - Start/stop
  - Input/output
  - Process instruction
  - Decision/conditional test
  - Connector
- Sequential structure
- Selection structure
  - Two outcomes
  - Multiple outcomes
- Repetition stucture
  - Test first
  - Test last
- Nesting
- Variable initialization versus variable input
- Process termination procedures

Problems

6.1 You have available a list of the heights of the 8 000 first-year students at your university. Draw a flowchart that will count the number of students whose height, h, is in each of the following categories. Heights must be input one at a time, and the height ranges and counts should be output.

categories on next page

$h < 5$ ft

5 ft $\leq h <$ 5 ft 6 in

5 ft 6 in $\leq h <$ 6 ft

6 ft $\leq h <$ 6 ft 6 in

$h \geq$ 6 ft 6 in

Price range	Tax (¢)
$0 < s \leq \$0.20$	1
$\$0.20 < s \leq \$0.40$	2
$\$0.40 < s \leq \$0.60$	3
$\$0.60 < s \leq \$0.80$	4
$\$0.80 < s \leq \$1.00$	5
$\$1.00 < s \leq \$1.20$	6
etc.	

6.2 Assume that the sales tax structure for your state is (s = sale value, $)
Produce a flowchart that will compute the appropriate sales tax and total price (sale value + tax) for any sale up to $1 000.

6.3 Draw a flowchart to calculate the areas and circumferences of N circles whose smallest diameter D is x cm (centimeters) and whose diameters increase in increments of z cm. A repetition structure is required. Provide the user with values of the diameter, area, and circumference of each circle. Allow the user to specify the number of circles, the smallest diameter, and the diameter increment.

6.4 The buckling load for a long column is given by Euler's equation:

$$F_B = \frac{n\pi^2 EA}{(L/r)^2}$$

where F_B = buckling load

E = modulus of elasticity

A = cross-sectional area

L = length of column

r = least radius of gyration

The factor n depends on the end conditions of the column as follows: both ends are hinged, $n = 1$; both ends are fixed, $n = 4$; one end is fixed, the other end is hinged, $n = 2$.

Draw a flowchart to compute the buckling load based on input values of E, A, L, r, and end conditions (H-H, F-F, or F-H).

6.5 Draw a flowchart to compute the approximate value of

$$\cos x = \left(1 - \frac{4x^2}{\pi^2}\right)\left(1 - \frac{4x^2}{3^2\pi^2}\right)\left(1 - \frac{4x^2}{5^2\pi^2}\right) \cdots$$

where x is the angle in radians. Design the flowchart so that the user specifies an angle in degrees and the number of terms and receives as output the angle in degrees and radians, the approximate value of cos x, and the value of the first product that is not included in the approximation.

6.6 A regular polygon of n sides can be inscribed in a circle of radius R. The area and perimeter of the polygon are given by

$$\text{Area} = \frac{1}{2} nR^2 \sin \frac{360^\circ}{n}$$

$$\text{Perimeter} = 2\pi R \sin \frac{180^\circ}{n}$$

Draw a flowchart that will compute the area and perimeter of a series of polygons beginning with $n = 3$ ($\Delta n = 1$) for a specified R. Processing should terminate either when the polygon area is within Z percent of the area of the associated circle or when the polygon perimeter is within Y percent of the perimeter of the circle. Y and Z must be input variables. There should be an indication of which condition resulted in the process termination.

6.7 The series for the hyperbolic cosine of a number x is given by

$$\cosh x = 1 + \frac{x^2}{2!} + \frac{x^4}{4!} + \frac{x^6}{6!} + \cdots \qquad (x^2 < \infty)$$

By using a repetition structure based on the general term of the series and by explicitly computing factorials, draw a flowchart to calculate the value of cosh x such that the last term included in the series has a magnitude less than a user-defined number. Provide output giving x, cosh x, and an indication that the last term meets the user's specifications.

6.8 For $z^2 < 1$ the inverse cosine of z is given by

$$\cos^{-1} z = \frac{\pi}{2} - \left[z + \left(\frac{1}{2}\right)\left(\frac{z^3}{3}\right) + \left(\frac{1 \times 3}{2 \times 4}\right)\left(\frac{z^5}{5}\right) + \left(\frac{1 \times 3 \times 5}{2 \times 4 \times 6}\right)\left(\frac{z^7}{7}\right) + \cdots \right]$$

Draw a flowchart to calculate $\cos^{-1} z$ for a user-specified value of z and the maximum magnitude of the first term to be dropped. Output z, $\cos^{-1} z$, the value of the first term dropped, and the number of terms that are included in the approximation.

6.9 Draw a flowchart that will produce a table of temperature in kelvins versus temperature in degrees Fahrenheit for temperatures ranging from −100 to 1 000°F in increments of T°F as specified by the user. Assume that T is an even divisor of 100 and 1 000.

6.10 The sum of a sinking fund, F, that results from an annual investment of A dollars at an interest rate of i (decimal) for n years is given by

$$F = A\left[\frac{(1+i)^n - 1}{i}\right]$$

Draw a flowchart that will produce a series of tables of F versus n $(0 \leq n \leq 20)$ for $A = A_1$ and A_2 and $i = i_1$, i_2, and i_3. Be sure each table is identified by the appropriate annual investment and interest rate.

6.11 Draw a flowchart that will input the exam scores (one score at a time) of 100 students taking an engineering problems course; count the numbers of grades (based on 100 percent) in the ranges from 90 to 100, 80 to 89, 70 to 79, 60 to 69, and < 60; and output the results as a table of grade ranges versus a count within the range. Also, compute and output the average score on the exam.

6.12 Draw a flowchart that will compute a student's grade point average for the term given the number of courses, the number of credits for each course, and the letter grade earned in each course. Assume A, B, C, D, and F grades in a 4-point system with $A = 4$, $B = 3$, $C = 2$, $D = 1$, and $F = 0$.

6.13 The current i flowing in a series circuit consisting of a resistor (resistance = R), inductor (inductance = L), and capacitor (capacitance = C) but having no voltage source is given by the following equations. Assume that the capacitor has an initial charge of Q at time $t = 0$.

$$i = \frac{2\pi f^2 Q e^{-Rt/2L}}{f_d} \sin 2\pi f_d t$$

where $f = \dfrac{1}{2\pi}\sqrt{\dfrac{1}{LC}}$

$$f_d = \frac{1}{2\pi}\sqrt{\frac{1}{LC} - \frac{R^2}{4L}}$$

Draw a flowchart to compute the current flow i for $0 \leq t \leq 0.1s$ with $\Delta t = 0.005s$. Allow the user to specify Q, R, L, and C. Produce a table of current versus time.

6.14 Swaging is a process in metal forming that changes the geometric shape of a part through a series of hammerlike blows produced via machine dies. For typical ductile materials it is possible to hold the following normal operating tolerances for tubes.

Diameter of tubes, in	Tolerance, in
$d \leq .5$	$\pm.001$
$.5 < d \leq 1.0$	$\pm.002$
$1.0 < d \leq 1.5$	$\pm.003$
$1.5 < d \leq 3.0$	$\pm.005$

Prepare a flowchart that will allow the user to input a diameter, d, and to learn the expected tolerance zone.

6.15 The method of interval halving can be used to find the root of a function written as $f(x) = 0$. (See Fig. 6.12.) The single real root must be between $x = a$ and $x = b$; that is, $f(a)$ and $f(b)$ must have opposite signs. If the midpoint of a to b is found as $x_m = (a + b)/2$, then $f(x_m)$ will be the same sign as either $f(a)$ or $f(b)$. If $f(x_m)$ is the same sign as $f(a)$, the root lies in the interval $(a + b)/2$ to b. If it has the same sign as $f(b)$, the root is in the interval a to $(a + b)/2$. One then discards the half interval where the root cannot be found and repeats the process until $|b - a|$ is less than some desired value.

Draw a flowchart to determine the root of $x^2 = 5x - 2$ between 2 and 6. Assure that the value is accurate to three significant figures. Output the root and a count of the number of trials that were needed.

6.16 Draw a flowchart to solve the quadratic equation $Ax^2 + Bx + C = 0$. The user may specify any (positive, negative, or zero) values of constants

Figure 6.12

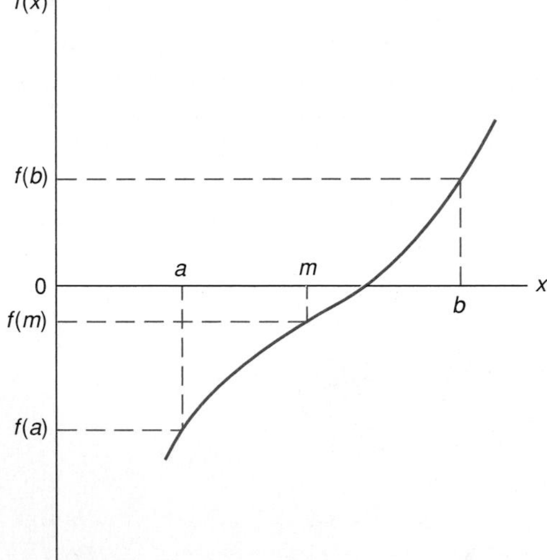

A, B, and C. Output the equation coefficients as well as the roots calculated.

6.17 The cotangent of an angle $|x| < \pi/2$ can be approximately computed from the series

$$\text{ctn } x \cong \frac{1}{x} - \frac{x}{3} - \frac{x^3}{45} - \frac{2x^5}{945} - \cdots - \frac{2^{2n}B_n x^{2n-1}}{(2n)!}$$

where B_n are the Bernoulli numbers ($B_1 = 1/6$, $B_2 = 1/30$, $B_3 = 1/42$, $B_4 = 1/30$, $B_5 = 5/66$, $B_6 = 691/2\,730$, $B_7 = 7/6$, $B_8 = 3\,617/510,\ldots$).

Draw a flowchart to calculate an approximation for ctn x for a given angle in degrees and the number of terms that are less than or equal to 7. Give the user an indication of the accuracy of the approximation by providing the value of the first term that is not included in the approximation.

6.18 An investment of P dollars produces an annual income of A dollars for n years. The rate of return i (equivalent interest rate per year as a decimal) can be found from the equation

$$P = A\left[\frac{(1 + i)^n - 1}{i(1 + i)^n}\right]$$

Using an iterative scheme—interval halving, for example—draw a flowchart to compute the rate of return for the given values of P, A, and n. Provide the output of P, A, n, and i as an annual percent.

6.19 The method of least squares gives the slope and intercept for a linear fit $y = mx + b$ to a set of data as

$$m = \frac{n(\Sigma\, x_i y_i) + (\Sigma\, x_i)(\Sigma\, y_i)}{n(\Sigma\, x_i^2) - (\Sigma\, x_i)^2}$$

$$b = \frac{\Sigma\, y_i - m(\Sigma\, x_i)}{n}$$

where n is the number of data pairs (x_i, y_i) and the summations are over all data pairs.

Draw a flowchart to read in n (user-defined) data pairs, one pair at a time, and to compute and output m and b.

6.20 Draw a flowchart that will read a set of n positive integers, one at a time; reorder them from the smallest to the largest; and output the ordered list, one at a time.

6.21 The expected linear tolerance for a casting operation depends on the size of the part and the care that is taken in the operation. With reasonable care, we can expect the tolerance values noted as practical. With less care, the tolerances in the column labeled economical are likely.

Dimension, in	Practical, in	Economical, in
$d \leq .675$	$\pm.003$	$\pm.010$
$.675 < d \leq 1.0$	$\pm.005$	$\pm.010$
$1.0 < d \leq 2.0$	$\pm.010$	$\pm.020$

Draw a flowchart that will accept as input the part dimension, d, and the care used in the operation and will produce the likely tolerance range.

6.22 Assume that for a specified Midwest campus, you have a data set containing the high and low temperatures in degrees Fahrenheit (°F) for each day of a calendar year. Prepare a flowchart that will find and output the highest temperature and the lowest temperature of the year. Be sure to account for the possibility of a leap year.

6.23 For the data set in Prob. 6.22, draw a flowchart to compute and output the high-temperature distribution in the following form:

Temperature range, °F	Number of days
Below 30	
30–50	
51–70	
71–90	
Over 90	

6.24 Use the data set in Prob. 6.22 and draw a flowchart that will determine the low-temperature distribution as follows:

Temperature range, °F	Number of days
Below 0	
1–20	
21–30	
31–40	
41–50	
Over 50	

7 Statistics

7.1 Introduction

Statistics, as used by the engineer, can most logically be called a branch of applied mathematics. It constitutes what some call the science of decision making in a world full of uncertainty. In fact, a degree of uncertainty exists in most day-to-day activities, from something as simple as the tossing of a coin to the results of an election, the outcome of a ball game, or the comparison of the efficiency of two production processes.

There can be little doubt that it would be virtually impossible to understand a great deal of the work done in engineering without having a thorough knowledge of statistics. Numerical data derived from surveys and experiments constitute the raw material upon which interpretations, analyses, and decisions are based; it is essential that engineers learn how to properly use the information derived from such data.

Everything concerned even remotely with the collection, processing, analysis, interpretation, and presentation of numerical data belongs to the domain of statistics. Figure 7.1 illustrates an engineering task that requires knowledge of statistics.

There exist today a number of different and interesting stories about the origin of statistics, but most historians believe it can be traced to two dissimilar areas: games of chance and political science.

During the eighteenth century, various games of chance involving the mathematical treatment of errors led to the study of probability and, ultimately, to the foundation of statistics. At approximately the same time, an interest in the description and analysis of the voting of political parties led to the development of methods that today fall under the category of *descriptive statistics,* which is basically designed to summarize or describe important features of a set of data without attempting to infer conclusions that go beyond the data. Descriptive statistics is an important part of the entire subject area; it is still used whenever a person wishes to represent data derived from observation.

In more recent years, however, statisticians have shifted their emphasis from methods that merely describe the data to methods that make generalizations about the data, called *statistical*

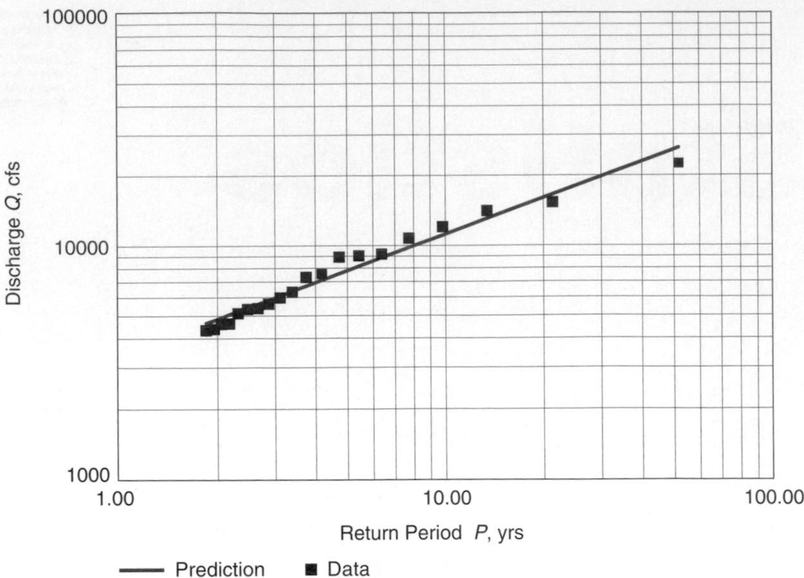

Figure 7.1 Statistical methods can aid in the prediction of river flows in flood conditions.

inference. To understand the distinction between descriptive statistics and statistical inference one must understand the terms *population* and *sample*. By definition, population includes all members of a specific group or all events of a particular description. For example, all residents of the state of Texas make up a population.

Normally populations contain a large number of members or events; however, populations may be large or small. The distinguishing characteristic of a population is not size but that it includes all members or events that meet the definition for membership.

In cases where populations are very large, it may not be practical to work with every member of the population. Instead a manageable subset of the population is selected which is called a sample.

A measure of a characteristic of a population is called a parameter. Later in this chapter, we will compute a measure called the mean or average. If we determine the mean age of all residents of the state of Texas, that mean would be a parameter (a characteristic of the population). On the other hand, a descriptive measure for a sample is called a statistic. Thus, if we draw a sample of Texas residents and determine the mean age for this sample, the mean age would be a statistic.

Inferential statistics consists of procedures for making generalizations about a population by studying a sample from that population. In other words, we make inferences about the parameters of the population from the statistics computed for the sample. For example, we might suggest the mean age for the sample of Texas residents should relate in some way to the mean age of all the residents of Texas.

To distinguish between descriptive measures for a population and for a sample we will use different symbols. For the mean or average, we will use mu (μ) to represent the mean of a population and xbar ($\overline{x}$) to represent the mean of a sample.

In statistics, it is not enough to consider only sets of data and calculated results when arriving at conclusions. Items such as control and authenticity of collected data and how the experiment or survey was planned are of major importance. Unless proper care is taken in the planning and execution stages, it may be impossible to arrive at valid results or conclusions.

7.2 Frequency Distribution

Various ways of describing measurements and observations, such as the grouping and classifying of data, are a fundamental part of statistics. In fact, when dealing with a large set of collected numbers, a good overall picture of the data can often be conveyed by proper grouping into classes. The following examples will serve to illustrate this point.

Consider the individual test scores received by students on the first major exam in their freshman computations course (see Tab. 7.1). Table 7.2 is a type of numerical arrangement showing scores distributed among selected classes. Some information such as the highest and lowest values will be lost once the raw data have been sorted and grouped.

The construction of numerical distributions as in this example normally consists of the following steps: select classes into which the data are to be grouped, distribute data into appropriate classes, and count the number of items in each class. Since the last two steps are essentially mechanical processes, our attention will be directed primarily toward the *classification* of data.

Two things must be considered when arranging data into classes: the number of classes into which the data are to be grouped and the range of values each class is to cover. Both these areas are somewhat arbitrary, but they do depend on the nature of the data and the ultimate purpose the distribution is to serve.

The following are guidelines that should be followed when constructing a frequency distribution.

Table 7.1

92	71	89	91	53	93	90	96	95
98	76	96	94	68	91	82	82	44
88	87	93	78	85	98	82	90	70
78	70	87	88	89	95	99	88	88
77	65	85	64	79	50	81	80	76

Table 7.2

Test scores	Tally	Frequency
41–50	||	2
51–60	|	1
61–70	~~||||~~	5
71–80	~~||||~~ |||	8
81–90	~~||||~~ ~~||||~~ ~~||||~~ |	16
91–100	~~||||~~ ~~||||~~ |||	13
	Total	45

1. Use no less than 6 and no more than 15 classes. The square root of n, where n is the number of data points, provides an approximate number of classes to consider.
2. Select classes that will accommodate all the data points.
3. Make sure that each data point fits into only one class.
4. Whenever possible, make the class intervals of equal length.

The numbers in the right-hand column of Tab. 7.2 are called the *class frequencies,* which denote the number of items that are in each class. Since frequency distributions are constructed primarily to condense large sets of data into more easily understood forms, it is logical to display or present that data graphically. The most common form of graphical presentation is called the *histogram*.

It is constructed by representing measurements or grouped observations on the horizontal axis and class frequencies along the graduated and calibrated vertical axis. This representation affords a graphical picture of the distribution with vertical bars whose bases equal the class intervals and whose heights are determined by the corresponding class frequencies. Figure 7.2 demonstrates a histogram of the test scores tabulated in Tab. 7.2.

7.3 Measures of Central Tendency

The solution of many engineering problems in which a large set of data is collected can be somewhat facilitated by the determination of single numbers that describe unique characteristics about the data. The most popular measure of this type is called the arithmetic mean or average.

The *arithmetic mean,* or mean of a set of n numbers, is defined as the sum of the numbers divided by n. In order to develop a notation and a simple formula for arithmetic mean, it is helpful to use an example.

Suppose that the average, or mean, height of a starting basketball team is to be determined. Let the height in general be

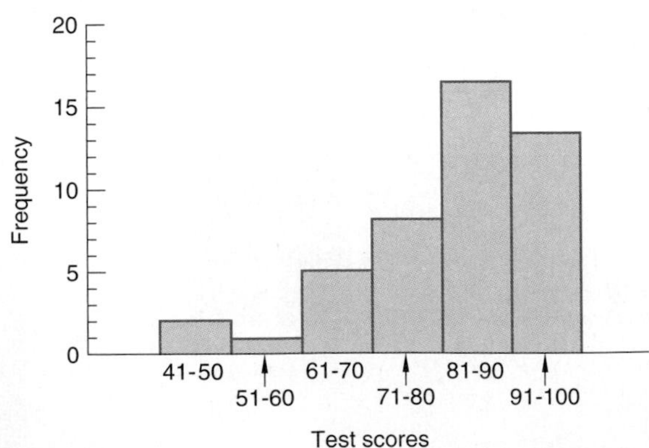

Figure 7.2 Test scores histogram.

represented by the letter x and the height of each individual player be represented by x_1, x_2, x_3, x_4, and x_5. More generally, there are n measurements that are designated $x_1, x_2, \ldots, x_n$. From this notation, the mean can be written as follows:

$$\text{Mean} = \frac{x_1 + x_2 + x_3 + \cdots + xn}{n}$$

A mathematical notation that indicates the summation of a series of numbers is normally written

$$\sum_{i=1}^{n} xi$$

which represents $x_1 + x_2 + x_3 + \cdots + x_n$. This notation will be written in the remainder of the chapter as Σx_i, but the intended summation will be from 1 to n.

As indicated earlier in the chapter, when a set of all possible observations is used, it is referred to as the *population;* when a portion or subset of that population is used, it is referred to as a *sample.* The notation for arithmetic mean will be $\overline{x}$ when the x values are representative of a random sample and not an entire population. For an entire population the symbol mu (μ) is used to represent the mean.

The standard notations discussed above provide the following common expressions for the arithmetic mean of the population and of a sample:

$$\mu = \frac{\Sigma x_i}{n} \text{ (population)} \qquad (7.1a)$$

$$\overline{x} = \frac{\Sigma x_i}{n} \text{ (sample)} \qquad (7.1b)$$

where for the population the sum is over all members of the population and for the sample, the sum is over just the members of the sample.

The mean is a popular measure of central tendency because (1) it is familiar to most people, (2) it takes into account every item, (3) it always exists, (4) it is always unique, and (5) it lends itself to further statistical manipulations.

One disadvantage of the arithmetic mean, however, is that any gross error in a number can have a pronounced effect on the value of the mean. To avoid this difficulty, it is possible to describe the "center" of a set of data with other kinds of statistical descriptions. One of these is called the *median,* which can be defined as the value of the middle item of data arranged in increasing or decreasing order of magnitude. For example, the median of the five numbers 15, 27, 10, 18, and 22 can be determined by first arranging them in increasing order: 10, 15, 18, 22, and 27. The median is 18.

If there are an even number of items, there is never a specific middle item, so the median is defined as the mean of the values of the two middle items. For example, the median of six numbers 5, 9, 11, 14, 16, and 19 is (11 + 14)/2, or 12.5.

The mean and median of a set of data rarely coincide. Both terms describe the center of a set of data, but in different ways. The median divides the data so that half of all items is greater than or equal to the median; the mean may be thought of as the center of gravity of the data.

The median, like the mean, has certain desirable properties. It always exists and is always unique. Unlike the mean, the median is not affected by extreme values. If the exclusion of the highest and lowest values causes a significant change in the mean, then the median should be considered as the indicator of central tendency of that data.

In addition to the mean and the median, there is one other average, or center, of a set of data, which we call the *mode*. It is simply the value that occurs with the highest frequency. In the following set of numbers 18, 19, 15, 17, 18, 14, 17, 18, 20, 19, 21, and 14 the number 18 is the mode because it appears more often than any of the other values.

An important point for a practicing engineer to remember is that there are any number of ways to suggest the middle, center, or average value of a data set. If comparisons are to be made, it is essential that similar methods be compared. It is only logical to compare the mean of brand A with the mean of brand B, not the mean of one with the median of the other. If one particular item, brand, or process is to be compared with another, the same measures must be used. If the average grade in one section of college calculus is to be compared with the average grade in other sections, the mean of each section would be one important statistic.

7.4 Measures of Variation

It is not likely that the mean values of the course grades of different sections of college calculus will be of equal magnitude. And the extent to which the means are dissimilar is of fundamental importance.

Measures of variation indicate the degree to which data are dispersed, spread out, or bunched together. Suppose that by coincidence two sections of a college calculus course have exactly the same mean grade values on the first hour exam. It would be of interest to know how far individual scores varied from the mean. Perhaps one class was bunched very closely around the mean, while the other class demonstrated a wide variation, with some very high scores and some very low scores. This situation is typical and is often of interest to the engineer.

It is reasonable to define this variation in terms of how much each number in the sample deviates from the mean value of the sample, that is, $x_1 - \bar{x}$, $x_2 - \bar{x}$, . . . , $x_n - \bar{x}$. If you wanted an average deviation from the mean you might try adding $x_1 - \bar{x}$ through $x_n - \bar{x}$ and dividing by n. But this does not give a useful result, since the sum of the deviations is always zero. The procedure generally followed is to square each deviation, sum the resulting squares, divide the sum by n, and take the square root. The resulting formula for the standard deviation of the entire population is

$$\sigma = \left[\frac{\Sigma(x_i - \mu)^2}{n}\right]^{1/2} \tag{7.2}$$

The formula for the standard deviation of a large sample is

$$s = \left[\frac{\Sigma(x_i - \bar{x})^2}{n}\right]^{1/2} \quad \text{(large sample)} \tag{7.3}$$

If you wish to use the standard deviation s of a sample to *estimate* the standard deviation σ of the population from which the sample is taken, you can obtain a good estimate from Eq. (7.3) as long as the sample size is 30 or larger.

It has been found, however, that for small samples ($n<30$) Eq. (7.3) underestimates the magnitude of the population standard deviation. Statisticians have shown that if n in Eq. (7.3) is replaced by $n - 1$, the resulting equation is more accurate for estimating the population standard deviation from the standard deviation of the small sample. Thus, for small samples, s is given by

$$s = \left[\frac{\Sigma(x_i - \bar{x})^2}{n - 1}\right]^{1/2} \quad \text{(small sample)} \tag{7.4}$$

An alternate form of Eq. (7.4) which is sometimes easier to use, is derived by expanding $(x_i - \bar{x})^2$, substituting for $\bar{x}$ from Eq. (7.1b), and reducing terms. It is

$$s = \left[\frac{n(\Sigma x_i^2) - (\Sigma x_i)^2}{n(n - 1)}\right]^{1/2} \tag{7.5}$$

Because Eq. (7.5) was obtained from Eq. (7.4), it is useful for small samples. For a sample size of 30, the value of s calculated by Eq. (7.5) differs from one calculated by Eq. (7.2) by less than 2 percent. As sample sizes grow, this difference becomes even less. Thus for large samples, either the n or $n - 1$ form of the standard deviation equations could be used with good results. Another common measure of variation is called the *variance;* it is the square of the standard deviation. Therefore, for small samples, the sample variance is given by

$$s^2 = \frac{\Sigma(x_i - \bar{x})^2}{n - 1} \tag{7.6}$$

Formulas giving the variance for large samples and for populations are obtained by squaring Eqs. (7.2) and (7.3), respectively.

Example problem 7.1 A midwestern university campus has 10 540 male students. Using a random selection process, 50 of these students were chosen and weighed to the nearest pound (pound-mass); the raw data were as recorded in Tab. 7.3. The data were then grouped (Tab. 7.4), and the histogram in Fig. 7.3 was constructed. Calculate the sample mean, sample standard deviation, and sample variance of the data.

Solution From Eq. (7.1b) the sample mean can be calculated (summary of computations shown in Tab. 7.5):

$$\bar{x} = \frac{\Sigma x_i}{n}$$

$$= \frac{8\,037}{50}$$

$$= 160.74 \text{ lbm (161 lbm after rounding)}$$

From Eq. (7.4) the sample standard deviation can be determined:

$$s = \left[\frac{\Sigma(x_i - \bar{x})^2}{n-1}\right]^{1/2}$$

$$= \left(\frac{4\,793.74}{49}\right)^{1/2}$$

$$= 9.89 \text{ lbm}$$

The sample standard deviation can also be determined from Eq. (7.5):

$$s = \left[\frac{n(\Sigma x_i^2) - (\Sigma x_i)^2}{n(n-1)}\right]^{1/2}$$

$$= \left[\frac{50(1\,296\,661) - (8\,037)^2}{50(49)}\right]^{1/2}$$

$$= 9.89 \text{ lbm}$$

The sample variance can be calculated from Eq. (7.6):

$$s^2 = \frac{\Sigma(x_i - \bar{x})^2}{n-1}$$

$$= \frac{4\,793.74}{49}$$

$$= 97.8 \text{ lbm}^2$$

Table 7.3

164	171	154	160	158	150	159	185	168	158
143	159	162	165	160	167	166	164	152	172
177	165	170	155	155	163	180	157	145	160
149	153	137	173	157	175	163	147	156	156
162	167	165	166	162	136	158	170	162	159

Table 7.4

Range	Frequency
136–140	2
141–145	2
146–150	3
151–155	5
156–160	13
161–165	11
166–170	7
171–175	4
176–180	2
181–185	1
	50

Table 7.5

Mass x_i, lbm	x_i^2	$x_i - \bar{x}$	$(x_i - \bar{x})^2$
164	26 896	3.26	10.63
143	20 449	−17.74	314.71
177	31 329	16.26	264.39
149	22 201	−17.74	137.83
156	24 336	−4.74	22.47
159	25 281	−1.74	3.03
8 037	1 296 661	0.00	4 793.73

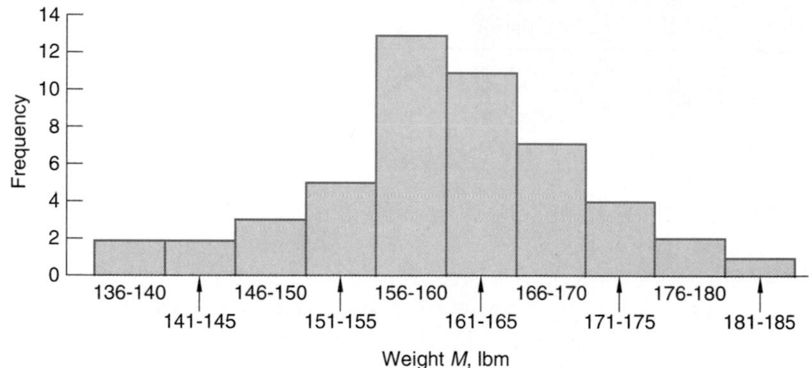

Figure 7.3
Histogram of weight data.

By examining the raw data in Example prob. 7.1 we can see the range in variation of values that occurs from a random sample. Certainly we would expect to find both larger and smaller values if all males at the university—that is, the entire population—were measured and recorded. If we were to select additional random values and develop a second sample from the population, we would expect to find a different sample mean and a different sample standard deviation. We would not expect, however, the differences in these measures of central tendency and variation to be significant if the two samples were truly random in nature.

Example problem 7.2 Engineers may be asked to suggest measures for controlling flooding near a river. Flow rate (discharge) in the river as a function of time must be known if reasonable alternatives are to be examined. Since flooding results from high flow rates, perhaps a data set consisting of the maximum flow rate occurring each year would be useful.

Table 7.6 contains the maximum annual discharge data (in cubic feet per second) for a small river in the Midwest.

For this data set, determine

(a) the sample mean
(b) the sample standard deviation
(c) the sample variance

Plot a histogram to graphically depict the data.

Table 7.6

Maximum Discharge, cfs		
14 000	5 950	2 380
9 150	2 780	4 330
2 480	8 900	5 360
5 320	5 110	7 530
2 730	12 000	2 350
4 620	14 000	15 300
4 200	3 420	3 320
6 300	7 300	2 300
4 600	2 690	3 300
3 800	4 470	9 000
5 600	10 700	3 750
1 150	3 350	2 100

Solution A spreadsheet was used to solve the problem. Figure 7.4 shows the programming used. The raw data were entered in column A, rows 2 through 37. The sum of the raw data was calculated in cell A40, Eq. (7.1) was used in cell A43 to obtain the sample mean, Eq. (7.4) was coded in cell A46 for the sample standard deviation, and Eq. (7.6) was placed in cell A49 to compute the sample variance. Cells A46 and A49 use the squares of the deviations computed in column B. Alternately, the mean, standard deviation, and variance are computed in cells B43, B46, and B49 respectively using statistical functions available in the spreadsheet.

The results of the computation are shown in Fig. 7.5. For these data, the sample mean is about 5 710 cfs (average maximum annual discharge over the 36 years of data available). Also, the sample standard deviation is nearly 3 690 cfs, and the sample variance is about 1.36×10^7 (cfs)^2.

The spreadsheet was used to prepare the histogram in Fig. 7.6. A range of 2 000 cfs was chosen. The counts are based on 0–1 999, 2 000–2 999, and so forth.

	A	B
1	Maximum Discharge, cfs	Deviation squared
2	14000	=(A2-$A$43)^2
3	9150	=(A3-$A$43)^2
4	2480	=(A4-$A$43)^2
5	5320	=(A5-$A$43)^2
6	2730	=(A6-$A$43)^2
7	4620	=(A7-$A$43)^2
8	4200	=(A8-$A$43)^2
9	6300	=(A9-$A$43)^2
10	4600	=(A10-$A$43)^2
11	3800	=(A11-$A$43)^2
12	5600	=(A12-$A$43)^2
13	1150	=(A13-$A$43)^2
14	5950	=(A14-$A$43)^2
15	2780	=(A15-$A$43)^2
16	8900	=(A16-$A$43)^2
17	5110	=(A17-$A$43)^2
18	12000	=(A18-$A$43)^2
19	14000	=(A19-$A$43)^2
20	3420	=(A20-$A$43)^2
21	7300	=(A21-$A$43)^2
22	2690	=(A22-$A$43)^2
23	4470	=(A23-$A$43)^2
24	10700	=(A24-$A$43)^2
25	3350	=(A25-$A$43)^2
26	2380	=(A26-$A$43)^2
27	4330	=(A27-$A$43)^2
28	5360	=(A28-$A$43)^2
29	7530	=(A29-$A$43)^2
30	2350	=(A30-$A$43)^2
31	15300	=(A31-$A$43)^2
32	3320	=(A32-$A$43)^2
33	2300	=(A33-$A$43)^2
34	3300	=(A34-$A$43)^2
35	9000	=(A35-$A$43)^2
36	3750	=(A36-$A$43)^2
37	2100	=(A37-$A$43)^2
38		
39	Sum	Sum
40	=SUM(A2:A37)	=SUM(B2:B37)
41		
42	Average	Average
43	=A40/36	=AVERAGE(A2:A37)
44		
45	Standard Deviation	Standard Deviation
46	=SQRT(B40/35)	=STDEV(A2:A37)
47		
48	Variance	Variance
49	=B40/35	=VAR(A2:A37)

Figure 7.4
Spreadsheet programming
Example prob. 7.2.

	A	B
1	Maximum Discharge, cfs	Deviation squared
2	14000	68687260.49
3	9150	11818316.05
4	2480	10447260.49
5	5320	153838.27
6	2730	8893649.38
7	4620	1192949.38
8	4200	2286816.05
9	6300	345482.72
10	4600	1237038.27
11	3800	3656593.83
12	5600	12593.83
13	1150	20813871.60
14	5950	56538.27
15	2780	8597927.16
16	8900	10161927.16
17	5110	362671.60
18	12000	39536149.38
19	14000	68687260.49
20	3420	5254282.72
21	7300	2521038.27
22	2690	9133827.16
23	4470	1543116.05
24	10700	24877927.16
25	3350	5580093.83
26	2380	11103704.94
27	4330	1910538.27
28	5360	124060.49
29	7530	3304316.05
30	2350	11304538.27
31	15300	91925482.72
32	3320	5722727.16
33	2300	11643260.49
34	3300	5818816.05
35	9000	10809482.72
36	3750	3850316.05
37	2100	13048149.38
38		
39	Sum	Sum
40	205640.00	476423822.22
41		
42	Average	Average
43	5712.22	5712.22
44		
45	Standard Deviation	Standard Deviation
46	3689.46	3689.46
47		
48	Variance	Variance
49	13612109.21	13612109.21

Figure 7.5 Spreadsheet results for Example prob. 7.2.

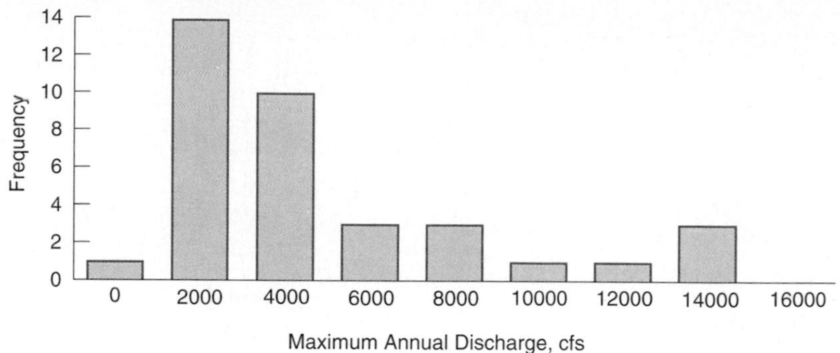

Figure 7.6
Histogram for Example prob. 7.2.

7.5 Continuous Distribution

Random variables are classified according to the values that the variable can assume. Discrete random variables may only take on a finite set of values. The flipping of a coin (two outcomes) or the number of automobiles that pass a certain location in a fixed time are examples of discrete random variables.

In contrast to discrete variables, a random variable is continuous when it can assume values on a continuous scale. Quantities such as time and temperature are examples of continuous variables.

Histograms, which were discussed earlier, can be used to determine the probability of a value falling into a given classification. Histograms permit examination of the area of the rectangle representing that classification. For example, one portion of the histogram from Fig. 7.3 is enlarged and shown in Fig. 7.7. The area of rectangle *ABCD* as a portion of the entire area of the histogram represents the probability that a male weighs between 151 and 155 lbm. It should also be apparent that the area of the rectangle *ABCD* is nearly equal to the shaded area under the continuous curve that could be constructed to represent the histogram.

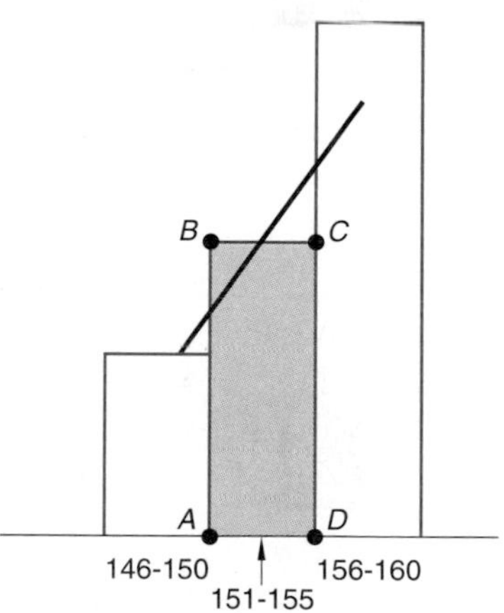

Figure 7.7
Segment of histogram.

More generally, if a histogram is approximated by means of a smooth curve (sometimes called a frequency distribution), the probability associated with any interval is related to the area under the curve bounded by the interval.

7.6 Normal Distribution

Among the many continuous distributions used in statistics, the normal distribution is by far the most useful.

The normal distribution is a theoretical frequency distribution for a specific type of data set. Its graphical representation is a bell-shaped curve that extends indefinitely in both directions. As can be seen in Fig. 7.8, the curve comes closer and closer to the

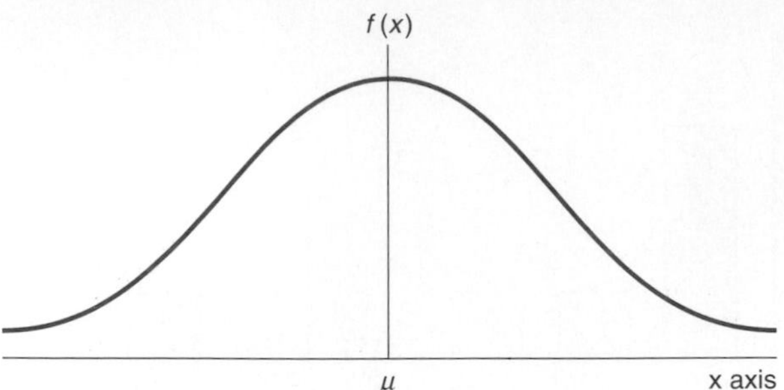

Figure 7.8
Normal distribution.

horizontal axis without ever reaching it, no matter how far the axis is extended from the mean (μ).

The location and shape of the normal curve can be specified by two parameters: (1) the population mean (μ), which locates the center of the distribution, and (2) the population standard deviation (σ), which describes the amount of variability or dispersion of the data.

Mathematically, the normal distribution is represented by Eq. (7.7):

$$f(x) = \frac{1}{\sigma\sqrt{2\pi}}\, e^{-(1/2)[(x-\mu)/\sigma]^2} \tag{7.7}$$

This expression can be used to determine the area under the curve between any two locations in the x-axis as long as we know the mean and standard deviation of the data:

$$\text{Area} = \int_{x_1}^{x2} f(x)\, dx \tag{7.8}$$

Since the evaluation of this expression is difficult, in practice we obtain areas under the curve either from a special table of values that was developed from this equation, from the functions built into a spreadsheet, or from a mathematics analysis program. What follows guides you through the use of the table.

As indicated previously, a normal curve is symmetrical about the mean; however, the specific shape of the distribution depends on the deviation of the data about the mean. As can be seen in Fig. 7.9, when the data are bunched around the mean, the curve drops rapidly toward the x-axis. However, when the data have a wide deviation about the mean, the curve approaches the x-axis more slowly. This presents a problem because if you examine the

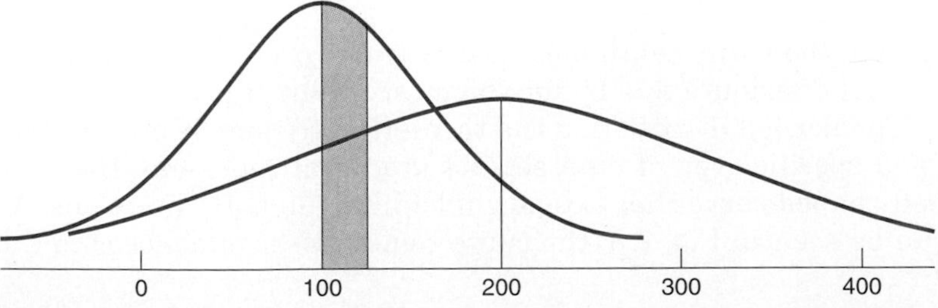

Figure 7.9
Normal curves having standard deviations and means of different magnitudes.

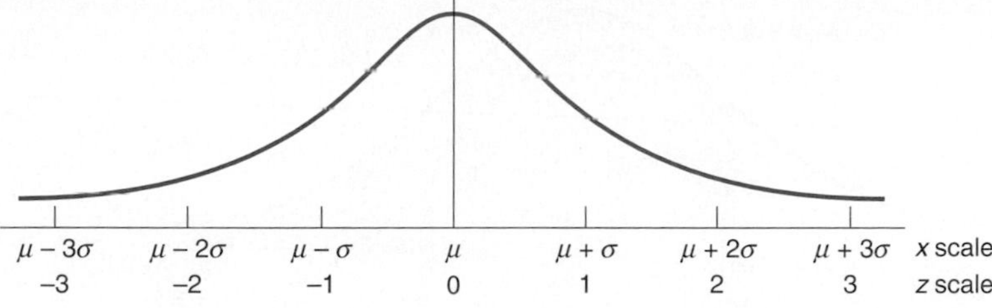

Figure 7.10
Normal curve with normalized distribution.

two curves in Fig. 7.9, the area under the curves between x values of 100 and 125 is not the same for the two distributions.

Thus with each different mean and standard deviation we would have to construct a separate table of normal-curve areas. To avoid having to use many tables, a transformation can be applied converting all curves to a standard form that has $\mu = 0$ and $\sigma = 1$ (see Fig. 7.10). Thus we can normalize the distribution by performing a change of scale that converts the units of measurement into standard units by means of the following equation:

$$z = \frac{x - \mu}{\sigma} \qquad (7.9)$$

In order to determine areas under a standardized normal curve, we must convert x values into z values and then use the table in App. D. Note that numbers in the table have no negative values. Because of the symmetry of the normal curve about the mean, this does not reduce the utility of the table.

Before looking at example problems that demonstrate the practical application of this concept, we will demonstrate the use of the normal-curve table. The total area under the curve to the left of $z = 0$ as well as the area to the right of $z = 0$ is equal to 0.500 0 because for this standardized normal curve, the area beneath it equals 1.000.

Referring to Fig. 7.11, we shall calculate the probability of getting a z value less than 0.85. The area under the curve for $z < 0.85$ is determined from 0.500 0 (area left of $z = 0$) plus a table value of 0.302 3 for $z = 0.85$ (area between $z = 0$ and $z = 0.85$). Adding these values gives us a probability of 0.802 3.

When working problems with this table, we must remember that the data should closely approximate a normal distribution, and we must know the population mean (μ) and the standard deviation (σ).

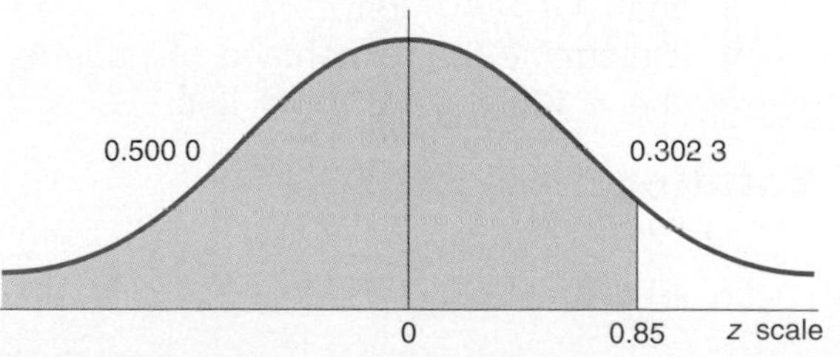

Figure 7.11
Normal curve, z scale.

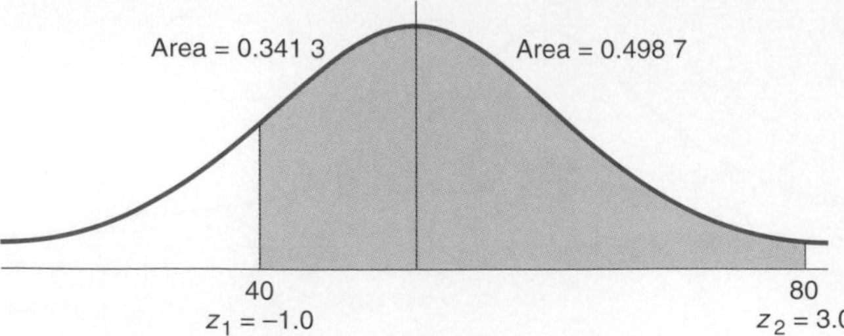

Figure 7.12

Example problem 7.3 A random variable has a normal distribution with $\mu = 5.0 \times 10^1$ and $\sigma = 1.0 \times 10^1$. What is the probability of the variable assuming a value between 4.0×10^1 and 8.0×10^1? (See Fig. 7.12.)

Solution Normalize the value of μ and σ:

$$z_1 = \frac{40 - 50}{10} = -1.0$$

$$z_2 = \frac{80 - 50}{10} = +3.0$$

From the z table in App. D,

Area = probability = 0.341 3 + 0.498 7 = 0.840 0

You may also solve this problem using a spreadsheet (the formula NORMDIST in EXCEL, for example). This is left as an exercise for you to try.

In most statistical applications we do not know the population parameters; instead we collect data in the form of a random sample from that population. It is possible to substitute sample mean and sample standard deviation, provided the sample size is sufficiently large ($n \geq 30$). A different theory called the *t distribution* is applicable to smaller sample sizes, but this distribution is not covered in our introduction to statistics.

Example problem 7.4 Assume that a normal distribution is a good representation of the data provided in Tab. 7.3, Example prob. 7.1.

(a) Determine the probability of a male student weighing more than 1.7×10^2 lbm.
(b) Determine the percentage of students who weigh between 1.4×10^2 and 1.5×10^2 lbm.

Solution

(a) $$z_1 = \frac{170 - 160.74}{9.89}$$

$$= 0.94$$

From the normal-curve table, area = 0.326 4. Since the area under the curve to the right of $z = 0$ is 0.500 0,

Probability = 0.500 0 − 0.326 4 = 0.173 6

$$\text{(b)}\quad z_1 = \frac{140 - 160.74}{9.89} = -2.10$$

$$z_2 = \frac{150 - 160.74}{9.89} = -1.09$$

For $z_1 = 2.10$,

Area = 0.482 1 (area between $z = -2.10$ and $z = 0$)

For $z_2 = 1.09$,

Area = 0.362 1 (area between $z = -1.09$ and $z = 0$)

The desired area is the difference, that is,

Probability = area = 0.482 1 − 0.362 1 = 0.120 0

Therefore, we would expect 12 percent of the males to weigh between 140 and 150 lbm. Try solving this problem with a spreadsheet function.

7.7 Linear Regression

There are many occasions in engineering analysis when the ability to predict or forecast the outcome of a certain event is extremely valuable. The difficulty with most practical applications is the large number of variables that may influence the analysis process. Regression analysis is a study of the relationships among variables. If the situation results in a relationship among three or more variables, the study is called *multiple regression*. There are many problems, however, that can be reduced to a relationship between an independent and a dependent variable. This introduction will limit the subject and treat only two-variable regression analyses.

Of the many equations that can be used for the purposes of prediction, the simplest and most widely used is a linear equation of the form $y = mx + b$, where m and b are constants. Once the constants have been determined, it is possible to calculate a predicted value of y (dependent variable) for any value of x (independent variable).

Before investigating the regression concept in more detail, we must examine how the regression equation is established.

If there is a reason to believe that a relationship exists between two variables, the first step is to collect data. For example, suppose x denotes the age of an automobile in years and y denotes the annual maintenance cost. Thus, a sample of n cars would reveal the age $x_1, x_2, x_3, \ldots, x_n$ and the corresponding annual maintenance cost $y_1, y_2, y_3, \ldots, y_n$.

The next step would be to plot the data on rectangular coordinate paper. The resulting graph is called a *scatter diagram*.

From the scatter diagram shown in Fig. 7.13, it may be possible to construct a straight line that adequately represents the data, in which case a linear relationship exists between the variables. In other cases, the line may be curved, and the relationship between variables is nonlinear in nature.

Ideally, we would hope to determine the best possible line (straight or curved) through the points. A standard approach to this problem is called the *method of least squares* and is explained as follows.

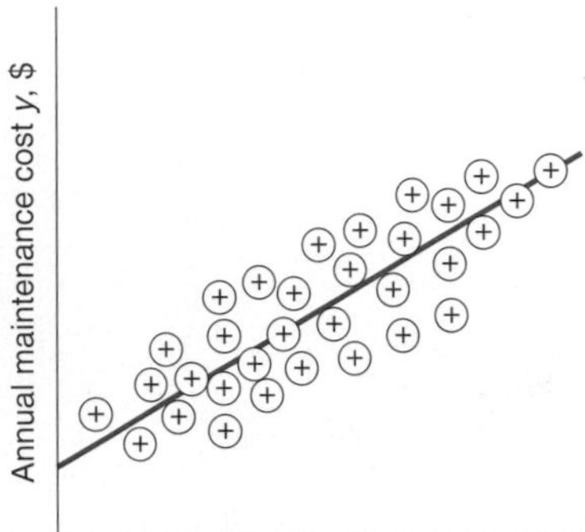

Figure 7.13
Scatter diagram.

To demonstrate how the process works, as well as to explain the concept of the method of least squares, consider the following situation. A class of 20 students is given a math test and the resulting scores are recorded. Each student's IQ score is also available. Both scores for the 20 students are shown in Tab. 7.7.

First, the data must be plotted on rectangular coordinate paper (see Fig. 7.14). As you can see by observing the plotted data, there is no limit to the number of straight lines that could be drawn through the points. In order to find the line of best fit, it is necessary to state what is meant by "best." The method of least squares requires that the sum of the squares of the vertical deviations, the residuals, from the data points to the straight line be a minimum.

To demonstrate how a least-squared line is fit to data, let us consider this problem further. There are n pairs of numbers (x_1, y_1), (x_2, y_2), . . . , (x_n, y_n), where here $n = 20$, with x and y being IQ and math scores, respectively. Suppose that the equation of the line that best fits the data is of the form

$$y' = mx + b \tag{7.10}$$

where the symbol y' (y prime) is used to differentiate between the observed values of y and the corresponding values calculated by means of the equation of the line. (Note that y' is sometimes used to represent a derivative in calculus, it is not a derivative

Table 7.7

Student #	Math score	IQ	Student #	Math score	IQ
1	85	120	11	100	130
2	62	115	12	85	130
3	60	100	13	77	118
4	95	140	14	63	112
5	80	130	15	70	122
6	75	120	16	90	128
7	90	130	17	80	125
8	60	108	18	100	140
9	70	115	19	95	135
10	80	118	20	75	130

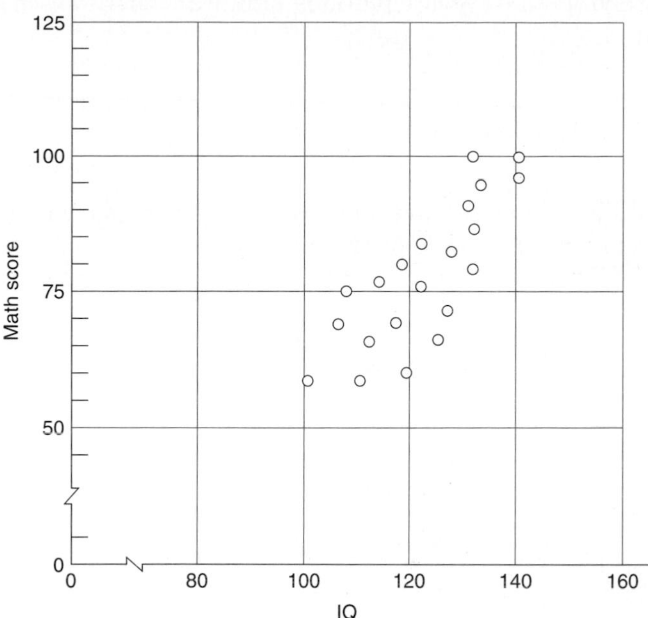

Figure 7.14
Math and IQ scores.

here, rather a computed value on the line being sought.) In other words, for each value of x, there exist an observed value (y) and a calculated value (y') obtained by substituting x into the equation $y' = mx + b$.

The least-squares criterion requires that the sum of all $(y - y')^2$ terms, as illustrated in Fig. 7.15, be the smallest possible. One must determine the constants m and b so that the difference between the observed and the predicted values of y will be minimized.

When this analysis is applied to the linear equation $y = mx + b$, it follows that we wish to minimize the summation of all deviations (residuals):

$$\text{SUM} = \Sigma[y_i - (mx_i + b)]^2 \tag{7.11}$$

From the calculus, to minimize SUM, the partial derivatives with respect to m and b must be zero as follows:

$$\frac{\partial(\text{SUM})}{\partial m} = \frac{\partial}{\partial m}\{\Sigma[y_i - (mx_i + b)]^2\} = 0 \tag{7.12}$$

$$\frac{\partial(\text{SUM})}{\partial b} = \frac{\partial}{\partial b}\{\Sigma[y_i - (mx_i + b)]^2\} = 0 \tag{7.13}$$

Performing these partial derivatives gives

$$nb + m\Sigma x_i = \Sigma y_i \tag{7.14}$$

$$b\Sigma x_i + m\Sigma x_i = \Sigma x_i y_i \tag{7.15}$$

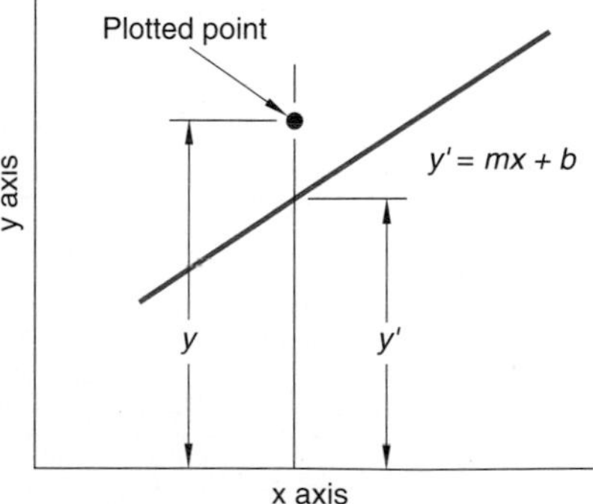

Figure 7.15
y-Axis deviation.

Solving these two equations simultaneously for m and b gives

$$m = \frac{n(\Sigma x_i y_i) - (\Sigma x_i)(\Sigma y_i)}{n(\Sigma x_i^2) - (\Sigma x_i)^2} \tag{7.16}$$

$$b = \frac{\Sigma y_i - m(\Sigma x_i)}{n} \tag{7.17}$$

Table 7.8 is a tabulation of the values necessary to determine the constants m and b for the math score–IQ problem. The independent variable is the IQ, and the dependent variable is the math score.

Substituting the values from Tab. 7.8 into Eqs. (7.16) and (7.17), we get the following values for the two constants:

$$m = \frac{20(198\,527) - (2\,466)(1\,592)}{20(306\,124) - (2\,466)^2}$$

$$= 1.081$$

$$b = \frac{1\,592 - (1.081)(2\,466)}{20}$$

$$= -53.7$$

The equation of the line relating math score and IQ using the method of least squares becomes

Math score = −53.7 + 1.08(IQ)

Table 7.8

Independent variable		Dependent variable		
IQ	(IQ)^2	Math score	(Math score)^2	(IQ) (Math score)
120	14 400	85	7 225	10 200
115	13 225	62	3 844	7 130
100	10 000	60	3 600	6 000
140	19 600	95	9 025	13 300
130	16 900	80	6 400	10 400
120	14 400	75	5 625	9 000
130	16 900	90	8 100	11 700
108	11 664	60	3 600	6 480
115	13 225	70	4 900	8 050
118	13 924	80	6 400	9 440
130	16 900	100	10 000	13 000
130	16 900	85	7 225	11 050
118	13 924	77	5 929	9 086
112	12 544	63	3 969	7 056
122	14 884	70	4 900	8 540
128	16 384	90	8 100	11 520
125	15 625	80	6 400	10 000
140	19 600	100	10 000	14 000
135	18 225	95	9 025	12 825
130	16 900	75	5 625	9 750
2 466	306 124	1 592	129 892	198 527

Interesting questions arise from this problem. Can IQ be used to predict success on a math exam and, if so, how well? Regression analysis or estimation of one variable (dependent) from one or more related variables (independent) does not provide information about the strength of the relationship. Section 7.8 will provide a method to determine how well an equation developed from the method of least squares describes the strength of the relationship between variables.

The method of least squares as just explained is a most appropriate technique for determination of the best-fit line. You should clearly understand that this method as presented is *linear regression* and is valid only for *linear* relationships. The technique of least squares can, however, be applied to power ($y = bx^m$) and exponential ($y = be^{mx}$) relationships if done correctly. The power function can be handled by noting that there is a linear relationship between $\log y$ and $\log x$ ($\log y = m \log x + \log b$, which plots as a straight line on log-log paper). Thus we can apply the method of least squares to the variables $\log y$ and $\log x$ to obtain parameters m and $\log b$.

The exponential function written in natural logarithm form is $\ln y = mx + \ln b$. Therefore, there exists a linear relationship between $\ln y$ and x (this plots as a straight line on semilog paper). The next examples will demonstrate the use of least squares method for power and experimental curves.

Example problem 7.5 The data in Tab. 7.9 were obtained from measuring the distance dropped by a falling body with time. We would expect that the distance should be related to the square of the time according to theory (neglecting air friction). Find the equation of the line of best fit.

Table 7.9

Time t, s	Distance s, m
0	0
1	4.9
2	19.6
3	44.1
4	78.4
5	122.5
6	176.4

Solution The expected form of the equation is $s = bt^m$ or in logarithmic form, $\log s = m \log t + \log b$. Therefore, if we use $\log t$ in place of x and $\log s$ in place of y in Eqs. (7.16) and (7.17), we can solve for m and $\log b$. (Note carefully that the parameters are m and $\log b$.) Refer to Tab. 7.10.

Table 7.10

		Independent variable		Dependent variable	
t	s	$\log t$	$(\log t)^2$	$\log s$	$(\log t)(\log s)$
1	4.9	0.000 0	0.000 0	0.609 2	0.000 0
2	19.6	0.301 0	0.090 6	1.292 3	0.389 0
3	44.1	0.477 1	0.227 6	1.644 4	0.784 5
4	78.4	0.602 1	0.362 5	1.894 3	1.140 6
5	122.5	0.699 0	0.488 6	2.088 1	1.459 6
6	176.4	0.778 2	0.605 6	2.246 5	1.748 2
		2.857 4	1.774 9	9.855 8	5.521 9

Substitute into Eq. (7.16):

$$m = \frac{6(5.521\,9) - (2.857\,4)(9.855\,8)}{6(1.774\,9) - (2.857\,4)^2}$$

$$= 2.00$$

Substitute into Eq. (7.17), using log *b*, not *b*:

$$\log b = \frac{9.855\,8 - (2.00)(2.857\,4)}{6}$$

$$= 0.690\,17$$

$$b = 4.9$$

The equation is then $s = 4.9t^2$

Example problem 7.6 Using the method of least squares, find the equation that best fits the data shown in Tab. 7.11.

Solution This data produces a straight line when plotted on semilog graph paper; therefore, its equation will be of the form $Q = be^{mv}$ or $\ln Q = mv + \ln b$.
An examination of the equation and the graph paper leads us to the following:

1. Since the line is straight, the method of least squares can be used.
2. Since the abscissa is a uniform scale, the independent variable values (velocity in this problem) may be used without adjustment.
3. Since the ordinate is a log scale, the dependent variable values (fuel consumption) must be the logarithms of the data, not the raw data.

Table 7.12 provides us with the needed values to substitute into Eqs. (7.16) and (7.17).
Substitute into Eq. (7.16):

$$m = \frac{7(1\,494.63) - (280)(33.665\,8)}{7(14\,000) - (280)^2}$$

$$= 0.052\,9$$

Table 7.11

Fuel consumption Q, mm^3/s	Velocity v, m/s
25.2	10.0
44.6	20.0
71.7	30.0
115	40.0
202	50.0
367	60.0
608	70.0

Table 7.12

Independent variable		Dependent variable	
v	v^2	$\ln Q$	$v(\ln Q)$
10	100	3.226 8	32.27
20	400	3.797 7	75.95
30	900	4.272 5	128.17
40	1 600	4.744 9	189.80
50	2 500	5.308 3	265.41
60	3 600	5.905 4	354.32
70	4 900	6.410 2	448.71
280	14 000	33.665 8	1 494.63

Substitute into Eq. (7.17), using ln b rather than b:

$$\ln b = \frac{33.665\,8 - (.052\,9)(280)}{7}$$

$$= 2.693\,4$$

$$b = 14.78$$

The equation becomes $Q = 15e^{0.053v}$

7.8 Coefficient of Correlation

The technique of finding the best possible straight line to fit experimentally collected data is certainly useful, as previously discussed. The next logical and interesting question is how well such a line actually fits. It stands to reason that if the differences between the observed y's and the calculated y's are small, the sum of squares $\Sigma(y - y')^2$ will be small; and if the differences are large, the sum of squares will tend to be large.

Although $\Sigma(y - y')^2$ provides an indication of how well a least-squares line fits particular data, it has the disadvantage that it depends on the units of y. For example, if the units of y are changed from dollars to cents, it will be like multiplying $\Sigma(y - y')^2$ by a factor of 10 000. To avoid this difficulty, the magnitude of $\Sigma(y - y')^2$ is normally compared with $\Sigma(y - \bar{y})^2$. This allows the sum of the squares of the vertical deviations from the least-squares line to be compared with the sum of squares of the deviations of the y's from the mean.

To illustrate, Fig. 7.16a shows the vertical deviation of the y's from the least-squares line, while Fig. 7.16b shows the deviations of the y's from their collective mean. It is apparent that where there is a close fit, $\Sigma(y - y')^2$ is much smaller than $\Sigma(y - \bar{y})^2$.

In contrast, consider Fig. 7.17. Again, Fig. 7.17a shows the vertical deviation of the y's from the least-squares line, and Fig. 7.17b shows the deviation of the y's from their mean. In the latter case, $\Sigma(y - y')^2$ is approximately the same as $\Sigma(y - \bar{y})^2$. This would seem to indicate that if the fit is good, as in Fig. 7.16, $\Sigma(y - y')^2$ is much less than $\Sigma(y - \bar{y})^2$; and if the fit is as poor as in Fig. 7.17, the two sums of squares are approximately equal.

The coefficient of correlation puts this comparison on a precise basis:

$$r = \pm\sqrt{1 - \frac{\Sigma(y_i - y')^2}{\Sigma(y_i - \bar{y})^2}} \tag{7.18}$$

If the fit is poor, the ratio of the two sums is close to 1 and r is close to zero. However, if the fit is good, the ratio is close to zero and r is close to $+1$ or -1. From the equation, it is obvious that the ratio can never exceed 1. Hence, r cannot be less than -1 or greater than $+1$.

The statistic is used to measure the strength of a linear relationship between any two variables. It indicates the goodness of fit of a line determined by the method of least squares, and this in turn indicates whether a relationship exists between x and y.

Although Eq. (7.18) serves to define the coefficient of correlation, it is seldom used in practice. An alternative form of the formula is

$$r = \frac{n(\Sigma x_i y_i) - (\Sigma x_i)(\Sigma y_i)}{\sqrt{n(\Sigma x_i^2) - (\Sigma x_i)^2}\sqrt{n(\Sigma y_i^2) - (\Sigma y_i)^2}} \tag{7.19}$$

Figure 7.16
Deviation from y' and y (good fit).

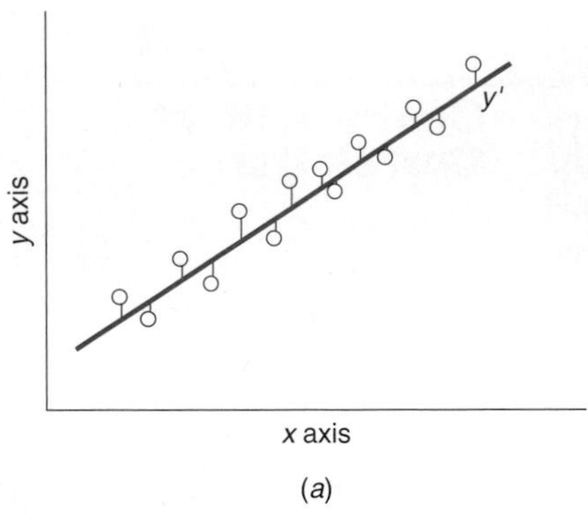

(*a*)

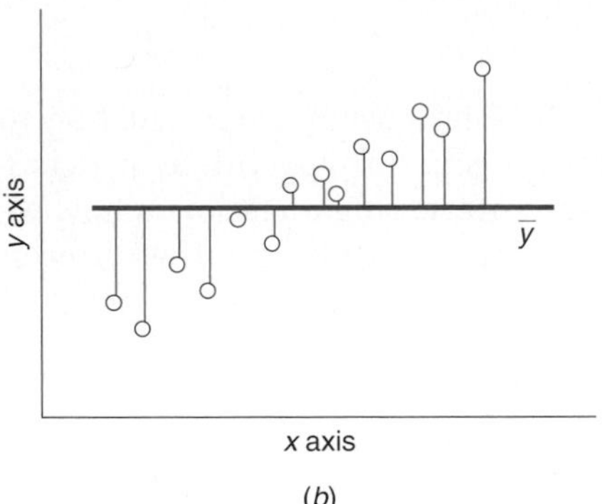

(*b*)

Figure 7.17
Deviation from y' and y (poor fit).

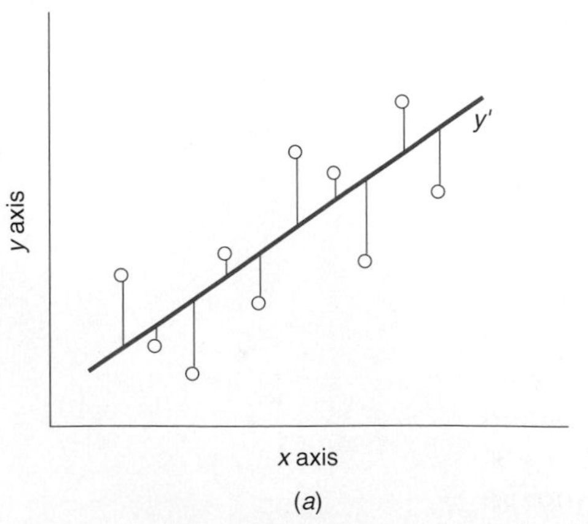

(*a*)

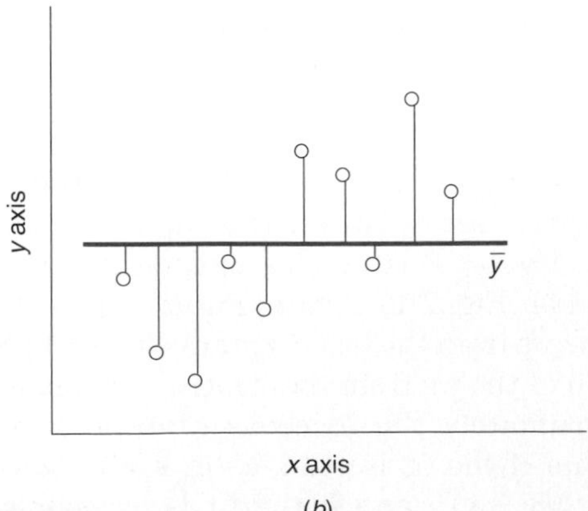

(*b*)

The interpretation of r is not difficult if it is ± 1 or zero: When it is zero, the points are scattered and the fit of the regression line is so poor that a knowledge of x does not help in the prediction of y; when it is $+1$ or -1, all the points actually lie on the straight line, so an excellent prediction of y can be made by using x values. The problem arises when r falls between zero and $+1$ or zero and -1.

General guidelines for interpreting the correlation coefficient are as follows:

Correlation Coefficient	Correlation Interpretation
0.9 to 1.0	Very high positive
0.7 to 0.9	High positive
0.5 to 0.7	Moderate positive
0.3 to 0.5	Low positive
−0.3 to 0.3	Little, if any
−0.5 to −0.3	Low negative
−0.7 to −0.5	Moderate negative
−0.9 to −0.7	High negative
−1.0 to −0.9	Very high negative

The physical interpretation of r can be explained in the following manner. If the coefficient of correlation is known for a given set of data, then $100r^2$ percent of the variation of the y's can be attributed to differences in x, namely, to the relationship of y with x. If $r = 0.6$ in a given problem, then 36 percent—that is, $100(0.6^2)$—of the variation of the y's is accounted for (perhaps caused) by differences in x values. The square of the correlation coefficient (r^2) is called the *coefficient of determination*.

Again consider the problem on IQ and math scores, substituting values from Tab. 7.8 into the equation for the correlation coefficient.

$$r = \frac{(20)(198\,527) - (2\,466)(1\,592)}{\sqrt{(20)(306\,124) - (2\,466)^2}\sqrt{(20)(129\,892) - (1\,592)^2}}$$

$$= 0.87$$

Computing the coefficient of determination and multiplying by 100 to obtain percent yields

$$100r^2 = 76 \text{ percent}$$

This would indicate that 76 percent of the variations in math scores can be accounted for by differences in IQ.

One word of caution when using or considering results from linear regression and coefficients of correlation and determination: There is a fallacy in interpreting high values of r or r^2 as implying cause-effect relations. If the increase in television coverage of professional football is plotted against the increase in traffic accidents at a certain intersection over the past 3 years, an almost perfect positive correlation (+1.0) can be shown to exist. This is obviously not a cause-effect relation, so it is wise to interpret the correlation coefficient and coefficient of determination carefully. The variables must have a measure of association if the results are to be meaningful.

Example problem 7.7 Data showing the annual peak flow (discharge) as a function of expected return period is given in Tab. 7.13. This is for a small watershed of about 440 mi^2 located in a midwestern state. Using the least-squares method, obtain an equation relating the annual peak discharge Q to the expected return period P. Compute the correlation coefficient and interpret the results.

Solution A spreadsheet is a convenient tool for this problem. The data were entered in columns A and B as seen in Fig. 7.18. Test plots were then done to determine visually whether linear, semilog, or log-log techniques would be most likely to produce a straight line. As shown in Fig. 7.18, the log-log plot, although not a perfect straight line, is more nearly so than either the linear plot or the semilog plot. Thus we will use the method of least squares to find an equation of the form $Q = bP^m$ ($\log Q = m \log P + \log b$).

The spreadsheet was then modified (see Fig. 7.19) by adding columns to compute $\log P$, $\log Q$, $(\log P)(\log Q)$, $(\log P)^2$ and $(\log Q)^2$. The sums of each of columns A through G were also computed. From these sums, the parameter m could be found (cells A27 and B27) and with m, $\log b$ was calculated in cells A29 and B29. The parameter $b = 10^{\log b}$ and is shown in cells A30 and B30.

The least-squares equation is then $Q' = 3.37 \times 10^3 P^{0.520}$ after rounding the coefficients. Column H was added to compute values

Table 7.13

Period P, yrs	Discharge Q, cfs
1.85	4 330
1.95	4 470
2.06	4 600
2.18	4 620
2.32	5 110
2.48	5 320
2.66	5 360
2.87	5 600
3.11	5 950
3.40	6 300
3.75	7 300
4.18	7 530
4.73	8 900
5.43	9 000
6.39	9 150
7.75	10 700
9.84	12 000
13.48	14 000
21.41	15 300
52.00	22 400

of the predicted discharge, Q', for each return period P. The original data as well as the prediction curve were plotted in Fig. 7.19.

The correlation coefficient was then computed in cell D32 and was found to be about 0.984. Squaring this value (coefficient of determination) and multiplying by 100 percent gives a result of 96.9 percent, which is an indication of a very good agreement between the data and the prediction line. It suggests that nearly 97 percent of the variation in the discharge can be accounted for by variation in the return period.

Spreadsheets also provide a mechanism for computing a least-squares curve fit (EXCEL calls this a trendline) to a data set. Figure 7.20 shows the result of this approach for Example problem 7.7.

Figure 7.18
Preliminary analysis for Example prob. 7.7.

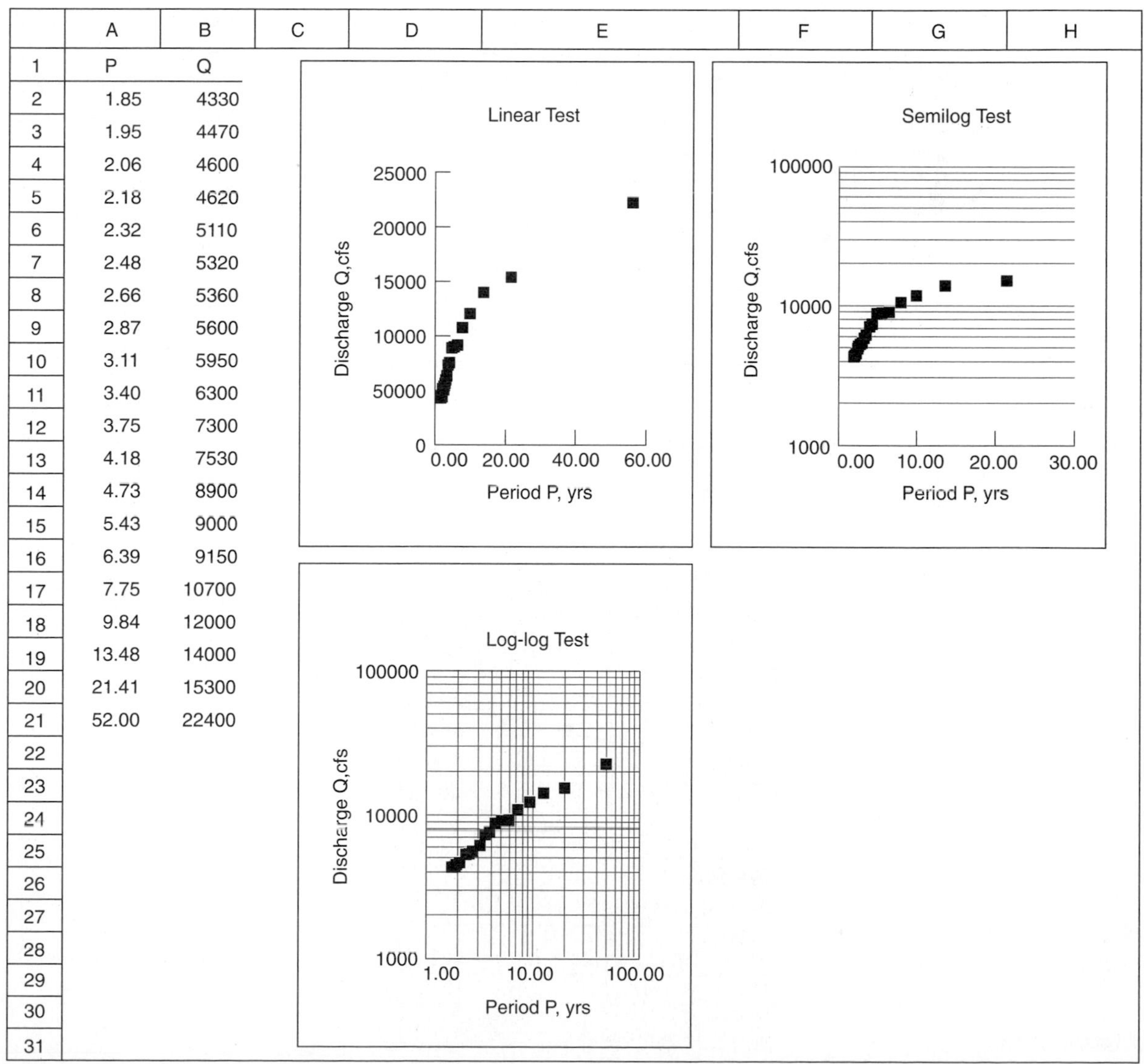

	A	B
1	P	Q
2	1.85	4330
3	1.95	4470
4	2.06	4600
5	2.18	4620
6	2.32	5110
7	2.48	5320
8	2.66	5360
9	2.87	5600
10	3.11	5950
11	3.40	6300
12	3.75	7300
13	4.18	7530
14	4.73	8900
15	5.43	9000
16	6.39	9150
17	7.75	10700
18	9.84	12000
19	13.48	14000
20	21.41	15300
21	52.00	22400
22		
23		
24		
25		
26		
27		
28		
29		
30		
31		

	A	B	C	D	E	F	G	H
1	P	Q	log P	log Q	(log P)(log Q)	(log P)^2	(log Q)^2	Q'
2	1.85	4330	0.2672	3.6365	0.9716	0.0714	13.2240	4641.74
3	1.95	4470	0.2900	3.6503	1.0587	0.0841	13.3247	4770.56
4	2.06	4600	0.3139	3.6628	1.1496	0.0985	13.4158	4908.65
5	2.18	4620	0.3385	3.6646	1.2403	0.1146	13.4296	5055.31
6	2.32	5110	0.3655	3.7084	1.3554	0.1336	13.7524	5221.60
7	2.48	5320	0.3945	3.7259	1.4697	0.1556	13.8824	5405.86
8	2.66	5360	0.4249	3.7292	1.5845	0.1805	13.9067	5606.45
9	2.87	5600	0.4579	3.7482	1.7162	0.2097	14.0489	5832.40
10	3.11	5950	0.4928	3.7745	1.8599	0.2428	14.2470	6081.12
11	3.40	6300	0.5315	3.7993	2.0193	0.2825	14.4350	6369.67
12	3.75	7300	0.5740	3.8633	2.2177	0.3295	14.9253	6702.60
13	4.18	7530	0.6212	3.8768	2.4082	0.3859	15.0295	7091.82
14	4.73	8900	0.6749	3.9494	2.6653	0.4554	15.5977	7562.63
15	5.43	9000	0.7348	3.9542	2.9056	0.5399	15.6360	8125.32
16	6.39	9150	0.8055	3.9614	3.1909	0.6488	15.6929	8843.08
17	7.75	10700	0.8893	4.0294	3.5833	0.7909	16.2359	9776.40
18	9.84	12000	0.9930	4.0792	4.0506	0.9860	16.6397	11068.73
19	13.48	14000	1.1297	4.1461	4.6838	1.2762	17.1904	13036.97
20	21.41	15300	1.3306	4.1847	5.5682	1.7705	17.5116	16582.71
21	52.00	22400	1.7160	4.3502	7.4650	2.9447	18.9247	26305.74
22	Sum P	Sum Q	Sum log P	Sum log Q	Sum (log P)(log Q)	Sum (log P)^2	Sum (log Q)^2	
23	153.84	167940	13.3454	77.4945	53.1639	11.7011	301.0502	
24								
25	Form: log Q = m log P + log b							
26								
27	m=	0.520						
28								
29	log b=	3.527756						
30	b=	3371						
31								
32	Correlation coefficient =			0.984				

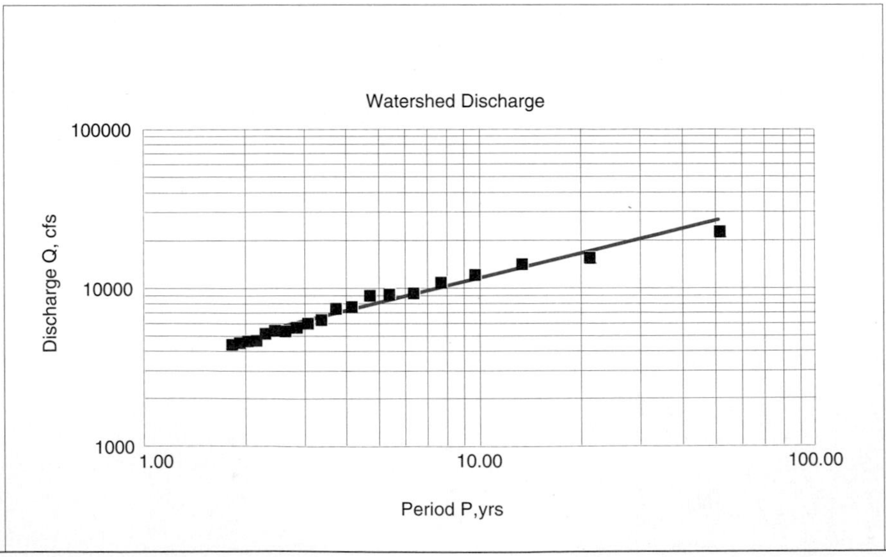

Figure 7.19
Spreadsheet for Example prob. 7.7.

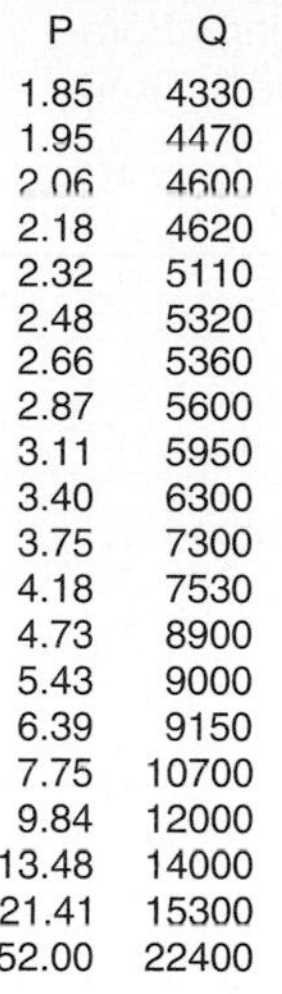

P	Q
1.85	4330
1.95	4470
2.06	4600
2.18	4620
2.32	5110
2.48	5320
2.66	5360
2.87	5600
3.11	5950
3.40	6300
3.75	7300
4.18	7530
4.73	8900
5.43	9000
6.39	9150
7.75	10700
9.84	12000
13.48	14000
21.41	15300
52.00	22400

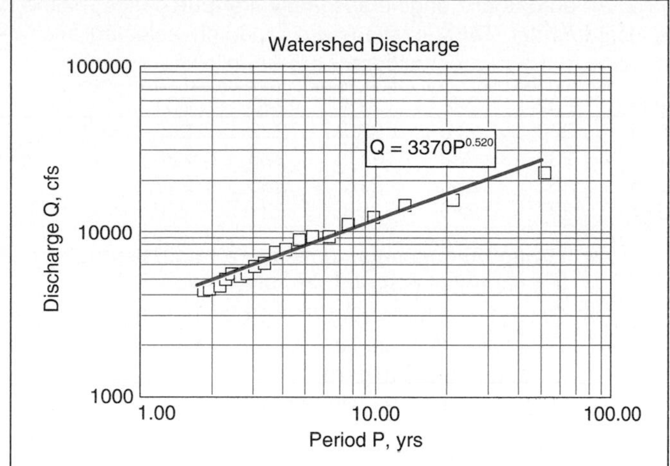

Figure 7.20
Alternate spreadsheet treatment for Example prob. 7.7.

7.9 Key Terms and Concepts

The following are some of the terms and concepts you should recognize and understand:

Descriptive statistics	Standard deviation
Statistical inference	Variance
Frequency distribution	Normal distribution
Classification of data	Scatter diagram
Class frequency	Linear regression
Histogram	Multiple regression
Arithmetic mean	Method of least squares
Median	Power relationship
Mode	Exponential relationship
Population	Coefficient of correlation
Sample	Coefficient of determination

Problems

7.1 Students in a housing complex selected random rooms and measured the width of student desks found in them to the nearest 0.1 inch.

56.7	64.9	61.4	66.9	61.5
71.2	66.6	68.2	68.9	63.6
67.8	67.5	67.6	71.3	56.2
64.2	55.7	68.8	68.5	65.9
67.5	72.0	63.9	64.3	73.0
61.4	63.4	59.6	68.7	64.4
70.3	68.2	68.2	69.2	65.6

(*a*) Group these measurements into a frequency-distribution table having five equal classes from 54.0 to 73.9.
(*b*) Construct a histogram of the distribution.
(*c*) Determine the median, mode, and mean of the data.
(*d*) Calculate the standard deviation.

7.2 An automobile engine assembly firm purchases engine castings from a local foundry. Thirty castings are randomly selected and weighed. Their masses to the nearest kilogram are as follows:

235	232	228	228	240	231
225	220	218	230	222	242
207	233	222	211	228	228
238	232	230	226	236	247
227	227	229	229	224	228

(*a*) Group the measurements into a frequency-distribution table having six equal classes from 206 to 247.
(*b*) Construct a histogram of the distribution.
(*c*) Determine the median, mode, and mean of the data.
(*d*) Calculate the standard deviation.

Use a spreadsheet to aid with this problem.

7.3 The burnout velocity of a test series of rockets was measured to the nearest 10 miles per hour.

980	860	950	1 010
930	880	870	960
850	1 020	970	890
970	900	1 030	950
1 000	940	970	900

(*a*) Group these measurements into a frequency distribution table having six equal classes that range from 800 to 1 099.
(*b*) Construct a histogram of the distribution.
(*c*) Determine the median, mode, and mean of the data.
(*d*) Calculate the standard deviation.

Perform this problem using pencil and calculator and repeat using a spreadsheet.

7.4 The following test scores were earned by a class of first-year engineering students on a physics test.

40	70	77	80	85	59	90	67	47	70
87	61	73	88	70	58	70	67	62	75
65	90	58	69	99	83	63	72	95	62
79	80	68	100	75	58	69	60	72	88
64	52	65	77	72	70	31	93	79	72

(*a*) Group these test scores into a frequency-distribution table.
(*b*) Construct a histogram of the distribution.
(*c*) Determine the median, mode, and mean of the data.
(*d*) Calculate the standard deviation.

Use a spreadsheet for this problem.

7.5 Survey at least 30 engineering students to obtain each student's total investment in calculators and computers. The investment figure should include all equipment, whether bought personally or received as a gift.

(*a*) Find the mean, mode, and median investment.
(*b*) Find the standard deviation.
(*c*) Assuming your sample to be a normal distribution, half of the engineering students at your school probably have invested between $ ________ and $ ________ each.

7.6 Find out from all members of this class how many credits toward graduation each will have at the end of this term.

(*a*) Find the mean, mode, and median of the data
(*b*) Find the standard deviation.
(*c*) Assuming that your class is representative of all of those taking this course this term, 12 percent of the students currently taking this course will have completed more than ________ credits at the end of this term. Fifty percent will have completed less than ________ credits. What percentage will have earned more than 30 semester credits?

7.7 The scores on the final exam for 300 students taking engineering computations produced a mean of 79 and a standard deviation of 13. If the distribution of the scores is approximately normal, determine

(*a*) The expected percentage of students earning 90 or higher.
(*b*) The number of students expected to have scores below 60.
(*c*) The expected number of Cs (70–79).
(*d*) The predicted maximum score of the lowest 15 percent of the class.

7.8 The quality control department measured the length of 100 pens randomly selected from a specified order. The mean length was found to be 14.76 cm, and the standard deviation was 0.01 cm. If the pen lengths are normally distributed, find:

(*a*) The percentage of pens shorter than 14.74 cm.
(*b*) The percentage of pens longer than 14.78 cm.
(*c*) The percentage of pens that meet the length specification of 14.75 $\pm$ 0.02 cm.
(*d*) The percentage of pens that are longer than the nominal length of 14.75 cm.

7.9 A sample of 40 resistors randomly taken from yesterday's production were tested with the following result: mean resistance = 985Ω and standard deviation = 45Ω. Assuming a normal distribution, compute the following:

(*a*) The percentage of resistors with resistance greater than 1 000Ω.
(*b*) The yield (that is, percentage good) if the acceptable values are 1 000Ω $\pm$ 10 percent.
(*c*) The yield if acceptable values include only 1 000Ω $\pm$ 5 percent.
(*d*) The yield if 1 000Ω $\pm$ 1 percent resistors are needed.

7.10 In Tab. 7.14 are grades that 20 students obtained on the midterm and final examination in a freshman graphics course:

(*a*) Using the method of least squares, determine the best linear relationship prediction of the final grade from the midterm grade.
(*b*) Plot the data on linear graph paper.
(*c*) Represent the equation from a on the graph from b.
(*d*) Calculate and interpret the coefficients of correlation and determination.
(*e*) Repeat steps a through c using a spreadsheet approach.

Table 7.14

Midterm exam *M*, %	Final exam *F*, %
88	78
75	85
97	91
68	82
86	81
91	75
53	64
84	91
77	78
92	83
62	52
83	73
36	50
51	40
81	83
91	87
82	80
74	70
85	89
96	98

Table 7.15

Time t, s	Velocity V, mph
1	15.1
2	32.2
3	63.4
4	84.5
5	118
6	139

Table 7.16

Time t, s	Velocity V, m/s
1.00	5.30
4.00	18.1
7.00	26.9
10.0	37.0
14.0	55.2

7.11 Table 7.15 has data from a vehicle acceleration test.

(*a*) Plot the data on rectilinear paper.
(*b*) Using the method of least squares, determine the equation of the line of best fit.
(*c*) Draw the line on the graph.
(*d*) Calculate and interpret the coefficient of correlation.

7.12 Table 7.16 lists values of velocities recorded for one of the participants during a drag race of 1930s stock automobiles.

(*a*) Plot the data on rectilinear paper.
(*b*) Using the method of least squares, determine the equation of the line of best fit.
(*c*) Draw the line on the graph.
(*d*) Calculate and interpret the coefficient of determination.

7.13 A furnace used in a manufacturing process produced the heat output per unit furnace volume shown in Tab. 7.17 when tested last month.

(*a*) Plot the data on log-log graph paper, with temperature as the independent variable.
(*b*) Using the method of least squares, find the equation of the line of best fit.
(*c*) Draw the line on the graph.
(*d*) Calculate the coefficient of correlation.

Table 7.17

Heat released H, 10^3 Btu/ft^3	Temperature T, °F
0.21	175
0.63	250
2.11	400
4.09	500
7.97	600
19.92	800
39.89	950
80.05	1 300

7.14 The capacity of a screw conveyor that is moving dry ground corn is expressed in liters per second and the conveyor speed in revolutions per minute. A test was conducted in Cleveland on the conveyor model JD172 last week. The results of the test are shown in Tab. 7.18.

(*a*) Plot the data on log-log graph paper.
(*b*) Using the method of least squares, find the equation of the line of best fit.
(*c*) Draw the line on the graph.
(*d*) Calculate the coefficients of correlation and determination.

Table 7.18

Capacity C, L /s	Angular velocity V, r/min
3.1	10
6.1	20
15.6	60
29.9	140
50.3	250
78.5	400
112.4	525

7.15 The resistance of a class of electrical conductor was tested over a wide range of sizes at constant temperature. The test was performed at Madison, Wisconsin, on April 4, 1995, at the Acme Electrical Labs. The test results are shown in Tab. 7.19. The resistance is expressed in milliohms per meter of conductor length.

(*a*) Plot the data on log-log graph paper.
(*b*) Using the method of least squares, find the equation of the line of best fit.
(*c*) Draw the line on the graph.
(*d*) Find the coefficient of correlation.

Table 7.19

Area A, mm^2	Resistance R, $m\Omega/m$
0.021	1 010
0.062	364
0.202	111
0.523	44
1.008	22
3.320	8
7.290	3.5

7.16 Table 7.20 is a collection of data from an iron-constantan thermocouple. Temperature is in degrees Celsius and the emf is in millivolts.

(*a*) Plot the data on rectilinear graph paper.
(*b*) Using the method of least squares, find the equation of the line of best fit.
(*c*) Draw the line on the graph.
(*d*) Calculate and interpret the coefficient of determination.

Table 7.20

Temperature t, °C	Voltage (emf), mV
50	1.3
100	3.3
150	4.4
200	5.6
300	8.5
400	11.3
500	13.0
600	16.3
700	18.9
800	20.5
900	24.0
1 000	27.6

7.17 A Sessions pump was tested to determine the power required to produce a range of discharge. The test was performed in Oak Park on July 1, 1995. The results of the test are shown in Tab. 7.21.

(*a*) Plot the data on rectilinear graph paper.
(*b*) Using the method of least squares, find the equation of the line of best fit.
(*c*) Draw the line on the graph.
(*d*) Calculate and interpret the coefficient of determination.

Table 7.21

Discharge Q, L/s	Power P, kW
6	57.0
14	67.6
20	78.2
27	86.4
34	96.0
40	103.6
50	120.0

7.18 A spring was tested in Des Moines last Thursday. The test of spring ZX-15 produced the data in Tab. 7.22.

(*a*) Plot the data on rectilinear graph paper and determine the equation that expresses the deflection to be expected under a given load. Use the method of least squares.
(*b*) Draw the line on the graph.
(*c*) Predict the load required to produce a deflection of 35 mm.
(*d*) Calculate and interpret the coefficient of determination.

Table 7.22

Deflection D, mm	Load L, N
1.1	35
6.0	80
10.0	120
14.0	160
17.5	200
22.5	250
27.5	300

7.19 A 90° triangular weir is commonly used to measure flow rate in a stream. Data on the discharge through the weir were collected and recorded in Tab. 7.23 for various heights.

(*a*) Plot the data on log-log graph paper.
(*b*) Using the method of least squares, find the equation of the line of best fit.
(*c*) Draw the line on the graph.
(*d*) Calculate the coefficient of correlation.
(*e*) What discharge would you predict for a water height of 5.5 m?

Table 7.23

Height *h*, m	Discharge *Q*, m^3/s
1.0	1.5
2.0	8.0
3.0	22
4.0	45
5.0	78
6.0	124
7.0	182
8.0	254

7.20 A Pitot tube is a device for measuring the velocity of flow of a fluid (see Fig. 7.21). A stagnation point occurs at point 2; by recording the height differential *h*, the velocity at point 1 can be calculated. Assume for this problem that the velocity at point 1 is known corresponding to the height differential *h*. Table 7.24 records these values.

(*a*) Plot the data on log-log graph paper.
(*b*) Using the method of least squares, find the equation of the line of best fit.
(*c*) Draw the line on the graph.
(*d*) Calculate the coefficient of determination.
(*e*) What is the expected velocity for a height differential of 0.5 m?

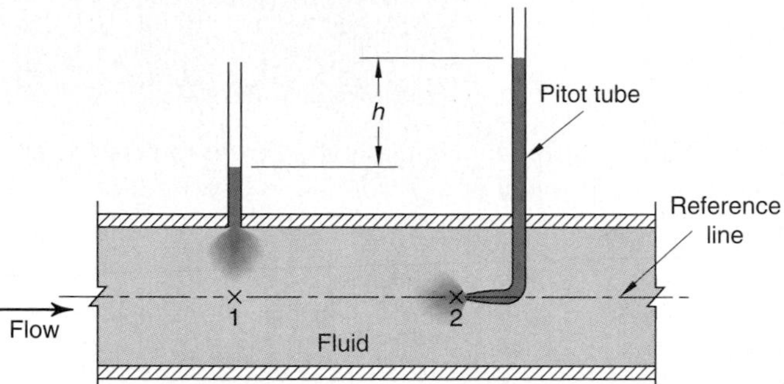

Figure 7.21 Pitot tube.

Table 7.24

Height *h*, m	Velocity *V*, m/s
0.10	1.40
0.20	2.00
0.40	2.80
0.60	3.40
0.80	4.00
1.00	4.40

7.21 The voltage across a capacitor during discharge was recorded as a function of time (see Tab. 7.25).

(*a*) Plot the data on semilog graph paper.
(*b*) Using the method of least squares, find the equation of the line of best fit.
(*c*) Draw the line on the graph.
(*d*) Calculate the coefficient of correlation.
(*e*) Predict the voltage at 25s, 70s, and 90s.

Table 7.25

Time t, s	Voltage V, V
12	98
20	62
34	23
50	9.5
64	3.5
76	1.9
84	1.3

7.22 When a capacitor is to be discharged, the current flows until the voltage across the capacitor is zero. This current flow when measured as a function of time resulted in the data given in Tab. 7.26.

(*a*) Plot the data on semilog graph paper.
(*b*) Using the method of least squares, find the equation of the line of best fit.
(*c*) Draw the line on the graph.
(*d*) Calculate the coefficients of correlation and determination.
(*e*) What current should be flowing at 2s?

Table 7.26

Time t, s	Current I, A
0.1	3.62
0.2	3.28
0.3	2.96
0.4	2.68
0.5	2.42
1.0	1.46

7.23 When fluid is flowing in the line, it is relatively easy to begin closing a valve that is wide open. But as the valve approaches a more nearly closed position, it becomes considerably more difficult to force movement. Visualize a circular pipe with a simple flap hinged at one edge being closed over the end of the pipe. The fully open position is $\theta = 0°$, and $\theta = 90°$ is the fully closed condition.

A test was conducted on such a valve by applying constant torque at the hinged position and measuring the angular movement of the valve. The test data are shown in Tab. 7.27.

(*a*) Plot the data on semilog graph paper.
(*b*) Using the method of least squares, find the equation of the line of best fit.
(*c*) Draw the line on the graph.
(*d*) Calculate the coefficient of determination.
(*e*) What torque would be required to hold the valve at 60°?

Table 7.27

Torque T, N·m	Movement θ, degrees
9	5.2
18	29.3
30	40.9
60	56.3
105	71.0
150	84.8

Table 7.28

Pressure P, MPa	Relative compression R, %
1.12	27.3
3.08	37.6
5.25	46.0
8.75	50.6
12.3	56.1
16.1	59.2
30.2	65.0

7.24 All materials are elastic to some degree. It is desirable that certain parts of some designs compress when a load is applied to assist in making the part air tight or watertight (for example, a jar lid). The test results shown in Tab. 7.28 resulted from a test made at the Herndon Test Labs in Houston on a material known as SILON Q-177.

(*a*) Plot the data on semilog graph paper.
(*b*) Using the method of least squares, find the equation of the line of best fit.
(*c*) Draw the line on the graph.
(*d*) Calculate the coefficient of correlation.
(*e*) What pressure should be applied to achieve a relative compression of 60 percent?

7.25 The rate of absorption of radiation by metal plates varies with the plate thickness and the nature of the source of radiation. A test was made at Ames Lab on October 1, 1994, using a Geiger counter and a constant source of radiation; the results are shown in Tab. 7.29.

(*a*) Plot the data on semilog graph paper.
(*b*) Using the method of least squares, find the equation of the line of best fit.
(*c*) Draw the line on the graph.
(*d*) Calculate the coefficients of correlation and determination.
(*e*) What minimum plot thickness is required to assure a radiation level below 1 000 counts per second?

Table 7.29

Plate thickness W, mm	Geiger counter C, counts per second
0.20	5 500
5.00	3 720
10.00	2 550
20.00	1 320
27.50	720
32.50	480

7.26 Survey at least 15 students (U.S. citizens) in your housing unit regarding the number of long-distance calls per year in which they have been involved, either as the caller or the receiver, and the number of miles from their homes to campus.

(*a*) Plot the data on the type of graph paper that appears to give the nearest to a straight-line relationship.
(*b*) Using the method of least squares, calculate the line of best fit.
(*c*) What is the correlation coefficient? What does it mean?

Perform the computations and plot the graphs either by hand or with a spreadsheet at your instructor's direction.

7.27 The annual low flows (lowest 1-day flow rate within a calendar year) of two tributaries of a river are expected to be linearly related. Data for a 12-year period are shown in Tab. 7.30. Use a spreadsheet to do the following:

(*a*) Compute the best-fit line using the method of least squares. Use data for the north tributary as the independent variable.

(*b*) Plot the data and the prediction equation on the same graph.
(*c*) Compute the correlation coefficient and interpret it.
(*d*) If the low flow for the north tributary is 200 cfs, what would you expect the low flow for the south tributary to be?

Table 7.30

Year	North tributary flow (cfs)	South tributary flow (cfs)
1	222	222
2	354	315
3	201	174
4	372	402
5	246	204
6	324	324
7	216	189
8	195	279
9	195	210
10	264	255
11	276	243
12	183	174

7.28 The data in Tab. 7.31 represent the flow through a channel as a function of the height of the water in the channel. Use a spreadsheet to accomplish the following:

(*a*) Determine whether a linear, exponential, or power function best represents the data.
(*b*) Using the method of least squares, find the best-fit equation.
(*c*) Compute and interpret the coefficient of determination.
(*d*) Plot the data and optimum curve on the same graph.
(*e*) Based on your calculations, how high would the water be if the discharge reached 500 cfs?

Table 7.31

Water height h, ft	Discharge Q, cfs
1.7	2.6
1.96	3.5
2.6	4.04
2.92	6.44
4.04	11.2
5.24	30.61
5.88	35.92
6.12	40.25
6.8	48.92
8.54	73.24
9.24	89.7
10.52	98.61
12.92	130.52
14.88	202.5
17.34	225.2
20.18	298.66

8

Total Quality: The First Step

8.1 Introduction

As a beginning engineering student busy with a variety of coursework and extracurricular activities you may not have the time or opportunity during your first year in engineering to experience a complete total quality process. In fact, it is not our intent in this introductory chapter to provide a complete detailed description of the entire process with sufficient information together with the necessary training to carry out a full-scale project. Rather the objective of this lesson is to provide an abbreviated overview that accomplishes these purposes:

1. To present a brief review of the background and history of Total Quality.
2. To develop an appreciation for the importance of Total Quality and its role in making the United States internationally competitive.
3. To provide an introduction to many definitions, unique terminology, and elements that apply to total quality.
4. To explore two important aspects of Total Quality that relate to other lower-division studies: problem-solving models and the team process.

We believe with this background you will better understand how Total Quality is destined to play an important role in your life both as an engineering student and eventually as a professional engineer.

Total Quality (TQ), Total Quality Management (TQM), and Continuous Quality Improvement (CQI) are three of the many titles given to the central concept of organizational behavior and management. In this chapter we will use the term *Total Quality,* or TQ, but it is the philosophy, not the title, that is fundamental.

As a student in engineering there are several ways this idea will impact your immediate life. As you read this chapter and begin the process of understanding TQ, it is likely that this subject area has been or is currently being integrated into your undergraduate curricular activities. This introduction to TQ could be followed with additional coursework and application or team involvement in a variety of other areas. A second way it may affect you is by virtue of the fact that you are a "customer" of some unit or agency that is striving to provide significantly improved customer satisfaction. A third way you could be affected would be to do some additional individual reading and apply total quality principles to your personal life, habits, and actions. One such reference for consideration would be *Quality Is Personal* by H. V. Roberts and B. F. Sergesketter.[1]

One ancillary objective or outcome of this introductory material is to help you understand that TQ is a subject that you can easily read and understand. It is not a complex or difficult subject to comprehend. If, however, you intend to effectively practice TQ, a totally different approach is required. A commitment to TQ demands a modification in the way you approach problem solving as an engineer; it involves a completely new and different way of thinking—a philosophically different attitude toward quality.

8.2 Background and History

The information summarized in Tab. 8.1 contains selected writings and authors who have made significant contributions in the last 100 years. Other references are contained in footnotes throughout the chapter.

Table 8.1 The Origin and Evolution of Total Quality

Date	Subject areas	Author/reference
1911	Principles of Scientific Research	Frederick W. Taylor
1924	Control Charts	W. A. Shewhart
1927	Human Relations Influence on Productivity	Hawthorne Research
1947	Administrative Behavior: Systems Theory	Herbert Simon
1950	Japanese Quality Movement	W. Edwards Deming
1951	Change Model	Kurt Lewin
1954	Quality Control Management in Japan	J. M. Juran
1962	Quality Circles	Kaoru Ishikawa
1962	Paradigms	Thomas S. Kuhn
1966	Open Systems Theory	Herman Kahn
1968	Cost of Quality	Armand Feigenbaum
1979	*Quality Is Free*	Philip Crosby
1980	If Japan Can . . . Why Can't We?	NBC Documentary
1982	*In Search of Excellence*	Tom Peters
1982	*Out of the Crisis*	W. Edwards Deming
1985	Organizational Culture	Edgar Schein
1985	Leadership	Warren Bennis and Burt Nanus
1985	Strategic Thinking and Visioning	John W. Zimmerman
1986	*The Transformational Leader*	Noel M. Tichy
1987	Empowerment	Peter Block
1988	National Quality Award	Malcolm Baldrige
1990	International Quality Requirements	ISO 1900
1990	*Quality or Else*	Lloyd Dobyns
1990	*The Fifth Discipline*	Peter M. Senge

1. H. V. Roberts and B. F. Sergesketter, *Quality Is Personal,* New York: The Free Press, 1993.

The amount of written material about TQ has significantly expanded in the past few years. From 1990 to the present there have been many books and numerous papers written on the subject. Only two are referenced here because they apply directly to the impact of TQ in higher education.

1992	*On Q: Causing Quality in Higher Education*	Daniel Seymour
1995	*Malcolm Baldrige National Quality Award*	

In the late 1970s and early 1980s international competition for market share was becoming a way of life—the United States was being out-performed by foreign companies in too many areas. Perhaps a simple way to illustrate this is by a brief example.

In the 1970s Japanese companies began to take market shares from many U.S. companies. David Kearn, who was at that time chairman of Xerox, tells it like this:[2]

> Whereas most American corporations were advancing 2 or 3 percent a year in productivity, we were achieving gains of 7 or 8 percent. But despite these gains, the Japanese continued to price their products substantially below us. We kept wondering: how were they doing it? Our team went over everything in a thorough manner. It examined all the ingredients of cost: turnover, design time, engineering changes, manufacturing defects, overhead ratios, inventory, how many people worked for a foreman, and so forth. When it got done with its calibrations, we were in quite a shock.

From this same reference another Xerox executive, Frank Pipp, remembers the results of the review as follows:

> Absolutely nauseating, it wasn't a case of being out in left field, we weren't even playing the same game.

The Xerox experience was, in fact, an example of what was happening with great frequency in the 1970s and 1980s. For corporate America this was a wake-up call. The manufacturing and service industries, long thought to be the envy of the entire planet, were no longer number one and in many cases were not even competitive.

American industry became justifiably alarmed by the lack of competitiveness of the United States in the global marketplace. In the eight critical industries shown in Fig. 8.1, the United States currently has a positive balance of trade in only two. The other six industries represent areas where the United States was once dominant, but subsequently foreign competition has eroded the U.S. position.

---

2. David T. Kearnes and David A. Nadler, *Prophets in the Dark: How Xerox Reinvented Itself and Beat Back the Japanese,* New York: Harper Business, 1992.

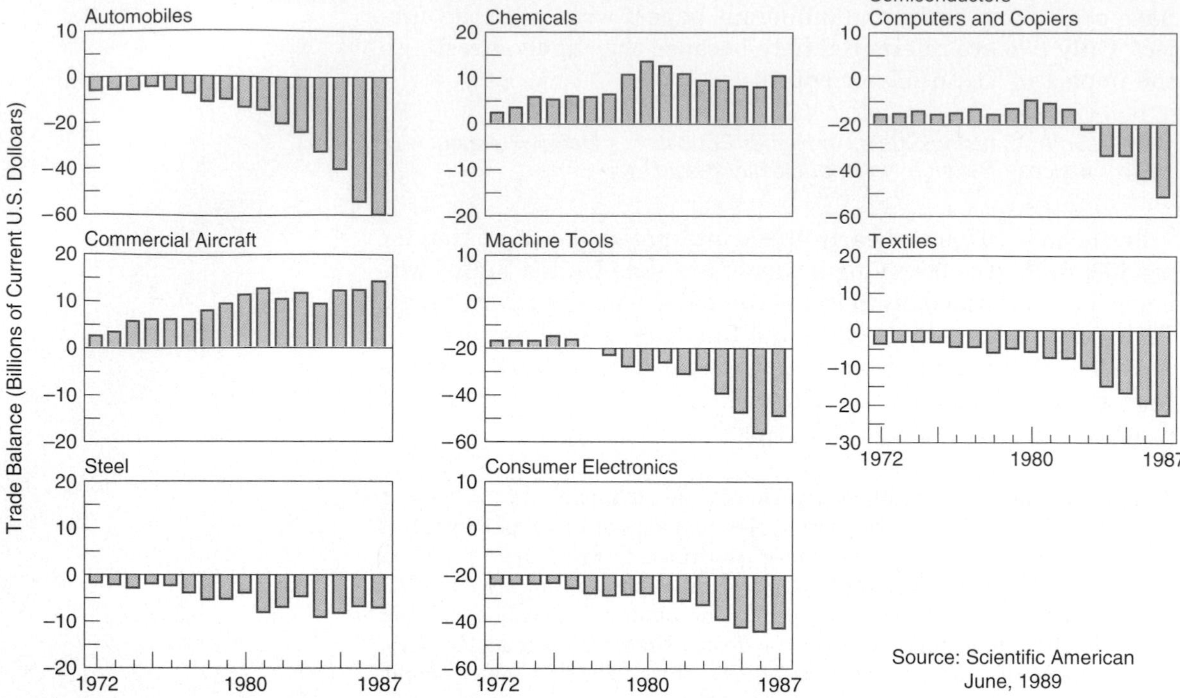

Figure 8.1
U.S. Balance of Trade in Key Industries

To better understand how this could happen, let's take a brief look at our history and how certain events allowed this country to lose global competitiveness in a number of areas.

Frederick Taylor (1856–1917)

Taylor is generally considered to be the father of modern time study in this country. His thoughts and ideas suggested that the work of each employee be planned by management and that each worker receive complete written instructions describing the task to be done in detail. Each job was to have a standard time for completion, which was to be fixed after time studies had been done by experts. In the timing process, Taylor advocated breaking up the work assignment into small dimensions of effort known as "elements." These were timed individually and their collective values used to determine the allowed time of a given task.

It is important to note that in the late 1800s this approach was needed because the work force was largely untrained and uneducated. Taylor asked the following specific question: "What is the best way to do a job?" He did not accept opinion; he demanded facts and evidence. As a mechanical engineer, he was the first to apply the principle of the scientific approach to the application of improving jobs. At that time he stated, "Any change the worker makes to the plan is fatal to success." The Taylor approach was a system that separated planning from execution but also tended to separate management from workers. Its major premise was that the supervisor and the worker lacked the education needed to plan how work should be done; hence planning was turned over to management and engineering.

Today this approach seems contrary to the TQ concepts, however we need to remember that in the late 1800s and early 1900s this approach significantly increased efficiency and productivity.

W. Edwards Deming

As a young man Deming worked for Walter Shewhart, who was the creator of statistical process control and the inventor of the control chart. In the 1940s Deming worked for the U.S. Bureau of the Census where he applied quality principles and taught courses in statistical process control. Therefore, his early contribution was as a statistician. He taught the value of statistics as a tool for measuring the output of a process within a system. He believed that organizations are systems with processes. What managers do is manipulate processes, consciously or unconsciously, skillfully or erratically, to get work done.

Deming had an enormous influence on the start and subsequent progress of the "quality revolution." His methods of measurement and process control resulted in significant improvement to the quality of American products prior to World War II. Following the war, the U.S. focus shifted from quality to quantity to meet postwar demands. Meanwhile Deming was invited to Japan to help rebuild their infrastructure. Deming, along with another American, Joseph M. Juran, was very influential in the 1950s, starting the quality improvement movement in Japan. Much of Deming's management philosophy is contained in what he calls his "14 points." Over the course of many years, both through practice and teaching, Deming modified and improved this list of 14 points. He believed these ideas to be an essential way of life for corporate America.

Deming's 14 points are listed in Tab. 8.2 because they are an important part of the TQ evolution. However, without considerable explanation they are difficult to totally understand. For you as a student they are included as an important reference.[3]

Philip B. Crosby

Crosby is chief executive of PCA, Inc., a quality management consulting and teaching firm he founded. Previously he was corporate vice president and director of quality for the ITT Corporation,

3. Deming, W. Edwards (1986), *Out of the Crisis,* MIT Center for Advanced Engineering Study, Cambridge, MA.

Table 8.2 W. Edwards Deming's 14 points

1. Create constancy of purpose for the improvement of product and service.
2. Adopt the new philosophy. We are in a new economic age. Poor workmanship and sullen service are no longer tolerable. Stay in business and provide jobs through innovation, research, and constant improvement.
3. Use modern science. Cease dependence on inspection to achieve quality. Quality comes not from inspection but from improvement of the process. Build quality into the product/process.
4. End the practice of awarding business on the basis of price tag. Keep total costs in mind, not partial costs. Seek the best quality, not the cheapest price.
5. Improve constantly and forever the system of production and service, to improve quality and productivity, and thus constantly decrease costs. Find problems. Management must work continually on the system.
6. Never stop training on the job.
7. Improve supervision of workers and managers. Institute leadership. The aim of leadership should be to help people, machines, and gadgets to do a better job.
8. Drive out fear so everyone may work effectively for the organization.
9. Break down barriers between departments. An organization must work as a team.
10. Eliminate slogans, exhortations, and targets for workers asking for better quality and productivity.
11. Eliminate work standards that have only numerical quotas. Quotas take account only of numbers, not quality or methods. Eliminate management by objectives. Eliminate management by numbers and numerical goals. Substitute leadership.
12. Remove barriers that stand between workers and their right to pride in workmanship.
13. Institute a vigorous program of education and self-improvement.
14. Put everybody in the organization to work to accomplish the transformation. The transformation is everybody's job.

responsible for worldwide quality operations. He is best known as the creator of the "zero defects" concept. Crosby explains zero defects as follows:

> Some American companies were in the habit of spending 15–20 percent of every sales dollar on reworking, scrapping, repeating, inspecting, testing, and other quality-related costs. This results from the traditional idea that quality control happens only on the manufacturing line. The idea that "quality is free" is meant to suggest that whatever you save from the fact that "unquality" costs money therefore increases your profit.

Crosby also developed 14 foundation blocks that he believes must be followed.[4] He stated that each step is complex and that no step can be eliminated. His 14 steps of quality improvement are included in Table 8.3.

---

4. Philip B. Crosby, *Quality Is Free,* McGraw-Hill Book Company, Copyright c 1979.

Table 8.3 Philip B. Crosby's 14 Steps of Quality Improvement

1. Management commitment: Management must be committed. They must be personally committed and provide leadership by example, in other words, "Walk the talk."
2. Quality improvement teams: Collect the correct individuals into a quality improvement team. Must be properly engaged.
3. Quality measurements: Quality measurements and current status are necessary to determine where corrective action is necessary and to document actual improvement.
4. Cost of quality evaluation: The cost of quality is not an absolute performance measure, it is an indication of where corrective action will be profitable.
5. Quality awareness: The process of sharing with employees the measurements and results of the amount nonquality is costing. Getting supervisors and employees in the habit of talking positively about quality.
6. Corrective action: An opportunity for workers to contribute to the solution set.
7. Establish an ad hoc committee for zero defects: The idea is that things should be done correctly the first time. Defects cost, and zero defects provide a better product with less cost.
8. Supervisor training: Training in total quality concepts and process is critical. Everyone must clearly understand the ideas and philosophy.
9. Zero defects day (ZD): Establishment of ZD as a performance standard of the company should be done in one day. New company attitude.
10. Goal setting: Supervisors should ask employees to establish 30, 60, and 90-day goals.
11. Error cause removal: Describe any problem that keeps them from performing error-free work.
12. Recognition: Award programs are established to recognize those who meet their goals.
13. Quality councils: Quality professionals and the team chairs should be brought together to exchange, learn, and redefine the quality program.
14. Do it over again: A typical quality team project takes from 12 to 18 months to complete. During that length of time many company employees may be transferred, leave, and so on—so DO IT AGAIN.

National Quality Awards

National quality awards in this country are a relatively new phenomenon. The Deming Prize, in honor of W. Edwards Deming, was established in 1950 by Japan and has generally become identified with the first country to recognize TQ as a national priority. Several countries have subsequently established national awards, including the United States.

On August 20, 1987, President Reagan signed documentation establishing the Malcolm Baldrige National Quality Award as a national law. The award was named after Malcolm Baldrige, secretary of commerce.

The first presentation of these awards was made by President Reagan on November 14, 1988. Up to two awards in each of three categories can be presented each year:

- manufacturing companies
- service companies
- small businesses

In 1988 the first three companies to receive this award were Motorola Inc; Commercial Nuclear Fuel, a Division of Westinghouse; and Globe Metallurgical Inc.

The award focuses on seven rigorous categories:

1. Leadership
2. Information and analysis
3. Strategy quality planning
4. Human resource development and management
5. Management of process quality
6. Quality and operational results
7. Customer focus and satisfaction

Total Quality Forum

In late 1989, David Kearns, chief executive officer of Xerox, convened several senior executives from other businesses, together with faculty and deans representing 20 schools of business into the first TQ forum. He believed that an understanding of TQ was an essential skill for industrial managers. In subsequent forums, colleges of engineering and education were invited.

Kearns believed that universities must be encouraged to teach students from within the disciples the concept and principles of TQ.

Joseph M. Juran

On February 23, 1990, Joseph M. Juran, a pioneer in the field of quality improvement, gave the summary address at "The Quest for Excellence," an executive conference featuring the 1989 winners of the Malcolm Baldrige National Quality Award. A portion of his remarks follow including his reference to the "Taylor system":

> A further observation on lessons learned relates to the use of human resources. Most of our companies still retain a considerable residue of the Taylor system of separating planning from execution. That system evolved about a century ago. Its major premise was that the supervisors and workers of that era lacked the education needed to plan how work should be done. Hence the planning was turned over to engineers. The supervisors and workers were left with the responsibility of carrying out the plans.
>
> During our century the education levels of the work force have increased sharply, and thereby destroyed the major premise of the Taylor system. The system has become obsolete, and it should be replaced by something else. However, despite much recent experimentation we are still not agreed on what should be that something else. Meanwhile we are failing to make use of our biggest underemployed asset—the education, experience, and creativity of the work force.
>
> The report we have heard at this conference included reference to use of self-managing teams. If I may be permitted another look into

> my fallible crystal ball, I suggest that self-managing worker teams will become the dominant successor to the Taylor system.

In the mid 1970s and early 1980s it was becoming apparent to much of corporate America that change, in fact immediate change, was essential. The excerpt from a speech presented by a Japanese CEO underscores the need for serious reexamination of the American industrial culture.

Konosuke Matsushita

Matsushita is the founder and executive advisor for Matsushita Electric Industrial Company. His company owns Panasonic, Universal Studios, and a host of other organizations. Speaking to an American audience in the late 1980s, Matsushita delivered a scathing speech on American business:

> We will win and you will lose. You cannot do anything about it because your failure is an internal disease. Your companies are based on Taylor's principles. Worse, your heads are Taylorized too. You firmly believe that sound management means executives on one side and workers on the other, on one side men who think and on the other side men who can only work. For you, management is the art of smoothly transferring the executives' ideas to the workers' hands.
>
> We have passed the Taylor stage. We are aware that business has become terribly complex. Survival is very uncertain in an environment filled with risk, the unexpected, and competition. Therefore, a company must have the commitment of the minds of all of its employees to survive. For us, management is the entire work force's intellectual commitment at the service of the company . . . without self-imposed functional or class barriers.
>
> We have measured—better than you—the new technological and economic challenges. We know that the intelligence of a few technocrats—even very bright ones—has become totally inadequate to face these challenges. Only the intellects of all employees can permit a company to live with the ups and downs and the requirements of its new environment. Yes, we will win and you will lose. For you are not able to rid your minds of the obsolete Taylorisms that we never had.

With this extremely brief history you can begin to see the situation that the United States found itself in during the early 1980s. One large American industry that decided to do something about declining market shares was Texas Instruments. Jerry R. Junkins, CEO of Texas Instruments, presented a summary of the actions taken by that company at a conference in Des Moines, Iowa, on February 25, 1993. The complete texts of Junkins' speech is presented in Sec. 8.5.

8.3 Definitions and Tools

TQ is a process, and any process is made up of its own unique array of terminology and definitions. This section provides an overview of the essential concepts associated with TQ.

8.3.1 Definition

A simplistic definition of TQ could be stated as:

> Meeting or exceeding the expectations of the customer.
>
> Other more comprehensive definitions follow.[5]

The Report of the Total Quality forum

> Total Quality is a people-focused management system that aims at continual increase of customer satisfaction at continually lower real cost. TQ is a total system approach, an integral part of organizational high-level strategy. It expands horizontally across functions and departments, it involves all employees—top to bottom—and extends backwards and forwards to include the supply chain and the customer chain. TQ stresses learning and adaptation to continual change as keys to organizational success.

Thomas Johnson

> The key feature of this paradigm is the idea that work is a process and that business is a system of processes that is aimed at exceeding customer expectations profitably.

Carla O'Dell

> A process is defined as an activity that is definable, repeatable, measurable, and predictable. Process management is a methodology that attempts continually to increase effectiveness and efficiency of processes.
>
> The process of identifying an organization's functions and breaking these functions into finer and finer subfunctions is called function analysis or function deployment.
>
> Processes are definable, predictable, repeatable actions that can be flow-charted. Processes are particular actions that deliver functions. For example, a key function of every organization is "to understand markets and customers"; hence every organization needs a process by which this is accomplished. From industry to industry and organization to organization, there are many different processes for delivering a given function.

Don Clausing

> Management by fact applied by teams to improve processes (as perceived by customers) with high expectations.

A common expression often heard when discussing Total Quality is that it begins with education and ends with education. Learning the process and system whether as an individual or as part of a team takes much effort, time, and commitment. This section introduces some of the elements and statistical tools used in the process.

---

5. *A Report of the Total Quality Leadership Steering Committee and Working Councils*, November, 1992, published by John K. Lowe Company, Cincinnati, Ohio.

8.3.2 Elements of Total Quality

Some of the critical elements of TQ that must be present or included in the process are discussed in the following sections.

Top-Level Management Commitment and Involvement

It is completely apparent to everyone if top management or the central administrative team of an organization is all "talk" and no "do." Commitment must be real and it must be visible. Jerry Junkins, CEO of Texas Instruments, stated that "Top management must be convinced. The idea that this concept of Total Quality is necessary for you but I'm already perfect does not work. Top administration must 'walk the talk.'"

Continuous Incremental Improvement

A statement overheard from an industrial CEO (to remain nameless) was as follows:

> "One of the problems that many organizations unfortunately face is that they cruise along in this state of uninformed euphoria and do not notice (or care) how out of touch or how bad things are becoming. One day the ax falls." When this is the case the resulting action normally means sweeping or radical change. No one likes or enjoys this outcome. Instead of experiencing the big changes, organizations should make hundreds of small improvements. People normally embrace and respect incremental change, particularly if employees are involved.

Focus on Process

A process is the steps that depict how the system achieves its objectives. To illustrate the concept of process we often use flow charts where a box represents a single step in that process. W. Edwards Deming discussed process as follows: "Every activity, every job is a process. A flow diagram of any process will divide the work into stages. The stages as a whole form a process. Each stage works with the next stage and with the preceding stage toward optimum accommodation."[6]

All organizations are composed of a multitude of processes (they can be stated and documented or unstated and undocumented) to get work done. Every organization that delivers a product or a service goes through a series of steps to produce something of value to a customer. This is why a focus on process is central. Organizational managers should carefully analyze the processes in their organization to understand how the work gets done.

The concept *kaizen* (improvement) is at the core of the way the Japanese manage their organizations. This philosophy has gen-

6. Deming, *Out of the Crisis,* 1986.

erated a process-oriented way of thinking, and they have developed strategies that assure continuous improvement through the involvement of people at all levels within the organizational hierarchy.

Think of any organization as the place where countless tasks are accomplished. Perhaps it is putting addresses on letters, tightening a machine screw, or removing burrs from a shearing process.

A process can be defined as each of the individual tasks grouped into a sequence necessary for a unique outcome. Every organization has thousands of these processes. If you can improve one or many of these processes, you can improve the outcome, which means better quality. People who view work as a process understand how the quality of what comes out is a function of the quality that goes in. Thus the need to review the product received and to insist that suppliers provide quality input.

Many times the TQ team will try to tackle an entire system that produces a product. Systems are composed of many processes. The production of a product involves thousands of interrelated processes, and it may be necessary for the team to narrow or focus on a specific portion.

Employee Involvement

Employees in today's organizational work force are educated, have job experience, are creative, and know their particular task better than others. They have ideas about how to improve the process if asked. Most employees are willing, perhaps eager, to share ideas. Management does not have all the answers. Management must work with employees to identify areas that need improvement, and then get the work force involved in providing suggestions for improvement. Once the people are engaged in the process, they are more committed to its implementation (their ideas) and to its success.

The Customer

Defining exactly who is the customer can often be complex. For example, if you are within an organization looking at the entire system, the customer may be defined as the individual that purchases a product. Most quality teams within an organization are only looking at a single process in a series of events within the larger system. Who then is their customer? How does a given customer define quality? Is it defined the same by all customers?

Consider the suppliers of materials as people in the organization who precede you in a series of tasks, and the customer as those who receive the product or service. If customers are the people who receive your work, only they can tell you what they expect in terms of quality.

Competitiveness

The late Kaoru Ishikawa, one of Japan's most distinguished quality gurus, believed that "95 percent of the quality problems in the workplace could be solved using what are known as the Seven Quality Control Tools." These seven tools establish the basis of the statistical quality control process which is a major component of TQ.

1. Check sheet/checklist: makes it easy to collect and use data.
2. Pareto diagram: helps to find the "vital few" causes that create the majority of the problems.
3. Histogram: describes the manner of dispersion of data.
4. Cause-and-effect diagram (Ishikawa diagram): arranges the cause-and-effect relationship.
5. Stratification: helps to find differences between data.
6. Scatter diagram: shows correlation between two variables.
7. Graph and control chart: helps to find a specific feature behind the data and to analyze and control the process.

Data

The TQ process and effective team results rely heavily on data. The eventual ability of the team to analyze and synthesize a solution requires a solid understanding of process together with an ability to visualize. These elements rely on a variety of tools.

Diagrams	Charts
Activity network diagram	Gantt chart
Affinity diagram	Control chart
Cause-and-effect diagram	Flow chart
Matrix diagram	Pareto chart
Scatter diagram	Radar chart
Tree diagram	Run chart

Application

As you have an opportunity to be selected as a TQ team member, either in school or after graduation, you will need to explore many of these tools learning their unique and special applications.

The Pareto Chart

The Pareto chart displays the relative importance of one item to another in a simple, visual format. Once the major problem has been identified and significant improvement realized, the team can pursue other problems in the quest for continuous quality improvement. The Pareto chart is based on the 20–80 percent rule; that is, 20 percent of the problem sources cause 80 percent of the problems. The key or fundamental reason for using a Pareto chart is to focus efforts of the team on the single problem that offers the greatest potential for improvement.

Table 8.4 Engineering Retention Survey Results

Problem area	Frequency	Percent (%)
Coursework related	160	38
Uncertainty as to "what is engineering?"	92	22
Motivation (attitude)	55	13
Personal problems	38	9
Poor pre-college preparation	21	5
Poor instruction	13	3
Poor advising	8	2
Others	33	8
	420	100

An example will illustrate how the Pareto chart works. The data in Tab. 8.4 were taken from an actual TQ team organized at Iowa State University to look at the issue of freshman engineering retention. The team conducted personal interviews with 50 students and 10 faculty advisors.

Once the data are collected, the next step is to create the Pareto chart. Each of the categories is listed in descending order from left to right on the horizontal axis (see Fig. 8.2). Next establish bars above each problem area corresponding in height to the frequency of occurrence on the y-axes. Last, construct the cumulative percentage line showing the portion of the total that each area represents.

Again, the tallest bar graphically illustrates the single largest contribution to the overall problem. It may be that the team can have absolutely zero impact on that particular problem area and

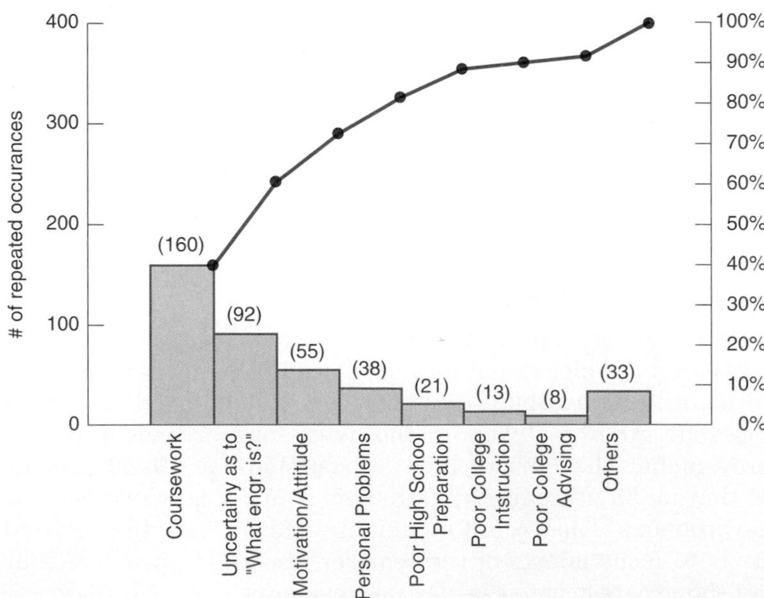

Figure 8.2 Example of a Pareto Chart.

will select the second or third problem area for action. The chart, however, guides the team toward solutions that will have a major impact on the problem.

Cause-and-Effect Diagram

This tool, often called the Ishikawa fishbone diagram, allows a team to identify and graphically display, in increasing detail, many of the possible "causes" related to a specific "effect" or condition.

To continue the retention example, the number one problem area identified was coursework. First, the team modified this concise problem statement to allow as much flexibility as possible. They agreed that the following statement summarized the problem area: "Students are not successful in first-year coursework."

What are the "causes" that produce this "effect"? The effect is placed in a box on the right-hand side of the diagram (see Fig. 8.3). Major causes are then developed and placed on the basic bone of the chart. There is no set number of categories.

Figure 8.4 is the result of a "brainstorming" session the team conducted using data they had collected as well as their personal experiences.

The Team

Why would we elect to form a team instead of having a talented individual solve a particular problem? Individuals can do great things, but rarely does a single person have the knowledge and experience needed to consider all elements in a process. Major gains in quality and productivity can be derived from a group of people pooling their talents and skills to tackle a complex problem.

As with most issues there are advantages and disadvantages to team versus individual effort. One disadvantage is that teams require time, patience, and the learning of new skills to deal effectively in a team environment, but there are advantages once a team is working together effectively. For example the mutual support received from other team members while working on a project that may take considerable time and effort.

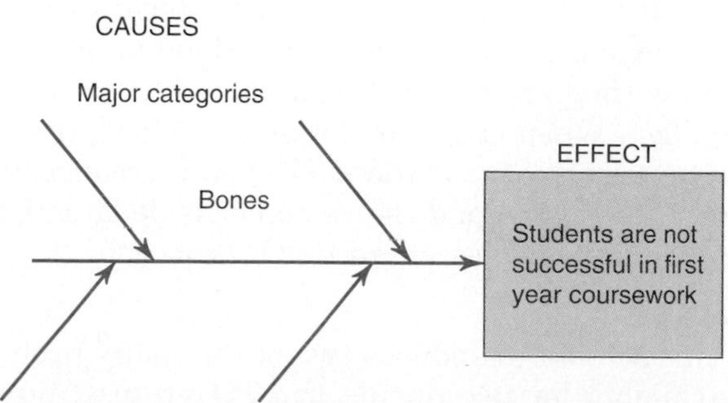

Figure 8.3
Cause-and-Effect, or Fishbone, Diagram Structure.

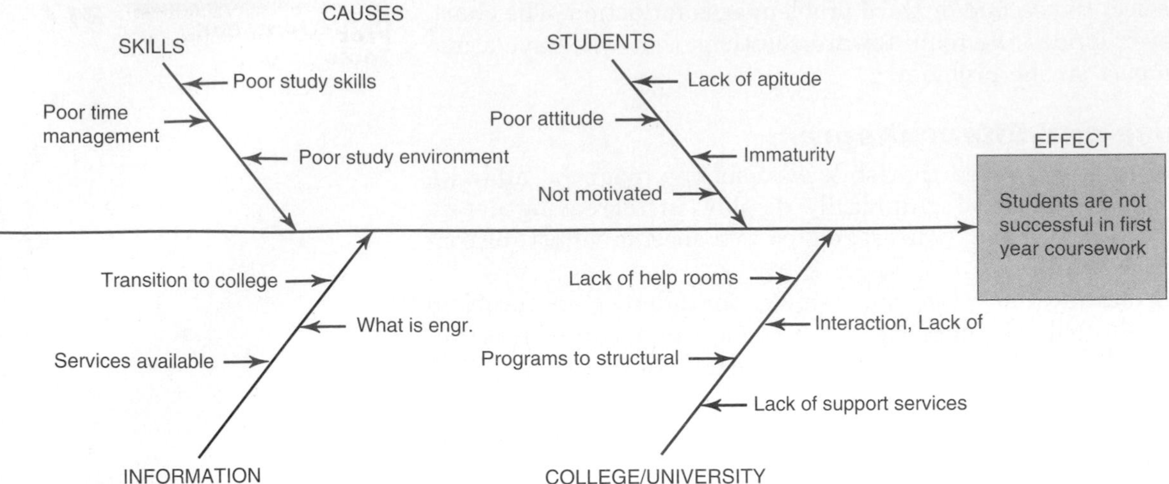

Figure 8.4
Example of Fishbone Diagram.

Scientific Approach

The central theme of quality improvement methods can be summed up in two words: scientific approach. Put simply this means that there are systematic ways for individuals and teams to learn about a process. It means making decisions based on facts derived from data rather than a hunch. It means to search for root causes of problems. A famous TQ expert stated during a TQ meeting: "In God we trust, others must have data."

Culture Change

Culture change involves a modification in the fundamental way we think and act, and it does not happen easily or quickly. The TQ process demands a change of culture within each individual and therefore throughout an organization, it means that the organization collectively thinks differently, reacts spontaneously. It means that the organization supports a totally different approach to the solution of problems, one that focuses on quality. It means every person is committed to quality. All events consist of processes, and all processes can be improved. Zero defects really means, do it right the first time. Total Quality means understanding customer needs and exceeding those expectations. It means never accepting conformity and always understanding that continuous incremental improvement is possible. It means all employees must be involved because they are the organization, and they must base all decisions on fact. When every employee is involved, the organizational environment is positive and healthy, the product or service is continually improving, and the customer is delighted, it is then the organization has experienced the TQ concept.

8.4 Problem Solving and the Team

This section introduces two of the many problem-solving models available for use during the TQ process, but the material presented will focus primarily on the team and not the model. We will present ideas about working in a group environment and include a look at personal style inventory.

8.4.1 Problem-Solving Models

A clearly defined problem-solving model in the TQ arena accomplishes three things:

1. It helps all team members understand the current situation.
2. It provides a detailed assessment of customer requirements.
3. It drives solutions by facts. The team relies on data analysis and measurement to justify selection of solutions.

Peter R. Scholtes in *The Team Handbook*[7] has developed a series of strategies or steps to follow. By using these strategies you can design a plan. He outlines the following 14 improvement strategies:

A. Strategies of the scientific approach

1. Collect meaningful data.
2. Identify root causes of problems.
3. Develop appropriate solutions.
4. Plan and make changes.

B. Strategies for identifying improvement needs

5. Identify customer needs and concerns.
6. Study the use of time.
7. Localize recurring problems.

C. Strategies for improving a process

8. Describe a process.
9. Develop a standard process.
10. Error-proof a process.
11. Streamline a process.
12. Reduce sources of variation.
13. Bring a process under statistical control.
14. Improve the design of a product or process.

Although these steps appear quite simple, most are more complex than they first appear and frequently lead to other meaningful events. For example, strategy 3, develop appropriate solutions, involves a 7-step process that is in fact very close to the traditional design process that engineers follow in the solution of open-ended problems.

---

7. Peter R. Scholtes, *The Team Handbook,* Madison, WI, Joiner Associates Inc., 1992.

Strategy 3: Develop Appropriate Solutions	Engineering Design Process
1. Describe the need.	1. Identify the need.
2. Define goals and criteria.	2. Define the problem.
3. Identify constraints.	3. Search.
4. Generate alternatives.	4. Criteria.
5. Evaluate alternatives.	5. Constraints.
6. Select best solution.	6. Alternative solutions.
7. Follow up.	7. Analysis.
	8. Decision.
	9. Specification.
	10. Communication.

Open-ended problems can be divided into two broad categories: the problem and the solution. A problem is a situation or condition that needs to be changed or improved. Problem solving on the other hand is the action taken to change the existing situation. Interactive problem solving is a process that allows people to work together productively to address issues and create opportunities. Most often the desired result is a consensus decision. Familiarity with open-ended problem-solving strategies is essential for individuals and teams to work well together, resolve issues, and make quality decisions.

Problem Space

In the beginning the team must develop a mutual understanding of all aspects of the problem. They must agree on a problem definition and identify root causes so that the eventual solutions can address these important causes rather than just symptoms. In other words, the team must agree on what the problem really is before beginning to work toward a solution. This will involve considerable discussion and a search for needed information. Proper identification of the problem is a critical step. The diamond shape shown in Fig. 8.5 represents three necessary phases that a team must move through as the problem is completely defined. Starting at location (1) the team must learn as much as possible about the problem. Phase A represents a large increase in the amount of information. Phase B begins the process of sorting and organizing information. It introduces a range of problem statements. After considerable discussion, agreement by the team is reached in phase C. Once the group has collectively decided on the best problem statement, location (2), they are ready to move into solution space.

Solution Space

Figure 8.6 illustrates the process of solution generation. A number of techniques are available to assist a team during this important phase. One common method is called "brainstorming." The diagram depicts the process of expanding solution space to include as many ideas as possible. Then the process of narrowing must

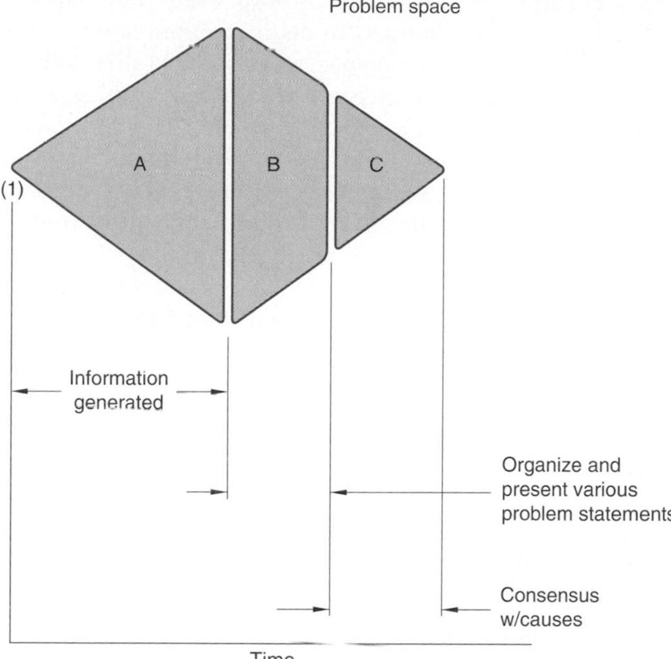

Figure 8.5
Visual diagram of problem space.

take place. This involves both evaluation of ideas against a predetermined list of criteria and the process of decision making.

The important point to be understood is that there are many adaptations and variations of the scientific method of problem solv-

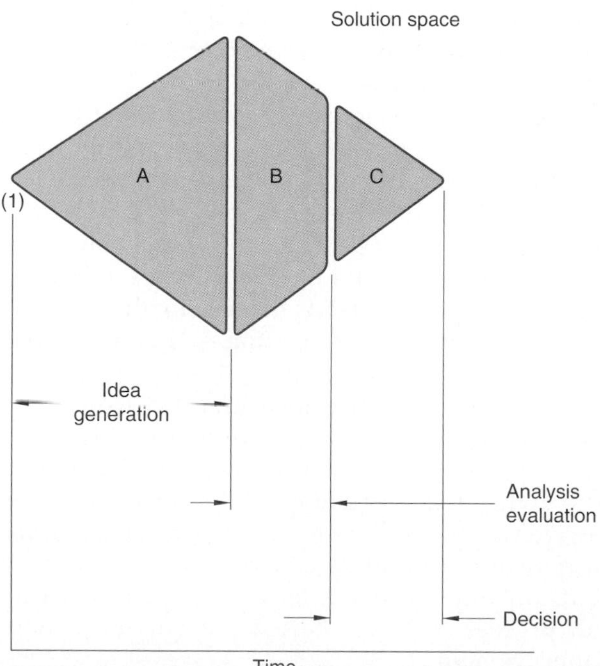

Figure 8.6
Visual diagram of solution space.

ing. Depending on the situation and what the goals or objectives of the team project happen to be, a problem-solving model can be or already has been developed to assist and direct this effort.

The undergraduate engineering educational process provides students with many fundamental engineering science courses that are essential to the professional engineer. However, the educational mission also includes practicing the process—that is, learning by application. It is through repetition that we develop optimum open-ended problem solving skills.

A different process or series of steps is suggested by a problem-solving model developed at Oregon State University. Their approach to the problem-solving process suggests the following steps for the team:

1. Interview customers.
2. Select issues and develop performance measures.
3. Diagram the process.
4. Diagram cause-and-effects performance measures.
5. Collect and analyze data on causes.
6. Develop solutions.
7. Benchmark.
8. Select and implement solutions.
9. Measure results and refine.

Each of these nine steps contains additional menus of items that need to be considered. For example, step 8 (select and implement solutions) recommends an additional process: plan-do-check-act. This process—called the P-D-C-A cycle—was used years ago by Walter Shewhart and is sometimes referred to as the Shewhart cycle.

Most of this terminology and many of these steps have been used by engineers for years. How and where they can be applied to TQ is an ongoing learning experience. A considerable amount of trial-and-error discovery as well as targeted research is currently underway in the TQ movement. Much of the work will be learning to improve current practice, but the single biggest change needed is not in learning or applying the process but in "behavior modification"—that is, changing the way we think.

Too many organizations and individuals currently operate with a mentality that if there is a problem, fix it and move on to something else; if no problem exists, there is no problem to fix. Today, the attitude must be to constantly and continuously strive to improve processes even when no problem exists. If we are aware of problems, naturally start with the big problems, but continue to search for better ways to continuously improve all processes. Never be complacent or satisfied—always seek improvement.

8.4.2 The Team

You will spend a high percentage of both your professional and personal life interacting with people. Some of these interactions will be one-on-one, but many will involve groups of individuals either self-selected or members that have been assigned to committees or teams. For example, in the school of engineering you will likely be assigned as a member of a design team or you may be a team member in a collaborative learning situation. Design-team membership is typically assigned by the instructor, although occasionally team membership can be self-selecting. Design teams typically consist of 4 to 6 members and these individuals work together for the duration of the academic term. The team and its membership in the TQ process is more prescribed. The structure consists of a sponsor, a team leader, a facilitator, a recorder, and the individual members.

Sponsor

The sponsor is responsible for translating the "company" mission and vision into appropriate action. This individual would typically be a company manager, but in the academic environment it would be your instructor.

Team Leader

The team leader has responsibility for both the content (what is to be accomplished) and the process (how it is to be accomplished). Leaders must receive sponsor sign-off before moving from one step to the next in the problem-solving model. The team leader may run the meeting or may wish to participate as a group member and therefore ask the group facilitator to conduct the meeting.

Facilitator

The facilitator is normally an individual from outside the team (another unit or department). This individual focuses on process. The facilitator helps the team leader prepare for meetings and assists in the implementation of the problem-solving model. He or she is the meeting chauffeur, a servant to the group. The facilitator is neutral and nonevaluative.

Recorder

The recorder is selected from the team. This responsibility is normally rotated from one team member to another. It is this individual's responsibility to record the "group memory," and, like the facilitator, the recorder is a neutral, nonevaluative servant of the group. Recorders write down in full view of the team (perhaps on a large chart) enough detail so that ideas can be preserved and reviewed at a later time.

Team Members

Individual members of a TQ team are normally selected due to a particular area of responsibility or expertise. They contribute both information and ideas. Effective team members are supportive and constructive because one important aspect of TQ focuses on the strength of the team as a group.

The possibility of you as an individual engineering student being part of team-related activities is high, so you should learn as much as possible about team interaction. Working together as a team on a project provides a wide array of advantages; however, the team environment is complex. If the team mix—that is, personality, chemistry, background, and knowledge—is proper, a great deal of enthusiasm, esprit de corps, and "fun," not to mention progress, can be derived from the team experience. On the other hand, if you have no knowledge of team dynamics, or if you have no training in facilitative leadership, you may not be an effective member. As a unit you will more than likely waste time on struggles for control and endless discussion, with little or no meaningful progress.

The more you can learn about what to expect or what to anticipate as you are placed on a team, the better equipped you will be to handle difficult situations. You must be able to recognize and avoid disruptive behavior and learn to work through those situations that cannot be avoided.

Team Dynamics

Most college freshmen have worked with other students in various team environments on specific projects. This may have occurred in secondary school, sports, church, or during some other social activities. Typically, when you are placed together as a college team, it is likely that the members will consist of a nonhomogeneous mixture of personalities and cultural backgrounds. For example, a design team may consist of someone from your own hometown, another individual from a different part of the state, an international student, or perhaps an adult student. This is both good and bad. It is good if you understand that the best solution to problems will be found from within a mixture of opinions, backgrounds, experiences, ideas, and cultural diversity. It is bad when you do not understand and accept the fact that input from other members may contrast your own ideas, thought processes, and beliefs. You must learn to appreciate the talent and value of all team members and open your mind to possible approaches that do not parallel your background and experience.

The moment you are placed within a team environment a series of events take place that are predictable. The following is adapted from work by Bruce W. Tuckman,[8] who suggests that team growth develops in the following four unique stages.

---

8. Bruce W. Tuckman, "Development Sequence in Small Groups," *Psychological Bulletin* Vol. #15, 1965.

Stage 1: Forming. During this stage members are tentative, exploring the limits of acceptable group behavior. The forming period includes excitement and anticipation, a feeling of pride that they have been selected for this assignment. However, it also includes serious suspicion, fear, and anxiety about the project. Teams begin the process of defining the task at hand and decide how it might be accomplished. They begin an informal search for acceptable group behavior and ponder how to deal with group problems. It is likely that discussion will be tentative, suggesting abstract thoughts, concepts, and issues. Because members are cautious and tentative, because individuals are trying to size-up other members within the team, the initial or forming stage does not result in much progress. If this happens, it is perfectly normal.

Stage 2: Storming. Typically the team transitions from little noticeable progress to downright trouble. Storming is the most difficult stage. A swimming analogy provides a good example. Team members believe they have been tossed in the water and think they are going to drown. Members thrash around because they see the task as formidable. They are not at all sure how deep the water is. They slowly begin to form a global vision of the task as a team but still rely on their own personal and professional experiences, resisting the many advantages of collaborative team work. The storming stage includes a resistance to the introduction of new tools available because these new approaches suggest methods that are at odds with your own background and experiences. There is sharp disagreement among members with frequent hostility. During this stage members tend to be defensive and competitive. It is very unlikely that much time or energy is being devoted to problem solving. This stage is typical and perhaps even necessary for team dynamics to evolve. Not much work is accomplished, but, once again, this is normal.

Stage 3: Norming. During this stage events start to crystallize. Members accept the "team ground rules." They begin to establish their roles as team members and slowly begin the process of helping each other and working together. Norming is the stage of accepting team membership, developing a belief that the project is possible, and expressing criticism constructively. Members become more confident in one another and develop a general sense of team cohesion, a common spirit and goal. As the team members work out differences, they spend more energy on problem resolution. They actually begin to see progress.

Stage 4: Performing. This last stage of team dynamics suggests that the team has finally resolved and established its relationships. The team arrives at the performing step by real problem solving. Members have discovered and accepted each other's strengths and shortcomings. The team is a cohesive unit capable of getting a lot of work done.

These stages are normal, and a member that understands all new teams must go through these steps in some fashion will be better equipped to deal with them.

Teams that have not received TQ training or teams that have not been involved in previous TQ projects are not as likely to be successful. Without prior training and planning a great deal of time and energy will be wasted. With this thought in mind we have provided the following material to help you get started in the team environment.

Team Formation

When you are initially placed on a team it can be difficult and sometimes stressful. You may not know any of the team members, and, lacking very specific instructions, it is not easy to know what to do or how to get started. To help you through this situation, let's review again the "forming" stage. During this stage members are cautious, they begin the process of defining the task at hand and discuss how it might be accomplished. It is unlikely that meaningful progress will be made during this stage, but this is perfectly normal. Remember, the team is a powerful unit when properly organized, motivated, empowered, and managed. Individual members that are collectively focused on a well-defined outcome can accomplish amazing results.

Team activity is normally divided into two components: content and process. Content is the "what" that is to be accomplished and process is the "how" it will get done. Team members are either working on a specific assignment as individual members or they are collected together in a team meeting.

Team Meetings

As a professional you will spend a large percentage of your time in meetings. Some of these meetings will be well organized and some will not. It has been estimated that throughout the United States in a single day there are 10 million meetings. As a new engineer, you may spend 25 percent of your time in meetings, and as you become a part of middle and top management you can spend as much as 50 percent of your time in meetings. It is safe to say meetings consume tremendous amount of time, money, and energy. Resources that should be carefully managed.

Let's return to the TQ team and use the organizational structure that we have developed to discuss team meetings. First, a successful team must have a team leader. To most people, the term *leader* typically means the one who is "in charge." Since we do not especially want the leader to be "in charge," a different title for this individual is used in the TQ domain: facilitative team leader. This individual is not "in charge" in the sense that they are the czar or dictator. The facilitator team leader is just as the name implies, he or she is to make sure the meeting is well organized. This person has a specific set of responsibilities just as the recorder and team members have specific duties. Every member of the team should acquire the training and skill to be an effective facilitative team leader, particularly if you are an aspiring engineering student.

The first important function of the facilitative team leader is to plan the meeting. There are different types of meetings that can be held. Three possible varieties are

Informational meeting: The purpose of this type of meeting is to distribute or gather information.

Action-oriented meeting: The purpose of this type of meeting is to take action by the group; that is, plan, problem solve, or make a decision.

Combination meeting: Many meetings are a combination of these two; however, it is important that team members know before the meeting if a decision will be made or an action item will take place.

Once the facilitative team leader has decided on the type of meeting, the next order of business is to plan the desired outcome—that is, to decide what will happen as a result of the meeting. It is important to establish a clear and concise statement of the desired outcome. Having defined desired outcomes provides direction for developing the meeting agenda. At the end of the meeting the team should be able to determine if they have accomplished the desired outcome. If the outcomes of a meeting are not clear, it is more than likely they will not be accomplished.

An effective facilitative team leader will prepare an agenda and distribute it to the team members with sufficient lead time for review and preparation by members. The facilitator who is assigned to assist the leader will arrange meeting space, as well as arrange for any special requirements such as flip charts, overhead projectors, and so on.

The more effective and complete the preparation, the more productive the meeting. In general, there is a one-to-one ratio between preparation and meeting time. If you are planning a two-hour meeting, it will require two hours of preparation.

A facilitative team leader may start a meeting as follows:

> I have convened this meeting to provide new information, gather some information from previous assignments of the membership, and to decide on a unique course of action from several possible actions listed on the flip chart at the front of the room (combination meeting). The desired outcome will be as follows . . . [whatever it is]. I'll be facilitating the meeting and I'd like to ask John Doe to serve as team recorder [that function is normally assigned and then rotated]. We'll make all decisions by consensus, and I'll be the fallback if we don't reach consensus in the time allowed.

It is extremely important to establish the decision-making process. Many teams do not do this and it is a big mistake. There are different ways that decisions can be made:

1. Decide and announce—If a team leader in an organization also happens to be the "boss" or has been delegated considerable authority, then this form of decision making may be appropriate.

2. Gather information from individuals or team members and then the leader decides.

3. Gather information from individuals or team members and then the team decides by majority vote.

4. Consensus.

5. Delegate decision to membership with criteria.

We will focus on the fourth method: decision making by consensus. This concept is not clearly understood by most people. Consensus does not mean everyone agrees, nor is it a vote whereby majority rules. Consensus means everyone agrees to support the group decision.

As stated by William G. Ouchi, author of *Theory Z,* Consensus has been reached when all team members can agree on a single solution or decision and each member can say:

- I believe that each of you have listened to and understand my point of view.
- I believe that I understand your point of view.
- I may or may not prefer this decision, but I will support it because it was reached openly and fairly.

In other words, a consensus decision is one that each team member can live with and actively support—it is not necessarily their number one choice.

This concept leads us to another critical issue for successful team meetings: the idea of "ground rules." Any well-functioning team must establish their own set of operational ground rules. The following is a set of typical "ground rules" for conducting a meeting:

1. All decisions will be by consensus.

2. Sensitive issues are confidential.

3. Listen to what others have to say.

4. Be prepared for the meeting.

5. Be on time.

6. Contribute actively.

7. Improve team relationships whenever possible.

8. Support the TQ process.

9. Keep records of your work.

10. Be flexible.

11. Do not dominate the meeting.

12. A minimum of four team members must be present to conduct a meeting.

It is important to understand the other responsibilities an effective facilitative team leader has beyond planning. Two additional key elements in any successful meeting are conducting the meeting itself and follow-up. Let's look at these two items separately.

Conducting the meeting. The facilitative team leader can run the meeting or can elect to be a participant during the meeting and invite the team "facilitator" to run the meeting. Recall that a TQ team has an external facilitator. Again, this is a person from outside the organization designated to assist the facilitative team leader with process.

As the meeting begins keep in mind that all teams will have their share of problems. Remember the second stage of team development was "storming." Recall that in the beginning a team's normal transition is from little noticeable progress to trouble. Storming is a difficult stage. The team facilitator is trained to help, but the individual team members are the ones that must make the team work. It may be necessary at different times, as a team, to work through a particular problem.

For example, providing feedback to other team members is necessary on occasion, but again this step must be done with care.

It is likely that the team will experience some or all of the following common problems, and there are methods and techniques developed to assist in their resolution. Common team or group problems:

- Stagnation
- Overbearing or dominating members
- Reluctant participants
- Unquestioned acceptance of opinion as fact
- Rush to accomplishments
- Feuding team members

All individuals on the team are different, and sometimes the most important thing to do is listen carefully to what others have to say. Only then can we begin to understand people and effectively deal with individuals that are not just like us.

Meeting Follow-up

Once the meeting is over there are a number of important activities that remain to be accomplished. It is extremely likely that during the course of the meeting there were a number of action items that need attention and follow-up. The minutes of the meeting must be distributed, and the team members may need reminders of things they were asked to do. The facilitative team leader may need to receive information or get approval from the sponsor.

8.4.3 Personal Style Inventory

We have indicated throughout this chapter, in fact stressed the idea, that all individuals are not the same. Various methods are available to explore these individual differences, and one very common approach is referred to as a personal style inventory assessment. A personal style inventory assessment allows each of us as individuals to understand our own personalities and how they differ from others in the team. The most beneficial information derived from personal inventories is a clear understanding that everyone does not think and act the same. Good, creative, hard working, ethical people think, act, and react differently.

In 1986, Robert W. Russell developed a personal inventory test which is a set of questions that can be answered to determine one's personality category. For example, do you like things to move with a "fast pace" or more "slow and deliberate"? Do you "tend to be impatient" or do you display "a great deal of patience"? Your response to a series of questions of this type tend to categorize you as either:

Task oriented	or	people oriented
Reflective	or	assertive

Individuals who are task oriented and reflective tend to be analyzers. Those who are people oriented and assertive are enthusiasts.

Glenn M. Parker, in his 1990 book *Team Players and Teamwork,* categorizes four team-player styles:

- Contributor

  They tend to provide all relevant knowledge, skills, and data. A contributor sees the team as a group of subject matter experts.

  A contributor checklist: dependable, responsible, organized, efficient, logical, clear, relevant, pragmatic, systematic, and proficient.

- Collaborator

  They seek an overall vision. The project goals are paramount. They are willing to work outside of a defined role to get the work done.

  A collaborator checklist: cooperative, flexible, confident, forward-looking, conceptual, accommodating, generous, open, visionary, and imaginative.

- Communicator

  They tend to be a process-oriented team member. They can be effective listeners and facilitators or function as participants focused on conflict resolution, consensus building, feedback, and the building of a relaxed climate.

A communicator checklist: supportive, encouraging, relaxed, tactful, helpful, friendly, patient, informal, considerate, and spontaneous.

- Challenger

They openly question process, vision, goals, and other team members.

A challenger checklist: candid, ethical, questioning, honest, truthful, outspoken, principled, adventurous, above board, and brave.

MBTI (Myers Briggs Type Indicator)

Katharine Briggs and Isabel Briggs Myers conceived the idea for a personality indicator in the early 1940s, and it has evolved into a reasonably sophisticated instrument. The actual MBTI consists of 200-plus questions that require considerable time to complete. Personality questionnaires such as the MBTI have been used for many years. They are a popular method of assessment.

R. Craig Hogan and David W. Champagne have derived a shorter version of the MBTI instrument that can be completed in much less time. This particular inventory is self-administering and self-scoring. It takes about 10 minutes to fill out and 5 minutes to score. Items are arranged in pairs (a and b), and each member of the pair represents a preference you may or may not hold. Individuals are asked to rate their preference for each item by giving it a score of 0 to 5 (0 meaning you *really* feel negative about it or strongly about the other member of the pair, 5 meaning you *strongly* prefer it or do not prefer the other member of the pair). The scores for a and b *must add up to* 5 (0 and 5, 1 and 4, 2 and 3, and so on).

A few sample pairs are listed here:[9]

I prefer:

1a. _______ Making decisions after finding out what others think.
1b. _______ Making decisions without consulting others.

2a. _______ Being called imaginative or intuitive.
2b. _______ Being called factual and accurate.

3a. _______ Making decisions about people in organizations based on available data and systematic analysis of situations.
3b. _______ Making decisions about people in organizations based on empathy, feelings, and understanding of their needs and values.

As a team member each of you bring a range of experiences and values to the team environment. In so doing you can promote ei-

9. The complete survey can be found in R. C. Hogan and D. W. Champagne, "Personal Style Inventory," *1980 Annual Handbook for Group Facilitators,* San Diego: University Associates, 1980, pp. 89–99. Although the material is copyrighted, it may be freely reproduced for educational/training/research activities.

ther congenial, comfortable, productive discussion or frustrating, conflicting, unproductive argument that reflects your needs and prejudices rather than the real issues of the project.

When people who interact with each other understand their own values and unique backgrounds and why these affect their thinking and interaction, they will likely be more constructive about the ideas and suggestions they advance, seeing them as ideas they value rather than as absolutes. They also will be more able to accept the ideas or actions of others that differ from their own.

The purpose of a personal style inventory is to enable individuals to identify their personality types so that they can learn to understand better the influence of personality style on their thoughts and actions and on the thoughts and actions of those with whom they interact. An inventory assessment will also help identify strengths and weaknesses in their own styles.

The MBTI personal style inventory classifies four pairs of personality traits:

Introversion—extroversion
Intuition—sensing
Feeling—thinking
Perceiving—judging

Every individual exhibits all eight in thought and action, but for each person one personality trait of each pair is used more often, is more comfortable, and has given rise to a greater number of beliefs and values than has the other member of the pair. The trait consequently characterizes the person's outlook and thought processes. As a result, a person's type of preference for four of these traits (one from each of the four pairs) can be determined, and predictions about values, beliefs, and behavior can be made based on the results.

When participants understand how profoundly these personality traits affect their values, choices, assumptions, beliefs, decisions, thoughts, and behavior and those of their spouses, colleagues, superiors, subordinates, students, and instructors, then they can begin to realize that the statements and actions that they and those around them live by are the result of different views of the world, not right or wrong thinking. Our own point of view and those of others can be more easily understood and accepted.

Personality Traits

These personality traits are present to some degree in all people. It is the extremes that are described here. Hogan and Champagne describe the pairs of traits as follows:

Introversion—extroversion. People who are more introverted than extroverted tend to make decisions somewhat independently of constraints and prodding from the situation, culture, people, or things around them. They are quiet, diligent at working alone, and socially reserved. They may dislike being interrupted while working and may tend to forget names and faces.

Extroverts are attuned to the culture, people, and things around them, endeavoring to make decisions congruent with demands and expectations. The extrovert is outgoing, socially adept, and interested in variety and in working with people. The extrovert may become impatient with long, slow tasks and does not mind being interrupted by people.

Intuition—sensing. Intuitive people prefer possibilities, theories, gestalts, the overall, invention, and the new; they become bored with nitty-gritty details, the concrete and actual, and facts unrelated to concepts. Intuitive people think and discuss in spontaneous leaps of intuition that may leave out or neglect details. Problem solving comes easily for these individuals, although there may be a tendency to make errors of fact.

The sensing types prefer the concrete, real, factual, structured, tangible here-and-now, becoming impatient with theory and the abstract, mistrusting intuition. The sensing type thinks in careful, detail-by-detail accuracy, remembering real facts, making few errors of fact but possibly missing a conception of the overall.

Feeling—thinking. The feelers make judgments about life, people, occurrences, and things based on empathy, warmth, and personal values. As a consequence, feelers are more interested in people and feelings than in impersonal logic, analysis, and things, and in conciliation with harmony more than in being on top or achieving impersonal goals. The feelers get along well with people in general.

The thinkers make judgments about life, people, occurrences, and things based on logic, analysis, and evidence, avoiding the irrationality of making decisions based on feelings and values. As a result, the thinkers are more interested in logic, analysis, and verifiable conclusions than in empathy, values, and personal warmth. The thinkers may step on others' feelings and needs without realizing it, neglecting to take into consideration the values of others.

Perceiving—judging. The perceivers are gatherers, always wanting to know more before deciding, holding off decisions and judgments. As a consequence, perceiving people are open, flexible, adaptive, nonjudgmental, able to see and appreciate all sides of issues, always welcoming new perspectives and new information about issues. However, perceivers are also difficult to pin down and may be indecisive and noncommittal, becoming involved in so many tasks that do not reach closure that they may become frustrated at times. Even when they finish tasks, perceivers will tend to look back at them and wonder whether they are satisfactory or could have been done another way. The perceiver wishes to roll with life rather than change it.

The judgers are decisive, firm, and sure, setting goals and sticking to them. The judgers want to close books, make decisions, and get on to the next project. When a project does not yet have closure, judgers will leave it behind and go on to new tasks and not look back.

The Team Mix

The purpose of this overview is to point out that each of the team members has different strengths. In the beginning you will perceive these differences as time-consuming irritations. Imagine someone challenging something you have stated or presented—the nerve of that individual!

The storming stage of team evolution can be less disastrous if we understand that we all have different styles. The categorization of these styles, however, is limited and somewhat arbitrary. Individuals are far too complex to be labeled. We do not wish to stereotype people too easily. There is no right or wrong style—all have strengths and weaknesses that make them no better or worse than each other. Styles are often situational and may change depending on mood, environment, and state of mind. To be successful, teams need a mix which provides a strength for task accomplishment and group process.

8.5 Texas Instruments and Total Quality

Mr. Junkins presented a summary of the actions taken by Texas Instruments at a Quality Coalition Conference in Des Moines, Iowa, on February 25, 1993. The complete text of Mr. Junkins' speech is presented below with permission of Texas Instruments.

> I appreciate the invitation to talk about Total Quality and what it means for American business, and particularly to focus on the experience we had with the Malcolm Baldrige process. I feel strongly about what it's doing for America, and I've seen firsthand the effect on our entire organization, and particularly our defense operation that won the Malcolm Baldrige Award last year.
>
> With all that's been written and talked about regarding quality, and all the different programs of the month, this has the potential to be a pretty dull story. But my message today is that Total Quality, and specifically the Malcolm Baldrige process, is the most powerful catalyst I've seen for driving the organizational and cultural changes we've got to make to compete in the environment we're living in today. This is the real thing.
>
> Before going into detail on our Total Quality journey and what we've learned, let me start with a brief description of Texas Instruments.
>
> Our principal business is semiconductors, which is about half of the company. Defense electronics is about 30 percent of the company, about $2 billion in revenues, and the balance is in computer software and hardware and materials and controls. There are about 60,000 of us in 30 countries around the world. About one-third of our people work outside the United States.

Figure 8.7
Jerry R. Junkins
As the immediate past chairman, president, and chief executive officer of Texas Instruments Incorporated, Mr. Jerry R. Junkins was recognized as a leader in the electronics industry. He initiated strategies that changed the electronics industry and played a significant role in revitalizing America's competitive position. Among these was Texas Instrument's leading effort to gain recognition for the true value of intellectual property.

Mr. Junkins began his career with Texas Instruments after graduating in 1959 from Iowa State with a degree in electrical engineering. As he progressed from an entry-level engineering position through the corporate ranks, Mr. Junkins demonstrated an outstanding ability to transform leading-edge technologies into high-volume production. He developed a manufacturing process for the Strike missile guidance system that converted the lab design into an affordable, mass-produced weapon system, and later used this expertise on the forward-looking infrared night sight that remains a key part of the U.S. arsenal today. Under his management, from 1975 until 1981, Texas Instruments' military defense operations grew more than 200 percent.

Mr. Junkins also held a master's in engineering administration from Southern Methodist University where he serves on the board of trustees. Among the many tributes to his achievements is an honorary doctorate of engineering conferred in 1989 by Rensselaer Polytechnic Institute. In 1988 Mr. Junkins was elected to the National Academy of Engineers and, in 1991, he received the Award of Excellence for Leadership in Management from the American Society of Engineering Management. Under his guidance, Texas Instruments has woven Total Quality Management practices into its fabric as evidenced by the 1992 presentation of a Malcolm Baldrige National Quality Award to the defense systems and electronics group in recognition of their superior management and customer service. A native of Montrose, Iowa, Mr. Junkins passed away May, 1996, and is survived by his wife Sally and two daughters.

Our principal focus is electronics. This is one of the most dynamic and competitive industries in the world. Electronics is forecast to be a \$1.5 to \$2 trillion market by early in the next decade—which will make it the largest industry in the world. And if you throw in the information services market, it could be another trillion dollars.

No other industry is growing this rapidly. But this has attracted a host of international competitors—in this business, the term "world-class" has real meaning. The market is truly global, and our customers will shop all over the world to do what's necessary to meet their requirements.

The semiconductor industry, where we participate, requires high investments. We've spent close to $2 billion in capital equipment over the past three years, and another billion and a half in research and development to stay at the leading edge of technology. And this was at a time when the market was not working in our favor, with the result that we were losing money while we were making these investments.

All of our businesses are going through tremendous changes, partly because of internal changes, but mostly driven by market forces. In semiconductors, most of the history of this business was in commodity products. Now, products developed for specific customer applications are replacing off-the-shelf parts. Cycle time is beginning to replace cost, quality and delivery as the Number One customer care-about.

In defense, global political changes are forcing us to adapt to a much smaller market—and at the same time, we have to develop the technologies for new, sophisticated weapon systems that can be used effectively in a shrinking defense market.

In computer systems, software and service are gaining in importance relative to hardware. And these changes are not just about products. In terms of skill sets, this means more software engineers and fewer hardware engineers; more service people and fewer machinists, and so on.

Across the board, the rate of technological change is accelerating, the level of competition is increasing, and the way we interact with customers is changing.

In this environment, paying close attention to your customers and satisfying their needs is the only way to survive, let alone prosper. As customers get more and more discriminating, the winners will be determined by how tightly we are tied with our customers and how well their care-abouts are satisfied. That is what makes Total Quality so important.

And this applies to all companies, big or small, global or domestic, manufacturing or service. Even though we may think of a particular business as local or regional, no business is immune from the effects of global competition, because we're all going to be affected by what happens to those companies that do compete in the international arena. It also applies to industry, government, and academia. Everyone has a customer, and the goal of all TQM activity is to make us more effective in serving that customer.

Like most companies, we have been involved in some form of quality and productivity improvement for years. In the 1950s we called it work simplification; in the 1960s we added employee attitude surveys and formalized a People and Asset Effectiveness program. But if you go back and look, much of our emphasis tended to focus on products as they rolled off the end of the assembly line.

In 1980—partly as a result of serious competition from Asian companies—we took another look at quality, and we identified quality as the real key to customer satisfaction and the competitive advantage that we're all seeking. During the early 1980s, more than 400 TI managers attended Crosby Quality College, and more than 40,000 TIers had Juran training. We added more teams; we added more training in team building and problem solving; we expanded the definition of quality to include every aspect of the relationship with customers and suppliers; and we expanded the responsibility for quality into the offices and design centers and work areas throughout the company.

Through these and other efforts, we made good progress and cut the cost of quality dramatically all across the company. However, in the mid- 1980s our rate of improvement leveled off. Before that, beginning in the early 1980s, one of our Japanese operations decided to begin the journey for the Deming Prize, and in 1985, our semiconductor factory in southern Japan became the first wholly owned U.S. subsidiary to win Japan's highest quality award. That stirred significant interest and had a dramatic effect on the rest of our semiconductor operations around the world.

But we still felt we needed something else to accelerate our progress. Then in 1988 the Malcolm Baldrige National Quality Award criteria appeared.

The Baldrige process gave some focus and coherence to the activities we had been engaged in across the corporation for all those years, and it gave us a common language across the different cultures and businesses of TI.

In 1989, we asked every organization, both operating and staff, to prepare a mock Baldrige application. These applications were reviewed by an internal team of people who had training in the Baldrige criteria. Based on this self-assessment, we concluded that we were a long way from being able to apply for and win the Baldrige Award. But having seen the five-year journey that our Japanese operation took toward the Deming Prize, we decided we would do an annual self-assessment, but if any operation wanted to apply formally, they were welcome to do so.

In 1990, our defense operation decided to apply for the award. The first time we submitted an application, we advanced to the second stage of evaluation. We did not receive a site visit, but we did receive very excellent feedback. We set up teams to respond to the feedback, with much of the improvement effort focused in the area of measurement.

We reapplied for the award in 1991. Again, we did not receive the award, but this time, we did get a site visit. In a Baldrige site visit, the examiners are trying to verify what you told them in your 75-page written application. Management has only a limited opportunity to talk to the examiners—I got to say hello, and the gentleman who runs our defense operation got to give a 20-minute welcome. But the examiners did go out and visit almost all of our sites, and they talked to more than a thousand of our people in the hallways, in the cafeteria, in their work groups. There's no way to anything but open the door, and they really see if you walk like you talk.

As a result of the feedback from the 1991 site visit, we renewed our focus on leadership and customer satisfaction. The result was that in 1992, we had the opportunity to accept the Malcolm Baldrige National Quality Award—the first time a defense contractor has received this award.

So—what have we learned from this process? And how has TI changed as a result of our Total Quality efforts? To answer that, let me talk a bit about the principles of our Total Quality process.

But before I do that, let me say that even though our defense business is in the spotlight because of the Baldrige Award, if you look across our company, you'll see many areas moving equally aggressively and making equal progress. Our defense group benchmarked our semiconductor operations on cycle time improvement, and so on. So there is learning taking place all across the company.

We built our total quality strategy on three principles: Customer focus, continuous improvement, and people involvement.

Customer focus is what every business should depend on, and the very best achieve it. At TI, we were talking about "market-driven strategies" in the early 1980s, and we piloted a special project in 1985 to talk to our customers to find out what it took to be more effective in doing business with them. But these were pretty much isolated programs. The Baldrige Award criteria gave us a tangible way to make "customer focus" the *centerpiece* of our daily activities.

We have to understand the customer's care-abouts, capture feedback, and improve the explicit and implicit commitments we make to our customers.

That leads to the second Total Quality principle, continuous improvement. Continuous improvement is essential because customer requirements and expectations are always climbing, so we must keep raising the bar on our own performance.

The Baldrige process relies on data that show continuous improvement over time. This was not new to TI. As an engineering company, we've always had a data-driven culture, and we've always tried to make things better. In semiconductors, that means putting more functions on a smaller chip for less cost year after year.

But sometimes, we weren't measuring the right thing, or we may have been measuring too much of the wrong thing. Even though we had data that showed we were improving, assessment against the Baldrige criteria showed that we were not improving nearly fast enough, and that our results were not always world-class when compared with out best competitors.

The results of the Baldrige process introduced two ideas to our culture that are beginning to make all the difference: benchmarking, and the idea of having stretch goals.

Benchmarking was important to us for two reasons. First, it deflated a lot of arrogance. Results that we thought were world class, compared with other organizations within TI, turned out to be only average, or slightly better than average, when compared with truly world-class companies. Let me tell you, there is nothing more sobering than patting yourself on the back for making 5 and 10 percent improvements every year, and then finding out that your best competitor is doing things twice as good as you do. It will wake you up in a big, big hurry.

Second, it made us better listeners. We were a company that took pride in doing everything ourselves. In benchmarking our processes with excellent companies such as Motorola, Xerox, L.L. Bean and others, we started getting excited about sharing our own best practices, and we learned ways to improve much faster than we could have done ourselves.

The key to benchmarking is to set goals that will get you to reach world-class levels quickly—not approach it at five to ten percent improvement per year. When we started looking at what the best companies were achieving, we found we'd been too easy on ourselves—our improvement goals weren't nearly aggressive enough. So we set what we called "stretch goals"—goals that could not be attained by 5 or 10 percent improvement, or by doing things organizationally or procedurally the same. That' s a significant point. If the goal is only a little bit ahead of where you are today, you'll find ways to get there without really changing the organization.

As we are buying into Motorola's pioneering work in Six Sigma, our current stretch goal is to achieve "six sigma" performance by 1995. Most industry processes today operate at about four sigma, or about 6200 defects per million opportunities. Six sigma says you've got to do better by a factor of 2,000, or *3.4* defects per million opportunities. The importance of this is, you have to make dramatic changes in design, so that a product can be *manufactured* at six sigma defect levels.

Along with our six sigma performance goals, we've also set some very aggressive goals for improving cycle times in all of our processes. I'm convinced that the biggest competitive advantage in the future will belong to those who can meet customer expectations in the shortest possible cycle times.

Our businesses have set aggressive cycle time goals to achieve breakthrough performance. To give you some idea of what this could mean, in the defense industry alone, because of the way the procurement process works, it is not unusual to take 5 or 6 years to develop a weapon system and get it into production. If we could achieve a 5-10X improvement in cycle times, we could cut that to one or two years. You can imagine the effect on the defense budget. In semiconductors, we're focusing on order fulfillment time for custom products—we expect to cut that time in half this year, and to one-tenth of its present level in three to five years. That will make a tremendous difference in cost, in customer satisfaction, and—most important—in the resources we would have available for new product development.

Achieving stretch goals requires radical change. And in the buzzwords of today, we think radical change requires re-engineering. We believe every organization that is going to survive and prosper has to re-engineer its business processes. We're finding every day that the organizational philosophy and methods of operation and management systems we invented and used in the past—that have been very successful—are no longer good enough, or maybe they no longer even apply. Every process has to be put on the table and questioned.

In the past, we've done things to get improvements in process that, in hindsight, we shouldn't have been doing at all. The best example of this is, we set up sophisticated systems to automate giant warehouses to dispatch parts to our assembly operations. We made outstanding improvements—until we realized we didn't need to put parts in the warehouse if we could get them scheduled and delivered directly to the line. These warehouses are monuments to our honest ignorance that existed ten or fifteen years ago.

If you're willing to attack it, re-engineering helps an organization to take out all unnecessary activities—activities that do not add value for the customer. You can then shorten the cycle times of the remaining processes. Re-engineering is driving revolution in the way companies do business. This revolution is most visible in its impact on our people.

This gets us to the third principle of our Total Quality thrust, after customer focus and continuous improvement, and that is people involvement. People are the key to quality. Customer focus depends on good relationships between people. Companies don't buy from companies; they buy from people. And continuous improvement relies on people for both ideas and execution.

Partly as a result of our adoption of the Malcolm Baldrige criteria, our approach to people involvement is entirely different than it was only five years ago. In the process, we are changing the nature of every job in the company:

- Flatter organizations are replacing the traditional hierarchical structure.
- "Supervisors" are giving way to "coaches" and "facilitators." And this is not an easy transition.
- Teams are replacing individuals as the basic performer of work.
- Cross-functional teams are cutting across our old stovepipe organizations.
- Self-directed work teams are absolutely incredible in terms of making improvements, but they are blurring the traditional distinctions between direct and indirect labor, and between salaried and hourly job definitions. That's bringing new challenges that I'll not pretend to have answers to. How do we measure and reward people? What do we do with the job classifications and pay rates? What about all of our standard procedures manuals that everyone has to sign off on? How much of that are you willing to let go to let the organization adapt to what's necessary to serve the customer?

Let me tell you what you can't do: You can't nail down three sides of the blanket and then tell everybody they're free to move the fourth corner. That won't work.

- And we are expecting people who used to wait for our instructions to make their own decisions.

The best definition I've heard of empowerment came out of the human resources director of our semiconductor group, who used to be the HR director in defense. He said that empowerment means that every TIer has the authority to do what is right for the customer without having to ask permission. Think about it.

As we go forward, we're focusing on developing a broader base of skills. The change from a supervisor to a facilitator calls for different kinds of leadership skills. Breaking down organizational barriers, working in cross-functional teams, having more teams that are self-directed—all of this calls for much more flexibility and a broader base of problem-solving and mathematical and reading skills. The versatility that lets a person move from job to job, or do different parts of the same job, implies the need for continuous, lifelong learning.

We are focusing a great deal of attention and resources to training and assessment across the entire company—and this also represents a change for us. We used to view training as a benefit or an expense—as something optional, that you could do in good times and cut back in harder times. It was seen as outside the normal work environment. We now understand that training is absolutely a prerequisite for achieving the kinds of process changes we're after.

[Comments on dinner with U.S. Department of Labor Secretary Reich and CEOs of 7–8 other companies; how to develop a long-term training vision. Unfortunately, most Labor Department-sponsored training is either an extension of the welfare system or an extension of severance benefits—they don't focus on continuous learning process or upgrade of skills necessary to have a world-class work force.]

Now—what lessons have we learned from our Total Quality journey? First, we must understand our customers and markets—both external and internal.

Second, leadership is vital—and we must lead by example. We learned this through our Baldrige application process, and by watching the way teams developed at TI. In the past, people on the shop floor, and maybe one or two levels of supervision, were involved in teams. Today we've got around 2,000 active teams. Teams became a way of life in our defense group when the top management got personally involved and began leading the group as a team.

Third, the process requires a major culture change within an organization. Teamwork, consensus building, buy-in and ownership or processes are time-consuming processes, and they require sensitivity.

Fourth, the Malcolm Baldrige focus on understanding who the customer is, finding out the customer's requirements, benchmarking the best in the world, and setting stretch goals to intersect that as quickly as possible has been—as I've said—the most powerful catalyst for corporate change I've seen in my career. But you cannot achieve rapid, radical change from the sidelines—you've got to "get in the game." I'm convinced that our defense business improved at twice the normal rate of change by continuing to apply formally and get feedback from the Baldrige Award, rather than go back to internal assessments until we were, quote, "ready."

Finally, we learned that the principles of Total Quality apply to all organizations—big or small, manufacturing or service, government, nonprofit organizations, and academia.

When we first examined the Baldrige Award criteria, there were skeptics who believed the criteria did not apply to the defense business. Our products, services and customers are unique, and there's no comparative data on how our missile rates against someone else's missile. Also, a lot of data about the performance of our products are classified. But we found that the criteria apply across the board: We all have customers. Our challenge now, internally, is to extend what we've learned to our entire worldwide operations, to the different national and business cultures across the entire company.

The focus on customer satisfaction through total quality, and the adoption of the Baldrige criteria, have made TI more efficient, improved TIers' morale, and led to better relationships with customers. There' s no question about it. We're winning business every day as a result of this focus.

This is also true for the United States in general. The focus on quality throughout American industry over the last 15 years or so has clearly made American industry more competitive. The press and the politicians don't seem to realize it, but there is a revolution in competitiveness that's taking place across American business. We're going to find out in the last half of this decade that we can stand very tall in terms of worldwide competitiveness and do very well as a result of these actions. The efficiency and improvement in productivity across all of American business is happening at a more rapid rate than anyone realizes. We're seeing improvement in the service sector that we haven't seen in the last two or three decades; we're seeing improvement in the manufacturing sector above and beyond what has been reasonable productivity improvement over the last decade. And the growth in the economy that we're seeing today is coming from this productivity improvement.

Unfortunately, we are paying a price. This is a major dislocating process. The downsizing and reorganizations are causing dislocations across the country. We have to live through that and reabsorb those people into the work force. But while we're doing that, we're building a very competitive American business environment.

I believe we could become even *more* competitive if schools, governments and other nonmanufacturing entities would get involved in the Baldrige process. Bob Galvin of Motorola estimates that the U.S. Gross Domestic Product would increase significantly—by half a percentage point or more—if every business seriously adopted the Baldrige criteria. Schools should not only apply the criteria to their own operations, but they should teach Total Quality disciplines in the classroom.

I'll end with a So what? First, I will bet you that in this room, there are organizations that, if you do some benchmarking, are doing something at world-class levels. But the person next to you doesn't know that.

If you want to try it, go to your people and ask, who are your customers—who is the machine shop's customer, the assembly operator, the clerk—who, specifically, is your customer, and what are their care-abouts, and are we satisfying those? If you go through each organization and ask every individual to identify who his or her customer is, you'll get some interesting answers. This is how to get started.

Then, find out who really does something the best. You don't have to study or analyze—you can go implement. And you will get step-function improvement just by doing that.

If you want a thrill, come with me and talk with one of our team leaders, who started in our assembly operations twenty or twenty-five years ago, and is so shy she keeps her head down all the time and just does her job. Listen to her as she introduces the other members of the team, and tells me that the defect levels are down by an order of magnitude, and that the costs are down 50 percent, and the cycle times are four-and-a-half days versus three weeks, and because cycle times are down, they have $750,000 less tied up in work in process. When you do this, you'll come away believing what I said at the beginning of this talk: This really is the real thing.

Thank you.

8.6 Key Terms and Concepts

Total Quality
Frederick Taylor
W. Edwards Deming
Philip Crosby
National Quality Award
Total Quality forum
Joseph Juran
Jerry Junkins
Malcolm Baldrige
Texas Instruments
The Malcolm Baldrige National Quality Award criteria
benchmarking
stretch goals
Continuous incremental improvement
The customer

Quality Control tools	Forming
Pareto diagram	Storming
Cause-and-effect diagram	Norming
The team	Performing
Cultural change	Team meetings
Problem space	Consensus
Solution space	Personal-style inventory
Team leader	Myers Briggs Type Indicator
Facilitator	Introversion—extroversion
Recorder	Intuition—sensing
Team members	Feeling—thinking
Team dynamics	Perceiving—judging

Problems

8.1 Class exercise: The student counseling unit on your campus will provide a short overview of the Myers Briggs Type Indicator and allow students to complete the questionnaire.

8.2 Class exercise: The short version of the MDTI (Hogan and Champagne) will be distributed by your instructor (see ref. 9). After completion and scoring, students will be placed in teams of 3 to 5 and asked to examine the description of all traits and discuss the results of the inventory with others in the group. The team will prepare a brief report summarizing the collective team results and suggest any impact MBTI differences may have on this group working together over a period of time.

8.3 Team projects: You will be placed in a team of 3 to 5 students and either assigned or asked to select one of the customer improvement areas listed below. In general the Total Quality process consists of two parts: problem space and solution space. Your assignment will be to complete all or a portion of the problem space depending on the time available.

Possible Areas for Improvement:

- Student parking on campus
- Transportation from residence to class locations
- Computer/printer access for students
- Used book market
- Team collaboration and study locations
- Assigned by instructor
- Allocation of student activity fee resources
- Food Service
- Area selected by student team.

Complete all of the following steps or those assigned:

Problem Space

Step 1

A. Identify customers
B. Develop interview questions
C. Interview customers

Step 2

D. Sort customer input
   Affinity diagram

E. Quantify customer concerns
   Pareto chart

F. Issue abatement

Step 3

G. Diagram the process
   Flow chart

Step 4

H. Diagram causes and effects
   Fishbone diagram

Step 5

I. Collect and analyze data on causes

Complete team report.

9 Mechanics

9.1 Introduction

Mechanics is the study of the effects of forces acting on bodies. The principles of mechanics have application in the study of machines and structures utilized in several engineering disciplines. Mechanics is divided into three general areas of application: rigid bodies, deformable bodies, and fluids.

Mechanics of rigid bodies is conveniently divided into two branches: statics and dynamics. When a body is acted upon by a balanced force system, the body will remain at rest or move with a constant velocity, creating a condition called equilibrium. This branch of mechanics is called *statics* and will be one focus of the discussion in this chapter. The study of unbalanced forces on a body, creating an acceleration, is called *dynamics*. The relationship between forces and acceleration is governed by F = ma, an expression of Newton's second law of motion.

Strength of materials, or *mechanics of materials*, is the branch of mechanics dealing with the deformation of a body due to the

Figure 9.1
This steel framework was designed specifically to withstand the forces of an earthquake. (Iowa State University)

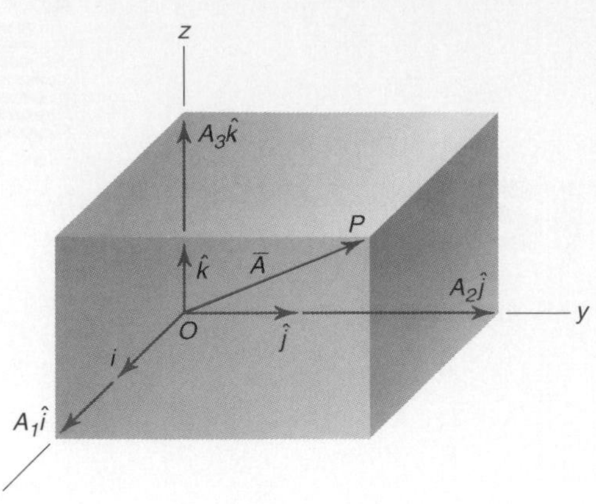

Figure 9.2 Graphical representation of a vector.

internal distribution of a system of forces applied to the body. A brief introduction to this branch of mechanics is included in this chapter.

Before developing concepts from statics and strength of materials, a discussion of coordinate systems and units is appropriate.

Coordinate Systems

The Cartesian coordinate system (shown as the *xyz*-axes in Fig. 9.2) is the basis of reference for describing the body and force system for mechanics. In this introduction to mechanics, we restrict our study to two-dimensional systems (bodies and forces lie in the *xy* plane) except for a brief introduction to concurrent three-dimensional force systems in Sec. 9.10.

Units

Two systems of units are commonly used in the United States. The first, the U.S. Customary System (formerly known as the British gravitational system) has the fundamental units, foot (ft) for length, pound (lb) for force, and second (s) for time. Mass (m) is a derived unit in this system and is called the slug. A slug is the mass that is accelerated one foot per second squared by a force of one pound. A slug therefore has units of $\text{lb}\cdot\text{s}^2/\text{ft}$.

The International System of Units (SI), has the fundamental units meter (m) for length, kilogram (kg) for mass, and second (s) for time. In the SI system, force is a derived unit and is called a newton (N). A newton has units of $\text{kg}\cdot\text{m/s}^2$.

9.2 Scalars and Vectors

A *scalar* is a physical quantity having magnitude but no direction. Examples of scalar quantities are mass, length, time, and temperature.

A *vector,* on the other hand, is a physical quantity having both a direction and a magnitude. Some familiar examples of vector quantities are displacement, velocity, acceleration, and force. Vector quantities may be represented by either graphical or an-

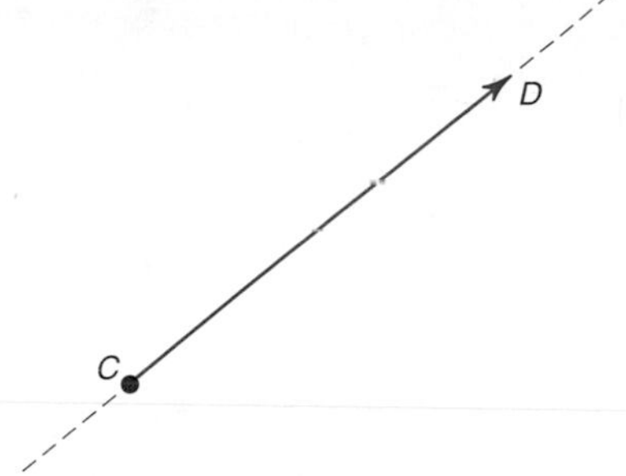

Figure 9.3
Vector shown in Cartesian coordinate system.

alytical methods. Graphically, a vector is represented by an arrow, such as CD in Fig. 9.3. Point C identifies the origin and D signifies the terminal point. The magnitude of the vector quantity is represented by selecting a scale and constructing the arrow to the appropriate length. The arrow positioned in either direction determines the *line of action* of the vector.

Analytically a vector can be represented in different ways. First, consider the familiar Cartesian coordinate system shown in Fig. 9.2. Graphically the vector is shown from O to P. Analytically the point O is at the (0, 0, 0) coordinate position, and the point P is at the (A_1, A_2, A_3) position. The vectors ($\hat{i}$, $\hat{j}$, $\hat{k}$) are called *unit vectors* in the *x, y, z* directions, respectively, and the vectors $A_1\hat{i}$, $A_2\hat{j}$, and $A_3\hat{k}$ are called *rectangular component vectors*. The unit vectors $\hat{i}$, $\hat{j}$, $\hat{k}$ establish the direction, while the magnitude of the rectangular components of the vector is given by A_1, A_2, and A_3, for the *x, y, z* directions respectively.

A typical notation for representing a vector quantity in printed material is a boldface type, for example, the vector **A**. However, since engineers generally solve problems using pencil and paper, the boldface type is difficult to duplicate. Hence, the convention of placing a bar or arrow over the vector symbol has been adopted, for example, $\bar{A}$ or $\vec{A}$.

The notation we will use in this text is to represent the vector as $\bar{A}$ and its magnitude as $|A|$ or A. The unit vectors, $\hat{i}$, $\hat{j}$, $\hat{k}$ are identified with the caret (^) symbol placed over the letter. The Cartesian vector $\bar{A}$ and its magnitude can then be written

$$\bar{A} = A_1\,\hat{i} + A_2\,\hat{j} + A_3\,\hat{k} \tag{9.1}$$

$$|A| = A = \sqrt{A_1^2 + A_2^2 + A_3^2} \tag{9.2}$$

Care must be exercised when working with vectors. Vector quantities normally add or subtract by the laws of algebra. For the most part we will work with the scalar components of the vectors. Remember, if you are representing a vector, both magnitude and direction must be specified. If you wish to represent the scalar component of a vector, then only magnitude need be specified.

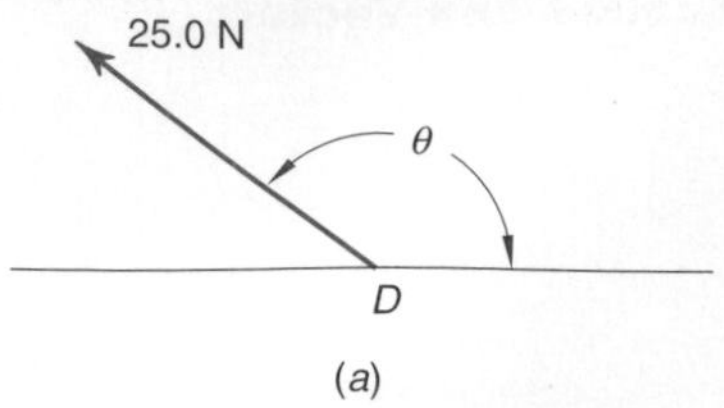

(a)

30.0 N
1
2
D

(b)

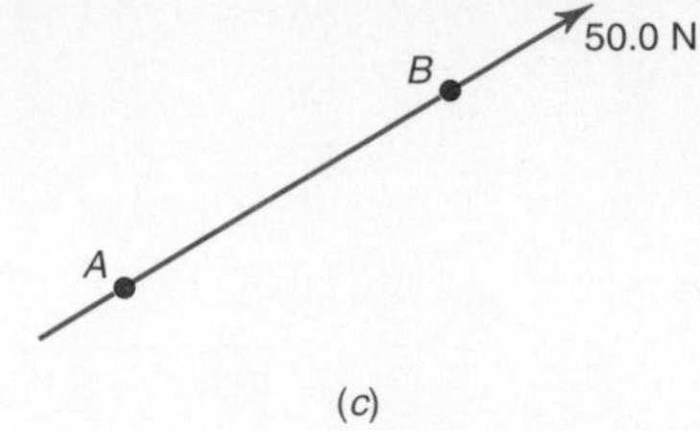

(c)

Figure 9.4 Methods of depicting a force vector.

9.3 Forces

The action of one body acting upon another body tends to change the motion of the body acted upon. This action is called a *force*. Because a force has both magnitude and direction, it is a vector quantity, and the previous discussion on vector notation applies.

Newton's third law states that if a body P acts upon another body Q with a force of a given magnitude and direction, then body Q will *react* upon P with a force of equal magnitude but opposite direction. Therefore, as indicated earlier, to describe a force you must give its magnitude, direction, and the location of at least one point along the line of action. Figure 9.4 indicates common ways of depicting a force. Although the reference line for the angle may be selected arbitrarily, Fig. 9.4a shows the standard procedure of measuring the angle counterclockwise from the positive horizontal axis. Instead of specifying the angle, one may indicate the slope, as shown in Fig. 9.4b.

An alternative method used to describe a force is illustrated in Fig. 9.4c. When two points along the line of action are known, the location and direction of the force can be identified. The vector is then represented as $\overline{AB}$. Incidentally, the reverse notation $\overline{BA}$ would specify a vector of equal magnitude but of opposite direction.

9.4 Types of Force Systems

Forces acting along the same line of action, as illustrated in Fig. 9.5a, are called *collinear forces*. The magnitude of collinear force can be added and subtracted algebraically. The three forces in Fig. 9.5a—$\bar{F}_1$, $\bar{F}_2$, and $\bar{F}_3$—can be replaced by a single resultant force $\bar{F}_4$.

$$|\bar{F}_4| = |\bar{F}_1| + |\bar{F}_2| - |\bar{F}_3| \tag{9.3}$$

The convenience of this addition and subtraction will be used to advantage when we resolve vectors into x and y components.

Forces that pass through the same point in space are called *concurrent forces*. The illustration in Fig. 9.5b represents a three-dimensional concurrent force system.

Forces that lie in the same plane, as seen in Fig. 9.5c, are known by definition as *coplanar*. Combinations of these force systems may also occur in problems. For example, the force system in Fig. 9.5d is both concurrent and coplanar.

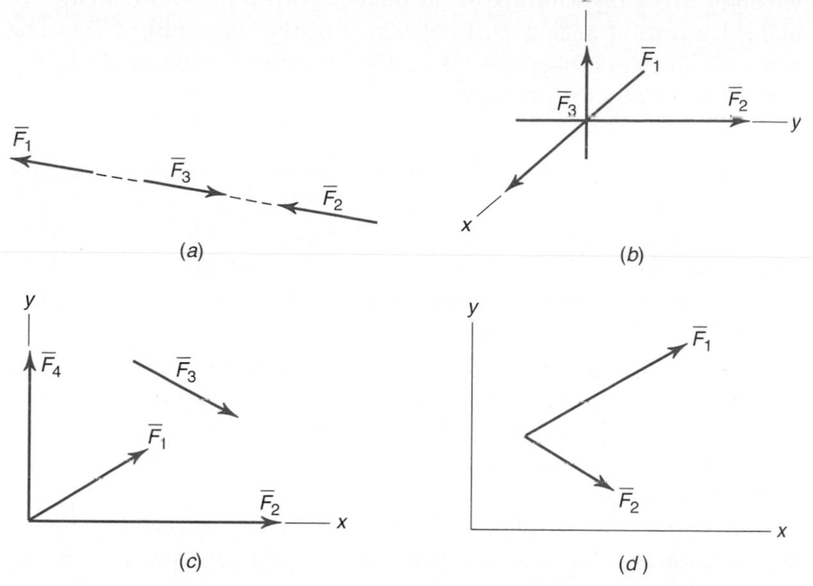

Figure 9.5
Force systems: (a) collinear forces; (b) concurrent forces; (c) coplanar forces; (d) concurrent, coplanar forces.

9.5 Transmissibility

The principle of transmissibility, as illustrated in Fig. 9.6, is an extension of the concept associated with collinear forces. From elementary physics we know that the gravitational attraction of the earth exerts a downward pull on a mass. To prevent the mass shown in the figure from falling, an equal force must be applied to the rope.

It is important to note that if we neglect the mass of the rope, the length of rope used for this application of force (pull) does not matter. That is, the force may be applied at points *A*, *B*, or *C*

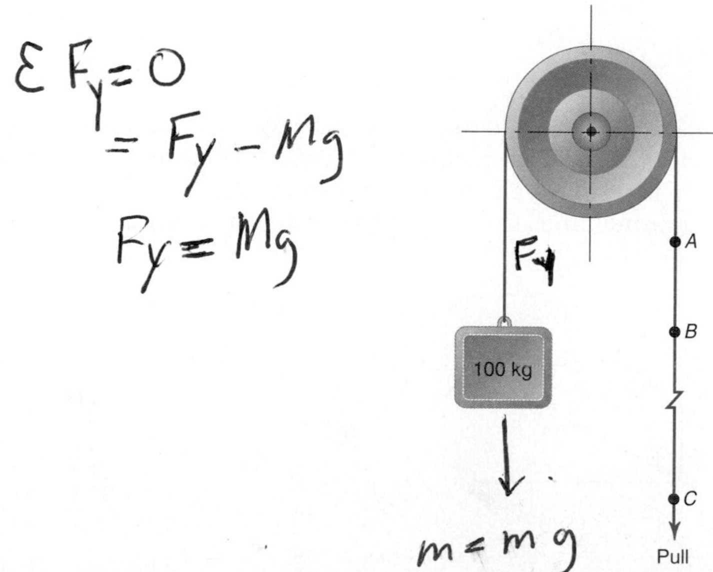

Figure 9.6
A force may be applied at different points along its line of action.

without affecting other external forces. Thus a force can be moved along its line of action without altering the external effect. The principle of transmissibility is applied frequently to simplify computations in statics problems.

9.6 Resolution of Forces

If a force is determined, then its magnitude, direction, and one point on the line of action are known. When calculating the effect of a force, quite often it is convenient to divide or resolve the force into two or more components. Consider the force $\bar{F}$ given in Fig. 9.7. It is acting through the point A and forms an angle of Θ with the positive x-axis.

In this chapter we will concentrate primarily on two-dimensional force systems; thus we will be working primarily in the xy coordinate system shown in Fig. 9.7. In most problems it will be convenient to work with the scalar components of forces in the x and y directions.

The force $\bar{F}$ can be replaced by its two vector components $\bar{F}_x$ and $\bar{F}_y$. This means that if A is a point on a rigid body, the net effect of a force $\bar{F}$ applied to that body is identical to the combined effect of its components $\bar{F}_x$ and $\bar{F}_y$ applied to that point. Had $\bar{F}_x$ and $\bar{F}_y$ been known initially, it would follow that $\bar{F}$ is the resultant of $\bar{F}_x$ and $\bar{F}_y$. The *resultant* of a force system acting on a body is the simplest system (normally a single force) that can replace the original system without changing the external effect on the body.

Expressed in mathematical terms, the scalar relationship of the quantities in Fig. 9.7 are

$$F_x = |\bar{F}| \cos \Theta \tag{9.4}$$

$$F_y = |\bar{F}| \sin \Theta$$

$$|\bar{F}| = \sqrt{F_x^2 + F_y^2} \tag{9.5}$$

$$= \tan^{-1} \frac{F_y}{F_x}$$

Equation (9.4) is used when the scalar components of a given vector are desired in the specified x and y directions. Equation (9.5) is used when the x and y components are known and the magnitude and direction of the resultant force are desired.

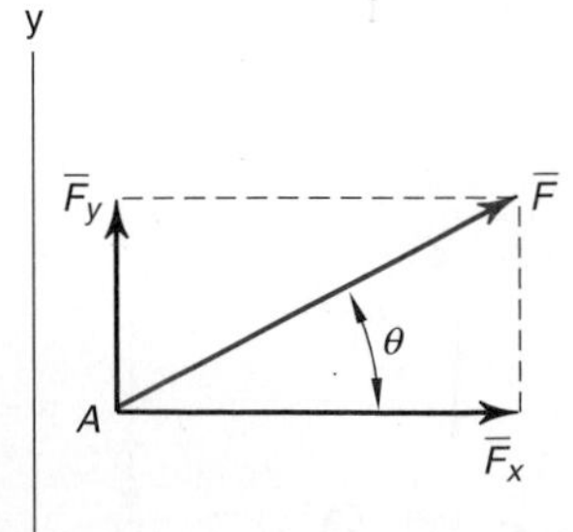

Figure 9.7 Resolution of a force into *x* and *y* components.

Example problem 9.1 In Fig. 9.7 the force vector $\bar{F}$ has a magnitude of 650 lb and acts through point A at a slope of 2 vertical to 5 horizontal. Determine the x and y components of $\bar{F}$.

Solution We can compute the angle θ in Fig. 9.7 from the tangent function.

$$\tan \theta = \frac{2}{5} = 0.40$$

$$\theta = 21.8°$$

$$F_x = |\bar{F}_x| = |\bar{F}| \cos 21.80° = 650\ (0.9285) = 604 \text{ lb}$$

$$F_y = |\bar{F}_y| = |\bar{F}| \sin 21.80° = 650\ (0.3714) = 241 \text{ lb}$$

We can write the actual forces $\bar{F}_x$ and $\bar{F}_y$ by showing a direction for each.

$$\bar{F}_x = 604 \text{ lb} \rightarrow \quad = 604\hat{i} \text{ lb}$$

$$\bar{F}_y = 241 \text{ lb} \uparrow \quad = 241\hat{j} \text{ lb}$$

An arrow indicates the direction in which the force acts. In this case the horizontal component of the force F acts in the positive x direction, and the vertical component acts in the positive y direction.

The force $\bar{F}$ can be written as

$\bar{F}$ = 650 lb (slope 2 vertical to 5 horizontal)

or

$\bar{F}$ = 650 lb ∠21.8°

The latter expression for the force vector $\bar{F}$ is written in polar coordinate notation.

In Cartesian vector notation the force $\bar{F}$ would be

$$\bar{F} = 604\hat{i} + 241\hat{j} \text{ lb}$$

As the number of force vectors within a two-dimensional system increases, the complexity of the problem dictates that an orderly procedure be followed. Extending the application of Eqs. (9.4) and (9.5) to a system composed of several vectors, we have

$$\rightarrow \Sigma F_x = R_x \tag{9.6}$$

$$\uparrow \Sigma F_y = R_y$$

$$|\bar{R}| = \sqrt{R_x^2 + R_y^2} \tag{9.7}$$

$$\theta = \tan^{-1} \frac{R_y}{R_x}$$

where $\bar{R}$ represents the resultant force. The arrows with the summation signs in Eq. (9.6) indicate the conventional positive directions for the scalar components.

Example problem 9.2 Given the two-dimensional, concurrent, coplanar force system illustrated in Fig. 9.8, determine the resultant, $\bar{R}$, of this system.

Solution

1. It is convenient to make a table of each force and its components assuming conventional positive *x* and *y* directions.

Force	Magnitude	*x*-component (→)	*y*-component (↑)
$\bar{F}_1$	86.0	86.0 cos 30.0° = 74.48 N	86.0 sin 30.0° = 43.00 N
$\bar{F}_2$	58.0	58.0 cos 180° = −58.0 N	58.0 sin 180° = 0
$\bar{F}_3$	72.0	$72.0\left(\frac{2}{\sqrt{20}}\right) = 32.2$ N	$72.0\left(\frac{-4}{\sqrt{20}}\right) = -64.4$ N

You should verify the computation of each of the components, taking particular note of the signs.

2. Applying Eq. (9.6)

$$\rightarrow \Sigma F_x = R_x = 74.48 - 58.0 + 32.2 = 48.68 \text{ N}$$

$$\uparrow \Sigma F_y = R_y = 43.00 - 64.4 = -21.4 \text{ N}$$

Figure 9.8

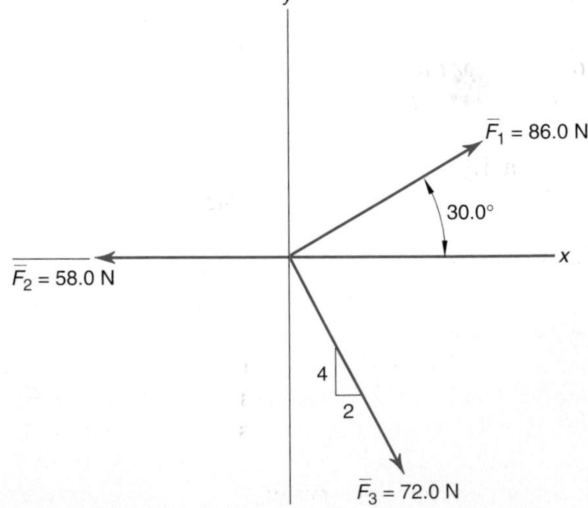

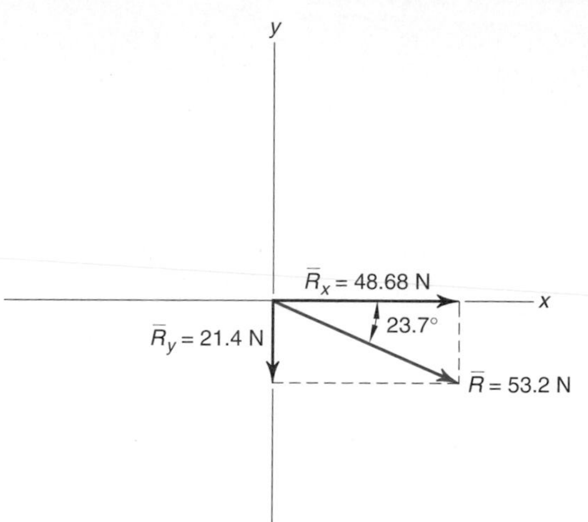

Figure 9.9

3. Applying Eq. (9.7)

$$|\bar{R}| = \sqrt{(48.68)^2 + (-21.4)^2}$$

$$= 53.2 \text{ N}$$

$$\theta_R = \tan^{-1} \frac{-21.4}{48.68}$$

$$= -23.7°$$

Alternatively, $\bar{R}$ can be expressed as

$$\bar{R} = 48.68\hat{i} - 21.4\hat{j} \text{ N}$$

The resultant force vector, $\bar{R}$, is shown in Fig. 9.9.

9.7 Moments and Couples

In order to solve problems involving complete force systems and their applications it is necessary to understand moments and couples. When was the last time you approached an exit in a public building and pushed on the panic bar of the door only to find that you had chosen the wrong side of the door, the side next to the hinges? No big problem. You simply move your hands to the side opposite the hinges and easily push open the door. You are thereby demonstrating the principle of the turning moment. By definition, the tendency of a force to cause rotation about a point is called the *moment* of the force relative to that point. The magnitude of the moment is the product of the magnitude of the force and the perpendicular distance from the line of action of the force to the point. With respect to the door just mentioned, the same force may have been exerted in both attempts to open the door. In the second case, however, the moment was greater owing to the fact that you increased the distance from the force to the hinges, the point about which the door turns.

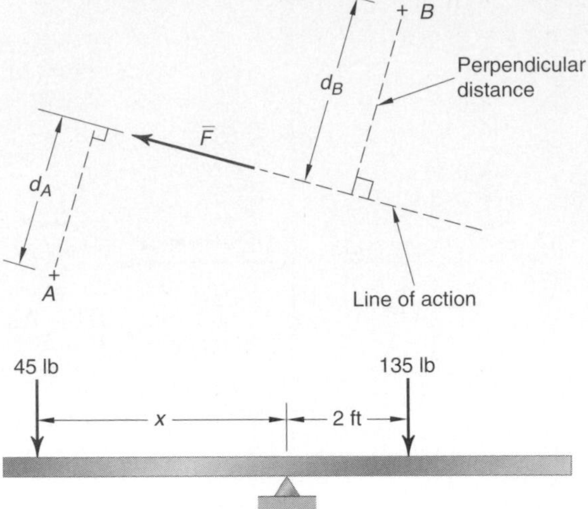

Figure 9.10
The force $\bar{F}$ and points A and B are in the same plane.

Figure 9.11
The forces are in the same vertical plane.

Figure 9.10 illustrates how a moment is evaluated for a specific problem. The magnitude of the moment of the force $\bar{F}$ about a point B is Fd_B. The distance d_B is the perpendicular distance, called the *moment arm,* from the point of application to the line of action of $\bar{F}$. The magnitude of the moment created by $\bar{F}$ about point A is Fd_A. The force $\bar{F}$ will tend to create a clockwise rotation about point B and a counterclockwise rotation about point A. The most common convention is to assign a positive sign to counterclockwise moments and a negative sign to clockwise moments.

Figure 9.11 illustrates the concept of a moment. If we neglect the mass of the beam, we can determine the distance x from the fulcrum that will keep the beam level under the two applied forces. To maintain balance the moments of the two forces must be equivalent and opposite in the tendency to rotate the beam. Thus the 45 lb force must be applied 6 feet to the left of the fulcrum to create a moment of 270 ft lb counterclockwise and balance the 270 ft lb clockwise moment of the 135 lb force.

A couple is similar to a moment. Figure 9.12 shows a hand wheel used to close and open a large valve. The two forces shown

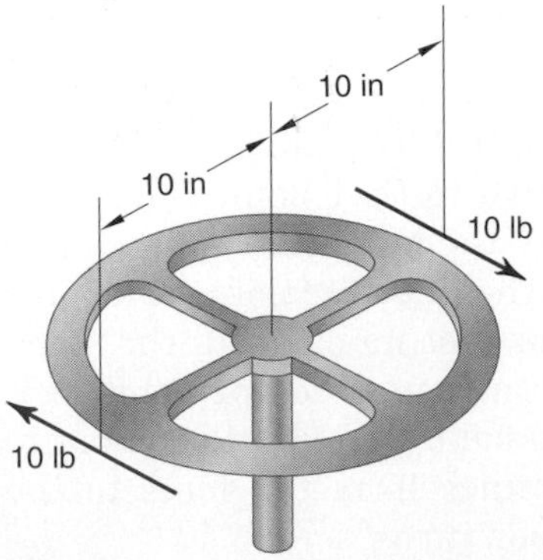

Figure 9.12
Illustration of a couple.

form a *couple* because they are parallel, equal in magnitude, and opposite in direction. The perpendicular distance between the two forces is called the *arm* of the couple. The equivalent moment of a couple is equal to the product of one of the forces and the distance between the forces. In Fig. 9.12, the couple has a magnitude of (10 lb) × (20 in), or 200 in lb.

9.8 Free-body Diagrams

The first step in solving a problem in statics is to draw a sketch of the body, or a portion of the body, and all the forces acting on that body. Such a sketch is called a *free-body diagram (FBD)*. As the name implies, the body is cut free from all others; only forces that act upon it are considered. In drawing the free-body diagram, we remove the body from supports and connectors, so we must have an understanding of the types of reactions that may occur at these supports.

Examples of a number of frequently used free-body notations are illustrated in Fig. 9.13. It is important that you become familiar with these so that each FBD you construct will be complete and correct.

Example problem 9.3 Construct a free-body diagram (FBD) for object *A*, shown in Fig. 9.14a. The surface is smooth (no friction). Object *A* is a homogeneous cylinder weighing 400 lb.

Solution A correct FBD will enable us to solve for unknown forces and reactions on an object. The steps to follow are:

(a) Isolate the desired object from its surroundings.
(b) Replace items cut free with appropriate forces.
(c) Add known forces, including weight.
(d) Establish a coordinate (*xy*) frame of reference.
(e) Add geometric data.

The result is shown in Fig. 9.14b. The weight of 400 lb is shown acting through the center of gravity. The cable is replaced with a tension force, *T*. The reaction of the smooth inclined plane on object *A* is shown by a normal force to the plane, N, acting through the center of gravity. The coordinate system and geometric characteristics complete the FBD.

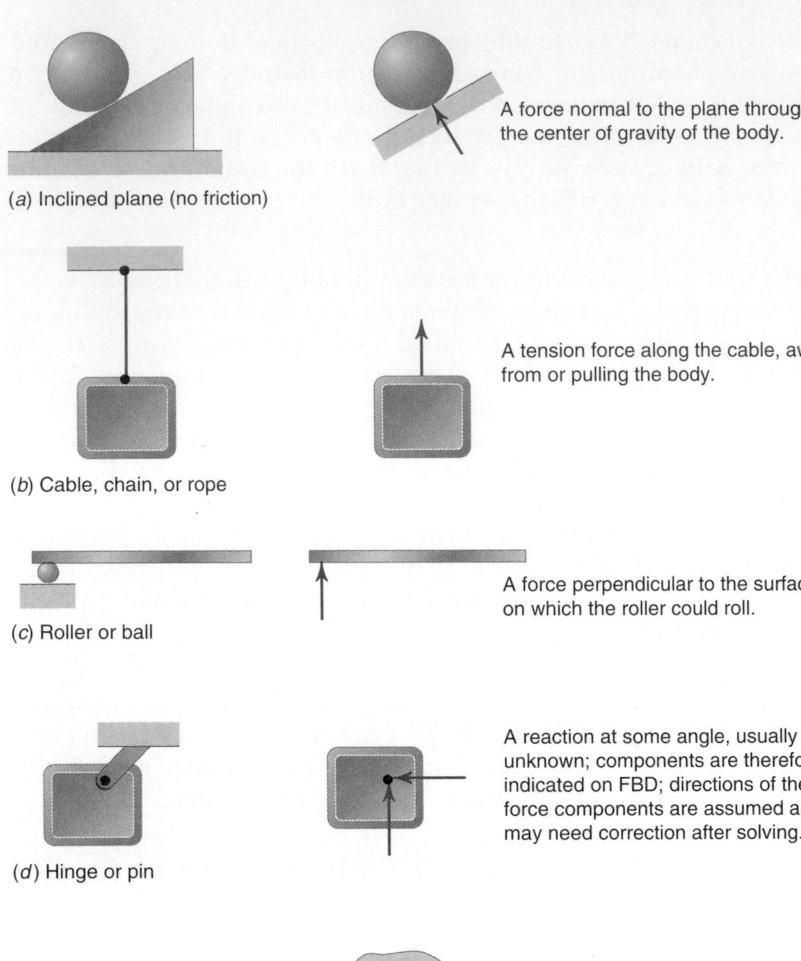

(a) Inclined plane (no friction)
A force normal to the plane through the center of gravity of the body.
(b) Cable, chain, or rope
A tension force along the cable, away from or pulling the body.
(c) Roller or ball
A force perpendicular to the surface on which the roller could roll.
(d) Hinge or pin
A reaction at some angle, usually unknown; components are therefore indicated on FBD; directions of the force components are assumed and may need correction after solving.
(e) Mass/earth
Mass multiplied by the earth's gravitational constant produces a force directed toward the center of the earth.

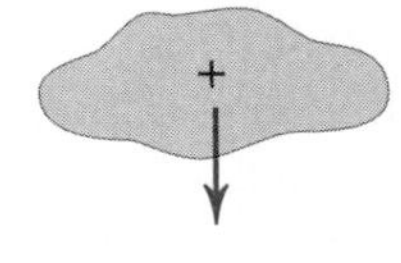

Figure 9.13
Free-body notations.

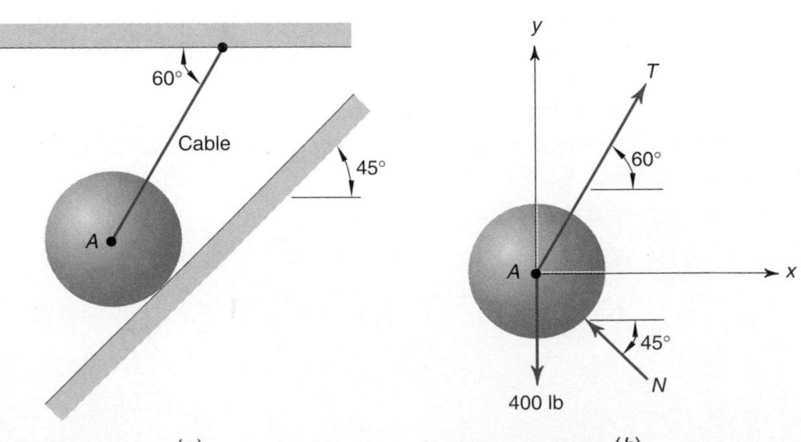

Figure 9.14

Equilibrium

Newton's first law of motion states that if the resultant force acting on a particle is zero, then the particle will remain at rest or move with a constant velocity. This concept is essential to statics.

Combining Newton's first law and the idea of equilibrium, we can state that a body will be in equilibrium when the sum of all external forces and moments acting on the body is zero. This requires the body to be at rest or moving with a constant velocity.

In this chapter, we will consider only bodies at rest. To study a body and the forces acting upon it, one must first determine what the forces are. Some may be unknown in magnitude and/or direction, and quite often these unknown magnitudes and directions are information that is being sought. The conditions of equilibrium can be stated in equation form as follows:

$$\begin{aligned} &\rightarrow \Sigma F_x = 0 \\ &\uparrow \Sigma F_y = 0 \\ &\circlearrowleft \Sigma M = 0 \end{aligned} \qquad (9.8)$$

In order to assist in the tabulation and summation of force components and moments, a convention is necessary. In this text, F_x is positive to the right ($\rightarrow$) and F_y is positive upward ($\uparrow$) and counterclockwise moments are positive ($\circlearrowleft$). When evaluating moments in a particular problem, the point from which the moment arms are measured is arbitrary. A point is usually selected that simplifies the computations.

The following example problems will illustrate the concept of equilibrium.

Example problem 9.4 A beam, assumed weightless, is subject to the load shown in Fig. 9.15a. Determine the reactions on the beam at supports *A* and *B* for the equilibrium condition.

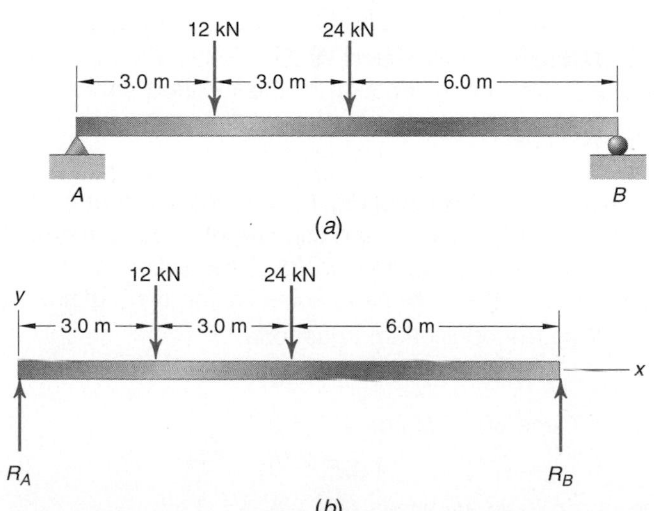

Figure 9.15

Solution This is an example of a beam problem, commonly occurring in buildings and bridges. The loads on the beam are expressed with the prefix k signifying thousands of newtons; thus the loads shown are 12 000 and 24 000 *N*, respectively. The force system is coplanar, and in this instance all forces are parallel. The solution is as follows:

1. Construct the free-body diagram (see Fig. 9.15b). The supports generate forces perpendicular to the length of the beam. Thus there are no forces in the *x*-direction. The directions for the R_A and R_B reactions are obvious in this example. However, the direction may be assumed either way, and the mathematics will produce the correct direction.

2. Write the equations of equilibrium

$$\uparrow \Sigma F_y = 0$$

$$R_A + R_B - 12 - 24 = 0$$

$$\circlearrowleft \Sigma M_A = 0$$

$$-12(3) - 24(6) + R_B(12) = 0$$

Note that we do not need to write $\Sigma F_x = 0$ because there are no *x*-direction forces. We have two equations and two unknowns which we can solve for R_A and R_B.

From the moment equation

$$12R_B = 24(6) + 12(3)$$

$$\overline{R_B} = 15 \text{ kN} \uparrow$$

Then from the sum of the vertical forces

$$R_A + 15 - 12 - 24 = 0$$

$$\bar{R}_A = 21 \text{ kN} \uparrow$$

Since the numerical values for R_B and R_A came out positive, the reactions shown on the FBD were in the correct direction.

Example problem 9.5 Solve Example prob. 9.3 for the cable tension and reaction of the inclined surface on the cylinder.

Solution

1. The FBD is shown in Fig. 9.14b. Observe that the force system on the cylinder is coplanar and concurrent. Since a moment equation will not produce any information for the equilibrium condition, we have two remaining equations to solve for the two unknowns T and N.

2. Write the equilibrium equations

$$\rightarrow \Sigma F_x = 0$$

$$\text{T} \cos 60° - \text{N} \cos 45° = 0$$

$$\uparrow \Sigma F_y = 0$$

$$\text{T} \sin 60° + \text{N} \sin 45° - 400 = 0$$

Solving for T from the *x*-direction equation

$$T = \frac{\cos 45^\circ}{\cos 60^\circ} N = 1.414 \text{ N}$$

Substituting this into the *y*-direction equation

$$1.414 \text{ N} (\sin 60^\circ) + \text{N} \sin 45^\circ - 400 = 0$$

$$1.932 \text{ N} = 400$$

$$\bar{N} = 207 \text{ lb} \quad 45^\circ$$

$$\bar{T} = 293 \text{ lb} \quad 60^\circ$$

Example problem 9.6 For the crane system shown in Fig. 9.16a, determine the reactions on the crane at pin *A* and roller *B*. Neglect the weight of the crane.

Solution

1. The FBD is shown in Fig. 9.16b. Note the two components for the pin reaction at *A*. The direction of each component was assumed.

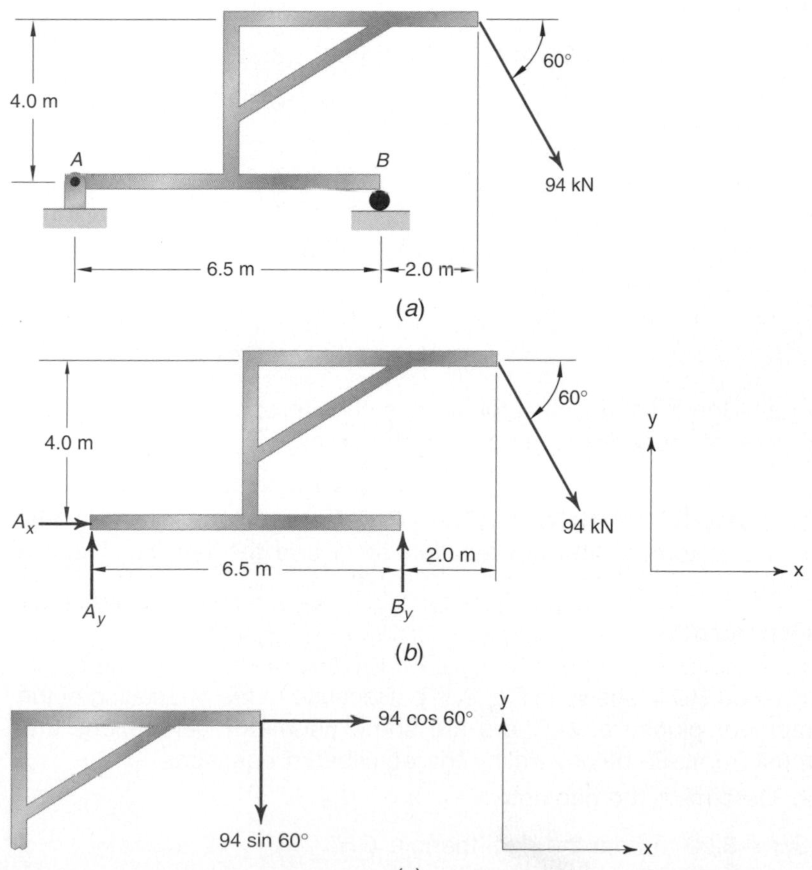

Figure 9.16

2. There are three unknowns and three equations of equilibrium that can be written.

$$\rightarrow \Sigma F_x = 0$$

$$A_x + 94 \cos 60° = 0$$

$$A_x = -47 \text{ kN}$$

$$\bar{A}_x = 47 \text{ kN} \leftarrow$$

The negative sign indicates that the initial direction chosen for A_x was opposite the actual direction of the *x*-direction force on the crane.

Before we write the sum of moments equation with respect to point *A*, we will show a convenient procedure for evaluating the moment of the 94 kN force. Finding the perpendicular distance from point *A* to the line of action of the 94 kN force requires trigonometry. It is less computation to first find the *x* and *y* components and apply these components at a convenient point on the line of action of the 94 kN force (principle of transmissibility). In this case the point where the force acts on the crane is the most convenient (see Fig. 9.16c). The moment arms of the horizontal and vertical components of the force with respect to the point *A* are 4.0 m and 8.5 m respectively.

$$\circlearrowright \Sigma M_A = 0$$

$$B_y (6.5) - (94 \cos 60°)(4.0) - (94 \sin 60°)(8.5) = 0$$

$$\bar{B}_y (6.5) = 188 + 692$$

$$B_y = 135 \text{ kN} \uparrow$$

$$\uparrow \Sigma F_y = 0$$

$$A_y + 135 - 94 \sin 60° = 0$$

$$A_y = -54 \text{ kN}$$

$$\bar{A}_y = 54 \text{ kN} \downarrow$$

Again, the initial direction for A_y was chosen opposite the actual direction of the *y*-direction force on the crane.

Example problem 9.7 For the structure shown in Fig. 9.17a, determine the pin reaction at *G* and the tension *T* in the cable.

Solution

1. The FBD is shown in Fig. 9.17b. Because of the orientation of the member pinned at *G*, there are some geometric calculations that must be made before writing the equilibrium equations.

2. Determine the geometry:

GH = 8.00 m from the right triangle *GHI*

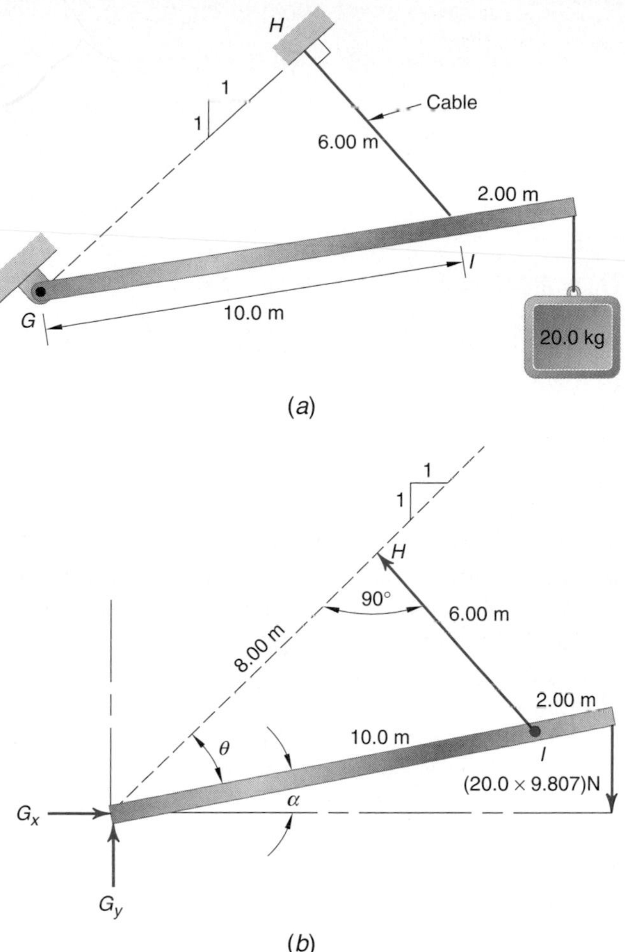

(a)

(b)

Figure 9.17

$$\theta = \measuredangle\ HGI = \tan^{-1}\frac{6}{8} = 36.87°$$

$$\alpha = 45 - \theta = 8.13°$$

3. Apply equations of equilibrium:

$\circlearrowright \Sigma M_G = 0$

$$-\ 20.0\ (9.807)\ (12 \cos 8.13°) + 8.00H = 0$$

$$H = 291.3\ \text{N}$$

$$\bar{H} = 291.3\ \text{N} \quad \nwarrow 45°$$

$$\rightarrow \Sigma F_x = 0$$

$$G_x - H \cos 45° = 0$$

$$G_x = 205.9\ \text{N}$$

$$\bar{G}_x = 205.9\ \text{N} \rightarrow$$

$$\uparrow \Sigma F_y = 0$$

$$G_y + H \sin 45° - 20.0\ (9.807) = 0$$

$$G_y = -9.840\ \text{N}$$

So

$\overline{G}_y = -9.840$ N ↑

or

$\overline{G}_y = 9.840$ N ↓

4. Combine vector components into resultant G using Eq. (9.7):

$\overline{G} = 206$ N

9.10 Three-dimensional Force Systems

Many applications encountered in mechanics involve forces with components in three directions. One such system, the non-coplanar, concurrent force system will be introduced here with an example. The complexity of three-dimensional systems makes it advantageous to use vectors to represent the forces and direction cosines to indicate orientation.

In Fig. 9.18 the vector $\overline{OA}$ represents a force which makes angles of Θ_x with the *positive* x-axis, Θ_y with the *positive* y-axis, and Θ_z with the *positive* z-axis. The cosines of these angles are called direction cosines based on the following relationships.

The length of the vector is

$$|\overline{OA}| = \sqrt{x^2 + y^2 + z^2}$$

From the geometry

$$\cos \Theta_x = \frac{x}{|\overline{OA}|}$$

$$\cos \Theta_y = \frac{y}{|\overline{OA}|} \qquad (9.9)$$

$$\cos \Theta_z = \frac{z}{|\overline{OA}|}$$

$$\cos^2 \Theta_x + \cos^2 \Theta_y + \cos^2 \Theta_z = \frac{x^2 + y^2 + z^2}{|\overline{OA}|^2} = 1$$

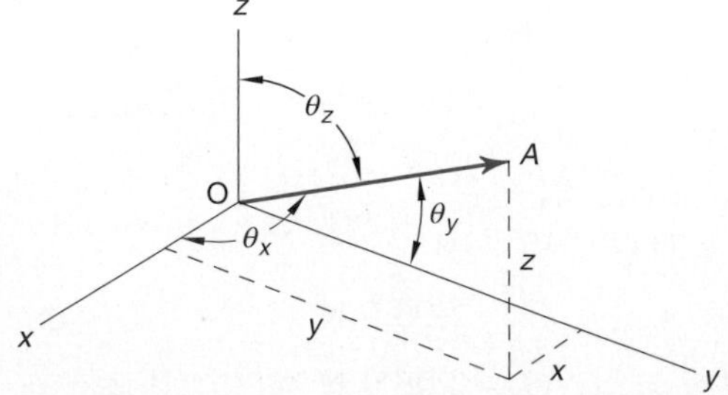

Figure 9.18 A force in three dimensions with direction indicated by direction cosine angles.

Thus a unit vector (length = 1) along the vector $\overline{OA}$ can be written as a Cartesian vector

$$\hat{n} = \cos\,\theta_x \hat{i} + \cos\,\theta_y \hat{j} + \cos\,\theta_z \hat{k}$$

and the vector $\overline{OA}$ can be written as

$$\overline{OA} = |\overline{OA}|\hat{n} \tag{9.10}$$

Similarly, Eq. (9.6) can be extended to three dimensions

$$\begin{aligned} \Sigma F_x &= R_x \\ \Sigma F_y &= R_y \\ \Sigma F_z &= R_z \end{aligned} \tag{9.11}$$

We will use Eqs. (9.9), (9.10), (9.11) to analyze non-coplanar, concurrent force systems.

Example problem 9.8 Express the 200 lb force vector, $\bar{F}$, in Fig. 9.19 as a Cartesian vector.

Solution From Eq. (9.9)

$$\cos^2\,\theta_x + \cos^2\,\theta_y + \cos^2\,\theta_z = 1$$

$$\cos^2\,(120) + \cos^2\,(60) + \cos^2\,\theta_z = 1$$

$$\cos\,\theta_z = \pm\,0.7071$$

$$\theta_z = 45°,\ 135°$$

Thus the force is, from Eq. (9.10)

$$\bar{F} = 200\,(\cos\,120\hat{i} + \cos\,60\hat{j} + \cos\,45\hat{k})$$

$$= -100\hat{i} + 100\hat{j} + 141\hat{k}$$

The small angle for θ_z is chosen because the vector lies above the x-y plane.

Example problem 9.9 Determine the resultant of the non-coplanar, concurrent force system in Fig. 9.20. Express the resultant as a Cartesian vector and its direction cosines.

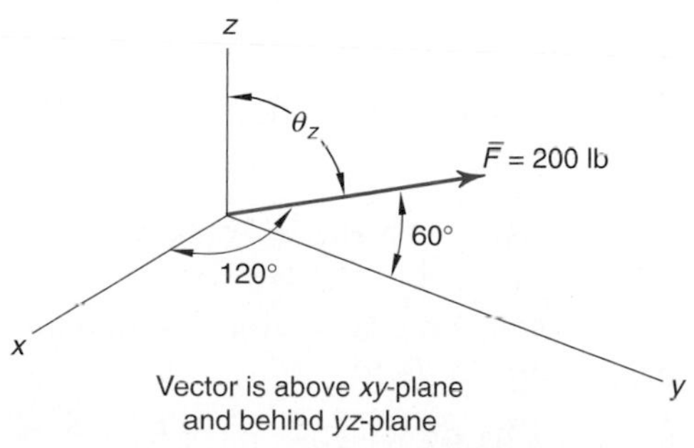

Figure 9.19

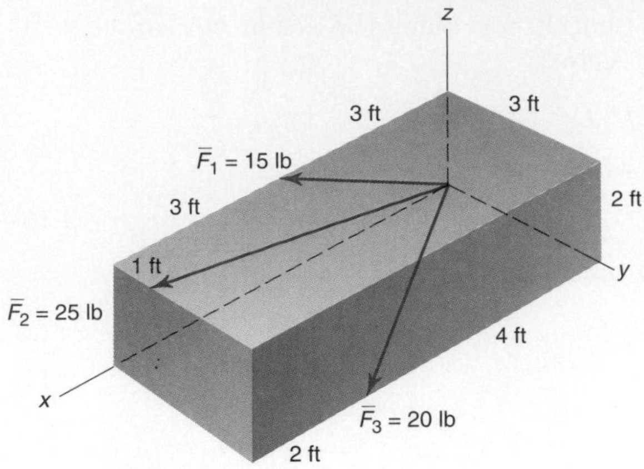

Figure 9.20

Solution

1. Find unit vectors, $\hat{n}_1$, $\hat{n}_2$, $\hat{n}_3$, for the directions of $\bar{F}_1$, $\bar{F}_2$, $\bar{F}_3$, respectively. We will have to use the geometry given since the direction cosines are not directly available.

$$\hat{n}_1 = \frac{3\hat{i} + 0\hat{j} + 2\hat{k}}{\sqrt{3^2 + 2^2}} = \frac{3\hat{i} + 2\hat{k}}{\sqrt{13}}$$

$$= 0.8321\hat{i} + 0.5547\hat{k}$$

$$\hat{n}_2 = \frac{6\hat{i} + \hat{j} + 2\hat{k}}{\sqrt{6^2 + 1^2 + 2^2}} = \frac{6\hat{i} + \hat{j} + 2\hat{k}}{\sqrt{41}}$$

$$= 0.9370\hat{i} + 0.1562\hat{j} + 0.3123\hat{k}$$

$$\hat{n}_3 = \frac{4\hat{i} + 3\hat{j} + 0\hat{k}}{\sqrt{4^2 + 3^2}} = \frac{4\hat{i} + 3\hat{j}}{5}$$

$$= 0.8000\hat{i} + 0.6000\hat{j}$$

2. Write the force vectors as Cartesian vectors. Use Eq. (9.10).

$$\bar{F}_1 = 15\hat{n}_1 = 12.48\hat{i} + 8.321\hat{k}$$

$$\bar{F}_2 = 25\hat{n}_2 = 23.43\hat{i} + 3.904\hat{j} + 7.809\hat{k}$$

$$\bar{F}_3 = 20\hat{n}_3 = 16.00\hat{i} + 12.00\hat{j}$$

3. Using Eq. (9.11), write the resultant vector $\bar{R}$. The *x, y, z* directions are $\hat{i}$, $\hat{j}$, $\hat{k}$.

$$\bar{R} = (12.48 + 23.43 + 16.00)\hat{i} + (3.904 + 12.00)\hat{j} + (8.321 + 7.809)\hat{k}$$

$$= 51.91\hat{i} + 15.90\hat{j} + 16.13\hat{k}$$

4. The direction cosines for the resultant vector $\bar{R}$ are found as follows using Eq. (9.9).

$$\cos\theta_x = \frac{51.91}{|\bar{R}|} = \frac{51.91}{\sqrt{51.91^2 + 15.90^2 + 16.13^2}}$$

$$= \frac{51.91}{56.64}$$

$$\theta_x = \cos^{-1} 0.9165$$

$$= 23.6°$$

$$\cos \theta_y = \frac{15.90}{56.64}$$

$$\theta_y = \cos^{-1} 0.2809$$

$$= 73.7°$$

$$\cos \theta_z = \frac{16.13}{56.64}$$

$$\theta_z = \cos^{-1} 0.2848$$

$$= 73.5°$$

9.11 Stress

In statics, concepts were limited to rigid bodies. It is obvious that the assumption of perfect rigidity is not always valid. Forces will tend to deform or change the shape of any body, and an extremely large force may cause observable deformation. In most applications, slight deformations are experienced, but the body returns to its original form after the force is removed. One function of the engineer is to design a structure within limits that allow it to resist permanent change in size and shape so that it can carry or withstand the force (load) and still recover.

In statics, forces are represented as having a magnitude or an intensity in a particular direction. Structural members are usually characterized by their mass per unit length or their size in principal dimension, such as width or diameter. Consider the effect on a wire dimension, such as width or diameter. Consider the effect on a wire that is 5.00 mm in diameter and 2.50 m in length. It is suspended from a well-constructed support and has a ball with a mass of 30.0 kg attached to its lower end (see Fig. 9.21). The force exerted by the mass is (30.0) (9.807) = 294 N. The wire has a cross-sectional area of $\pi\,(5^2/4)$, or 19.6 mm^2. If it is assumed that every square millimeter equally shares the force, then each square millimeter supports 294 ÷ 19.6, or 15.0 N. This fact can be stated another way: The stress is 15.0 N/mm^2 (15.0×10^6 N/m^2 or 15.0 MPa); that is, the force in the wire is literally trying to separate the atoms of the material by overcoming the bonds that hold the material together.

Figure 9.21
The circular wire is under a tensile stress due to the suspended mass.

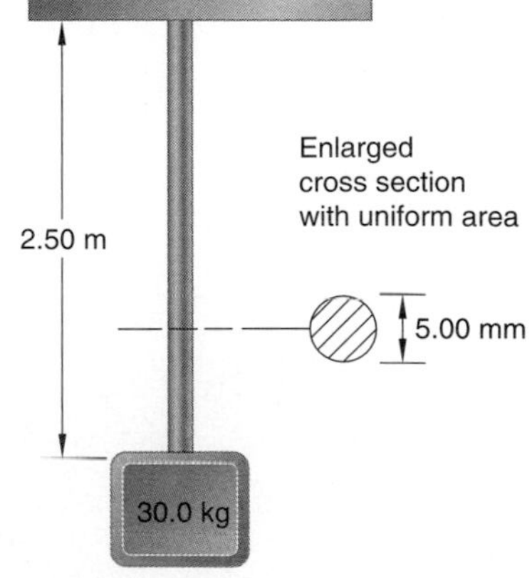

The relationship above is normally expressed as

$$\sigma = \frac{F}{A} \tag{9.12}$$

where σ = stress, Pa

F = force, N

A = cross-sectional area, m^2

Since stress is force per unit area, it is obvious that the stress in the wire (ignoring its own weight) is unaffected by the length of the wire. Such a stress is called *tensile stress* (tending to pull the atoms apart). If the wire had been a rod of 5.00 mm diameter resting on a firm surface and if the mass had been applied at its top, the force would have produced the identical stress, but it would be termed *compressive,* for obvious reasons.

The simple or direct stress in either tension or compression as described earlier results from an applied force (load) that is in line with the axis of the member (axial loading). Also the cross-sectional area in both examples was constant lengthwise. If you have an axial load but the cross section varies, the stresses in separate cross sections are different because the areas are different.

A third type of stress is called *shear.* While tension and compression attempt to separate or push atoms together, shear tries to slide layers of atoms in the material across each other. (Imagine removing the top half of a stack of sheets of plywood without lifting.) Consider the pin in Fig. 9.22a as it resists the force of 1.00 (10^5) N.

The average shear stress in the pin is

$$\tau = \frac{F}{A} = \frac{1.00 \times 10^5}{(\pi)\,(2.00 \times 10^{-2})^2(2)/4}$$

$$= 159 \text{ MPa}$$

Note: This is an example of double shear in that two cross sections of the pin resist the force; hence, the area is the cross-sectional area of two points. The two pin shear surfaces are indicated in Fig. 9.22a.

Figure 9.22

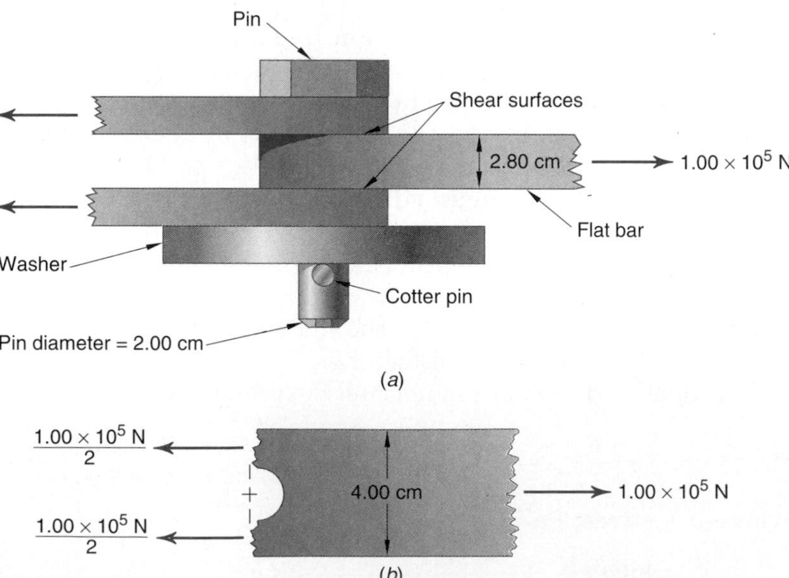

To complete the computation, the average tensile stress in the bar at the critical section through the pin hole, as shown in Fig. 9.22b, is

$$\sigma = \frac{F}{A} = \frac{1.00 \times 10^5}{(2.00 \times 10^{-2})\,(2.80 \times 10^{-2})}$$

$$= 179 \text{ MPa}$$

The actual stress in both cases is somewhat greater than the average because stresses tend to concentrate at the edges of the pin and hole. For this reason, engineers apply a factor of safety in the design process.

9.12 Strain

As an engineer you may be called upon to design both structures and mechanisms. In design work, it is important to consider not only the external forces but also the strength of each individual part or member. It is critical that each separate element be strong enough, yet not contain an excessive amount of material. Thus in the solution of many problems a knowledge of the properties of materials is essential.

One important test that provides designers with certain material properties is called the *tensile test*. Figure 9.23 illustrates a schematic of a tensile-test specimen. When this specimen is loaded in an axial tensile-test machine, the force applied and the corresponding increase in material length can be measured. This increase in length is called *elongation*. Next, in order to permit comparisons with standard values, the elongation is converted to a unit basis called *strain*.

Strain (ϵ) is defined as a dimensionless ratio of the change in length (elongation) to the original length:

$$\epsilon = \frac{\Delta l}{l} = \frac{\delta}{l}$$

where

ϵ = strain, mm/mm (9.13)

δ = deformation, mm

l = length, mm

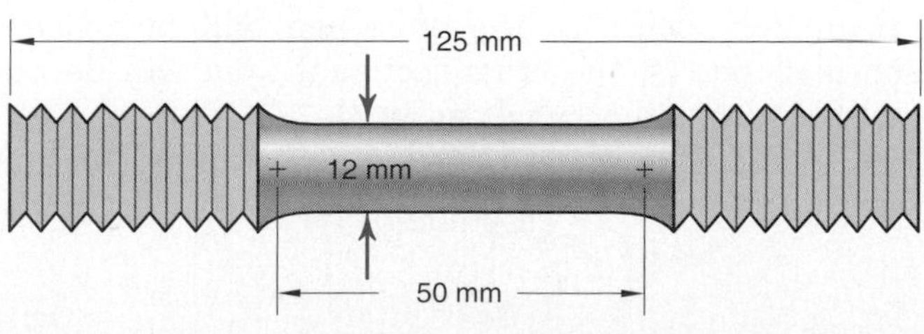

Figure 9.23

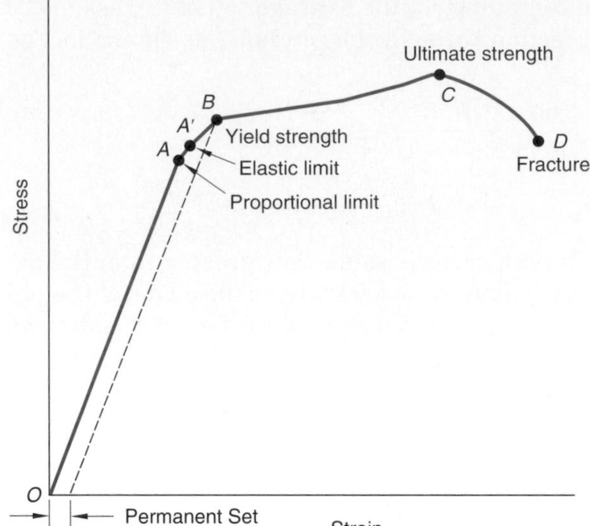

Figure 9.24 Stress-strain diagram.

A *stress-strain diagram* is a plot of the results of a tensile test (see Fig. 9.24). The shape of this diagram will vary somewhat for different materials, but in general there will first be a straight-line portion *OA*. Point *A* is the proportional limit—the maximum stress for which stress is proportional to strain.

At any stress up to point *A′*, called the *elastic limit,* the material will return to its original size once the load has been removed. At stresses higher than *A′*, permanent deformation (set) will occur. For most materials, points *A* and *A′* are very close together.

If the load is increased to the stress and strain at *B*, and then returned to zero, the stress-strain curve will follow the dotted line, leaving a permanent deformation (strain) in the material called a permanent set. The stress at *B* in Fig. 9.24 that causes a permanent set of 0.05 to 0.3 percent (depending on material) is termed the *yield strength.* The corresponding strain is 0.0005 to 0.003.

Point *C,* called the *ultimate strength,* is the maximum stress that the material can withstand. Between points *B* and *C,* a small increase in stress causes a significant increase in strain. At approximately point *C,* the specimen will begin to neck down sharply; that is, the cross-sectional area will decrease rapidly, and fracture will occur at point *D.*

The Fig. 9.25 photo shows a typical steel specimen prior to test and the specimen after it was pulled apart.

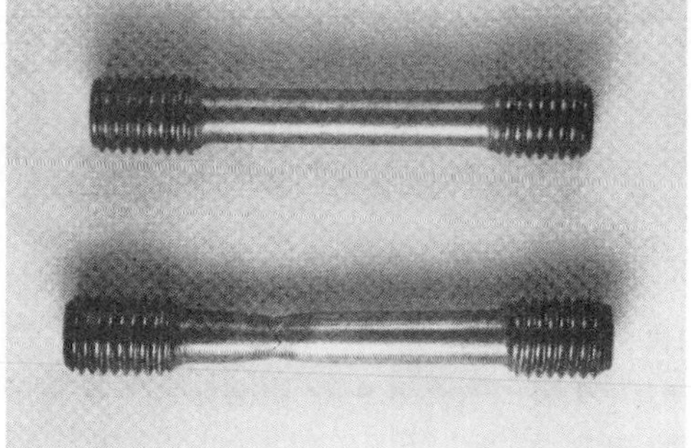

Figure 9.25
The lower portion of the photo shows the actual "necking-down" effect that occurs during a tensile test just prior to failure of the steel specimen.

9.13 Modulus of Elasticity

Approximately 300 years ago, Robert Hooke recognized the linear relationship between stress and strain. For stresses below the proportional limit, Hooke's law can be written

$$\epsilon = K\sigma \qquad (9.14)$$

where K is a proportionality constant. The modulus of elasticity E (the reciprocal of K) rather than K is commonly used, yielding

$$\sigma = E\epsilon \qquad (9.15)$$

Values of E for selected materials are given in Tab. 9.1.

Table 9.1 Modulus of elasticity for selected materials.

	E, psi	E, GPa
Cold-rolled steel	30×10^6	210
Cast iron	16×10^6	110
Copper	16×10^6	110
Aluminum	10×10^6	70
Stainless steel	27×10^6	190
Nickel	30×10^6	210

9.14 Design Stress

Obviously, most products or structures that engineers design are not intended to fail or become permanently deformed. The task facing the engineer is to choose the proper type and size of material that will perform correctly under the conditions likely to be imposed. Since the safety of the user and the liability of the producer (including the engineer) are dependent on valid assumptions, the engineer typically selects a design stress that is less than the yield strength. The ratio of the yield strength to the design stress is called the *safety factor.* For example, if the yield strength is 210 MPa and the design stress is 70 MPa, the safety factor, based on yield strength, is 3. Care must be exercised in reporting and interpreting safety factors because they are expressed in terms of both yield strength and ultimate (tensile) strength. Table 9.2 lists typical values used in structural design.

Table 9.2 Ultimate and yield strength.

	Ultimate strength		Yield strength	
	psi	MPa	psi	MPa
Cast iron	45×10^3	310	30×10^3	210
Wrought iron	50×10^3	345	30×10^3	210
Structural steel	60×10^3	415	35×10^3	240
Stainless steel	90×10^3	620	30×10^3	210
Aluminum	18×10^3	125	12×10^3	85
Copper, hard drawn	66×10^3	455	60×10^3	415

It should be noted that the United States still lists most of its standards in the English system. Conversions in this area will be necessary for some time to come.

Example problem 9.10 A round bar is 40.0 cm long and must withstand a force of 20.0 kN. What diameter must it have if the stress is not to exceed 140.0 MPa?

Solution

$$\sigma = \frac{F}{A}$$

$$A = \frac{F}{\sigma} = \frac{20.0 \times 10^3 \text{ N} \mid 10^6 \text{ mm}^2}{140.0 \times 10^6 \text{ N/m}^2 \mid 1 \text{ m}^2} = 143 \text{ mm}^2$$

$$= \frac{\pi d^2}{4}$$

$$143 = \frac{\pi d^2}{4}$$

$$d = 13.5 \text{ mm}$$

Example problem 9.11 Assume that the bar in Example prob. 9.10 is made from cold-rolled steel and is permitted to elongate 0.125 mm. Determine the required diameter.

Solution

1. We must use the quantities that define the relationship between stress and strain.

From Eqs. (9.12), (9.13), and (9.15)

$$\text{E} = \frac{\sigma}{\epsilon} = \frac{F/A}{\Delta \text{l}/\text{l}}$$

Therefore

$$\Delta \text{l} = \frac{\text{Fl}}{\text{AE}}$$

Δl is usually written as δ for deflection or elongation.

$$\delta = \frac{Fl}{AE}$$

and the necessary area for the given material and loading becomes

$$A = \frac{Fl}{\delta E}$$

2. Substituting the numerical quantities; E is 21×10^4 MPa from Tab. 9.1.

$$A = \frac{20 \times 10^3 \text{ N} | 0.400 \text{ m} | \text{m}^2 \qquad | 10^6 \text{ mm}^2}{125 \times 10^{-6} \text{ m} | \qquad | 21 \times 10^{10} \text{ N} | \text{m}^2}$$

$$= 304.8 \text{ mm}^2 = \frac{\pi d^2}{4}$$

$$d = 19.7 \text{ mm}$$

Example problem 9.12 Given the configuration in Fig. 9.26a, calculate the load that can be supported under the following design conditions:

(a) The pin at point *R*, enlarged in Fig. 9.26b, is 10.0 mm diameter. What load can be supported by the pin if the ultimate shear strength of the pin is 195 MPa and a safety factor of 2.0 based on ultimate strength is required?
(b) Using the load condition from (a) size cable *ST* if it has a design stress of 207.5 MPa.

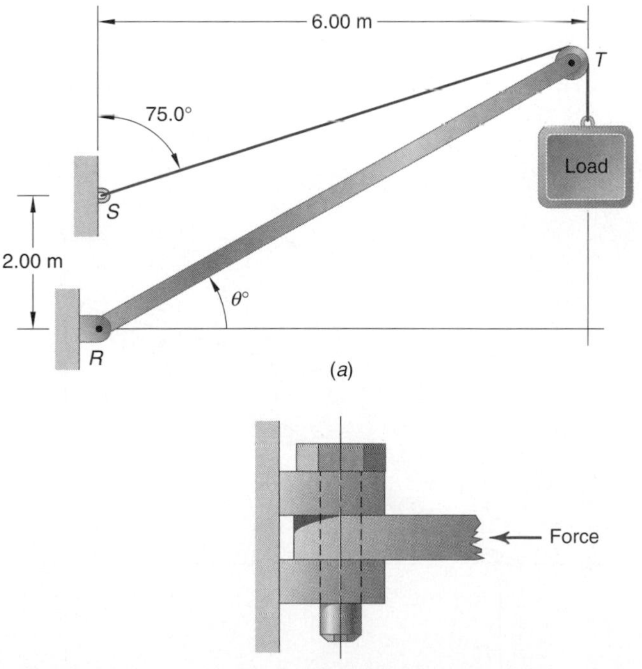

Figure 9.26

Solution (a)

1. Inspection of the FBD in Fig. 9.27 reveals four unknown forces and three equations of equilibrium. We will write the cable tension S, and the pin reactions, R_x and R_y in terms of the load, L, and then use the pin design conditions to find the largest permissible L.

2. Determine the geometry, H and θ

$$\tan 15° = \frac{H}{6.00}$$

$$H = 1.608 \text{ m}$$

$$\tan \theta = \frac{2 + H}{6}$$

$$\theta = 31.02°$$

3. Apply the equations of equilibrium

$$\circlearrowleft \Sigma M_S = 0$$

$$2.00\ R_x - 6.00\ L = 0$$

$$R_x = 3.00\ L$$

$$\circlearrowleft \Sigma M_T = 0$$

$$(2.00 + 1.608)R_x - 6.00\ R_y = 0$$

$$(3.608)\ (3L) - 6.00\ R_y = 0$$

$$R_y = 1.804\ L$$

$$R = \sqrt{R_x^2 + R_y^2}$$

$$= \sqrt{(3L)^2 + (1.804L)^2}$$

$$= 3.501L$$

Figure 9.27

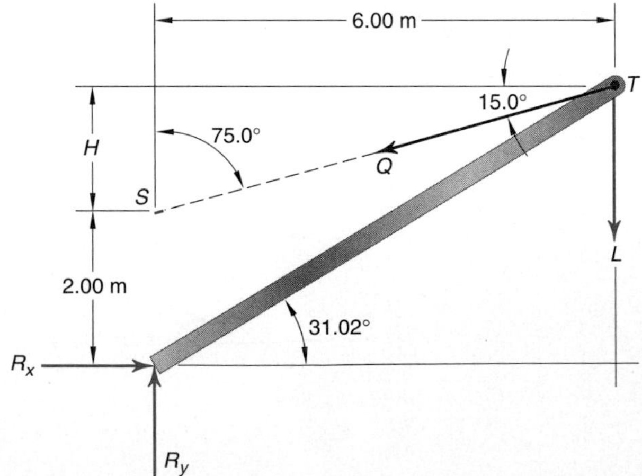

4. Determine pin limits from design conditions. Note that the pin is in double shear.

$$\text{Design stress} = \frac{\text{Ultimate strength}}{\text{Safety factor}}$$

$$= \frac{195 \text{ MPa}}{2.0} = 97.5 \text{ MPa}$$

$$\tau = \frac{R}{2A}$$

$$R = 2\tau A = 2\tau\left(\frac{\pi d^2}{4}\right)$$

$$= \frac{2|97.5 \times 10^6 \text{ N}|\pi|(10 \text{ mm})^2|\text{m}^2}{|\text{m}^2 \quad |4| \quad |10^6 \text{ mm}^2}$$

$$= 15\ 300 \text{ N}$$

$$L = \frac{R}{3.501} = \frac{15\ 300}{3.501} = 4.37 \text{ kN}$$

Solution (b)

1. Determine cable load Q from part (a)

$$\circlearrowleft \Sigma M_R = 0$$

$$-6.00\ L + 2Q \cos 15° = 0$$

$$Q = 3.106\ L$$

$$= 13.57 \text{ kN}$$

2. Calculate cable size. The design stress is 207.5 MPa.

$$\sigma = \frac{F}{A}$$

$$A = \frac{F}{\sigma} = \frac{13\ 570 \text{ N}|\text{m}^2}{|207.5 \times 10^6 \text{ N}}$$

$$= 6.54 \times 10^{-5} \text{ m}^2$$

$$\frac{\pi d^2}{4} = 6.54 \times 10^{-5} \text{ m}^2$$

$$d^2 = 8.32 \times 10^{-5} \text{ m}^2$$

$$d = 9.12 \times 10^{-3} \text{ m}$$

$$= 9.13 \text{ mm}$$

A practical choice would be the next larger size that is commercially available.

9.15 Key Terms and Concepts

Statics
Mechanics of materials
Scalars
Vectors
Forces
  Collinear forces
  Coplanar forces
  Concurrent forces
Transmissibility
Resolution of forces
Moments
Couples
Free-body diagram (FBD)
Equilibrium conditions
Stress
Strain
Modulus of elasticity
Hooke's law
Design stress
Safety factor

Problems

For Problems 9.1 through 9.4, find the x and y components of the force F. The angle θ is measured positive counterclockwise from the positive x-axis. Include a sketch of the force F and its components.

	F	θ
9.1	350 lb	50°
9.2	1200 N	135°
9.3	68 lb	225°
9.4	85 kN	310°

For Problems 9.5 through 9.7, find the resultant of two concurrent forces: F, which makes an angle θ with respect to the positive x-axis, and G, which makes an angle ϕ with respect to the positive x-axis. Show a sketch of $\bar{F}$ and $\bar{G}$ and the resultant.

	F	θ	G	ϕ
9.5	850 N	120°	480 N	60°
9.6	240 lb	75°	320 lb	330°
9.7	150 kN	90°	75 kN	225°

For Problems 9.8 through 9.10, find the resultant of two concurrent forces; $\bar{R}$ and $\bar{S}$ for which the direction is specified by slope, expressed as rise and run values. Show a sketch of $\bar{R}$ and $\bar{S}$ and the resultant.

	R	Rise	Run	S	Rise	Run
9.8	36 lb	3	4	52 lb	−3	4
9.9	725 N	−5	−12	435 N	2	−3
9.10	322 lb	4	−5	248 lb	2	2

For Problems 9.11 through 9.13, find the resultant of the system of concurrent, coplanar forces shown in Fig. 9.28. Force $\bar{A}$ acts vertically. Express the resultant in each problem in Cartesian vector notation.

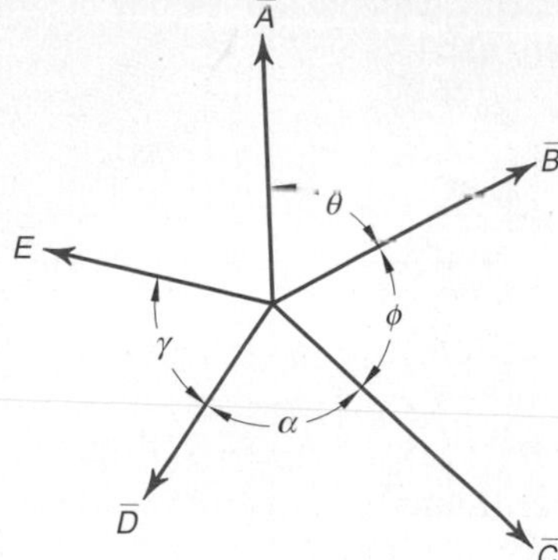

Figure 9.28

	A	*B*	*C*	*D*	*E*	θ	φ	δ	γ
9.11	55 lb	45 lb	72 lb	32 lb	38 lb	60	60	60	60
9.12	1.5 kN	1.3 kN	2.2 kN	1.1 kN	1.3 kN	45	70	90	50
9.13	8.0 lb	7.0 lb	12.0 lb	5.0 lb	6.0 lb	30	90	60	45

For Problems 9.14 through 9.20, the force $\bar{F}$ goes through point O and makes an angle Θ with the horizontal, as shown in Fig. 9.29. Calculate the moment of $\bar{F}$ about points *A* and *B*, assigning positive values to counterclockwise moments.

	F	Θ
9.14	150 lb	0°
9.15	75 lb	90°
9.16	325 lb	45°
9.17	925 N	30°
9.18	760 N	180°
9.19	1.55 kN	270°
9.20	3.20 kN	120°

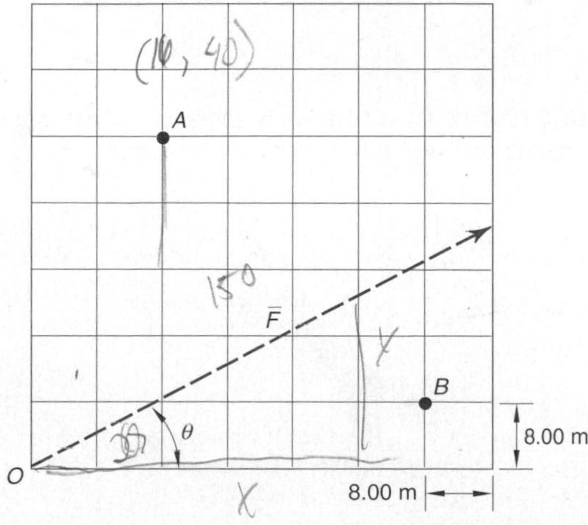

Figure 9.29

9.21 Construct a free-body diagram for the joint *R* in Fig. 9.30. *RP* and *RS* are cables.

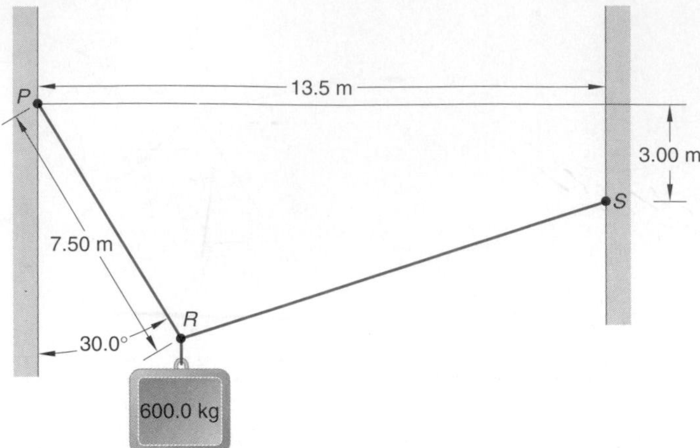

Figure 9.30

9.22 Construct a free-body diagram of the joint *K* in Fig. 9.31. The suspended block has a mass of 42 kg.

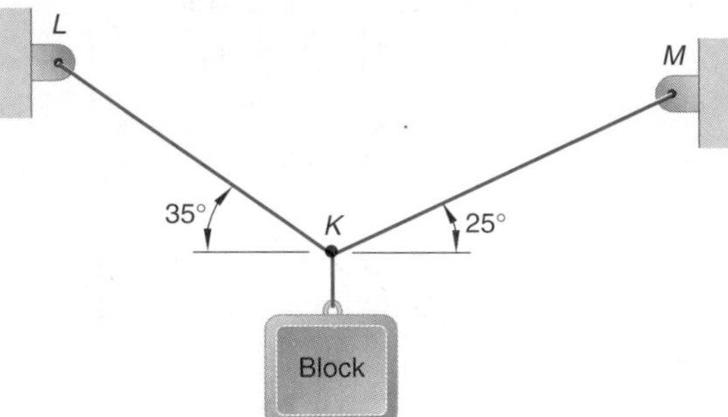

Figure 9.31

9.23 Construct a free-body diagram of the sphere *S* in Fig. 9.32 resting against two smooth surfaces. The sphere has a mass of 350 kg.

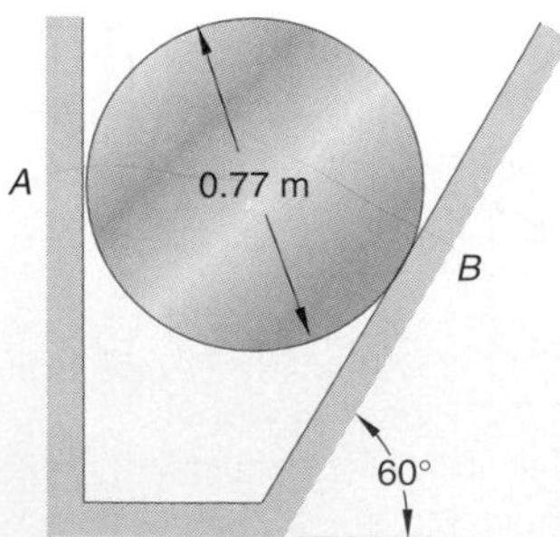

Figure 9.32

9.24 Construct a free-body diagram for each of the cylinders A and B in Fig. 9.33 supported at the contact surfaces R, S, and T. The cylinders each have a mass of 25 kg. All surfaces are smooth.

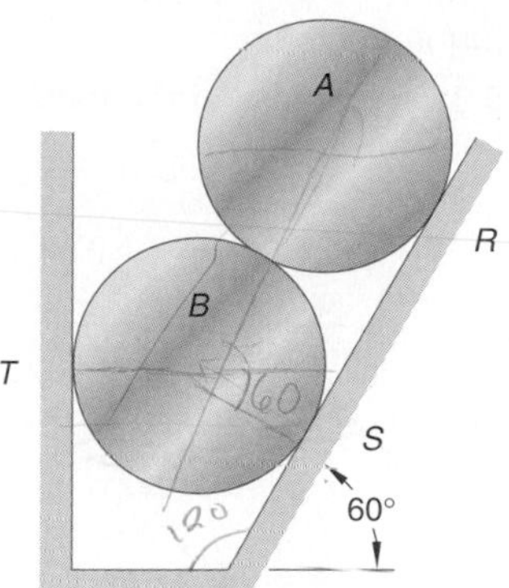

Figure 9.33

9.25 Construct a free-body diagram of the beam shown in Fig. 9.34. Neglect the mass of the beam.

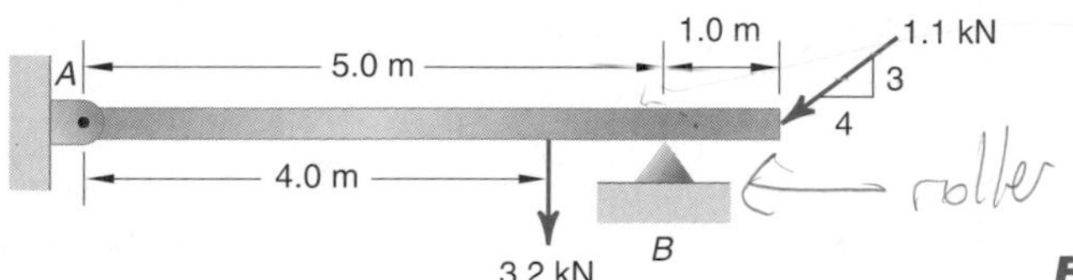

Figure 9.34

9.26 Construct a free-body diagram of the beam shown in Fig. 9.35. The beam weighs 15 lb per foot of length. Assume that the weight of the beam acts at the center.

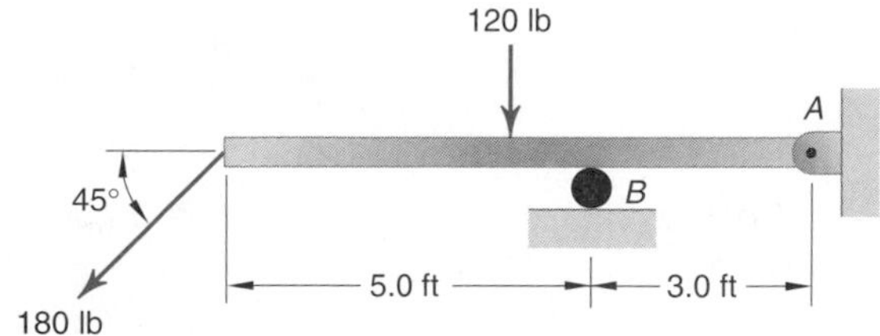

Figure 9.35

9.27 Find the tension in the cables RP and RS in Fig. 9.30.

9.28 Find the tension in the cables KL and KM in Fig. 9.31. The block weighs 440 lb.

9.29 Find the reactions of the smooth surfaces at A and B on the sphere in Fig. 9.32. The sphere weighs 825 lb.

9.30 Find the reactions of the smooth surfaces at R, S, and T on the cylinders A and B in Fig. 9.33. The cylinders each weigh 1 200 lb.

9.31 Find the reactions at A and B for the beam in Fig. 9.34. Assume the beam is massless.

9.32 Find the reactions at A and B for the beam in Fig. 9.35. The beam weighs 35 lb per foot of length. Assume that the weight of the beam acts at the center.

9.33 The mass of the frame in Fig. 9.36 may be neglected. Find the reactions at A and B.

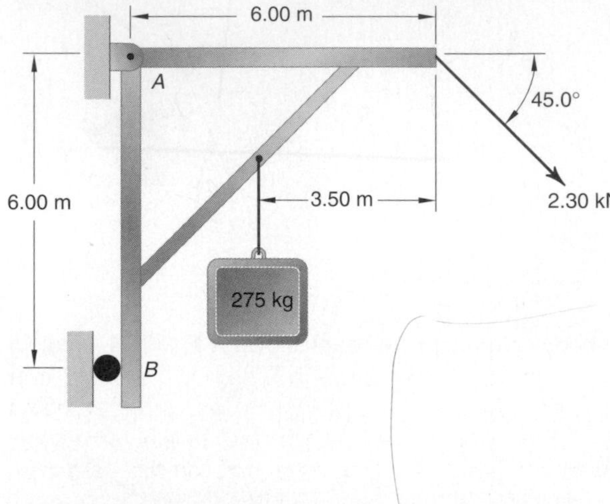

Figure 9.36

9.34 For the structure in Fig. 9.37, determine the tension in the cable ST and the pin reaction at R. The beam weighs 21 lb per foot of length. Assume the weight of the beam acts at the center.

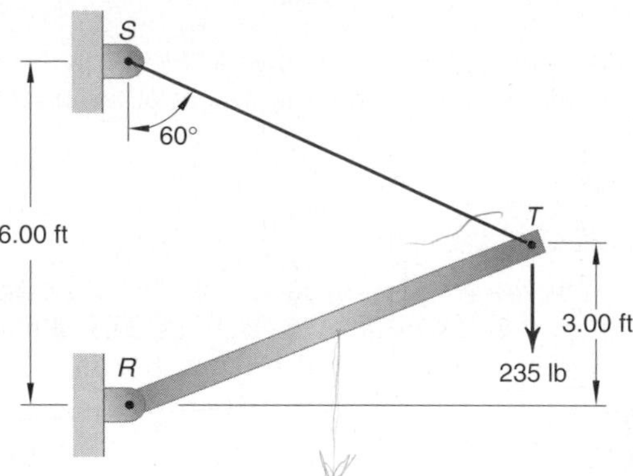

Figure 9.37

9.35 A hot-air balloon has a lifting force of 650 lb. A single rope is used to hold it in a gentle breeze causing the rope to make an angle of 23° with the vertical. Find the tension in the rope.

9.36 Find the resultant of the force system in Fig. 9.38 and express it in Cartesian vector notation and determine the angles θ_x, θ_y, and θ_z that the resultant makes with the x, y, and z axes, respectively.

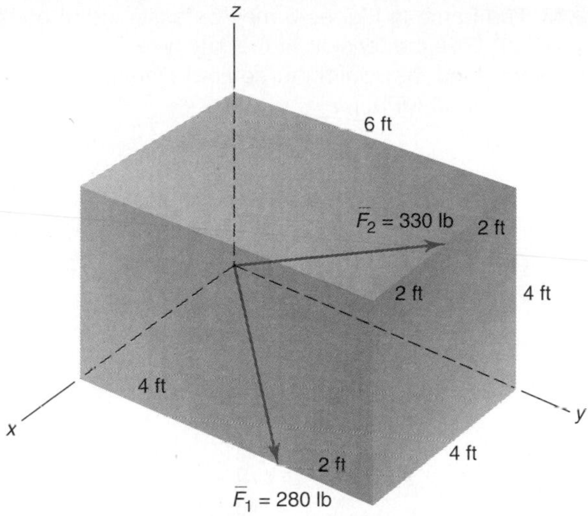

Figure 9.38

9.37 A hot-air balloon is tethered by three ropes *A, B,* and *C* all tied to a ring on the bottom of the gondola 15 meters above the ground (coordinates 0, 0, 15). The cables *A, B,* and *C* are tied to ground-level tethers at coordinates (6, 4, 0), (−3, 4, 0), (0, −6, 0). The lifting force on the balloon is 1060 lb.

(*a*) Write the Cartesian unit vectors for each of the ropes.

(*b*) Write the force vectors for the tension in terms of the unit vectors and unknown magnitudes.

(*c*) Apply the conditions of equilibrium for a three-dimensional concurrent force system.

$\Sigma F_x = 0$ (sum of $\hat{i}$ components)

$\Sigma F_y = 0$ (sum of $\hat{j}$ components)

$\Sigma F_z = 0$ (sum of $\hat{k}$ components)

(*d*) Solve the resulting system of equations for the tensions.

9.38 A round reinforcing rod 36 inches long is required to support a tensile load of 4200 lb. Determine the minimum diameter if the stress cannot exceed 25 000 psi (lb per square inch) or elongate more than 0.015 in.

9.39 The short cold-rolled steel tube in Fig. 9.39 is used as a support member. Determine the allowable load *P* that can be applied if the compressive stress is limited to 20 000 psi and the deformation to 0.010 in.

9.40 What wall thickness for the tube in Fig. 9.39 would allow both maximum stress and maximum deformation to occur at the same time? For what load *P* does this occur?

9.41 Refer to Fig. 9.26b. If the force is 4 700 lb and the allowable shear stress is 18 000 psi, determine the minimum required pin diameter.

9.42 A pin made from structural steel is 18 mm in diameter and is in double shear. What load can be supported by the pin if the ultimate shear strength is 195 MPa and the factor of safety is 4.0?

Figure 9.39

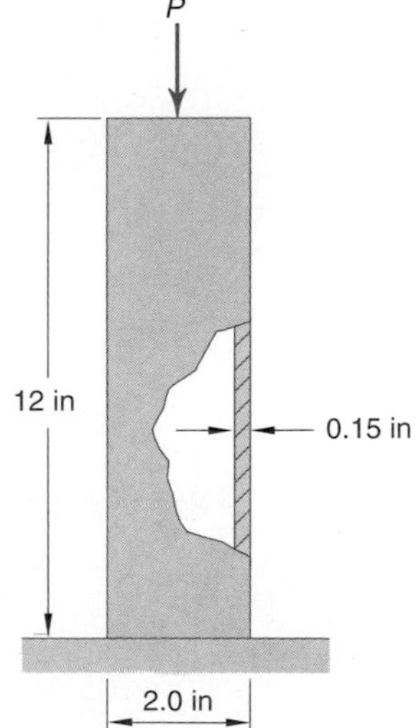

9.43 The frame in Fig. 9.40 may be assumed to be massless.

(*a*) Find the tension in the guy wire.

(*b*) Find the minimum required diameter of wire if the design stress is 95 MPa.

Figure 9.40

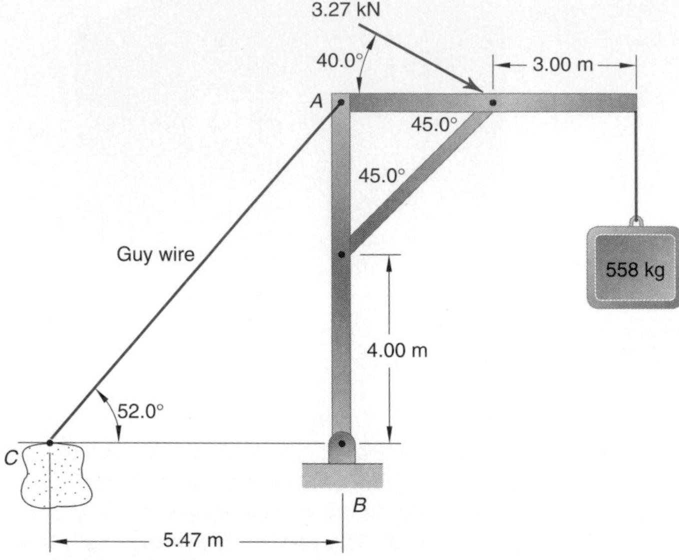

9.44 The beam in Fig. 9.41 has no mass.

(*a*) Find the tension in the cable.

(*b*) If the design stress is 25 MPa, what diameter cable is needed?

(*c*) Find the pin reaction at *P*.

(*d*) If the design shear stress is 15 MPa, and the pin is in single shear, what diameter is required?

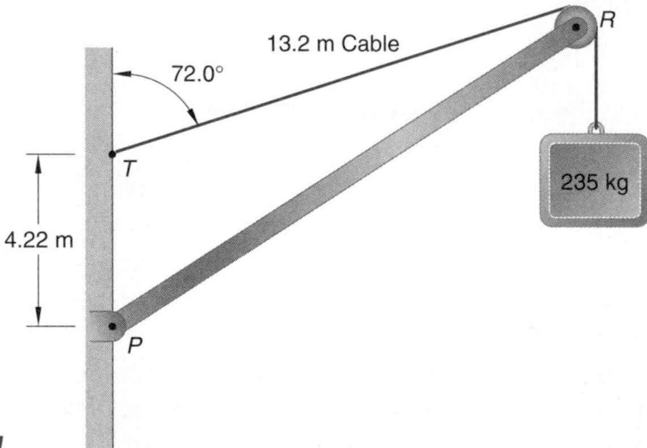

Figure 9.41

9.45 A business wants to hang a sign from the end of a horizontal pole. The pole will be attached to a wall by a pin and supported by a cable, as shown in Fig. 9.42. The pole and cable are each 8.0 m long. The sign has a mass of 100.0 kg. Ultimate strength is 1.0×10^9 Pa. A safety factor of 5 should be employed when designing the cable. At what point should the cable be attached to the pole in order to minimize the required cable diameter D? You should:

(*a*) Show the derivation of D as a function of X.
(*b*) Plot D vs. X.
(*c*) Find the value of X which minimizes D and state the minimum D.

As an alternate approach, write a computer program that permits the user to input the mass of the sign, the lengths of the cable and pole, the ultimate strength, and the safety factor. The program should print a table of values of D for corresponding values of X. In addition, the program should calculate the value of X and D to some specified accuracy where D is a minimum.

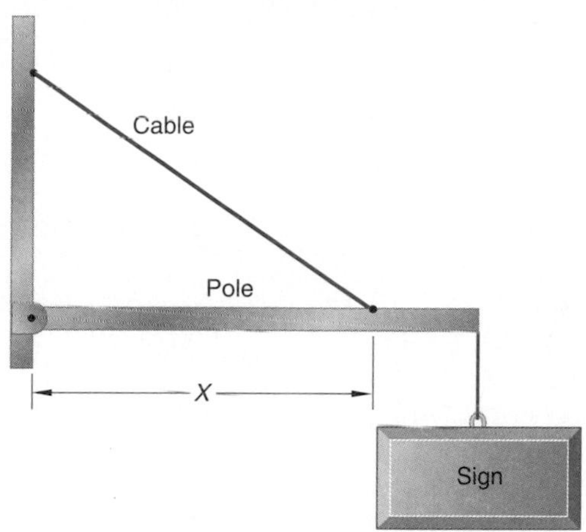

Figure 9.42

9.46 The boom *AB* (Fig. 9.43) is raised and lowered by the cable *BC*. The boom has a mass of 850 N. For the condition that *AB* is 9.5 m long and the weight, *W*, is 4 800 N

(*a*) Plot the pin reaction at *A* and the tension in the cable *BC* for $0 \leq \theta \leq 75°$ in 5° increments.
(*b*) What size cable is needed if the cable design stress is 180 MPa?

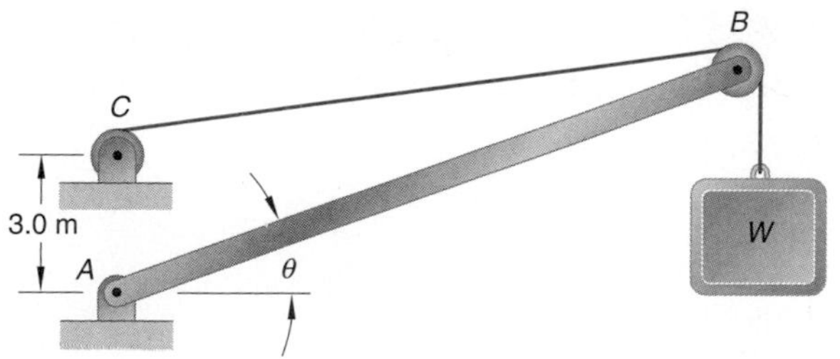

Figure 9.43

9.47 The beam AB in Fig. 9.44 has a weight of 480 N that may be assumed to act at the center.

(*a*) For a force F of 1 800 N, plot the reactions at A and B from $0 \leq x \leq 13$ in increments of 1.0 m.

(*b*) Determine the value of F which makes the reactions at A and B equal for $x = 7.6$ m.

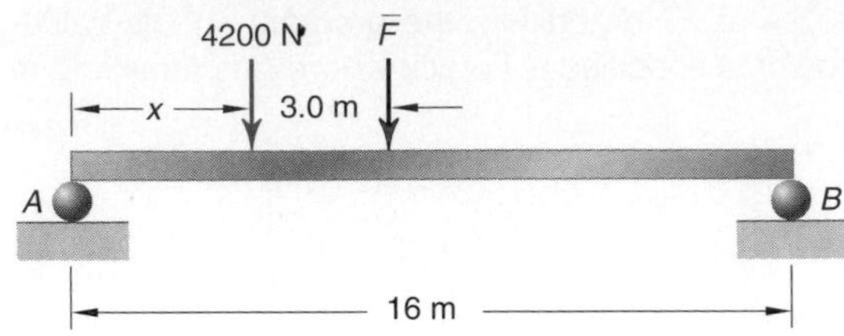

Figure 9.44

10 Material Balance

10.1 Introduction

We depend a great deal on industries that produce food, household cleaning products, energy for heating and cooling homes, fertilizers, and many other products and services. These process industries, as they are called, are continually involved with the distribution, routing, blending, mixing, sorting, and separation of materials. (See Fig. 10.1 for one example of a processing system.)

A typical process problem that an engineer might be called on to solve is exemplified by the drying process, shown schematically in Fig. 10.2. You as a process engineer designing a system to dry grain would most likely know the percent moisture (on a mass basis) of the wet grain, the desired moisture content for the dried grain, and the amount of grain to be dried in a specific amount of time. You would then have to calculate the flow rate of dry heated air required to be forced through the grain. Knowing the air flow rate, you could then design the mechanical system of heaters, motors, blowers, and ducting.

To perform computations involving material flow in a process, you will use an engineering analysis technique called *material balance,* which is based on the principle of conservation of mass.

Figure 10.1
Highway construction materials form several streams through this machine. (Kerry B. Gibson)

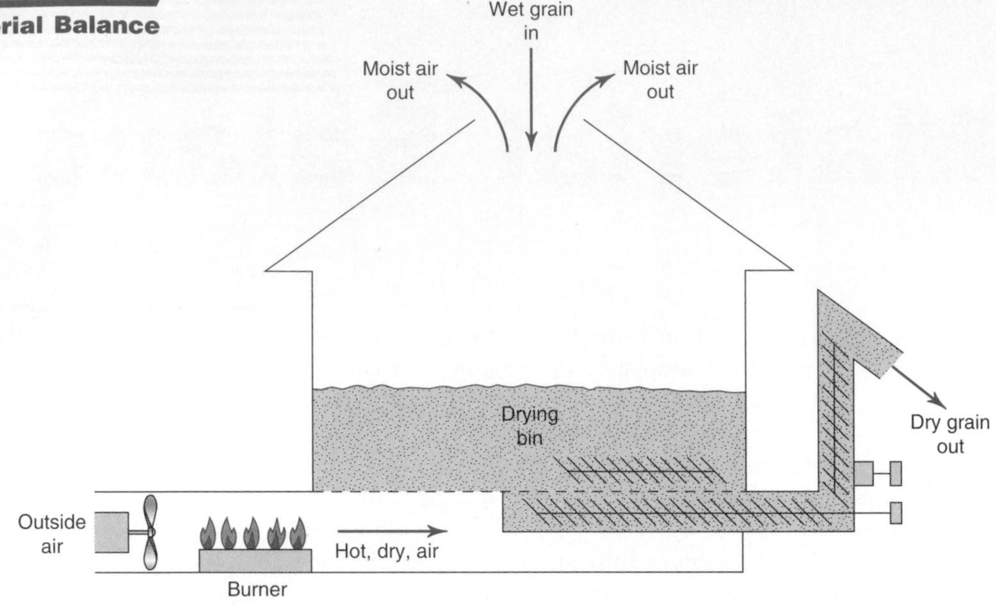

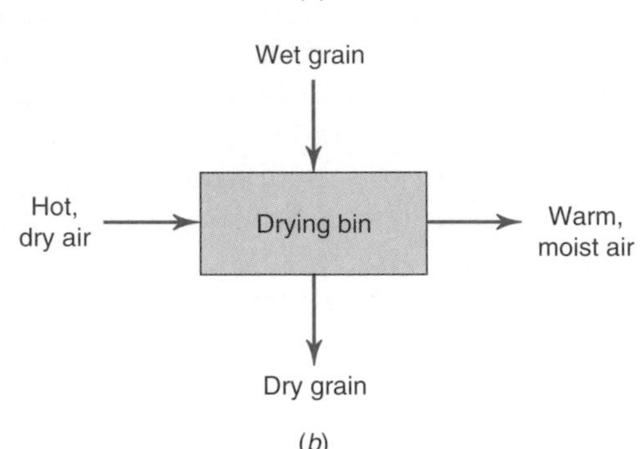

Figure 10.2
(a) Cross section of a typical grain-drying process; (b) working schematic used to depict the actual process so that the necessary calculations can be performed.

10.2 Conservation of Mass

The conservation-of-mass principle is a very useful concept in the field of engineering analysis. Simply stated, it says that, excluding nuclear reactions, mass is neither created nor destroyed. We know that mass is converted to energy in a nuclear reactor so the conservation-of-mass principle does not apply to the reaction itself.

Before we apply the conservation-of-mass principle to a material balance problem, additional concepts and terminology will be introduced. Figure 10.3 illustrates a number of these terms.

A *system* can be any designated portion of the universe with a definable boundary. Whenever mass crosses the boundary either

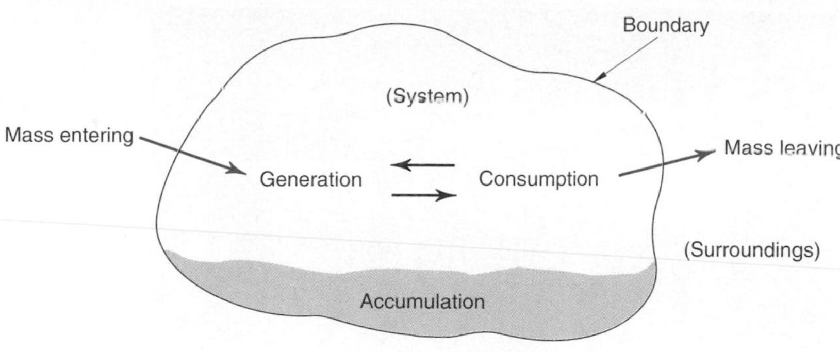

Figure 10.3
Illustration of a typical system with a defined boundary.

into or out of the system, it must be considered. In certain situations, the amount of mass entering the system is greater than the amount leaving. In the absence of chemical reactions, this results in an increase of mass within the system called *accumulation.* If the mass leaving the system is greater than that entering, the accumulation is negative.

When chemical reactions are included within the system boundaries, chemical reactants are consumed in the formation of reaction products. A simple example would be the dehydrogeneration of ethane in a reactor—this is, $C_2H_6 \rightarrow C_2H_4 + H_2$. One constituent is consumed and two others are generated. Thus, if chemical reactions occur, it is necessary to account for the consumption of some elements or compounds and the generation of others. It is important to understand that, even considering chemical reactions, mass is conserved. In the preceding example ($C_2H_6 \rightarrow C_2H_4 + H_2$), the number of atoms of carbon and hydrogen remains constant.

Before using the conservation-of-mass principle to analyze a system, you must understand the terminology and be able to account for all the constituents that enter, leave, or change within the system boundaries.

Consider a familiar example to more clearly visualize the definition of a system. Let the system boundary be the city limits of your city and let the population represent mass. A person moving to the city could be analogous to mass entering while someone moving away would represent mass leaving. If a person is born within the city, we have generation and if someone dies, we have consumption of population. The net of persons moving in and out and of births and deaths is the change in population of the city, or accumulation.

When all considerations are included, the conservation-of-mass principle applied to a system or to system constituents can be expressed as

$$\text{Input} + \text{generation} - \text{output} - \text{consumption} = \text{accumulation} \qquad (10.1)$$

Figure 10.4
This arrangement of ductwork directs the air flow in a large building. It illustrates the complicated network in an actual situation (Stanley Consultants).

10.3 Processes

Two types of processes typically analyzed are the batch process and the rate process. In a *batch process,* materials are put or placed into the system before the process begins and are removed after the process is complete. Cooking is a familiar example. Generally, you follow a recipe that calls for specific ingredients to be placed into a system that produces a processed food.

A *rate-flow process* involves the continuous time rate of flow of inputs and outputs. The process is performed continuously as mass flows through the system. An example of a rate-flow process is a pipe delivering water to a tank at the rate of 2.0 kg/s.

Rate processes may be classified as either *uniform* or *nonuniform, steady* or *unsteady.* A process is uniform if the flow rate is the same throughout the process, which means the input rate must equal the output rate. It is steady if rates do not vary with time. Solution of material balance problems involving nonuniform and/or unsteady flows may require the use of differential equations. However, many important processes either are or are nearly uniform and steady and thus can be analyzed with algebraic techniques.

Since this chapter is intended to be an introductory look at conservation of mass, we will now make a number of simplifying assumptions.

Many engineering problems involve chemical reactions, but if we assume no such reactions, then Eq. (10.1) can be reduced to

$$\text{Input} - \text{output} = \text{accumulation} \tag{10.2}$$

If we assume for a batch process that we take out at the end of the process all of the mass we placed into the system at the

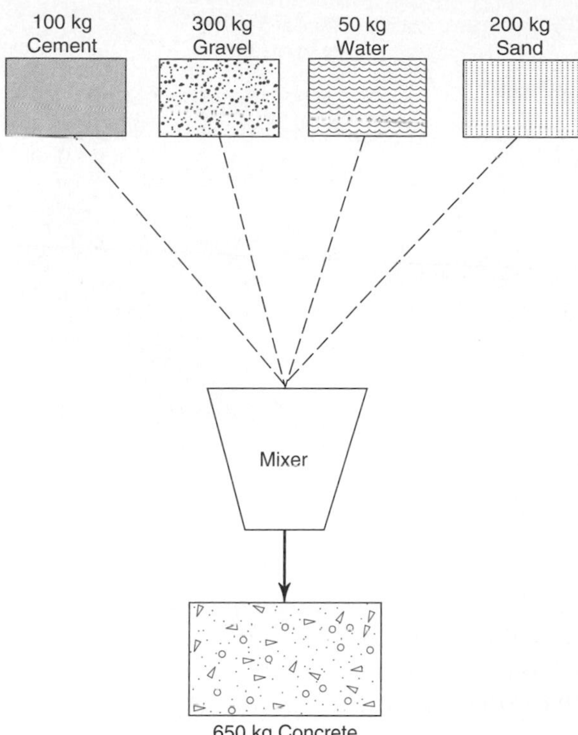

Figure 10.5
The constituents of a proper batch mix of concrete.

beginning, then the accumulation term is zero and Eq. (10.2) can be written

$$\text{Total input} = \text{total output} \tag{10.3}$$

Figure 10.5 is an example of a batch process for a concrete mixer.

For continuous flow, if we assume a uniform, steady rate process, the accumulation term is also zero, so Eq. (10.2) written on a rate basis reduces to

$$\text{Rate of input} = \text{rate of output} \tag{10.4}$$

Figure 10.6 illustrates a rate process for a coal dewatering process.

Although Eqs. (10.3) and (10.4) seem so overly simple as to be of little practical use, application to a given problem may be complicated by the need to account for several inputs and outputs as well as for many constituents in each input or output. The simplicity of the equation is in fact the advantage of a material balance approach, because order is brought to seemingly disordered data.

10.4 A Systematic Approach

Material balance computations require the manipulation of a substantial amount of information. Therefore, it is essential that a systematic procedure be developed and followed. If a systematic

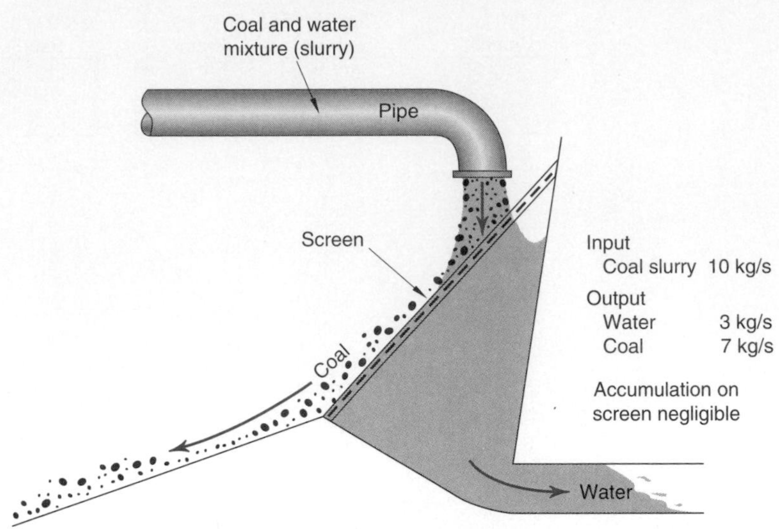

Figure 10.6
A coal dewatering system shown as a rate process. Sampling the downstream coal and water flows would provide a more detailed material balance than the single one shown because the coal would carry some water with it and the water side would contain some fine coal particles.

approach is used, material balance equations can be written and solved correctly in a straightforward manner. The following list of steps is recommended as a procedure for solving material balance problems.

1. Identify the system(s) involved.

2. Determine whether the process is a batch or rate process and whether a chemical reaction is involved. If no reaction occurs, it is evident that the material balance involves compounds. If a reaction is to occur, elements must be involved and must be balanced. In a process involving chemical reactions, additional equations based on chemical composition may be required in order to solve for the unknown quantities.

3. Construct a diagram showing the feeds (inputs) and products (outputs) schematically.

4. Label known material quantities or rates of flow.

5. Identify each unknown input and output with a symbol.

6. Apply Eq. (10.3) or (10.4) for each constituent as well as for the overall process. Note that not all equations written will be algebraically independent.

7. Solve the equations (selecting an independent set) for the desired unknowns and express the result in an understandable form.

Example problem 10.1 Drinking water can be obtained from saltwater by partially freezing the saltwater to create salt-free ice and a brine solution. If saltwater is 3.50 percent salt by mass and the brine solution is found to be an 8.00 percent concentrate by mass, determine how many kilograms of saltwater must be processed to form 2.00 kg of ice.

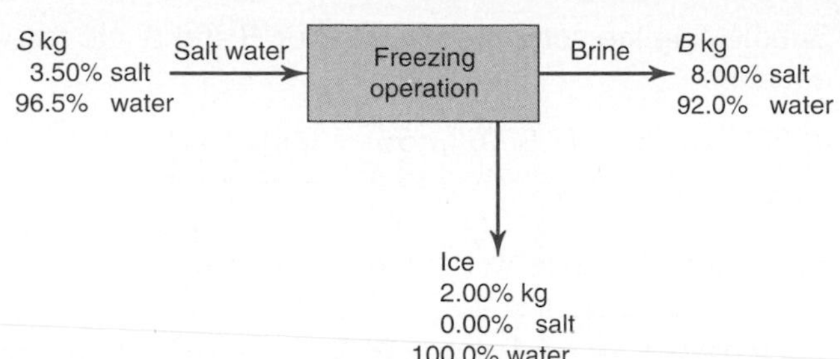

Figure 10.7
Schematic diagram of the saltwater freezing operation.

Solution

1. The system in this example problem involves a freezing operation.

2. The freezing operation is a batch process because a fixed amount of product (ice) is required. There are no chemical reactions.

3. A diagram of the process is shown in Fig. 10.7.

4. Saltwater is the input to the system, with brine and ice taken out at the end (see Fig. 10.7).

5. Appropriate symbols are used to identify unknown quantities (see Fig. 10.7).

6. The material balance equation for each constituent as well as for the overall process is written. It is important to understand that the material balance equation (Eq. [10.3]) is applicable for each constituent as well as for the overall process. In this example, three equations are written, but only two are independent. That is, the overall balance equation is the sum of the salt and water balance equations. Thus we have a good method of checking the accuracy of the equations we have written.

Equation	Input = output
Overall balance	$S = B + 2.00$
Salt balance	$0.035S = 0.08B$
Water balance	$0.965S = 0.920B + 2.00$

7. The equations are solved by substitution for S from the overall balance equation into the salt balance equation.

$$0.035(B + 2.00) = 0.08B$$

$$0.045B = 0.070$$

$$B = 1.56 \text{ kg}$$

Since $S = B + 2.00$, then

$$S = 3.56 \text{ kg}$$

The water balance equation was not used to solve for B and S but can serve as a check of the results (that is, it should balance).

Substituting the computed values for B and S into the water balance equation:

$$(0.965)\,(3.56) \stackrel{?}{=} (0.920)\,(1.56) + 2.00$$

$$3.4354 = 3.4352$$

which does balance within the roundoff error.

Example problem 10.2 A process to remove water from solid material consists of a centrifuge and a dryer. If 35.0 t/h of a mixture containing 35.0 percent solids is centrifuged to form a sludge consisting of 65.0 percent solids and then the sludge is dried to 5.00 percent moisture in a dryer, how much total water is removed in 24-hour period?

Solution There are three possible systems involved in this problem: the centrifuge, the dryer, and the combination (see Fig. 10.8). The operation in this system is a continuous flow process. There are no chemical reactions and the process is steady and uniform, thus

Rate of input = rate of output

The following equations are written for the process illustrated in Fig. 10.8. The overall process is illustrated in Fig. 10.8a, with subsystem diagrams for the centrifuge and the dryer shown in Fig. 10.8b and c, respectively. The overall balance equation for a selected system is the sum of the constituent balance equations for that system. This means that the set of equations written for a selected system

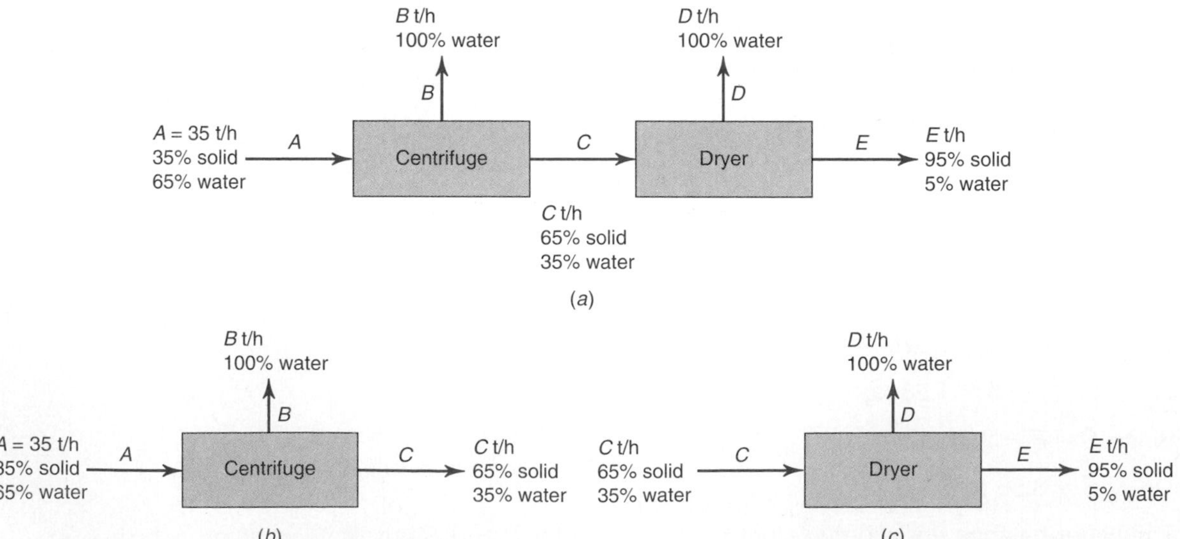

Figure 10.8 Schematic diagram illustrating flow process inputs and outputs depending on system boundaries.

are not all independent when the overall balance equation and the constituent balance equations are included.

For the entire system (Fig. 10.8a):

1. Solid balance $0.35\ (35) = 0.95E$
2. Water balance $0.65\ (35) = B + D + 0.05E$
3. Overall balance $35 = B + D + E$

For the centrifuge (Fig. 10.8b):

4. Solid balance $0.35\ (35) = 0.65C$
5. Water balance $0.65\ (35) = B + 0.35C$
6. Overall balance $35 = B + C$

For the dryer (Fig. 10.8c):

7. Solid balance $0.65C = 0.95E$
8. Water balance $0.35C = D + 0.05E$
9. Overall balance $C = D + E$

Solve for rate of mass out of centrifuge (C) from Eq. (4):

$$C = \frac{0\ 35\ (35)}{0.65}$$

$$= 18.85 \text{ t/h}$$

Solve for rate of water out of centrifuge (B) from Eq. (6):

$$B = 35 - C$$

$$= 35 - 18.85$$

$$= 16.15 \text{ t/h}$$

Solve for rate of mass out of dryer (E) from Eq. (7):

$$E = \frac{0.65C}{0.95}$$

$$= \frac{0.65\ (18.85)}{0.95}$$

$$= 12.90 \text{ t/h}$$

Solve for rate of water out of dryer (D) from Eq. (9):

$$D = C - E$$

$$= 18.85 - 12.90$$

$$= 5.95 \text{ t/h}$$

Check the results obtained in the three balance equations for the entire system, Eqs. (1), (2), and (3).

$$(0.35)(35) \stackrel{?}{=} (0.95)(12.90)$$

$$12.25 = 12.25 \qquad \text{(checks)}$$

$$(0.65)(35) \stackrel{?}{=} 16.15 + 5.95 + (0.05)(12.90)$$

$$22.75 = 22.75 \qquad \text{(checks)}$$

$$35 \stackrel{?}{=} 16.15 + 5.95 + 12.90$$

$$35 = 35 \qquad \text{(checks)}$$

All the results check so then calculate total water removed in 24 h:

$$\text{Total water} = (B + D)\,24$$
$$= (16.15 + 5.95)\,24$$
$$= (22.10)\,24$$
$$= 5.30 \times 10^2 \text{ t}$$

A general problem that would involve typical material balance consideration is a standard evaporation, crystallization, recycle process. Normally this type of system involves continuous flow of some solution through an evaporator. Water is removed, leaving the output stream more concentrated. This stream is fed into a crystallizer where it is cooled, causing crystals to form. These crystals are then filtered out, with the remaining solution recycled to join the feed stream back into the evaporator. This system will be more clearly illustrated in the following example problem.

Example problem 10.3 A solution of potassium chromate (K_2CrO_4) is to be used to produce K_2CrO_4 crystals. The feed to an evaporator is 2.50×10^3 kilograms per hour of 40.0 percent solution by mass. The stream leaving the evaporator is 50.0 percent K_2CrO_4. This stream is then fed into a crystallizer and is passed through a filter. The resulting filter cake is entirely crystals. The remaining solution is 45.0 percent K_2CrO_4. The 45 percent solution that passed through the filter is recycled. Calculate the total input to the evaporator, the feed rate to the crystallizer, the water removed from the evaporator, and the amount of pure K_2CrO_4 produced each hour.

Solution See Fig. 10.9. There are different ways the system boundaries can be selected for this problem—that is, around the entire system, around the evaporator, around the crystallizer-filter, and so on. There are no chemical reactions that occur in the process, and the process is steady and uniform, thus

Rate of input = rate of output

The following balance equations can be written, not all of which are independent.

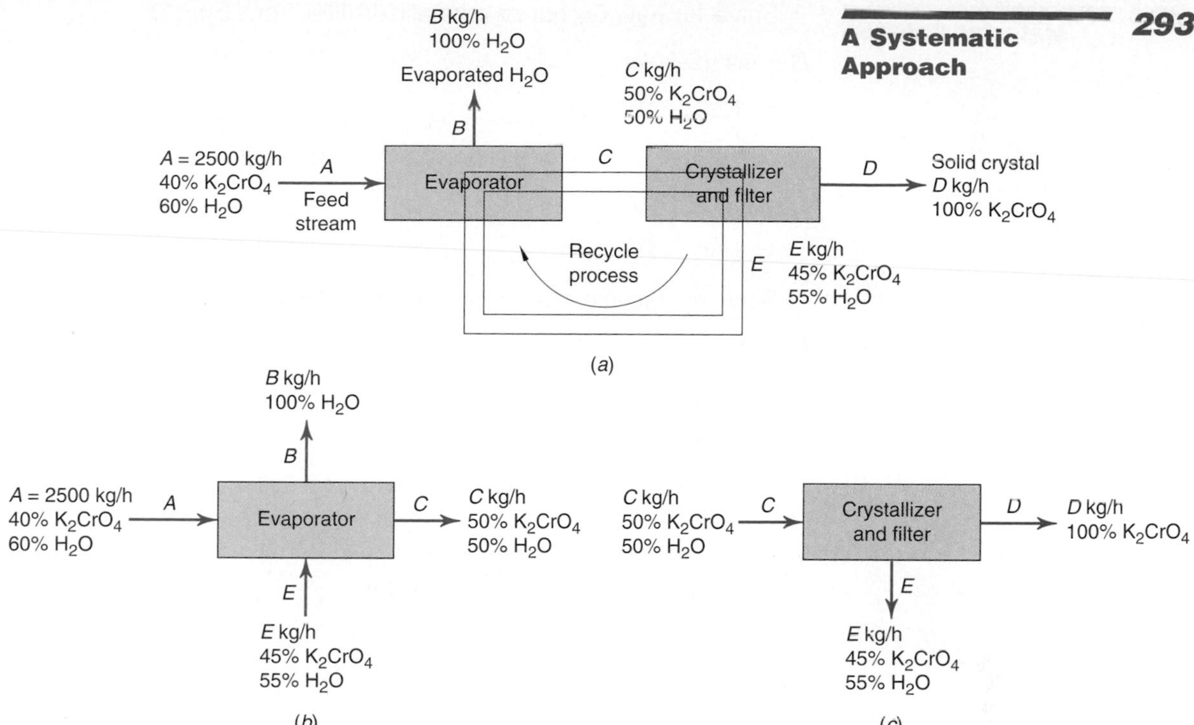

Figure 10.9
(a) Schematic of the overall system; (b) and (c) are diagrams where boundaries are selected around individual components.

For the entire system (Fig. 10.9a):

1. K_2CrO_4 balance $\quad 0.40\ (2\ 500) = D$
2. H_2O balance $\quad 0.60\ (2\ 500) = B$
3. Overall balance $\quad 2\ 500 = B + D$

For the evaporator (Fig. 10.9b):

4. K_2CrO_4 balance $\quad (0.40)\ (2\ 500) + 0.45E = 0.50C$
5. H_2O balance $\quad (0.60)\ (2\ 500) + 0.55E = B + 0.50C$
6. Overall balance $\quad 2\ 500 + E = B + C$

For the crystallizer-filter (Fig. 10.9c):

7. K_2CrO_4 balance $\quad 0.50C = D + 0.45E$
8. H_2O balance $\quad 0.50C = 0.55E$
9. Overall balance $\quad C = D + E$

We will first develop the solution by selecting appropriate equations from the preceding set and solving them by hand. Then we will place a set of four equations in four unknowns in a math solver to obtain (and verify in this case) the solution we have found by hand.

Solve for K_2CrO_4 out of crystallizer-filter from Eq. (1):

$$D = 0.4\ (2\ 500)$$
$$= 1.00 \times 10^3 \text{ kg/h}$$

Solve for H_2O out of evaporator from Eq. (2):

$$B = 0.6\ (2\ 500)$$
$$= 1.50 \times 10^3 \text{ kg/h}$$

Solve for recycle rate from crystallizer-filter from Eqs. (9) and (7):

$$C = D + E \quad (9)$$
$$= 1\ 000 + E$$
$$0.50C = D + 0.45E \quad (7)$$
$$0.50\ (1\ 000 + E) = 1\ 000 + 0.45E$$
$$E = 10\ 000 \text{ kg/h}$$

Calculate total input for evaporator:

$$\text{Total input} = E + A$$
$$= 10\ 000 + 2\ 500$$
$$= 1.25 \times 10^4 \text{ kg/h}$$

Calculate feed rate for crystallizer-filter from Eq. (6):

$$C = 2\ 500 + E - B$$
$$= 2\ 500 + 10\ 000 - 1\ 500$$
$$= 1.10 \times 10^4 \text{ kg/h}$$

To use the math solver, we will select four independent equations from the nine that were written to solve for the four unknowns *B, C, D,* and *E.* The math solver requires an initial guess. Each of the four unknowns were somewhat arbitrarily given the same initial guess. The equations selected were (other combinations could also have been selected): (4), (5), (7), and (8). The results are shown in Fig. 10.10.

From these values for *B, C, D,* and *E,* the total input for the evaporator and the feed rate for the crystallizer-filter can be found as before.

To solve for *N* unknowns from a system of linear equations, *N* independent equations are needed. From an engineering point of view, this means that each equation of the set must bring new information to the set. In preparing the system of equations, if you choose an equation that can be obtained from the ones already present (for example, the sum of two of the current equa-

Example Problem 10.3

Four equations containing 4 unknowns are entered into a math solver, in this case Mathcad. The Find function is used to determine the solution to the simultaneous equations once an initial guess for each unknown is given.

Initial guesses for the unknowns

$$b := 1200$$
$$c := 1200$$
$$d := 1200$$
$$e := 1200$$

The given section defines the simultaneous equations

Given $1000 + .45\cdot e = .5\cdot c$

$$.5\cdot c = d + .45\cdot e$$

$$.5\cdot c = .55\cdot e$$

$$1500 + .55\cdot e = b + .5\cdot c$$

The find section then provides the solutions

$$\text{Find}(b,c,d,e) = \begin{bmatrix} 1.5\cdot 10^3 \\ 1.1\cdot 10^4 \\ 1\cdot 10^3 \\ 1\cdot 10^4 \end{bmatrix}$$

Figure 10.10 Mathcad solution of simultaneous equations in Example prob. 10.3.

tions), this new equation does not provide any new information. Likewise, if the new equation is simply one of the current ones multiplied by a constant, no new information is provided.

The consequences of not having a linearly independent set of N equations to solve for N unknowns is that you will not be able to obtain a unique solution. There will be many sets of the N unknowns that will satisfy the set of equations rather than the single set you are seeking.

To show one way to check for linear independence, consider the four equations in unknowns B, C, D, and E just solved in the math solver. Organize them as follows:

$$0B + 0.5C + 0D - 0.45 = 1\,000$$

$$1B + 0.5C + 0D - 0.55 = 1\,500$$

$$0B + 0.5C - 1D - 0.45 = 0$$

$$0B + 0.5C + 0D - 0.55 = 0$$

or in matrix form:

$$\begin{bmatrix} 0 & 0.5 & 0 & -0.45 \\ 1 & 0.5 & 0 & -0.55 \\ 0 & 0.5 & -1 & -0.45 \\ 0 & 0.5 & 0 & -0.55 \end{bmatrix} \begin{bmatrix} B \\ C \\ D \\ E \end{bmatrix} = \begin{bmatrix} 1\,000 \\ 1\,500 \\ 0 \\ 0 \end{bmatrix}$$

For the set of equations to be linearly independent, the determinant of the coefficient matrix must be nonzero. Here

$$\begin{vmatrix} 0 & 0.5 & 0 & -0.45 \\ 1 & 0.5 & 0 & -0.55 \\ 0 & 0.5 & -1 & -0.45 \\ 0 & 0.5 & 0 & -0.55 \end{vmatrix} = -0.05$$

Since the determinant is nonzero, the equations are independent and a unique solution can be expected.

Note that with the equations in this form, many math solvers, spreadsheets, or modern calculators can be used to solve the set using matrix procedures.

10.5 Key Terms and Concepts

The following are some of the terms and concepts you should recognize and understand:

Material balance
Conservation-of-mass principle
System
Boundary
Input
Output
Accumulation
Consumption
Batch process
Rate flow process
Uniform flow
Nonuniform flow
Steady flow
Unsteady flow

Problems

For each problem, follow the steps of Sec. 10.4—that is, (1) identify the system, (2) determine whether the process is batch or rate, and so on.

10.1 A quantity of pure dry salt is added to 325 lbm of brine that is 11.3 percent salt. How much dry salt must be added to produce a 13.62 percent brine?

10.2 If 16.71 kg of water is evaporated from a vat containing 123.2 kg of a 12.3 percent brine and 2.37 kg of pure dry salt is added, what percentage of salt does the resulting brine contain?

10.3 A vat contains 2 240 lbm of brine that is 7.62 percent salt. If 466 lbm of water is evaporated from the vat, how much pure dry salt must be added to produce a 9.82 percent brine?

10.4 Sap from a maple tree contains 5.12 percent sugar and 94.88 percent water. Your evaporator will boil away 1.263 kg of water per minute. How many hours must you operate the evaporator in order to produce 450.0 kg of maple sugar containing 20.5 percent moisture?

10.5 A commercial dryer receives 4 500 kg/h of a wet wood pulp. If the pulp contains 36 percent water before it enters the dryer and 12 000 kg of water is removed in an 8-h day, what is the final moisture content of the output pulp?

10.6 A drilling mud contains 60.0 percent water and 40.0 percent special clay. The driller wishes to increase the density of the mud, and a curve shows that 48 percent water will give the desired density. Calculate the mass of bone-dry clay that must be added per metric ton of original mud to give the desired composition.

10.7 Syrups *A*, *B*, and *C* are mixed together. It is known that the quantity of *A* is equal to 40.0 percent of the total. Syrup *A* is 3.53 percent sugar, *B* is 4.27 percent sugar, and *C* is 5.76 percent sugar. To this mixture is added 17.34 kg of pure dry sugar, while 123 kg of water is boiled away. This results in 1 963 kg of syrup that is 5.50 percent sugar. How much of each syrup (*A*, *B*, and *C*) was added initially? Suggest use of a math solver.

10.8 Cod-liver oil is produced by an extracting process in which ether dissolves the oil from the livers. In one process, the livers are fed into the extractor at the rate of 1 120 kg/h. These livers consist of 29.7 percent oil; the rest is inert material. The solvent is mostly ether (97.9 percent) with 2.10 percent oil. It is fed into the extractor at the rate of 1 975 kg/h. The extract leaves the extractor at the rate of 1 820 kg/h and consists of 17.3 percent oil and 82.7 percent ether. Determine the flow rate and composition of the product (processed livers). Do by hand and also with a math solver.

10.9 Machine parts are often coated with a thick layer of grease when in storage. Before use, they must be degreased. Kerosene, with a specific gravity of 0.809, is added to a large vat containing 1.73 t of grease-coated parts. The average coating is 2.76 kg of grease for each 96.5 kg of parts (metal plus grease). The used kerosene, containing 8.76 percent grease, is withdrawn and sent to a separator. How many liters of kerosene were used? *Note:* 1 t = 1 000 kg.

10.10 Construction engineers choose different "mixes" of stone to produce desired strengths of concrete. One such mix consisted of 37.5 percent stones (by mass) between 16 and 25 mm in diameter and 26.8 percent between 10 and 15 mm; the remaining percentage of stones was less than 10 mm in diameter. The engineer decides to use this combination as a base supply to create a new mix by screening out all of the stones less than 10 mm in diameter and adding some stones between 16 and 25 mm. The resulting mixture has 74.6 percent of the largest size, and the total mix has a mass of 46.8 t. How much of the first mix was used? Use a math solver.

10.11 A bottle of cleaning solution contains 1.00 gal and is 11.38 percent alcohol (by mass). The solution has a specific gravity of 0.978. This is mixed with 0.750 gal of a second solution that is 39.6 percent alcohol (by mass) and has a specific gravity of 0.920. Based on mass, what is the percentage of alcohol of the mixture?

10.12 If the mixture resulting in Prob. 10.11 has 1.50 lbm of water removed and 0.212 qt of pure (100 percent) alcohol (specific gravity = 0.790) added, what is the alcoholic content of the resulting mixture?

10.13 A company makes alcohol from corn to blend with gasoline (gasohol). Officials are not satisfied with the production so they analyze the operation. They feed 8.00×10^2 kg/h into the still. This feed has been tested and it contains 11.3 percent alcohol, 83.9 percent water, and some inert material. From the still, a vapor is drawn off and passed through a condenser where it is cooled. The finished product is 12.6 percent of the feed and contains 73.1 percent alcohol, 26.2 percent water, and 0.7 percent inert materials. What is the quantity and composition of the bottoms—that is, the waste from the bottom of the still? Solve by hand and with a math solver.

10.14 Because of environmental concerns, your plant must install an acetone recovery system. Your task is to calculate the size of the various components of the system, which includes an absorption tank into which is fed 1 250 kg of water per hour and 7 000 kg/h of air containing 1.63 percent acetone. The water absorbs the acetone and the purified air is expelled. The water and acetone solution go to a distillation process where the solution is vaporized and then to a condenser. The resulting product is 98.9 percent acetone and 1.1 percent water. The bottoms (waste) of the distillation process contains 4.23 percent acetone and 95.77 percent water. To determine the volume of a holding tank, calculate how much product is generated in kilograms per hour.

10.15 Fish is used as animal feed by removing the fish oil and then drying the remainder into a cake that is mixed with other feed. One operation feeds 2 640 kg/h of fish that contains 5.27 percent oil, 73.82 percent water, and the balance dry fish cake. The fish cake produced has 0.123 percent oil and 12.7 percent water. How much water is evaporated during the process if the fish oil contains 1.25 percent water?

10.16 It is desired to produce a 60.0 percent solution of nitric acid (NHO_3) and water. The solution is produced by beginning with a dilute acid containing 22.3 percent HNO_3, and adding a quantity of 94.6 percent HNO_3. How many kilograms of the stronger acid must be added to 245 kg of the dilute acid?

10.17 An industry must clean up 3.2×10^3 kg of its by-product containing both toxic and inert materials. The toxic content is 11.2 percent; the remainder is inert material. Treatment with 3.60×10^4 kg of solvent results in dirty solvent containing 0.35 percent toxic material and a discard composed of 1.2 percent toxic material and all the inert materials. Determine the quantity of dirty solvent, the percentage of solvent in the discard, and the percentage of toxic substance removed in the process. Use a math solver.

10.18 Benzene, toluene, and xylene can be separated by distillation. When 562 kg of a mixture containing 45 percent benzene, 32 percent toluene, and 23 percent xylene is separated into three streams, stream *A* contains 98.9 percent benzene and 1.1 percent toluene; stream *B* contains 95.8 percent toluene, 2.7 percent benzene, and 1.5 percent xylene; and stream *C* contains 93.6 percent xylene and 6.4 percent toluene. Find the mass of the three streams.

10.19 Water is often used to wash ore into a separator. In one such installation, water transports a mixture of dirt and pure iron into the separator (the mixed ore contains three times the percentage of iron ore as dirt) from which two streams emerge; the washed ore stream has 62.8 percent iron ore, 3.8 percent dirt, and 33.4 percent water; the waste stream has 3.6

percent iron ore, 33.7 percent dirt, and 62.7 percent water. The amount of iron ore contained in the washed ore stream is 27.2 t/h. Determine the flow rate of all three streams.

10.20 A stream of fluid feeds 9.27 t/h of a mixture containing 41.3 percent ethane, 29.2 percent propane, and 29.5 percent butane into a still. Three streams are drawn off; stream *A* is 93.7 percent ethane, 5.12 percent propane, and 1.18 percent butane; stream *B* is 92.1 percent propane, 0.82 percent ethane, and 7.08 percent butane; and stream *C* is 94.2 percent butane and 5.80 percent propane. What is the quantity of each of the streams?

10.21 Leftover acid from a nitrating process contains 24.0 percent nitric acid (HNO_3), 55.0 percent sulfuric acid (H_2SO_4), and 21.0 percent water (H_2O) (mass percents). The acid is to be concentrated (strengthened in acid content) by adding sulfuric acid with 92.0 percent H_2SO_4 and nitric acid containing 89.0 percent HNO_3. The final product is to contain 28.0 percent HNO_3 and 61.0 percent H_2SO_4. Compute the mass of the initial acid solution and the mass of the concentrated acids that must be combined to obtain 1.00×10^2 kg of the desired mixture. Use a math solver.

10.22 A syrup contains 6.27 percent sugar. If some of the water is boiled away and 12.9 kg of dry sugar is added, leaving 873.6 kg of syrup that is 8.92 percent sugar, how much syrup was in the initial mixture and how much water was removed?

10.23 The water analysis in a flowing stream shows 1.80×10^2 PPM (parts per million) of sodium sulfate. If 11.0 lbm of sodium sulfate is added to the stream over a 1-hour period, and the analysis downstream where the mixing is complete indicates 3.30×10^3 PPM of sodium sulfate, how many gallons of water are flowing per hour? Neglect the effect of sodium sulfate on the fluid density.

10.24 A very sweet syrup is made by combining some beet syrup and some corn syrup. The beet syrup is 12.34 percent sugar (the remainder is water) and the corn syrup is 7.89 percent sugar. They are mixed, and 13.62 kg of pure dry sugar is added while 456.7 kg of water is boiled away. This leaves 891.2 kg of syrup that is 16.78 percent sugar. How much beet syrup and how much corn syrup did you have to start the process?

10.25 Two brine solutions are mixed. Brine *A* is 68.7 percent of the total and brine *B* is 31.3 percent. Brine *A* is 42.1 percent salt (the remainder is water), and brine *B* is 15.8 percent salt (the remainder is water). Some of the water is then removed from the mixture, leaving 852.6 kg of a mixture (brine) that is 43.62 percent salt.

(*a*) How much of brine *A* and of brine *B* was mixed initially?
(*b*) How much water was removed?

11 Electrical Theory

11.1 Introduction

Electricity is universally one of our most powerful and useful forms of energy. It impacts our world in many useful ways. For example, it affects our lives through communication systems, computer systems, control systems, and power systems. Certainly electrical engineers, but to some extent all engineers, must understand how to design, analyze, and maintain such systems.

Electrical and computer engineering are very large and diverse fields of study. This chapter provides an introduction to one small aspect called circuit theory. It is important to the study of engineering because almost any product or system that is designed involves the application of electrical theory. It is an area fundamental to computer and electrical engineers but extremely important to all engineering disciplines.

Circuit theory does many things for us. It provides simple solutions to practical problems with sufficient accuracy to be useful. It allows us to reduce the analysis of large systems to a series of smaller problems that we can conveniently handle. It provides a means of synthesizing (building up) complex systems from basic components.

In this chapter, we will review a number of the elementary concepts of electricity first learned in physics. We will also introduce and apply some of the fundamental circuit-analysis equations such as Ohm's law and Kirchhoff's laws. Applications, however, will be restricted to those involving steady-state direct current (DC). Methods of analyzing transients and alternating current (AC) circuits can be found in other electrical engineering textbooks.

11.2 Structure of Electricity

Matter consists of minute particles called molecules. Molecules are the smallest particles into which a substance can be divided and still retain all the characteristics of the original substance. Each of these particles will differ according to the type of matter to which it belongs. Thus, a molecule of iron will be different from a molecule of paper.

Looking more closely at the same molecule, we find that it can be divided into still smaller parts called atoms. Each atom has a central core, or nucleus, that contains both protons and neutrons. Moving in a somewhat circular motion around the nucleus are particles of extremely small mass called electrons. In fact the entire mass of the atom is practically the same as that of its nucleus, since the proton is approximately 2.0×10^3 times more massive than the electron.

To understand how electricity works bear in mind that electrons possess a negative electric charge, and protons a positive electric charge. Their values are opposite in sign but numerically they possess the same magnitude. The neutron is considered neutral, being neither positive nor negative.

The typical atom in its entirety has no electric charge because the positive charge of the nucleus is exactly balanced by the negative charge of the surrounding electron cloud. That is, each atom contains as many electrons orbiting the nucleus as there are protons inside the nucleus.

The actual number of protons depends on the element of which the atom is a part. Thus, hydrogen (H) has the simplest structure, with one proton in its nucleus and one orbital electron. Helium (He) has two protons and two neutrons in the nucleus; and since the neutrons exhibit a neutral charge, there are two orbital electrons (see Fig. 11.1). More complex elements have many more protons, electrons, and neutrons. For example, gold (Au) has 79 protons and 118 neutrons, with 79 orbital electrons. As the elements become more complex, the orbiting electrons arrange themselves into regions, or "shells," around the nucleus.

The maximum number of electrons in any one region is uniquely defined. The shell closest to the nucleus contains two electrons, the next eight, and so on. There are a maximum of six shells, but the last two shells are never completely filled. Atoms can therefore combine by sharing their outer orbital electrons and thereby fill certain voids and establish unique patterns of molecules.

Figure 11.1
Schematic representation of a helium atom.

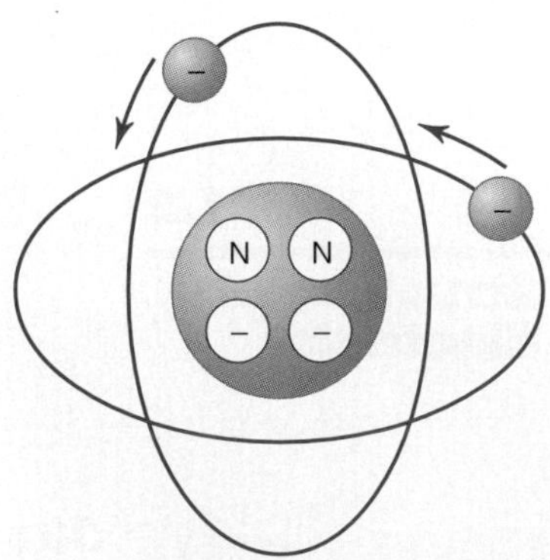

Atoms are extremely minute. In fact, it may be very difficult for one to imagine the size of an atom, since a grain of table salt is estimated to contain 10^{18} atoms. However, it is possible to understand the relation between the nucleus and the orbital electrons. Assume for purposes of visualization that the diameter of the hydrogen nucleus is a 1.0 mm sphere. To accurately scale the electron and its orbit, the electron would revolve at an average distance of 25 m from the nucleus. Although the relative distance is significant, this single electron is prevented from leaving the atom by an electric force of attraction that exists because the proton has a positive charge and the electron has an opposite but equal negative charge.

How closely the millions upon millions of atoms and molecules are packed together will determine the state (solid, liquid, or gas) of a given substance. In solids, the atoms are packed closely together, generally in a very orderly manner. The atoms are held in a specific, lattice structure but vibrate around their nominal positions. Depending on the substance, some electrons may be free to move from one atom to another.

It is therefore the nature of the molecular structure and the availability of free electrons that results in electricity.

11.3 Static Electricity

History indicates that the word *electricity* was first used by the Greeks. They discovered that after rubbing certain items together, the materials would exert a force on one another. It was concluded that during the rubbing process, the bodies were "charged" with some unknown element, which the Greeks called electricity. For example, they concluded that by rubbing silk over glass, electricity was added to the substance. We realize today that during the process of rubbing, electrons are displaced from some of the surface atoms of the glass and added to the surface atoms of the silk.

The branch of science concerned with static, or stationary, charged bodies is called electrostatics.

The charge on an electron can be measured, but the amount is extremely small. A larger unit, the coulomb, has been selected to denote electric charge. Charles Augustin Coulomb (1736–1806) was the first individual to measure such an electric force. In recognition of his work, the SI unit of charge is called the *coulomb* (C). A coulomb is defined in terms of the force exerted between unit charges. A charge of one coulomb (C) will exert on an equal charge, placed 1 m away in air, a force of about 8.988×10^9 N. The magnitude of this force however is very large and, equivalent to the weight of approximately 15 million people. Using this unit of measure, the charge on a proton is $+1.6x10^{-19}$ C.

11.4 Electric Current

Earlier we noted that electrons are prevented from leaving the atom by the attraction of the protons in the nucleus. It is entirely possible, however, for an electron to become temporarily separated from an atom. These free electrons drift around randomly in the space between atoms. During their random travel, many of them will collide with other atoms; when they do so with sufficient force, they dislodge electrons from those atoms. Since electrons are frequently colliding with other atoms, there can be a continuous movement of free electrons in a solid. If the electrons drift in a particular direction instead of moving randomly, there is movement of electricity through the solid. This continuous movement of electrons in a direction is called an electric current. If the electron drift is in only one direction, it is called direct current (DC). If the electrons periodically reverse direction of travel, then we have alternating current (AC).

The ease with which electrons can be dislodged by collision as well as the number of free electrons available varies with the substance. Materials in which the drift of electrons can be easily produced are good conductors; those in which it is difficult to produce an electron drift are good insulators. For example, copper is a good conductor, whereas glass is a good insulator. Practically all metals are good conductors. Silver is a very good conductor but it is expensive. Copper and aluminum are also good conductors and are commonly used in electric wire.

11.5 Electric Potential

Both theory and experimentation suggest that like charges repel and unlike charges attract. Consequently, to bring like charges together, an external force is necessary, and therefore work must be done. The amount of work required to bring a positive charge near another positive charge from a large distance is used as a measure of the electric potential at that point. This amount of potential is measured in units of work per unit charge or joules per coulomb. By definition, one joule per coulomb is one volt, a unit of electric potential.

Certain devices, such as electric batteries or generators, are capable of producing a difference in electric potential between two points. Such devices are rated in terms of their ability to produce a potential difference, and this difference in potential is measured in volts. When these devices are connected to other components in a continuous circuit, electric current flows. You could also think of batteries as devices that change chemical energy to electric energy while generators convert mechanical energy to electric energy.

11.6 Simple Electric Circuits

When electric charge and current were initially being explored, scientists thought that current flow was from positive to negative. They had no knowledge of electron drift. By the time it was discovered that the electron flow was from negative to positive, the idea that current flow was from positive to negative had become so well established that it was decided not to change the convention.

When two battery terminals are connected to a conducting material, the battery or generator creates a potential difference across the load measured as a voltage. It follows that the random movement of the negatively charged electrons will have a drift direction induced by this potential difference. The resulting effect will be the movement of electrons away from the negative terminal of the battery or generator. Electrons will travel in a continuous cycle around the circuit, reentering the battery or generator at its positive terminal.

The speed by which any individual electron moves is relatively slow, less than 1 mm/s. However, once a potential difference is connected into a circuit, the "flow" of electrons starts almost instantaneously at all points. Individual electrons at all locations begin their erratic movement around the circuit, colliding frequently with other atoms in the conductor. Because electron ac-

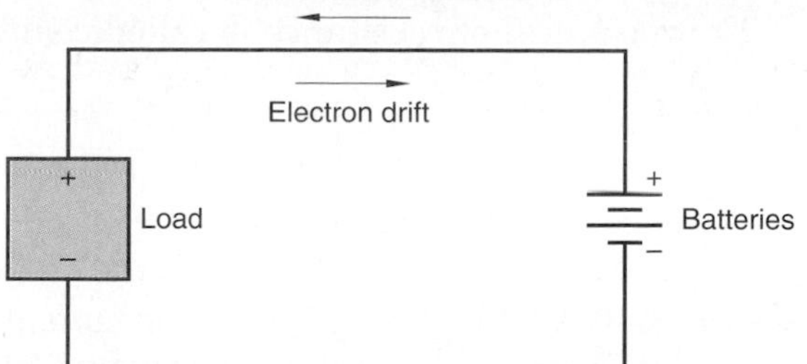

Figure 11.2
A simple DC electric circuit.

tivity starts at all points practically simultaneously, electric current appears to travel about 3×10^8 m/s.

Electric current is really nothing more than the rate at which electrons pass through a given cross section of a conductor. The number of electrons that migrate through this cross section is gigantic in magnitude in that approximately 6.28×10^{18} electrons pass a point per second per ampere. Since this number is so very large, it is not convenient to use the rate of electron flow as a unit of current measurement. Instead, current is measured in terms of the total electric charge (coulomb) that passes a certain point in a unit of time (second). This unit of current is called the *ampere*. That is, one ampere equals one coulomb per second.

11.7 Resistance

Another critical component of circuit theory is *resistance*. George Simon Ohm (1789–1854), a German scientist investigating the relation between electric current and potential difference, found that for a metal, the potential difference across the conductor was directly proportional to the current. This important relationship has become known as Ohm's law, which is stated as:

At constant temperature, the current I *in a conductor is directly proportional to the potential difference between its ends,* E.

The ratio E/I describes the "resistance" to electron flow and is denoted by the symbol R:

$$R = \frac{E}{I} \tag{11.1}$$

or

$$\text{Resistance, or } R(\text{ohms, } \Omega) = \frac{\text{potential, or } E \text{ (volts, V)}}{\text{current, or } I \text{ (amperes, A)}}$$

This is one of the simplest but most important relations used in electric circuit theory. A conductor has a resistance R of one ohm when the current I through the conductor is one ampere and the potential difference E across it is one volt.

The reciprocal of resistance is called *conductance* (G):

$$G = \frac{1}{R} \tag{11.2}$$

Conductance is measured in siemens (S).

Example problem 11.1 The current in an electric aircraft instrument heater is 2.0 A when connected to a battery with a potential of 60.0 V. Calculate the resistance of the heater.

Solution

$$R = \frac{E}{I}$$

$$= \frac{60.0}{2.0}$$

$$= \underline{\underline{3.0 \times 10^1\ \Omega}}$$

11.8 DC Circuit Concepts

A considerable amount of information about electric circuits can be presented in a compact form by means of circuit diagrams. Figure 11.3 illustrates three typical symbols that are frequently used in such diagrams.

Circuits may have resistors that are connected either end to end or parallel to each other. When resistors are attached end to end, they are said to be connected in series, and the same current flows through each. When several resistors are connected between the same two points they are located in parallel.

For the series circuit illustrated in Fig. 11.4, there will be a potential across each resistor, since a potential must exist between the ends of a conductor if current is to flow. This potential difference E is related to current and resistance by Ohm's law. For each unknown voltage (potential), we can write

$$E_1 = R_1 I \qquad E_2 = R_2 I \qquad E_3 = R_3 I$$

Since the total voltage drop E_T across the three resistors is the sum of the individual drops, and since the current is the same through each resistor, then

$$\begin{aligned} E_T &= E_1 + E_2 + E_3 \\ &= IR_1 + IR_2 + IR_3 \\ &= I(R_1 + R_2 + R_3) \end{aligned}$$

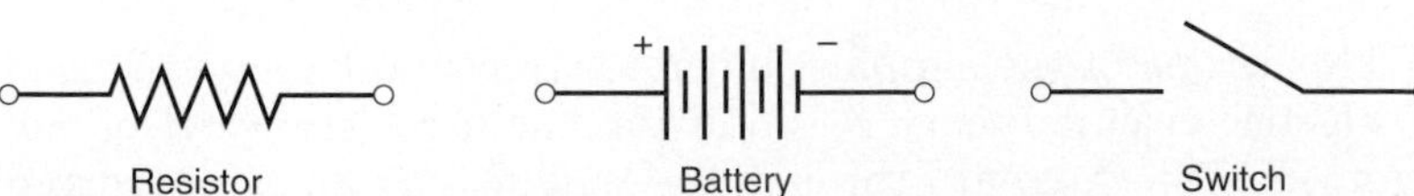

Figure 11.3 Symbols.

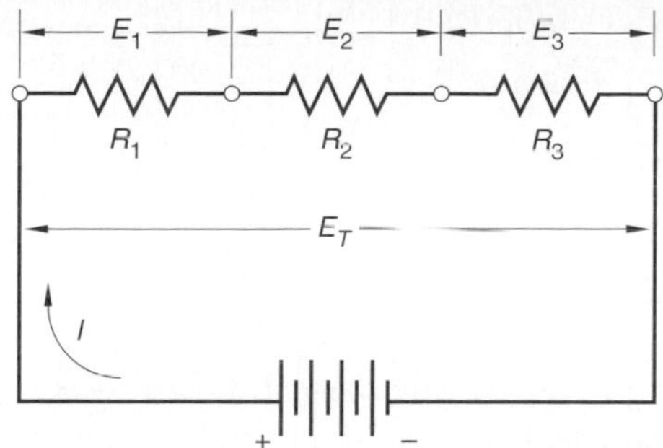

Figure 11.4
Series circuit.

or

$$R_1 + R_2 + R_3 = \frac{E_T}{I}$$

Since E_T is the total potential difference across the circuit and I is the circuit current, then the total resistance must be

$$R_T = \frac{E_T}{I}$$

Therefore,

$$R_T = R_1 + R_2 + R_3 \tag{11.3}$$

This total resistance R_T is sometimes called the equivalent resistance R_E because it is exactly equal to R_1, R_2, and R_3.

These steps demonstrate that when any number of resistors are connected in series, their combined resistance is the sum of their individual values.

Example problem 11.2 The circuit in Fig. 11.5 has three resistors connected in series with a 12.0 V source. Determine the line current I and the voltage drop across each resistor (E_1, E_2, and E_3).

Solution For resistors in series,

$$R_T = R_1 + R_2 + R_3$$

$$R_T = 4.0 + 8.0 + 12$$

$$= \underline{\underline{24\Omega}}$$

Ohm's law gives

$$E = RI$$

$$I = \frac{E_T}{R_T} = \frac{12.0}{24} = 0.50\text{A}$$

Then

$$E_1 = 0.50(4.0)$$

$$= \underline{\underline{2.0 \text{ V}}}$$

Figure 11.5

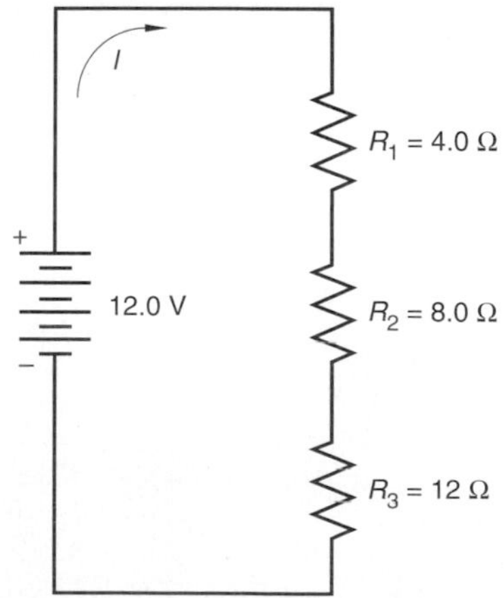

$$E_2 = 0.50(8.0)$$
$$= \underline{\underline{4.0 \text{ V}}}$$
$$E_3 = 0.50(12)$$
$$= \underline{\underline{6.0 \text{ V}}}$$

Check: $E_T = 12.0 = 2.0 + 4.0 + 6.0$

Figure 11.6
Parallel circuit.

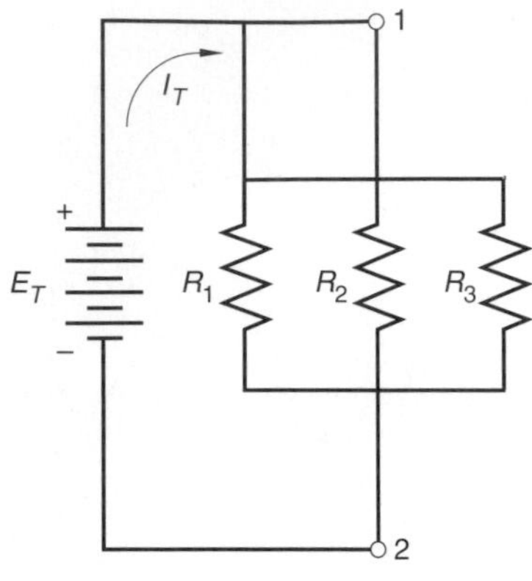

When several resistors are connected between the same two points, they are in parallel. Figure 11.6 illustrates three resistors in parallel.

The current between points 1 and 2 divides among the various pathways formed by the resistors. Since each resistor is connected between the same two points, the potential difference across each of the resistors is the same.

In analyzing the problem, let I_T be the total current passing through the points 1 and 2 with I_1, I_2, and I_3 representing the branch currents through R_1, R_2, and R_3, respectively.

Using Ohm's law we can write

$$I_1 = \frac{E_T}{R_1} \qquad I_2 = \frac{E_T}{R_2} \qquad I_3 = \frac{E_T}{R_3}$$

or

$$I_1 + I_2 + I_3 = E_T\left(\frac{1}{R_1} + \frac{1}{R_2} + \frac{1}{R_3}\right)$$

But we know that

$$I_T = I_1 + I_2 + I_3$$

Applying Ohm's law to total circuit values reveals that

$$I_T = \frac{E_T}{R_T}$$

Therefore,

$$\frac{E_T}{R_T} = E_T\left(\frac{1}{R_1} + \frac{1}{R_2} + \frac{1}{R_3}\right)$$

or

$$\frac{1}{R_T} = \left(\frac{1}{R_1} + \frac{1}{R_2} + \frac{1}{R_3}\right) \tag{11.4}$$

This equation indicates that when a group of resistors are connected in parallel, the reciprocal of their combined resistance is equal to the sum of the reciprocals of their separate resistances.

Example problem 11.3 In Fig. 11.7 three resistors are connected in parallel across a 6.0×10^1 V battery. What is the equivalent resistance of the three resistors and the line current?

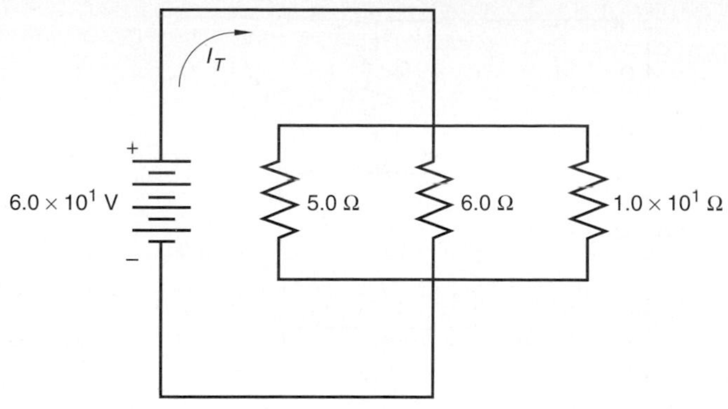

Figure 11.7

Solution For resistors in parallel,

$$\frac{1}{R_T} = \frac{1}{R_1} + \frac{1}{R_2} + \frac{1}{R_3}$$

$$= \frac{1}{5.0 \times 10^0} + \frac{1}{6.0 \times 10^0} + \frac{1}{1.0 \times 10^1}$$

$$= \frac{28}{6.0 \times 10^1}$$

$$R_T = \frac{6.0 \times 10^1}{2.8 \times 10^1}$$

$$= \underline{\underline{2.1\Omega}}$$

From Ohm's law

$$E = RI$$

$$E_T = R_T I_T$$

so,

$$I_T = \frac{E_T}{R_T}$$

$$= (6.0 \times 10^1)\,\frac{2.8 \times 10^1}{6.0 \times 10^1}$$

$$= \underline{\underline{28\text{ A}}}$$

Many electric circuits involve combinations of resistors in parallel and series. The next example problem demonstrates a solution of that nature.

Example problem 11.4 Determine the line current I, the circuit equivalent resistance R_T, and the voltage drop E_4 across the resistor R_4 for the circuit in Fig. 11.8. What resistance should be substituted for R_1 to reduce the line current by one-half?

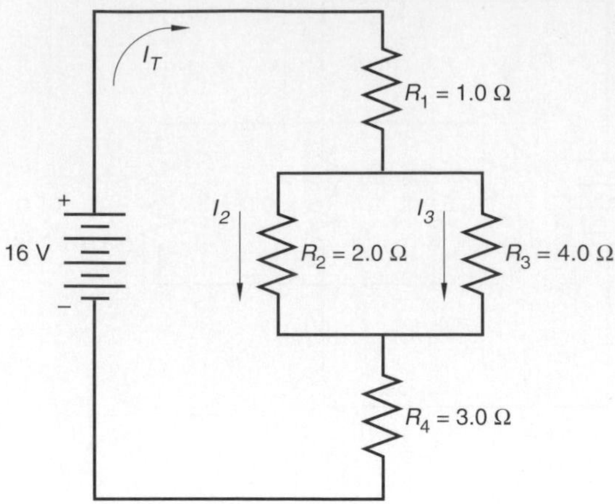

Figure 11.8

Solution Ohm's law: $E = RI$

Resistors in series: $R_T = R_1 + R_2 + \ldots + R_N$

Resistors in parallel: $\dfrac{1}{R_T} = \dfrac{1}{R_1} + \dfrac{1}{R_2} + \ldots + \dfrac{1}{R_N}$

For parallel resistors,

$$\frac{1}{R_T} = \frac{1}{R_2} + \frac{1}{R_3}$$

$$= \frac{1}{2.0} + \frac{1}{4.0}$$

$$R_T = \underline{\underline{\frac{4.0}{3.0}\ \Omega}}$$

For series resistors,

$$R_E = R_1 + R_T + R_4$$

$$= 1.0 + \frac{4.0}{3.0} + 3.0$$

$$= \frac{16}{3.0}$$

$$= \underline{\underline{5.3\ \Omega}}$$

Then the line current is

$$E = RI$$

$$I = E\left(\frac{1}{R_E}\right)$$

$$= 16\left(\frac{3.0}{16}\right)$$

$$= \underline{\underline{3.0\ \text{A}}}$$

and the voltage drop is

$$E_4 = R_4 I$$
$$= 3.0(3.0)$$
$$= \underline{\underline{9.0\ \text{V}}}$$

If we reduce the line current by one-half the new line current is 3.0/2 = 1.5 A. The new equivalent resistance must then be

$$R_E = \frac{E}{} = \frac{16}{1.5} = 10.667\ \Omega$$

But

$$R_E = R_1 + R_T + R_4$$

so the new R_1 is

$$R_1 = R_E - R_T - R_4 = 10.667 - \frac{4.0}{3.0} - 3.0 = \underline{\underline{6.3\Omega}}$$

11.9 DC Electric Power

Consider the following illustration of a simple DC circuit. A common household lantern, as illustrated in Fig 11.9, is represented

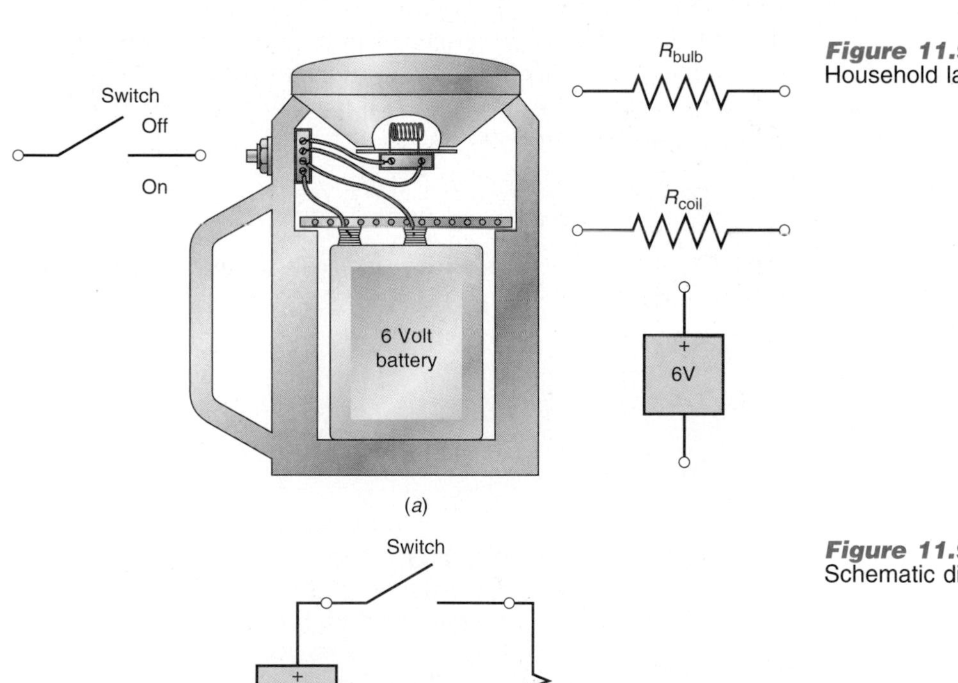

Figure 11.9(a) Household lantern.

Figure 11.9(b) Schematic diagram.

by a section view (part a) and a circuit diagram (part b). As the switch is closed, current (I) flows for some period of time (t), that is, until the switch is opened. Since there is a potential difference across the battery (E), there will be an amount of electrical energy produced. The bulb converts the electrical energy into heat and light energy and would continue to do so until the source is dissipated.

$$\text{Energy} = E\,I\,t \tag{11.5}$$

This equation is developed understanding that,

$$E\ (\text{volt}) = \frac{\text{joule}}{\text{coulomb}}$$

$$I\ (\text{ampere}) = \frac{\text{coulomb}}{\text{second}}$$

Energy, commonly measured in joules (J), can be in the form of work, heat, or its potential to produce that result, such as the potential energy of the battery. You can imagine the potential energy of a gallon of gasoline or the kinetic energy of a speeding locomotive.

Power, on the other hand, is the time rate at which energy is supplied or consumed. It is expressed in joules per second, but this unit has been given a special name after the Scottish engineer James Watt (1736–1819). Thus, one watt equals one joule per second.

Power can be expressed as

$$\text{Power} = \frac{\text{energy}}{\text{time}} = \frac{E\,I\,t}{t} = E\,I \tag{11.6}$$

By applying Ohm's law, we can express power in two other convenient forms:

$$P = \frac{E^2}{R} \tag{11.7}$$

and

$$P = I^2R \tag{11.8}$$

Example problem 11.5 The lantern in Fig 11.9 contains a lamp with a measured resistance of 2.0 Ω. When the switch is closed, what is the power consumed by the lamp?

Solution

$$P = \frac{E^2}{R}$$

$$= \frac{(6.0)^2}{2.0}$$

$$= \underline{\underline{18\ \text{W}}}$$

Example problem 11.6 A gallon of gasoline has the potential energy of approximately 132 000 kJ. Assume a gasoline engine is driving a generator with a 20 percent efficiency and the generator is supplying electricity to a 100 W bulb.

(a) How long will the bulb provide light from one gallon of gasoline?
(b) If the system operates at 120 volts, what is the bulb current?
(c) How much electric charge passes through the bulb in 10 seconds?

Solution

(a) $131.8 \times 10^6 \text{ J} \times 0.2 = 26.36 \times 10^6 \text{ W·s}$

$$\frac{26.36 \times 10^6 \text{ W·s}}{} = 0.2636 \times 10^6 \text{ s} = 73\text{h} - 13 \text{ min} - 20 \text{ s}$$

(b) $I = \dfrac{P}{E} = \dfrac{100}{120} = 0.833 \text{ A}$

(c) $0.833 \text{ A} = 0.833 \text{ C/s}$ or 8.33 C in 10 seconds

11.10 Terminal Voltage

Figure 11.10 is a basic circuit in which there can be different sources of electric potential. The storage battery and the electric generator are two familiar examples.

The potential does work of amount E in joules per coulomb on charges passing through the voltage source from the negative to the positive terminal. This results in a difference of potential E across the resistor R, which causes current to flow in the circuit. The energy furnished by the voltage source reappears as heat in the resistor.

Current can travel in either direction through a voltage source. When the current moves from the negative to the positive terminal, some other form of energy (such as chemical energy) is converted into electric energy. If we were to impose a higher potential in the external circuit—for example, forcing current backward through the voltage source—the electric energy would be converted to some other form. When current is sent backward through a battery, electric energy is converted into chemical energy (which can, by the way, be recovered in certain types of batteries). When current is sent backward through a generator, the device becomes a motor.

A resistor, on the other hand, converts electric energy into heat no matter what the direction of current. Therefore, it is impossible to reverse the process and regain electric energy from the heat. Electric potential always drops by the amount IR as current travels through a resistor. This drop occurs in the direction of the current.

In a generator or battery with current flowing negative to positive, the positive terminal will be E above the negative terminal

Figure 11.10 Basic circuit.

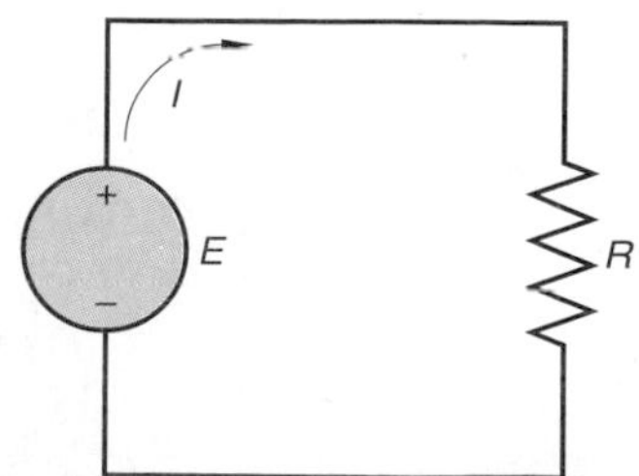

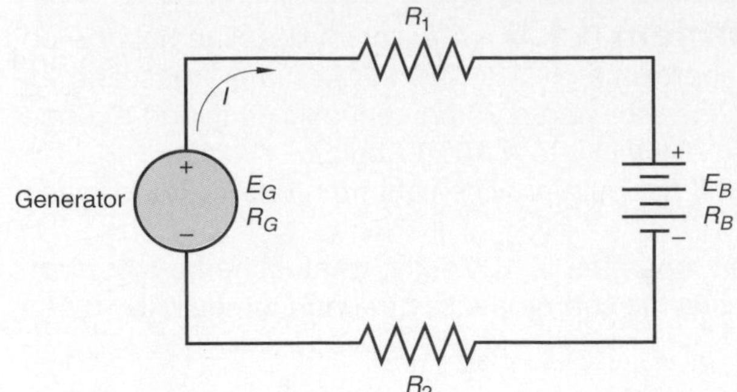

Figure 11.11
Generator charging a battery.

minus the voltage drop due to internal resistance between terminals. There will always be some energy converted to heat inside a battery or generator no matter which direction the current.

When a battery or a generator is driving the circuit, the internal current passes from the negative to the positive terminal. Each coulomb of charge gains energy E from chemical or mechanical energy but loses IR in heat dissipation. The net gain in joules per coulomb can be determined by $E - IR$.

For a motor or a battery being charged, the opposite is true, because the internal current passes from the positive to the negative terminal. Each coulomb loses energy E and IR. The combined loss can be determined by $E + IR$.

In the case of a motor, the quantity E is commonly called *back-emf* (the electromotive force, or potential), since it represents a voltage that is in a direction opposite the current flow.

Figure 11.11 shows a circuit wherein a battery is being charged by a generator. E_G, E_B, R_G, and R_B indicate the potentials and internal resistances, respectively, of the generator and battery. Each coulomb that flows around the circuit in the direction of the current I gains energy E_G from the generator and loses energy E_B in the form of chemical energy to the battery. Heat dissipation is realized as IR_G, IR_1, IR_B, and IR_2.

11.11 Kirchhoff's Laws

We can consider that a circuit has been analyzed when the voltage across and the current in every component is known. For the lantern circuit shown in Fig 11.9 there are five unknowns: I_s, I_b, I_c, E_b, and E_c. The voltage source, E_s, was given as 6V.

From Ohm's law we can write three equations:

$$E_s = I_s R_s$$

$$E_b = I_b R_b$$

$$E_c = I_c R_c$$

Since we have three equations and five unknowns it would appear to be a mathematical impossibility.

Fortunately, Gustav Kirchhoff observed that when components are placed in a circuit the interconnection between voltage and current places constraints on the system. These constraints have become known as Kirchhoff's laws. Kirchhoff's current law (KCL):

> The algebraic sum of all the currents at any node in a circuit equals zero.

In a circuit with n nodes, exactly $(n - 1)$ independent current equations can be derived from KCL.

Kirchhoff demonstrated that the remaining equation could be derived from his voltage law. Kirchhoff's voltage law (KVL):

> The algebraic sum of all the voltages around any path in a circuit equals zero.

Example problem 11.7 Given the circuit illustrated in Fig. 11.12, determine the current I_y.

Solution From Kirchhoff's current law we can write:

$$I_y = I_x + I_z$$

Applying Kirchhoff's voltage law around both closed paths, we find:

$$I_y\ (2) + 14 - I_x\ (4) = 0$$

$$12 - I_z\ (6) - I_y\ (2) = 0$$

By substitution of $I_x + I_z$ for I_y we find:

$$(I_x + I_z)(2) + 14 - I_x\ (4) = 0$$

$$12 - I_z\ (6) - (I_x - I_z)\ (2) = 0$$

$$I_z = 1\ \text{A}\uparrow$$

$$I_x = 2\ \text{A}\downarrow$$

$$I_y = 3\ \text{A}\downarrow$$

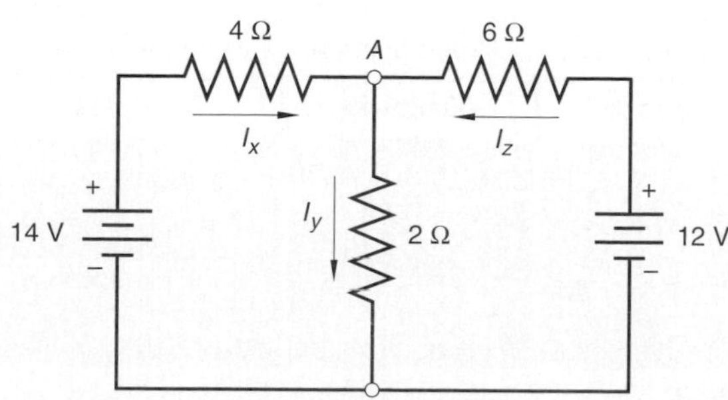

Figure 11.12 Application of Kirchhoff's laws.

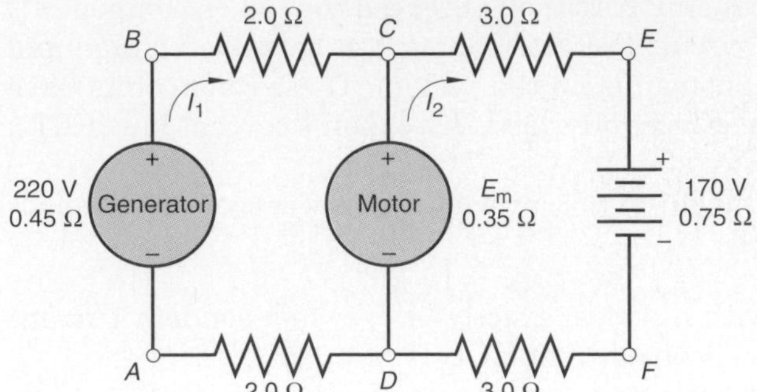

Figure 11.13
Application of Kirchhoff's laws.

Example problem 11.8 A 220 V generator is driving a motor drawing 8.0 A and charging a 170 V battery. Determine the back-emf of the motor (E_m), the charging current of the battery (I_2), and the current through the generator (I_1).

Solution Since the current through the motor is given as 8.0 A, we can see by applying Kirchhoff's current law to junction *C* that the current out of the junction must be $I_2 + 8.0$ if the current through the battery is I_2. That is, $I_1 = I_2 + 8.0$.

Kirchhoff's voltage law (KVL) dictates that we select a beginning point and travel completely around a closed loop back to the starting point, thereby arriving at the same electric potential. As a path is selected and followed, note carefully all changes in potential. Once a loop has been completely traveled and all potential changes noted, the sum is set equal to zero.

The following voltage summation for the circuit ABCDA in Fig. 11.13 results from Kirchhoff's voltage law:

$$(E_{A-B})_{\text{emf}} + (E_{A-B})_{\text{loss}} + E_{B-C} - (E_{C-D})_{\text{emf}} + (E_{C-D})_{\text{loss}} + E_{D-A} = 0$$

By applying correct algebraic signs according to the established convention, we get the results given in Tab. 11.1.

Table 11.1 Voltage Summation for Wvp ABCDA

Symbols	Quantities	Notes
$(E_{A-B})_{\text{emf}}$	$+$ 220 V	Potential of generator
$(E_{A-B})_{\text{loss}}$	$-0.45(I_2 + 8.0)$	Loss in generator
(E_{B-C})	$-2.0(I_2 + 8.0)$	Loss in line
$(E_{C-D})_{\text{emf}}$	$-E_m$	Back-emf of motor
$(E_{C-D})_{\text{loss}}$	$-0.35(8.0)$	Loss in motor
(E_{D-A})	$-2.0(I_2 + 8.0)$	Loss in line

Table 11.2 Voltage Summation for Loop DCEFD

Symbols	Quantities	Notes
$(E_{D-C})_{emf}$	$+E_m$	Back-emf of motor
$(E_{D-C})_{loss}$	$+0.35(8.0)$	*IR* rise in motor
(E_{C-E})	$-3.0I_2$	Loss in line
$(E_{E-F})_{emf}$	-170 V	Drop across battery
$(E_{E-F})_{loss}$	$-0.75I_2$	Loss in battery
(E_{F-D})	$-3.01I_2$	Loss in line

Substituting the values from Tab. 11.1, the equation becomes

$$220 - 0.45(I_2 + 8.0) - 2.0(I_2 + 8.0) - E_m - 0.35(8.0) - 2.0(I_2 + 8.0) = 0$$

Simplifying, we get

$$-4.45I_2 - E_m + 181.6 = 0$$

This equation cannot be solved because there are two unknowns. However, a second equation can be written around a different loop of the circuit.

$$(E_{D-C})_{emf} + (E_{D-C})_{loss} + E_{C-E} + (E_{E-F})_{emf} + (E_{E-F})_{loss} + E_{F-D} = 0$$

From this we can develop Tab. 11.2 and, therefore, a second equation:

$$E_m + 0.35(8.0) - 3.0I_2 - 170 - 0.75I_2 - 3.0I_2 = 0$$

$$E_m - 6.75I_2 - 167.2 = 0$$

Solving these two equations simultaneously, we obtain the following results:

$$I_2 = 1.3 \text{ A}$$

$$E_m = 1.8 \times 10^2 \text{ V}$$

Then

$$I_1 = I_2 + 8.0 = 1.3 + 8.0 = 9.3 \text{ A}$$

These values can be checked by writing a third equation around the outside loop.

By the preceding procedure, a set of simultaneous equations may be found that will solve any similar problem, provided the number of unknowns is not greater than the number of circuit paths or loops.

The following general procedure is outlined as a guide to systematically applying Kirchhoff's laws.

1. Sketch a circuit diagram and label all known voltages, and so on. Show + and − signs on potentials.
2. Assume a current direction in each branch of the circuit. If the direction is not known, choose a direction. A negative current solution will indicate that the current is flowing in the opposite direction.
3. Assign symbols to all unknown currents, voltages, and so forth.
4. Apply Kirchhoff's first law to circuit loops and Kirchhoff's second law at junctions to obtain as many independent equations as there are unknowns in the problem.
5. Solve the resulting set of equations.

11.12 Mesh Analysis

Individual elements or components can be connected to form unique circuits. The interconnectivity of each element can be described in terms of nodes, branches, meshes, paths, and loops.

A node is a specific point or location within a circuit where two or more components are connected.
A branch is a path that connects two nodes.
A mesh is a loop that does not contain any other loops within itself.

A mesh current is defined as a current that exists only in the perimeter of the mesh. Mesh currents are selected clockwise for each mesh. A mesh current is considered to travel all the way around the mesh.

Notice that Fig 11.14 is the same circuit diagram as Fig 11.12 without specific values. Referring to Fig. 11.14, we can apply Kirchhoff's voltage law around the two meshes expressing voltages across each component in terms of the mesh currents I_a and I_b.

$$E_1 - I_a R_1 - (I_a - I_b) R_3 = 0$$

and

$$- E_2 - (I_b - I_a) R_3 - I_a R_2 = 0$$

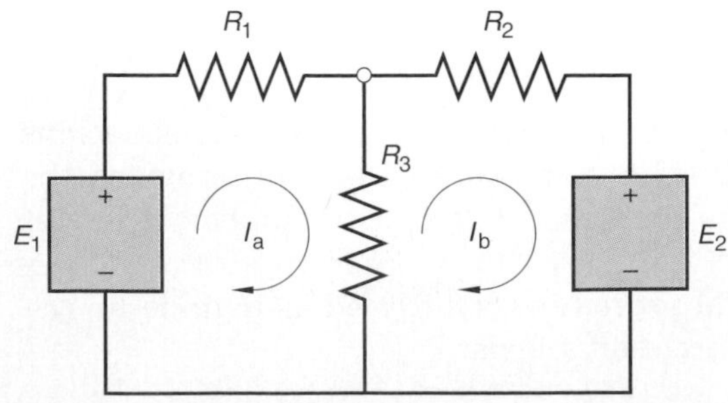

Figure 11.14 Application of a mesh analysis.

Collecting and rearranging these two equations gives us,

$$E_1 - I_a\,(R_1 + R_3) + I_b\,R_3 = 0$$

and

$$-\,E_2 + I_a\,R_3 - I_b\,(R_3 + R_2) = 0$$

By comparing Figs. 11.12 and 11.14 the branch currents can be expressed in terms of the mesh currents.

$$I_1 = I_a$$

$$I_2 = I_b$$

$$I_3 = I_a - I_b$$

Once we know the mesh currents, we know the branch currents, and once we know the branch currents, we can compute any voltage.

Example problem 11.9 Substitute the specific values given in Fig. 11.12 into the mesh current equations and verify the answers obtained in Example prob. 11.7. Writing KVL for the two loops gives:

$$E_1 - I_a\,(R_1 + R_3) + I_b\,R_3 = 0$$

$$-\,E_2 + I_a\,R_3 - I_b\,(R_3 + R_2) = 0$$

Substituting values we see that,

$$14 - 6\,I_a + 2\,I_b = 0$$

$$-\,12 + 2\,I_a - 8\,I_b = 0$$

$$I_b = 1A\uparrow,\ I_a = 2A\downarrow$$

but using KCL

$$I_3 = I_a - I_b = 2A\downarrow + 1A\uparrow = 3A\downarrow$$

11.13 Key Terms and Concepts

The following terms are basic to the material in Chapter 11. You should be able to define these terms and to be able to interpret them into various applications.

Electricity
Electric current
Direct current
Alternating current
Ampere
Coulomb
Resistor
Resistance
Ohm
Ohm's Law
Volt
Voltage
Voltage-drop
Kirchhoff's Laws
Circuit
Steady-state

Electric potential
Conductance
Potential difference
Electric energy
Joule
Generator
Motor
EMF
Battery

Problems

11.1 What is the average current through a conductor that carries 3 675 C during a 8.00 min time period?

11.2 The measured current through a battery is a constant 3.25 A. How many coulombs are supplied by the battery in 12 h?

11.3 Assuming that a current flow through a conductor is due to the motion of free electrons, how many electrons pass through a cross section normal to the conductor in 3 h if the current is 850 A?

11.4 Five 1.5 V batteries in series are required to operate an 8.0 W portable radio. What is the current flow? What is the equivalent circuit resistance?

11.5 A small vacuum cleaner designed to be plugged into the cigarette lighter of an auto has a 12.6 V DC motor. It draws 4.0 A in operation. What power must the auto battery deliver? What size resistor would consume the same power?

11.6 When the leads of an impact wrench are connected to a 12.0 V auto battery, a current of 15 A flows. Calculate the power required to operate the wrench. If 75 percent of the power required by the wrench is delivered to the socket, how much energy in joules is produced per impact if there are 1 100 impacts per minute?

11.7 A portable electric drill produces 1.2 hp at full load. If 85 percent of the power provided by the 9.6 V battery pack is consumed, what is the current flow? How much power is converted into heat?

11.8 An ideal 25 V DC power supply is connected to four resistors in series: $R_1 = 5.0$ ohm, $R_2 = 12\ \Omega$, $R_3 = 25\ \Omega$, and $R_4 = 75\ \Omega$.

(*a*) Draw the circuit diagram.
(*b*) Determine the equivalent circuit resistance.
(*c*) Calculate the line current.
(*d*) Find the voltage drop across each resistor.
(*e*) Compute the power produced by the 25 V supply and the power consumed by each resistor.

11.9 Three resistors 2.0 M Ω, 8.0 M Ω, and M Ω are connected in series to a 75 V ideal DC voltage source.

(*a*) Draw the circuit diagram.
(*b*) Determine the equivalent circuit resistance.
(*c*) Calculate the line current.
(*d*) Find the voltage drop across each resistor.
(*e*) Compute the power consumed by each resistor.

11.10 A 12.0 V supply is connected to three resistors that are wired in parallel: $R_1 = 15$ ohm, $R_2 = 25$ ohm, and $R_3 = 1.0$ k ohm.

(*a*) Draw the circuit diagram.
(*b*) Calculate the equivalent circuit resistance.
(*c*) Find the current through the voltage supply.
(*d*) Determine the current through each resistor and the power consumed by each.

11.11 A battery has a measured voltage of 12.0 V when the circuit switch is open. The internal resistance of the battery is 0.35 ohms. The circuit contains resistors of 15, 45, and 65 ohms connected in parallel.

(*a*) Draw the circuit diagram.
(*b*) Calculate the equivalent circuit resistance as seen by the battery.
(*c*) Find the current flow through the battery when the switch is closed.
(*d*) Compute the current through the 15 ohm resistor and the power consumed by it.
(*e*) Calculate the rate at which heat (watt) must be removed from the battery if it is to maintain a constant temperature.

11.12 Given the circuit diagram and values in Fig. 11.15, determine the current through the 15 ohm resistor and the voltage drop across it. Find the fraction of the power produced by the battery that is consumed by the 10 Ω resistor.

11.13 Add a 5.0 ohm resistor in parallel with the 10 ohm and 15 ohm resistors in the circuit shown in Fig. 11.15 and repeat Prob. 11.12.

11.14 A 24 V battery has an internal resistance of 0.15 ohm. When connected to a load, a current of 45 A flows.

(*a*) What voltage would you expect to measure across the battery terminals when the load is connected?
(*b*) How much power is delivered to the load?
(*c*) At what rate is heat produced in the battery?

11.15 A DC motor has an internal resistance of 0.25 ohm. When connected to a 75 V source, it draws 15 A.

(*a*) What is the back-emf of the motor?
(*b*) How much power is required to drive the motor?
(*c*) At what rate is heat generated in the motor?
(*d*) What power is produced by the motor?

11.16 Refer to Fig. 11.16 and compute the circuit current when $R_1 = 12$ ohm and $R_2 = 15$ ohm. Determine the potential at points *T*, *U*, and *V* if the potential at point *S* is 0 V. What percentage of the power available at terminals *S* and *T* is actually charging the battery?

11.17 The 220 V generator in Fig. 11.17 is charging a 75 V battery and driving a motor. Determine the battery charging current, the current through the motor, and the back-emf of the motor. Assume no internal resistance in the generator, motor, and battery.

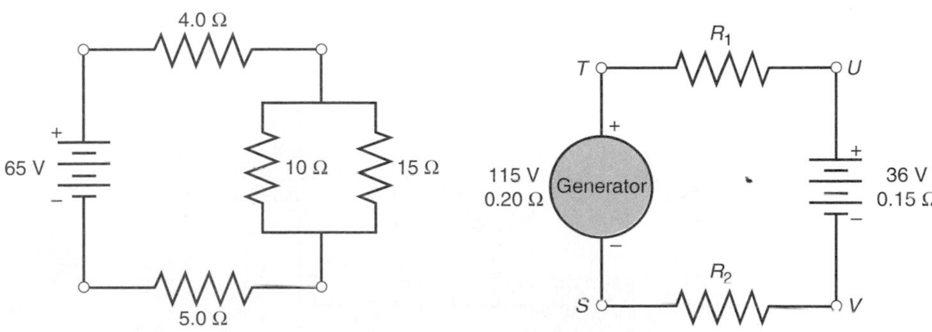

Figure 11.15

Figure 11.16

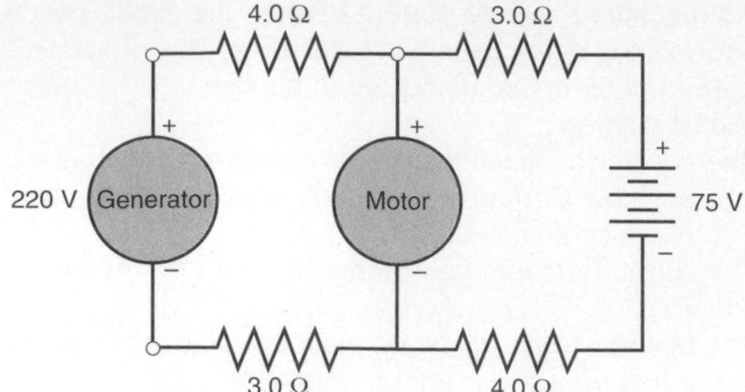

Figure 11.17

11.18 Figure 11.18 shows a resistance circuit driven by two ideal (no internal resistance) batteries. Determine:

(*a*) The currents through each resistor.
(*b*) The power delivered to the circuit by the 110 V battery.
(*c*) The voltage across the 25 ohm resistor.
(*d*) The power consumed by the 15 ohm resistor.

Figure 11.18

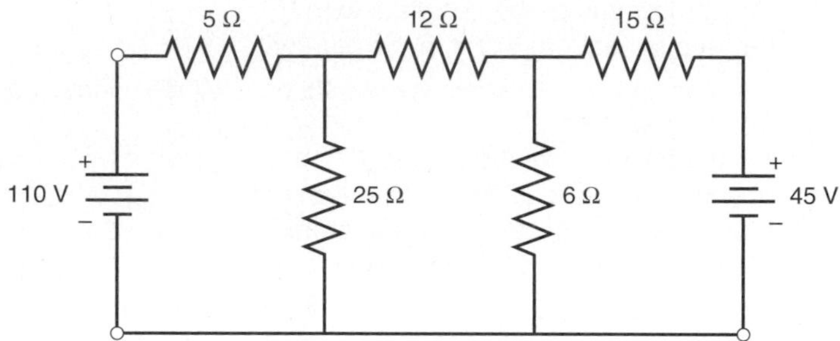

11.19 A 75 V generator is driving a motor and charging a 24 V battery (see Fig. 11.19). The measured current through the motor is 8 A. Calculate:

(*a*) Current I_A and I_B.
(*b*) The back-emf of the motor.

Figure 11.19

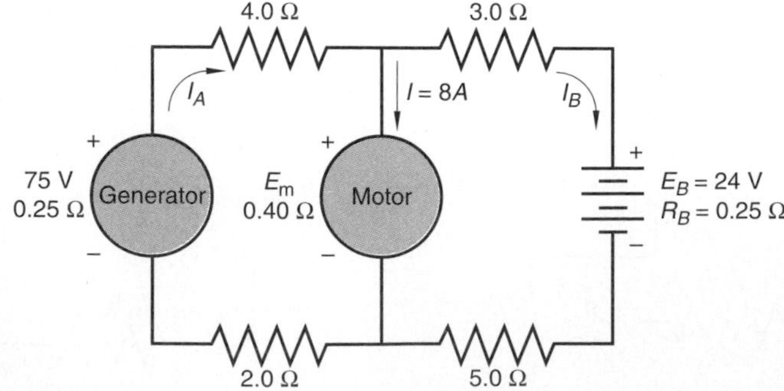

(*c*) The power delivered by the generator to the circuit.
(*d*) The power converted to heat in the motor.
(*e*) The voltage across the motor terminals.

11.20 An ideal 14 V generator and an ideal 12.6 V battery are connected in a circuit as shown in Fig. 11.20.

(*a*) For $R_A = 2.0\ \Omega$ and $R_B = 3.0\ \Omega$, compute the current through each circuit component (resistors, battery, and generator). Is the battery being charged or discharged?

(*b*) With $R_B = 1.0$ ohm, find the current through the battery for 0.50 ohm $< R_A <$ 10.0 ohm using a ΔR_A of 0.50 ohm. Plot the battery current versus R_A with R_A as the independent variable. Repeat the process with $R_B = 5.0$ ohm and 10.0 ohms, placing all three curves on the same graph. *Note:* Use a computer program if approved by your instructor.

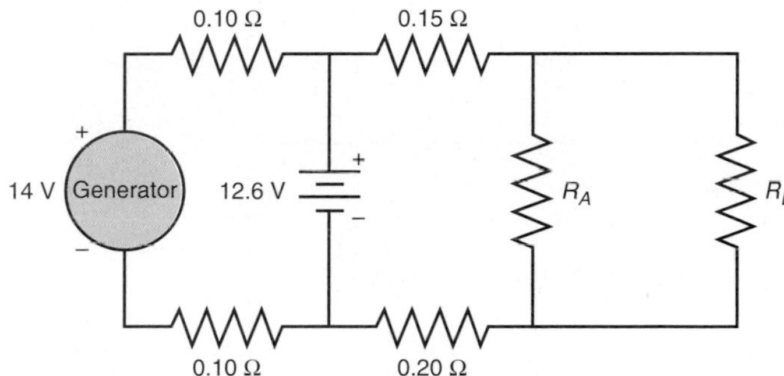

Figure 11.20

11.21 For the circuit shown in Fig. 11.21,

(*a*) Calculate the currents, I_1, I_2, and I_3 when the switch is placed at *A*. How much power is consumed by each of the three resistors?

(*b*) Repeat the problem with the switch placed at *B*.

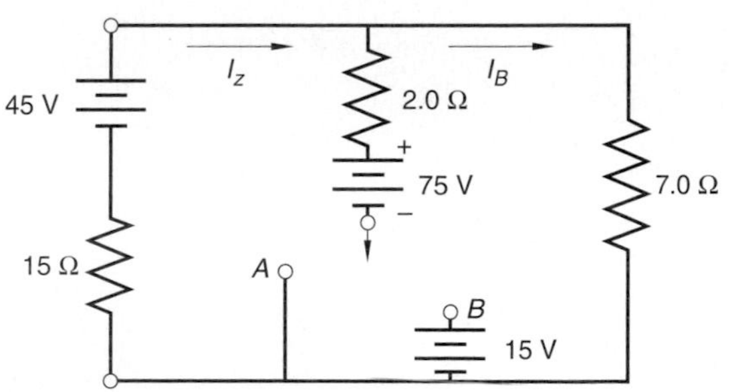

Figure 11.21

11.22 Compute the current through each resistor and the battery shown in Fig. 11.22. Determine the power supplied by the battery and the power consumed by the 25 ohm resistor.

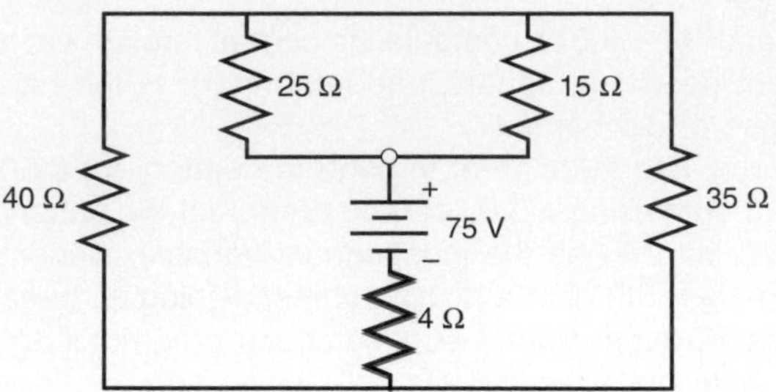

Figure 11.22

11.23 Determine the current through each component of the circuit in Fig. 11.23. Find the power delivered to the 6.0 V battery, the voltage across the 5.0 ohm resistor, and the power consumed by the 10.0 ohm resistor.

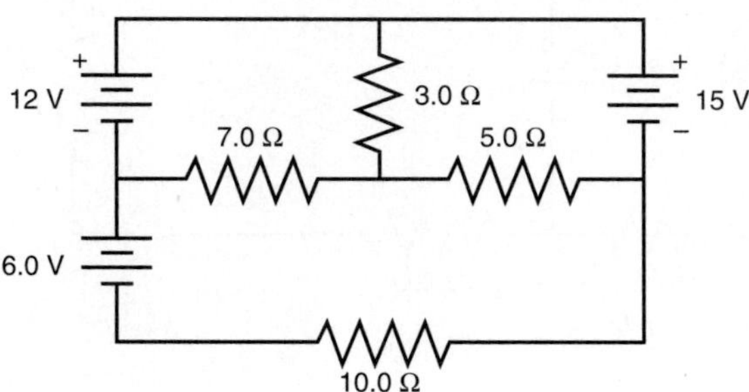

Figure 11.23

12 Energy

12.1 Introduction

Energy is the world's most important commodity. One way of characterizing the development of society during the past 200 years would be in the substitution of machine power for muscle power. This transformation has been helped by the rapid development of the uses of natural sources of energy, namely, fossil fuels, water, solar, and the atom. Energy from these natural sources must be converted into forms that can be transported, stored, and applied at the appropriate time and place. The degree of industrial development of society can be determined by the extent of energy usage. There is an excellent correlation between productivity of a nation and its capability to generate energy.

What is energy? Energy cannot be seen; it has no mass or defining characteristics; it is distinguished only by what it can produce. In a broad sense, *energy* may be defined as an ability to produce an effect (change) on matter. Energy may be within an object or may move from one object to another. Thus, energy is usually spoken of as being either stored or in transit.

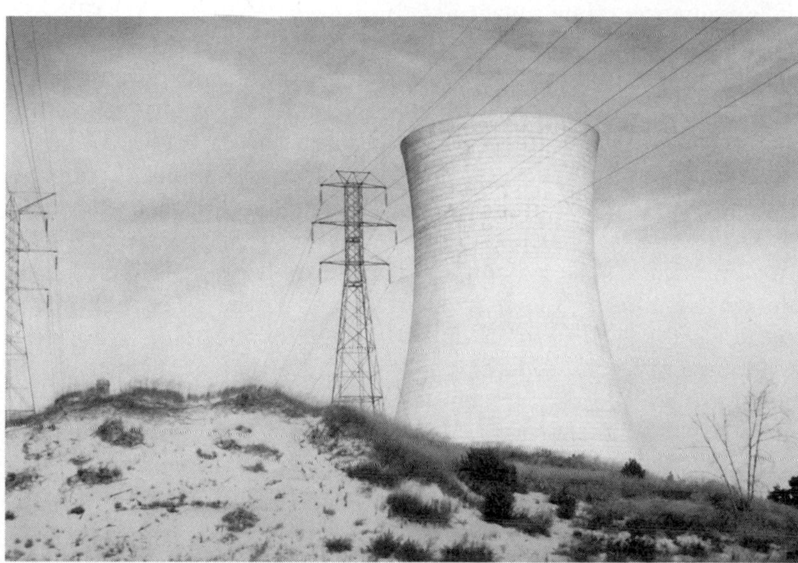

Figure 12.1
Cooling towers from a nuclear power plant release the exhaust gases high into the atmosphere, minimizing the environmental impact.

When supplied, energy can transform natural resources into products and services beneficial to society. We observe the effects of energy when physical changes occur in objects.

12.2 Stored Energy

Stored energy exists in five distinct forms: potential, kinetic, internal, chemical, and nuclear. The ultimate usefulness of stored energy depends on how efficiently the energy is converted into a form that produces a desirable result.

When an object or mass m is elevated to a height h in a gravitational field, a certain amount of work must be done to overcome the gravitational pull. Energy is the capacity to do work. (Work will be considered further in Sec. 12.3.) The object may be said to possess the work required to elevate it to the new position. Its potential energy has been increased. The quantity of work done is the amount of force multiplied by the distance moved in the direction in which the force acts. It is a result of a given mass going from one condition to another. *Potential energy* is thus stored-up energy. It is derived from force and height from a datum plane, not from the means by which the height was attained. Mass m stores up energy when being elevated to height h and does work when it comes back to its starting point. When an object is raised, the force needed is that to overcome the effect of gravity (weight) and the distance is the change in elevation. Thus,

$$\begin{aligned} \text{PE} &= (\text{weight})(\text{height}) \\ &= Wh \\ &= mgh \end{aligned} \tag{12.1}$$

If the units for mass (m) are kilograms, those for acceleration of gravity (g) are meters per second squared, and those for height (h) are meters, then the units for potential energy, or PE, are newton-meters, or joules.

It is unnecessary in most cases to evaluate the total energy of an object, but it is essential to evaluate the energy changes when there is a change in the physical state of the object. In the case of potential energy, this is accomplished by establishing a datum plane (see Fig. 12.2) and evaluating the energy possessed by objects in excess of that possessed at the datum plane. Any convenient location, such as sea level, may be chosen as the datum plane. The potential energy at any other elevation is then equal to the work required to elevate the object from the datum plane.

Figure 12.2 Potential energy

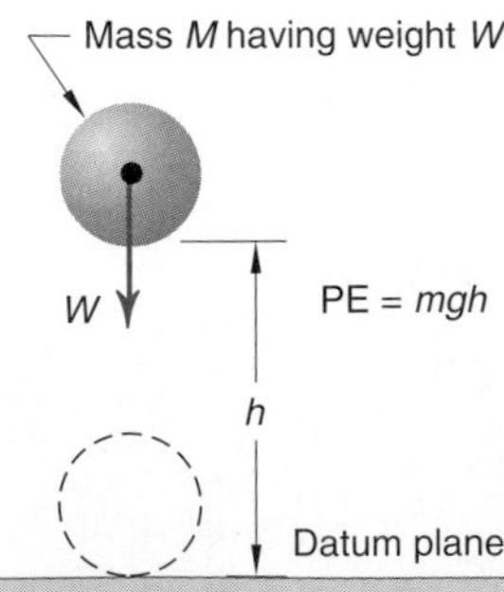

The energy possessed by an object by virtue of its velocity is called *kinetic energy*. It is equal to the energy required to accelerate the object from rest to its given velocity (considering the earth's velocity to be the datum plane). In equation form,

$$\text{KE} = 1/2\ mv^2 \tag{12.2}$$

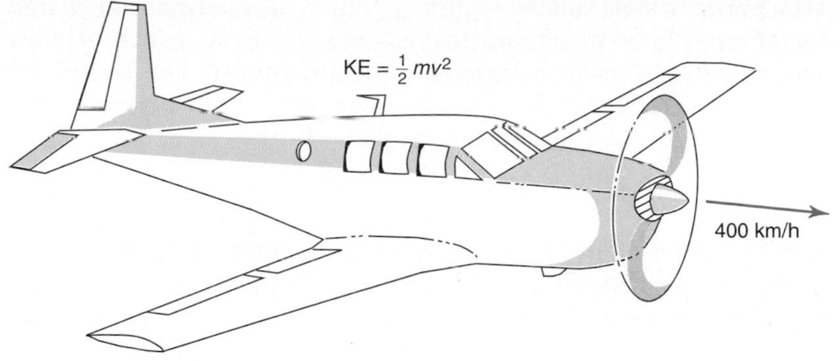

Figure 12.3
Kinetic energy

where

m = mass, kg

v = velocity, m/s

KE = kinetic energy, J

See Fig. 12.3.

In many situations in nature, an exchange of the forms of energy is common. Consider a ball thrown into the air. It is given kinetic energy when thrown upward. When the ball reaches its maximum altitude, its vertical velocity is zero, but it now possesses a higher potential energy because of the increase in altitude (height). Then as it begins to fall, the potential energy decreases and the kinetic energy increases until it is caught. If other effects such as air friction are neglected and the ball is caught at the same altitude from which it was thrown, there is no change in energy. This phenomenon is called *conservation of energy,* which is discussed in Sec. 12.4.

All matter is composed of molecules that, at finite temperatures, are in continuous motion. In addition, there are intermolecular attractions which vary as the distance between molecules changes. The energy possessed by the molecules as a composite whole is called *internal energy,* designated by the symbol U, which is largely dependent on temperature.

When a fuel is burned, energy is released. When food is consumed, it is converted into energy that sustains human efforts. The energy that is stored in a lump of coal or a loaf of bread is called *chemical energy.* This energy is transformed by a natural process called photosynthesis, that is, a process that forms chemical compounds with the aid of light. The stored energy in combustible fuels is generally measured in terms of heat of combustion, or heating value. For example, gasoline has a heating value of $47.7(10^6)$ J/kg, or $20.5(10^3)$ Btu/lbm.

Certain processes change the atomic structure of matter. During the processes of *nuclear fission* (breaking the nucleus into two parts, which releases high amounts of energy) and *fusion* (combining light-weight nuclei into heavier ones, which also releases energy), mass is transformed into energy. The stored energy in atoms is called *nuclear energy,* which will play an ever-increasing role in the technology of the future.

12.3 Energy in Transit

Energy is transferred from one form to another during many processes, such as the burning of fuels to run engines or the converting of electric energy to heat by passing a current through a resistance. Like all transfer processes, there must be a driving force or potential difference in order to effect the transfer. In the absence of the driving force or potential difference, a state of equilibrium exists and no process can take place. The character of the driving force enables us to recognize the forms of energy in transit, that is, work or heat.

Energy is required for the movement of an object against some resistance. When there is an imbalance of forces and movement against some resistance, mechanical work is performed according to the relationship

$$W = (\text{force})(\text{distance})$$

$$= Fd \tag{12.3}$$

where the force is in the direction of movement. If force has units of newtons and distance is expressed in meters, then work has units of joules. Electric energy is another form of work. This form of energy is transferred through a conducting medium when a difference in electric potential exists in the medium.

Other forms of energy transfer which are classified as work include magnetic, fluid compression, extension of a solid, and chemical. In each case a driving function exists that causes energy to be transferred during a process.

Heat is energy that is transferred from one region to another by virtue of temperature difference. The unit of heat is the joule. The large numerical values occurring in energy-transfer computations has led to the frequent use of the megajoule (10^6 J). The symbol used for heat is Q.

The relationship between the energy forms of heat and work during a process is given by the first law of thermodynamics. In addition, every form of work carries with it a corresponding form of friction which may change some of the work into heat. When this happens, the process is irreversible, meaning the energy put into the process cannot be totally recovered by reversing the process. Another way of stating this is that heat is a low-grade form of energy and cannot be converted completely to another form such as work. This concept is basic to the second law of thermodynamics. The first and second laws of thermodynamics are fundamentals to the study of processes involving energy transfer.

Figure 12.4
A solar wafer is shown under test conditions. This wafer can serve as an energy source for recharging storage batteries. (McGraw-Edison.)

Several examples follow of computations involving energy quantities. Particular attention should be given to the units and unit conversions in the examples.

Example problem 12.1 A boulder with mass of 1.000 (10^3) kg rests on a ledge 200.0 m above sea level. What type of energy does the object possess and what is its magnitude?

Solution The object possesses potential energy, so from Eq. (12.1),

$$\begin{aligned} PE &= mgh \\ &= 1.000\ (10^3)\ \text{kg}(9.807\ \text{m/s}^2)(200.0\ \text{m}) \end{aligned}$$

Note that 1 N = 1 kg · m/s^2, so

$$\begin{aligned} PE &= 1.961(10^6)\text{kg} \cdot \text{m}^2/\text{s}^2 \\ &= 1.961(10^6)\text{Nm} \\ &= 1.961(10^6)\text{J} \\ &= 1.961\ \text{MJ} \end{aligned}$$

Example problem 12.2 A 2.00 (10^4) lbm semi-trailer truck is traveling at sea level with a speed of 50.0 mi/h. Determine the energy form and magnitude possessed by the truck. Express the magnitude in SI units.

Solution Note that the units are in the English system and must be converted to SI units so that energy can be expressed in joules.

Converting to SI units, we get

$$2.00(10^4)\text{lbm} = 2.00(10^4)\text{lbm}(0.453\ 6\ \text{kg/lbm}) = 9\ 072\ \text{kg}$$

$$50.0\ \text{mi/h} = \frac{50.0\ \text{mi}}{1\ \text{h}} \Bigg| \frac{1\ 609\ \text{m}}{1\ \text{mi}} \Bigg| \frac{1\ \text{h}}{3\ 600\ \text{s}} = 22.3\ \text{m/s}$$

The truck possesses kinetic energy; therefore, from Eq. (12.2),

$$\begin{aligned} KE &= 1/2\ mv^2 \\ &= 1/2(9\ 072\ \text{kg})(22.3\ \text{m/s})^2 \\ &= 2.26(10^6)\ \text{kg} \cdot \text{m}^2/\text{s}^2 \\ &= 2.26(10^6)\ \text{Nm} \\ &= 2.26\ \text{MJ} \end{aligned}$$

The 2.26 MJ is the amount of energy that must be absorbed by the truck brakes in order to bring the vehicle to zero velocity in a required stop. If the stop is gradual, some of the energy can be absorbed by the engine by gearing down and road friction; in an emergency stop situation nearly all of the energy must be absorbed in the brakes and road friction.

Example problem 12.3 A mass of water is heated from 10.0 to 20.0 °C by the addition of $5.00(10^3)$Btu of energy. What is the final form of the energy? Express the final form of the energy in megajoules.

Solution The final form of the heat energy added appears as increased internal energy of the water. If state 1 of the water is prior to heating and state 2 is after heat has been added, then $U_2 - U_1 =$ 5 000 Btu.

Converting to SI, we get

$$\begin{aligned} U_2 - U_1 &= 5.00\ (10^3)\ \text{Btu} = (5.00 \times 10^3\ \text{Btu})(1\ 005\ \text{J/Btu}) \\ &= 5.28\ (10^6)\ \text{J} \\ &= 5.28\ \text{MJ} \end{aligned}$$

Example problem 12.4 A force of $2.00\ (10^2)$N acting at an angle of 20.0° with the horizontal is required to move block *A* along the horizontal surface (see Fig. 12.5). How much work is done if the block is moved $1.00\ (10^2)$m?

Figure 12.5
An applied force doing work

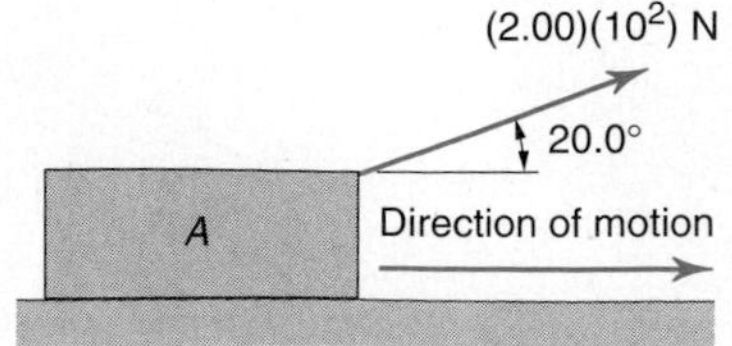

Solution Work is computed as the product of the force in the direction of motion and the distance moved, as in Eq. (12.3):

$$\begin{aligned} W &= Fd \\ &= (2.00 \times 10^2\ \text{N})(\cos 20.0°)(1.00 \times 10^2\ \text{m}) \\ &= 18\ 800\ \text{Nm} \end{aligned}$$

= 18 800 J

= 18.8 kJ

Where did the energy of the work go? There is no increase in potential energy, since height was not changed. There is no velocity change, so kinetic-energy change is zero. The energy of the work in this example is dissipated as heat of friction between the block and surface.

12.4 Conservation of Energy

The law of the conservation of energy states simply that energy can never be created or destroyed but only transformed (in nonnuclear processes). In effect, energy is converted from one form to another without loss. Careful measurements have shown that during energy transformations there is a definite relationship in the quantitative amounts of energy transformed. This relationship—the first law of thermodynamics—is a restatement of the principle of the conservation of energy.

12.5 Concepts from Thermodynamics

Thermodynamics is one of the major areas of engineering science. It is usually introduced to engineering students in a one-quarter or one-semester course, with students in energy-related disciplines continuing with one or more advanced courses. But it is necessary in this introduction to energy to introduce briefly some concepts that will help you solve some basic problems involving the transformation and consumption of energy. Our discussion is limited to the first law for closed systems and the second law, which governs efficiency and power. These terms will be explained in the following sections.

12.5.1 First Law of Thermodynamics

When applying the first law to substances undergoing energy changes, it is necessary to define a system and write a mathematical expression for the law. In Fig. 12.6a, a generalized closed system is illustrated. In a closed system, no material may cross the boundaries, but the boundaries may change shape. Energy, however, may cross the boundaries and/or the system may be moved intact to another position.

Some applications involve analysis of an open system in which mass crosses the defined boundaries and a portion of the energy transformation is carried in or out of the system with the mass (Fig. 12 6b) An example is an air compressor, which takes atmospheric air, compresses it, and delivers it to a storage tank. (Because of the complexity of analysis, problems involving open systems will not be considered here.)

For the generalized closed system, the first law is written as

Energy in + energy stored at condition 1 = energy out + energy stored at condition 2

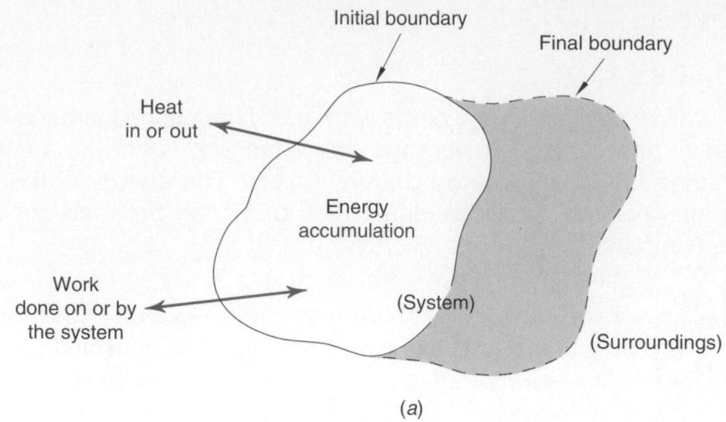

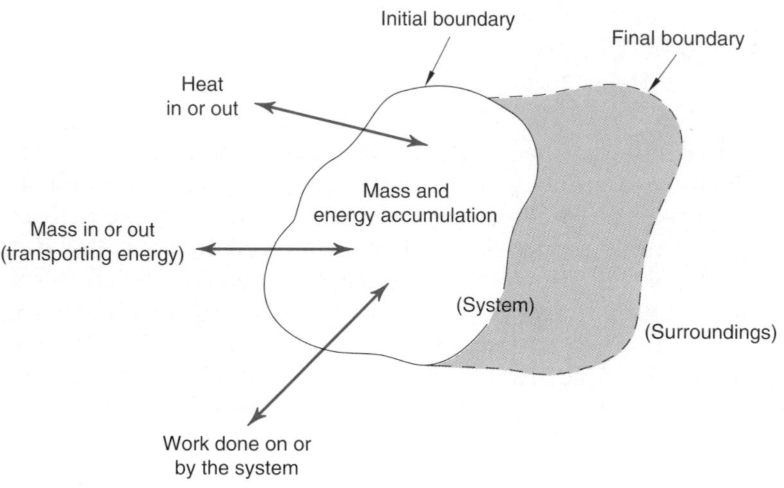

Figure 12.6
(a) Closed and (b) open thermodynamic systems

where conditions 1 and 2 refer to initial and final states of the system.

Energy in any of its forms may cross the boundaries of the system. If the assumptions are made that there is no nuclear, chemical, or electrical energy involved and that no changes will occur in potential and kinetic energy, then the first law becomes

Heat in + work done on system + internal energy at condition 1 = heat out + work done by the system + internal energy at condition 2

or combining heat, work, and internal energy quantities at conditions 1 and 2:

$$ {}_1Q_2 = U_2 - U_1 + {}_1W_2 \tag{12.4} $$

where

$_1Q_2$ = heat *added* to system

$_1W_2$ = net mechanical work *done by* system

$U_2 - U_1$ = change in internal energy

Example problem 12.5 The internal energy of a system decreases by 108 J while 175 J of work is done by the system on the surroundings. Determine if heat must be added to or taken from the system.

Solution

$$_1Q_2 = U_2 - U_1 + {}_1W_2$$

$$= -108 + 175$$

$$= +67 \text{ J} \qquad \text{(heat is added to system)}$$

Example problem 12.6 Analyze the energy transformations that can take place when the piston in Fig. 12.7 moves in either direction and/or heat is transferred.

Solution The air is considered to be the system. If the position shown is condition 1, there must be a force applied to the piston toward the left to hold its position. The magnitude of the force is

$$F = \text{(pressure of air) (area of piston)}$$

$$= PA \qquad \text{(in newtons)}$$

At this position, with no change in volume of the air, heat could be added or removed, which would increase or decrease, respectively, the internal energy of the system. In terms of the first law,

$$_1Q_2 = U_2 - U_1$$

Note the absence of the work term. Work cannot take place without action of a force through a distance. The force (PA) is present, but no movement takes place; thus no work is transferred.

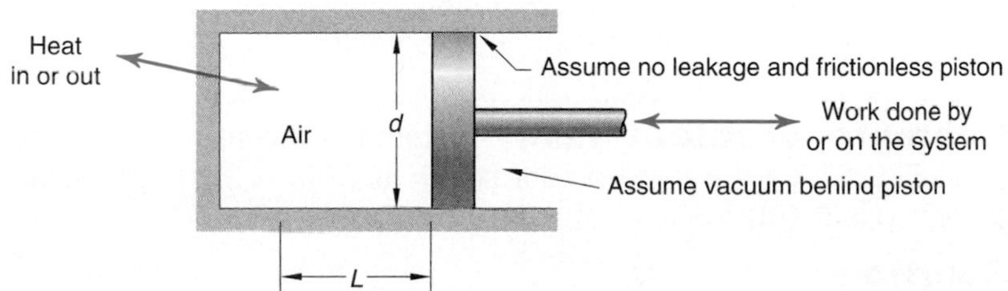

Figure 12.7
Piston-cylinder combination

If the force to the left is increased slightly, then the piston will move left. Work is done because movement occurs as a result of a force. The air temperature will change depending on how much heat is removed from the air. The first law for this situation is written

$${}_1Q_2 = U_2 - U_1 + {}_1W_2$$

The ${}_1W_2$, equal to $P \cdot A \cdot L$ if the piston travel is L, will be negative, since work is being done on the system by the piston.

For the case of the piston moving to the right, the first law is still

$${}_1Q_2 = U_2 - U_1 + {}_1W_2$$

but ${}_1W_2$ will be positive.

If no heat is allowed to transfer across the boundaries, then the first law becomes

$${}_1W_2 + U_2 - U_1 = 0$$

This process, where no heat crosses the boundary, is called an *adiabatic process.*

Many engineering applications utilize the assumption that air and other gases behave ideally, that is, the pressure, volume, and temperature obey an equation of state derived from the laws of Boyle and Charles. One characteristic of a perfect gas is that internal energy is a function of temperature only. If the ideal-gas assumption is made, then a constant-temperature (*isothermal*) process implies that there is no change in internal energy, and therefore the first law for an isothermal process becomes

$${}_1Q_2 = {}_1W_2$$

Example problem 12.7 Using the previous example and Fig. 12.7, let the piston be moved so that the volume of air is reduced to one-half its original value. During this process, 116 MJ of heat leave the air. If the process is isothermal, how much work is done on the system?

Solution

$${}_1Q_2 = {}_1W_2$$

$$-116 \text{ MJ} = {}_1W_2$$

The negative sign indicates that work has been done on the air by the piston.

Example problem 12.8 Determine the pressure of the air in Fig. 12.7 if the force on the piston is $1.00(10^2)$ N and the diameter d is 0.100 m.

Solution

$$F = PA$$

$$P = \frac{F}{A}$$

$$= \frac{1.00(10^2)}{\pi d^2/4}$$

$$= \frac{1.00(10^2)4}{0.01\pi}$$

$$= 1.27\ (10^4)\text{N/m}^2$$

$$= 1.27\ (10^4)\ \text{Pa}$$

$$= 12.7\ \text{kPa}$$

12.5.2 Second Law of Thermodynamics

The first law states that heat is a form of energy; it may be turned into work, and conversely, work can be returned to heat. However, there is a limitation in the conversion of heat into work. Stated simply, heat cannot perform work except when it passes from a higher to a lower temperature. In other words, heat will not flow spontaneously from a colder to a hotter substance. This limitation on heat conversion forms the basis of the second law of thermodynamics. There are many statements of the second law. Clausius stated it as: "It is impossible for a self-acting machine, unaided by any external agency, to convey heat from one body to another at higher temperature." Another statement is: "No device can completely and continuously transform all of the heat supplied to it into work." The heat that can be transformed into work is called *available energy*; the remaining portion is termed *unavailable energy*. As indicated in Sec. 12.2, whenever work of any form is performed, there is friction present that will downgrade some available energy to unavailable energy. This unavailable energy is heat that is at too low a temperature to perform work under the conditions specified by the system and its surroundings. For a given system and surroundings, the available and unavailable energies can be computed. The procedures for this computation are beyond the scope of this text.

Another application of the second law is the determination of the maximum efficiency of any device that converts heat into work. Such devices, called heat engines, cannot attain 100 percent efficiency because of the second law. In 1824 Sadi Carnot proposed an ideal engine cycle that had the highest attainable efficiency within thermodynamic laws. In reality, an engine following the Carnot cycle cannot be constructed, but the theory provides a basis of comparison for practical engines. The Carnot efficiency can be shown to be

$$\text{Carnot efficiency} = 1 - \frac{T_2}{T_1} \tag{12.5}$$

where T_1 is the absolute temperature at which the engine receives heat (high temperature) and T_2 is the absolute temperature at which the engine rejects heat (low temperature) after performing work. Absolute temperatures are determined by adding 273° to a reading in °C or 460° to a reading in °F. Temperatures on the absolute scales are in degrees Kelvin (°K) for Centigrade-size degrees, and in degrees Rankine (°R) for Fahrenheit-size degrees.

Example problem 12.9 A steam engine is designed to accept steam at 300 °C and exhaust at 100 °C. What is its maximum possible efficiency?

Solution

$$\text{Carnot efficiency} = 1 - \frac{T_2}{T_1}$$

$$= 1 - \frac{100 + 273}{300 + 273}$$

$$= 1 - \frac{373}{573}$$

$$= 35 \text{ percent}$$

12.5.3 Efficiency

All the energy put into a system does not end up producing useful results. According to the second law, a certain amount of energy is unavailable for productive work. In addition, the available energy does not perform an equivalent amount of work because of losses incurred during the transfer of energy from one form to another. An automobile engine converts chemical energy in gasoline to mechanical energy at the axle; however, some of the energy is lost through bearing friction, incomplete combustion, cooling water, and other thermodynamic and mechanical losses. The engineer designing units to convert heat into work must be concerned with the overall efficiency of the proposed unit. This is an estimate of actual performance, not ideal performance. The overall efficiency for a heat engine may be written as

$$\text{Overall efficiency} = \frac{\text{useful output}}{\text{total input}} \tag{12.6}$$

From Eq. (12.5), it is clear that the maximum (Carnot) efficiency can be increased by lowering the exhaust temperature T_2 and/or increasing the input temperature T_1. In theory this is true, but practical design considerations must include available materials for construction of the heat engine. The engineer thus attempts to obtain the highest possible efficiency using existing technology. Examples of overall efficiency are 17 to 23 percent for automobile engines (gasoline), 26 to 38 percent for diesel engines, and 20 to 33 percent for turbojet aircraft engines. Thus, in the case of the gasoline automobile engine, for every 80 L (21 gal) tank of gasoline, only 20 L (5.3 gal) ends up moving the automobile.

Care must be exercised in the calculation and use of efficiencies. To illustrate this point, consider Fig. 12.8, which depicts a steam power plant from the burning of fuel for steam generation to driving an electric generator with a turbine. Efficiencies of each stage or combinations of stages in the power plant may be calculated by comparing energies available before and after the particular operations. For example, combustion efficiency can be

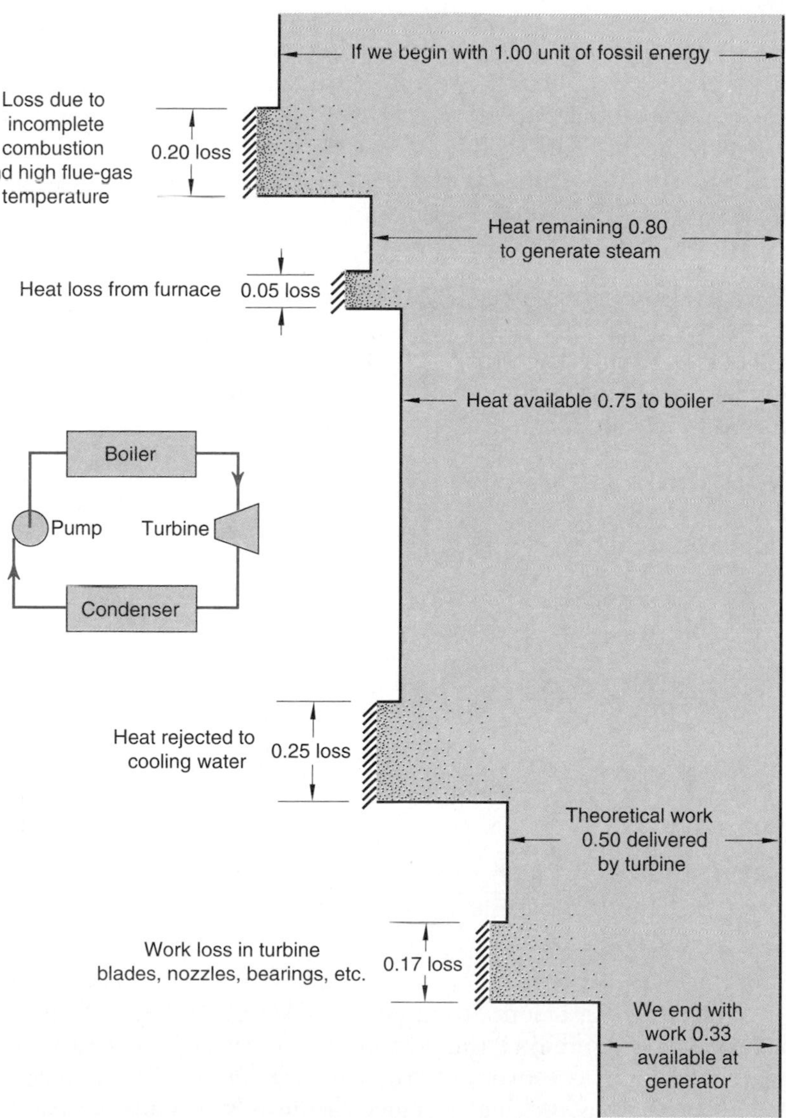

Figure 12.8
Energy losses in a typical steam power plant

calculated as 0.80/1.00, or 80 percent. The turbine efficiency is 0.33/0.50, or 66 percent. The overall power plant efficiency up to the electric generator is 0.33/1.00, or 33 percent. By no means is the 33 percent a measure of the efficiency of generation of electricity for use in a residential home. There will be losses in the generator and line losses in the transmission of the electricity from the power plant to the home. The overall efficiency from fuel at the power plant to the electric oven in the kitchen may run as low as 25 to 30 percent. It is perhaps ironic to note that the cycle is complete when the oven converts electricity back into heat, which is where the entire process began.

12.5.4 Power

Power is the rate at which energy is transferred, generated, or used. In many applications it may be more convenient to work with power quantities rather than energy quantities. The SI unit of power is the watt (one joule per second), but many problems will have units of horsepower (hp) and/or foot-pound-force per second (ft·lbf/s) as units, so conversions will be necessary.

Example problem 12.10 A steam turbine at a power plant produces 3 500 hp at the shaft. The plant uses coal as fuel (12 000 Btu/lbm). Using Fig. 12.8 to obtain an overall efficiency, determine how many metric tons of coal must be burned in a 24 h period to run the turbine.

Solution From Fig. 12.8, efficiency is 0.33/1.00, or 33 percent. From Eq. (12.6),

$$\text{Total input} = \frac{\text{useful output}}{\text{overall efficiency}}$$

$$= \frac{3\,500 \text{ hp}}{0.33} \left| \frac{2\,545 \text{ Btu}}{1 \text{ hp}\cdot\text{h}} \right.$$

$$= 2.7(10^7)\text{Btu/h}$$

The amount of coal needed for 1 day of operation is therefore

$$\text{Coal required} =$$

$$\frac{2.7(10^7)\text{Btu}}{1 \text{ h}} \left| \frac{1 \text{ lbm}}{12\,000 \text{ Btu}} \right| \frac{1 \text{ kg}}{2.205 \text{ lbm}} \left| \frac{1 \text{ t}}{10^3 \text{ kg}} \right| \frac{24 \text{ h}}{1 \text{ d}}$$

$$= 24 \text{ t/day}$$

A measure of efficiency commonly used in the refrigeration and air-conditioning fields is called the *energy efficiency ratio*, abbreviated *EER*. In essence, a refrigerating machine is a reversed-heat engine; that is, heat is moved from a low-temperature region to a high-temperature region, requiring a work input to the reversed-heat engine. The expression for efficiency of a reversible-heat engine is

$$\text{Refrigeration efficiency} = \frac{\text{refrigerating effect}}{\text{work input}} \tag{12.7}$$

Refrigeration efficiency is also called *coefficient of performance* (*CP*). The numerical value for CP may be greater than 1.

Industrial-size refrigeration units are measured in tons of cooling capacity. The ton unit originated with early refrigerating machines and was defined as the amount of refrigeration produced by melting 1 ton of ice in a 24 h period. If the latent heat of ice is taken into account, then a ton of refrigeration will be equivalent to 12 000 Btu/h. Home-size units are generally rated in British thermal units per hour of cooling capacity.

The EER takes into account the normal designations for refrigerating effect and work input and expresses these as power rather than energy. Thus, EER is the ratio of a refrigerating unit's capacity to its power requirements:

$$\text{EER} = \frac{\text{refrigerating effect, Btu/h}}{\text{power input, W}}$$

Example problem 12.11 Compute the cost of running a 36 000 Btu/h air-conditioning unit an average of 8 h/d for 1 month if electricity sells for 6.8 ¢/kWh. The unit has an EER of 8.

Solution

$$\text{EER} = \frac{\text{Refrigerating effect}}{\text{power input}}$$

$$8 = \frac{36\ 000\ \text{Btu/h}}{\text{power input}}$$

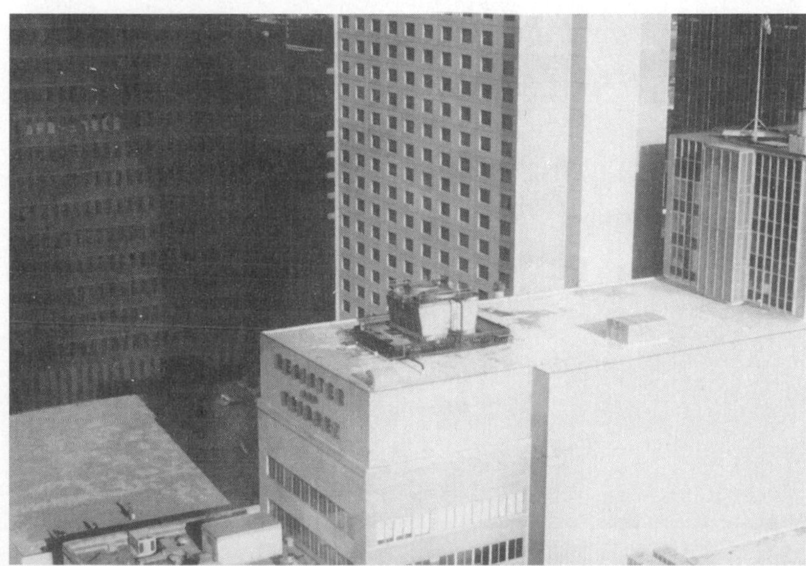

Figure 12.9 The air-conditioning system on a high rise building is located on the roof for more efficient distribution.

Power input = 4 500 W

$$\frac{\$}{\text{month}} = \frac{4\,500\text{W}}{} \left| \frac{8\text{ h}}{1\text{ d}} \right| \frac{30\text{ d}}{1\text{ month}} \left| \frac{\$0.068}{1\text{ kWh}} \right| \frac{1\text{ kW}}{10^3\text{W}}$$

$$= \$73.44/\text{month}$$

12.6 Energy Sources in the Future

The engineer will continue to design devices that will perform work by conversion of energy. However, the source of energy used to produce work and the conversion efficiency of the energy into work will play a much greater role in the design procedure.

Much has been written in the popular press concerning the "energy problem" and the development of energy sources to solve the problem. Many opinions are given with regard to how we should proceed to seek new energy sources and how we should use existing sources. Environmental concerns about energy acquisition and consumption become stronger by the day. The engineer must be able to discern the facts that may not be clear in some of the opinions. The following discussion provides an insight into some energy sources that will be utilized in the future.

12.6.1 Wind Power

Heated equatorial air rises and drifts toward the poles. This phenomenon coupled with the earth's rotation results in a patterned

Figure 12.10 A large university consumes enormous amounts of energy and in many instances provides its own power and heat. (Iowa State University.)

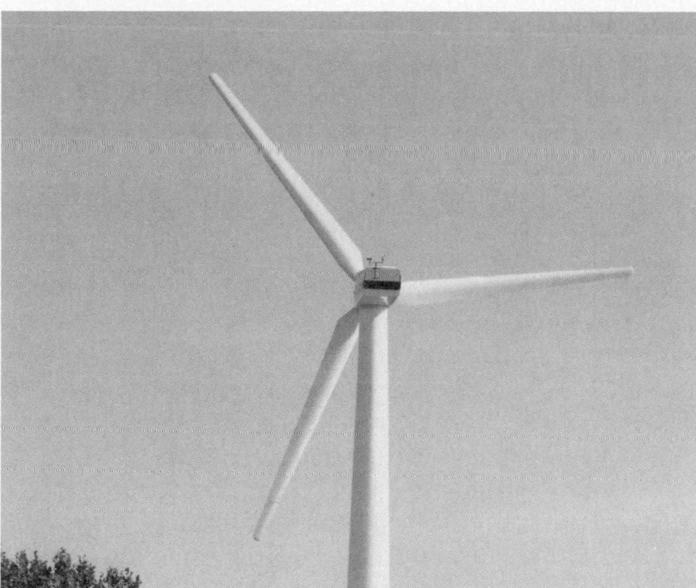

Figure 12.11
Generation of electricity from windmills is increasing in areas that have high average wind speeds

air flow. In the United States, we see this pattern as weather systems moving from west to east across the country. These weather systems possess enormous energy, but the energy is diffused, variable, and difficult to capture.

We are all familiar with the historical use of the wind to drive ships, to pump water, and to grind grain. These units are still present, even though the advent of inexpensive fossil fuels and rural electrification made them uneconomical for a period of time. Today, a great deal of research is under way in an attempt to capture energy from the wind in an economically feasible manner.

Engineers have learned that the power output from a windmill is approximately proportional to the square of the blade diameter and to the cube of the wind velocity. This suggests that to generate high power output, the windmills must be large and must be located in areas where the average wind velocity is high. Thus, coastal regions and the Great Plains are promising locations. Efforts continue toward reducing the costs of wind-power installations and toward overcoming the supply-demand problem. For example, we could use excess power generated when the wind is blowing to pump water to an elevated storage area and use the resulting potential energy to drive a water turbine for power generation when the wind is not blowing.

12.6.2 Water Power

There are a number of ways in which water can provide energy to be used by humankind. We can obtain energy, for example, from rivers, tides, ocean currents, waves, and thermal gradients.

Rivers

Generation of electricity by use of the energy of falling water provides less than 15 percent of the electricity requirements in the United States (see Fig. 12.12). In most areas of the country, less than one-half of the potential for hydroelectric power has been developed. Hydroelectric power facilities will remain an important factor in energy generation far into the future. However, because of the environmental concerns for the protection of scenic rivers and wildlife, new facility construction will be limited. Hydroelectric power thus will furnish a smaller fraction of the total power production.

Tides

The difference between the elevation of the ocean at high tide and low tide varies from 1 to 2 m in most places to 15 to 20 m in some locations. The idea of using this energy source is not new, and any sailor can attest to the power potentially available. The technical feasibility has been proved and plants are in use in France and in the former Soviet Union. The economic feasibility is still in serious doubt and is not yet even claimed by responsible sources. The toughest problems seem to be that vast storage volumes are required and suitable basins are rare.

Ocean Currents

We are all aware of the warming effect that the Gulf Stream provides for the British Isles and Western Europe. This and other ocean currents possess massive amounts of kinetic energy even though they move at very slow velocities. It has been proposed that a series of large (170 m in diameter) turbines be placed in

Figure 12.12 Electricity generated from water power continues to be important. The Folsom Dam and power plant provide electricity for parts of California (U.S. Bureau of Reclamation.)

the Gulf Stream off Florida. Ten such turbines could produce power equivalent to that produced by one typical coal-fired power plant. Detractors warn that such installations may reduce the stream velocity to the point that the warming of Western Europe would be lost.

Waves

No doubt you have watched the surf smash into the beach and have been in awe of the obvious power being displayed. Machines have been made that produce power from the wave action. But to be successful, the installations must be located where the magnitude of the wave action is high and somewhat uniform, and it is difficult and expensive to design such installations against major storms. The best sites in the United States are on the coasts of Washington and Oregon.

Thermal Gradients

Earlier we discussed the Carnot principle in which the theoretical efficiency of a heat engine is defined by Eq. (12.5). It should be possible to employ the difference in temperature of the warm water at the ocean surface and the cold water at considerable depths to drive a heat engine. It is not uncommon to have temperatures of 75 °F at the surface and 40 °F at a depth of about 1 500 ft. Such a temperature difference would produce a Carnot efficiency of about 6 percent. An efficiency of about 3 percent might then be realized for an actual engine.

There are major problems to be overcome if a steam cycle is used with these temperatures. Very low pressures (less than 0.5 lbf/in^2 absolute) would need to be provided by a vacuum pump in order to produce steam at 75 °F. Even lower pressure, perhaps 0.1 lbf/in^2 absolute, would be needed at the turbine outlet to produce useful work. Because of the low pressures, the steam would occupy a large volume, meaning that all the equipment would have to be much larger than is normally used in a steam power plant. Also, the turbine work output would be very small in comparison to that produced by a conventional system per pound-mass of steam flowing through the turbine.

Perhaps some fluid with a lower boiling temperature, say ammonia, could be employed to better utilize the low temperature of the ocean surface. Other problems to consider include the corrosive action of seawater, protection of the plant from weather, and the difficulties of transmitting the electrical energy produced.

Example problem 12.12 How many megawatts of electric power can be obtained from the water of the Zambezi River over Victoria Falls if 1.0 $(10^5) m^3/s$ of water passes over the falls and drops 1.0 (10^2) m. Assume a conversion efficiency of 80 percent and that the water turbines can be placed at the bottom of the falls.

Solution Equation (12.6) may be used with an estimated efficiency of 0.80:

output = (0.80)(input).

The input may be calculated as the potential energy of the water that would be converted to kinetic energy during the 100 m drop. Equation (12.1) is used, with the mass equal to the volume multiplied by the density of the water. Conversion of input energy to power is accomplished by dividing by 1 s.

$$\begin{aligned} \text{PE} &= mgh \\ &= Vp\,(gh) \\ &= (10^5\ \text{m}^3)(10^3\ \text{kg/m}^3)(9.81\ \text{m/s}^2)(100\ \text{m}) \\ &= (9.81)(10^{10})\text{kg·m}^2/\text{s}^2 \\ &= 9.81(10^{10})\text{J} \end{aligned}$$

The input power is then

$$\begin{aligned} \text{Input} &= \frac{9.81(10^{10})\text{J}}{1\ \text{s}} \\ &= 9.81(10^{10})\text{W} \end{aligned}$$

and the output is

$$\begin{aligned} \text{Output} &= (0.80)(9.81)(10^{10})\text{W} \\ &= (7.8)\ (10^4)\text{MW} \end{aligned}$$

12.6.3 Geothermal Power

It is commonly accepted that the earth's core is molten rock with a temperature of 10 000 to 12 000 °F. This source has the potential to provide a large portion of our energy needs. The earth's crust varies in thickness from a few hundred feet to perhaps 20 mi. It is composed mostly of layers of rock—some solid, some porous, and some fractured. Engineers are now exploring the use of the earth's heat (geothermal energy) to produce power from the hot water, steam, and heated rock.

There are many areas throughout the world where large amounts of hot water are available from the earth. The hot water varies in many ways—in quantity, temperature, salinity, and mineral content. You will recall that we earlier discussed the reduced work produced and large volumes required for steam at low temperatures and pressures. These factors explain why much larger geothermal power plants (relative to coal-fired plants) are needed to produce the same amount of power. Since large volumes of hot water must be withdrawn from the earth, there is concern about the possibility of subsidence of the earth. Hence,

Figure 12.13
Geothermal energy is a potential source of power for areas where this energy exists. (Phillips Petroleum Company.)

the water used will probably be returned to the formation from which it was withdrawn.

At only a few locations in the world is steam available in sufficient quantities to be used to produce electricity. The only area in the United States producing large quantities of steam is in northern California. Electricity has been produced there for about 30 years at a cost that is less than that produced by fossil-fuel plants. Because of the vast amount of energy available from this source, new plants are being constructed regularly. This steam is captured by drilling from 500 to 10 000 ft deep, and it has a temperature of about 350 °F but at low pressure. Besides the corrosive nature of the steam, it contains several gases, including ammonia and hydrogen sulfide, that have objectionable odors and are poisonous.

In almost all areas we can consider drilling deep enough to tap the heat in the dry, heated rock of the earth. Water could be introduced by one set of pipes and steam removed by another. Since the steam would be at relatively low pressure and temperature, large amounts will be required. This fact points to the need for a large surface area of exposed rock to accomplish the heat transfer. It has been suggested that the rock can be broken by explosions or by hydraulic fracturing to create more exposed surface.

It is clear that using geothermal energy has great potential, but many problems need to be solved and solutions will probably come only after large-scale pilot plants have been in operation long enough to obtain reliable efficiency data and operating costs.

12.6.4 Solar Power

The sun is an obvious source of energy that we have employed in different ways since the beginning of time. It supplies us with many, many times as much energy as we need. Our problems in using this energy lie with collecting, converting, and storing the inexhaustible supply. Our efforts are being directed primarily toward direct heating and steam power plants. We are also experimenting with crop drying and metallurgical furnaces. You are no doubt aware of the increasing use of solar energy in home and building heating and in domestic hot-water supplies. This use will surely continue to increase, particularly in new installations. But we will have to improve on current technology or have tax relief and low-interest loans or other incentives to make solar installations economically feasible.

Solar (photovoltaic or PV) cells have been shown to work on space vehicles but are extremely costly. We must reduce their cost to between 1 and 2 percent of current costs in order for this application to come into general use.

Example problem 12.13 Estimate the area over which solar energy must be collected to provide the electric energy needed by a small community of 4 500 people in a 24 h period. Assume a conversion efficiency of 8 percent.

Solution This example requires some assumptions in order to obtain a meaningful result.

1. Each home consumes about 13 000 kWh of electricity on the average each year.

Figure 12.14 A solar-heated home (Iowa State University.)

2. An average of three people live in each home.
3. The sun shines an average of 8 h/d.
4. The typical solar heat transfer rate is 1 kW/m^2.

Equation (12.6) may be used.

$$\text{Input} = \frac{\text{output}}{\text{efficiency}}$$

$$= (\text{area})(\text{solar heat-transfer rate})$$

$$= \text{area } (1 \text{ kW/m}^2)$$

Therefore, the area can be computed as

$$\text{Area} = \frac{}{0.08}\ \frac{1 \text{ m}^2}{1 \text{ kW}}\ \frac{13\,000 \text{ kWh}}{\text{year-home}}\ \frac{1 \text{ home}}{3 \text{ people}}\ \frac{4\,500 \text{ people}}{}\ \frac{1 \text{ d}}{8 \text{ h}}\ \frac{1 \text{ year}}{365 \text{ d}}$$

$$= 84\,000 \text{ m}^2$$

$$= 8.4 \text{ ha}$$

12.6.5 Nuclear Power

As mentioned earlier in this chapter, nuclear power may be generated from two processes, fission and fusion. Fission is the splitting of an atom of nuclear fuel, usually uranium 235 (U-235 or Pu 239). The splitting process releases a great quantity of heat, which can then be converted to other energy forms. Nearly all reactors in use today producing electric energy are fission reactors using U-235. However, U-235 comprises only 0.7 percent of the natural uranium supply and thus would quickly become a scarce resource if there were total energy dependence on it. Other iso-

Figure 12.15
The Duane Arnold Energy Center, is a nuclear-powered electric generating station. (Iowa Electric Light and Power Company.)

topes, such as U-238 and thorium 232, are relatively abundant in nature. The breeder reactor is designed to use these isotopes that are generally a waste product in current reactors.

The radioactive waste material from nuclear power generation has created some disposal problems. The half-life of radioactive materials can be 1 000 years or more, thus creating a perpetual need for secure disposal of the waste materials from a nuclear power–generation facility. Radiation leaks and other environmental concerns have led to the shutdown of some facilities. This has made nuclear power generation a politically sensitive issue. Nuclear engineers and scientists continue to perform research to solve the complex problems associated with this source of energy. As these problems are solved and the general public becomes aware of the potential afforded by nuclear energy, increased use of nuclear power is likely.

Example problem 12.14 If 1 g of U-235 is capable of producing the same amount of heat as 3 500 kg of coal, how much water would have to be stored $1.0(10^3)$ m above sea level to provide the same energy as 1 kg of U-235? Assume that coal has a heating value of 30.0 MJ/kg.

Solution A kilogram of U-235 would possess the same energy as $3.5\ (10^6)$ kg of coal and therefore would have a total energy value of

$$(30.0\ \text{MJ/kg})(3.5 \times 10^6\ \text{kg}) = 105(10^6)\text{MJ}$$

Setting this equivalent to the potential energy of the water,

$$mgh = 105(10^{12})\text{J}$$

$$\begin{aligned} m &= \frac{105(10^{12})\text{J}}{(9.81\ \text{m/s}^2)(10^3\ \text{m})} \\ &= 10.7(10^9)(\text{N}\cdot\text{m}\cdot\text{s}^2)/\text{m}^2 \\ &= 10.7(10^9)(\text{kg}\cdot\text{m/s}^2)(\text{m}\cdot\text{s}^2)/\text{m}^2 \\ &= 10.7(10^9)\ \text{kg} \\ &= 10.7\ \text{Tg} \end{aligned}$$

12.7 Key Terms and Concepts

Stored energy
Potential energy
Kinetic energy
Internal energy
Chemical energy
Nuclear energy
Energy in transit
Work
Heat
Conservation of energy law
First law of thermodynamics
Closed system
Second law of thermodynamics
Available energy
Unavailable energy

Efficiency
Carnot efficiency
Power
Energy efficiency ratio
Refrigeration efficiency
Wind power
Water power
Geothermal power
Nuclear power

Problems

12.1 Conduct a study on one or more of our sources of fossil fuels. Find out estimated annual usage of the fuels on a national and international basis, the geographical locations of the sources, and estimated time until the sources are exhausted at current usage rates. Compare fossil fuel usage rates in the United States with those in Japan, England, Germany, China, India, and South Africa. Working in teams of two, develop a written report or oral presentation according to your instructor's directions for

(*a*) coal
(*b*) oil
(*c*) natural gas

12.2 Prepare a report or oral presentation on the status of nuclear power in the United States. Include usage data, public opinion on the use of nuclear power, and regulations on the industry. Work in teams of two.

12.3 Prepare a report or oral presentation on the status of one of the following alternative sources of power in the United States. Include usage data, geographic locations of these sources, and ultimate form of energy that is consumed from these sources. Work in teams of two.

(*a*) water power
(*b*) geothermal power
(*c*) wind power
(*d*) solar power

12.4 A minivan with a mass of 3 340 lbm is being driven on a mountain road 4 500 ft above sea level at 45 mi/h. Determine the total energy possessed by the vehicle using sea level as the datum for potential energy. Express the answer in joules.

12.5 A truck with a mass of 9 300 kg is traveling at a speed of 86 km/h. If the truck had to stop in a distance of 75 m, what constant braking force would be required? Neglect air and road friction.

12.6 Referring to the truck in prob. 12.5, if instead of braking, it is allowed to coast up a grade of 9.2 percent, what distance in meters will it travel along the grade? Neglect air and road friction.

12.7 An automobile has a kinetic energy of 396 kJ and is traveling at 65 mi/h. Determine its mass in kilograms.

12.8 An automobile weighs 4 150 lbm and has a kinetic energy of 573 kJ. Determine its velocity in miles per hour.

12.9 A sled and passenger, total mass 220 lbm, coasts down an 11.5 percent grade. Neglecting air and snow friction, what is the speed in m/s after the sled has traveled 385 m?

12.10 An airplane weighing 15 tons is moving at 380 mi/h at an altitude of 9 480 ft. What is the total energy in joules?

12.11 A cliff diver, mass 62 kg, dives from a 29 m cliff into the sea. Neglecting air friction,

(*a*) What is the diver's total energy in joules at the following heights; 29 m, 15 m, surface of the water?
(*b*) What is the diver's velocity in ft/s at the first contact with the water?

12.12 A fly ball in a baseball game reaches a peak height of 210 ft. Neglecting air friction, with what speed does the baseball hit the centerfielder's glove when the ball is caught 6.5 ft above the ground?

12.13 A father pulls his two children in a wagon for 15.5 minutes covering 0.85 miles on a level sidewalk. He pulls with an average force of 25 lb at an angle of 35° to the sidewalk.

(*a*) How much work is done (in joules)?
(*b*) What is the average horsepower required?

12.14 How much work in ft·lbf is done when raising a mass of 39 slugs to an altitude of 89 m above the surface of the moon? The gravitational field on the surface of the moon is about one-sixth that on earth at sea-level.

12.15 If the mass in Prob. 12.14 fell back to the surface of the moon, with what velocity, in ft/s, would it strike the surface?

12.16 Air is heated in an 8.2 in diameter cylinder until it takes a force of 16 000 lb to hold the piston. What is the pressure exerted by the air on the piston (in lbf/ft^2).

12.17 A piston/cylinder device is 6.35 in in diameter and is rated at an internal design pressure of 7.50 MPa. Prior to shipping, the cylinder is tested by increasing the internal pressure to 2.0 times the design pressure (safety factor of 2.0). What maximum force in newtons must be applied to the piston to hold it stationary during the test?

12.18 If 5 960 ft lbf of work is done on a closed system while 52 Btu of heat is added, what is the change in internal energy (joules) of the system?

12.19 During a process, 46.6 MJ of heat is removed from a closed system that contains 63.0 kg of fluid. If the internal energy is increased by 18.6 MJ, how much work in Btu was done on the gas?

12.20 In an adiabatic process on 38 kg of fluid, 82 kJ of work is done on the fluid. Express the change in internal energy (Btu) per pound of mass of fluid.

12.21 Heat at the rate of 1.05 MJ/min is added to a closed system that produces work at the rate of 3.85 hp.

(*a*) How much heat (joules) is lost in 5 minutes from the system?
(*b*) What is the system's overall efficiency?

12.22 An internal combustion engine burns fuel at the rate of 6.5 gallons per hour. The fuel has a heating value of 36.7 kJ/mL. Calculate the power output (in kilowatts) if the overall efficiency is 36 percent.

12.23 Compute the Carnot (ideal) efficiency of a heat engine operating between 350 and 60 °C.

12.24 An inventor claims to have developed an engine that produces 150 hp if it receives 350 hp at 575 °C and rejects heat at 250 °C. Is this a reasonable claim and why?

12.25 A Carnot engine produces 31 kW and rejects $2.1(10^3)$ Btu/min to a 38 °C low-temperature reservoir. What is the temperature (°C) of the high-temperature source?

12.26 Fuel with a heating value of 36 MJ/L is burned at the rate of 9.4 L/hr in an internal combustion engine. The engine puts out 46 kW and drives an electric generator that produces 150 amps at 240 volts DC. Determine:

(*a*) engine efficiency
(*b*) generator efficiency
(*c*) overall efficiency

12.27 A system consisting of a water pump and electric motor has an overall efficiency of 56 percent. What power output of the motor is required (in kilowatts) to pump 2 500 gallons of water from a lake to a storage tank 44 ft above the lake surface in 19 minutes?

12.28 A pump is 65 percent efficient and is rated at 32 kW. It is used to pump a grain-slurry mixture. An electric motor, 83 percent efficient, drives the pump. Determine:

(*a*) the input horsepower to the pump
(*b*) the power, in kilowatts, required for the motor.

12.29 A 0.75 hp electric motor, 85 percent efficient, drives a water pump, 68 percent efficient.

(*a*) How much power, in watts, does the motor require?
(*b*) How much water, in gallons, could this combination lift 7.5 ft in one hour?

12.30 Estimate the number of households that could receive electric energy from 35 kg of U-235. Assume the average household requires 17 000 kWh per year, that the conversion of U-235 to electricity is 68 percent efficient, and that the energy equivalent of U-235 is 10^8 MJ/kg.

12.31 Using the data given in Prob. 12.29, determine the surface area (in hectares) of solar collectors operating at an efficiency of 12 percent to supply electric energy to the equivalent number of houses. Assume an average of 8 hours of sunlight per day and a transfer rate of one kW per square meter.

12.32 Write a computer program or produce a spreadsheet that will produce values for solar collector area in acres needed to supply electric energy for cities with populations of 5 000, 20 000, 50 000, 100 000, 1 000 000, and 5 000 000. Provide results for conversion efficiencies of 5 to 15 percent in increments of 1 percent. Interpolate the data to determine collector area for your city. Use the same assumptions as used in Example prob. 12.13.

12.33 A 36 000 Btu/h air conditioner is installed in a home. What fraction of the unit's capacity is needed to remove the heat generated by four 150 W lamps, six 100 W lamps, four 75 W lamps, a 500 W refrigerator, an 800 W microwave oven, and four people watching TV (125 W)? The average heat output of a person at rest is about 480 Btu/h. If the EER of the unit is 9.85, what power (kilowatts) is used by the unit when operating? How many kilowatt-hours of electric energy are used during each hour for these conditions?

12.34 Write a computer program or produce a spreadsheet that will determine the average monthly cost of operating five-hundred 24 000 Btu/h home air conditioners and plot this cost versus EER. Consider EER values from 7.0 to 11.0 in increments of 0.5. The electrical rate structure is as follows:

First 250 kWh = 9.05¢/kWh

Next 750 kWh = 7.85¢/kWh

Next 9 000 kWh = 6.35¢/kWh

Over 10 000 kWh = 5.25¢/kWh

Assume that the units run an average of 6 hours per day on weekdays and 12 hours per day on weekends. Use a 31-day month beginning on a Monday.

12.35 A typical refrigerator uses about 1 800 kWh per year. An energy-efficient refrigerator uses about 800 kWh per year. If the cost of electricity is 9.5¢/kWh, how many energy efficient refrigerators replacing the others would it take to create an annual energy savings of $1 million?

12.36 Assume an average annual mileage of 20 000 miles for a typical family car that gets 23 miles per gallon of gasoline. How many of these cars would have to be replaced with cars getting 26 miles per gallon to generate an annual savings of 15 million gallons of gasoline?

13 Engineering Economy

13.1 Introduction

The role of engineers in company management has increased because of the importance of technology in many business decisions. Engineers serve as managers or executive officers of businesses and therefore are required to make financial as well as technical decisions. Even in companies where the managers are not engineers, engineers serve as advisers and provide reports and analyses that influence decisions. Also, the amount of capital investment (money spent for equipment, facilities, and so on) in many industries represents a significant part of the cost of doing business. Thus, estimates of the cost of new equipment, facilities, software, processes, etc., must be carefully done if the business is to be successful and earn a profit on its products and services.

A couple of examples will be used to illustrate how an engineer might be involved in the financial decision-making process. Suppose a manufacturing company has decided to upgrade its computer network. The network will connect all parts of the company, such as engineering design, purchasing, marketing, manufacturing, field sales, accounting, and so on. Ten different vendors are invited to submit bids based on the specifications prepared by engineers. The bids will include hardware, software, installation, and maintenance. As an engineer, you will then do an analysis of the bids submitted and rank the 10 vendors' proposals based on predetermined criteria.

This assignment is possible once a method of comparison such as the equivalent annual cost or the equivalent present worth method is selected. These methods of analysis are discussed later in the chapter. However, they only allow us to compare tangible costs. Other intangible items, for example, safety, must also be considered.

Another example of a common engineering task is justification of the purchase of a new machine to reduce the costs of manufacture. This justification is usually expressed as either a rate of return on investment or a *payback period.* A simplified definition of *rate of return* is profit divided by investment or

$$\text{Annual rate of return} = \frac{\text{annual profit}}{\text{investment}}$$

Often the annual profit comes from the reduction of production costs as a result of the new machine.

The other approach mentioned above is to use the cost reduction to compute the number of months or years to pay back the investment in the machine (payback period).

Since any venture has some risk involved and the cost reductions expected are only estimated, companies will not choose to invest in new equipment unless there is a promise of a much greater return than could be realized by less risky investments such as bank deposits or the purchase of treasury notes or government bonds.

In addition to the application of engineering economy methods in your professional life, you will also have applications in your personal life. Major purchases such as a vehicle or a home, as well as investments you may make using your own funds, require understanding of the principles you will learn in this chapter. As an example, Fig. 13.1 shows an annual investment of $1 200 ($100 per month) for 10 years beginning at age 22. This is left at 8 percent annual compounded interest until age 65. Compare that with starting the same yearly investment at age 32 and continuing until age 65. You may draw your own conclusions.

13.2 Simple Interest

The idea of interest on an investment is certainly not new. The New Testament refers to banks, interest, and return. History records business dealings involving interest at least 40 centuries ago. Early business was largely barter in nature with repayment in kind. It was common during the early years of the development of the United States for people to borrow grain, salt, sugar, animal skins, etc., from each other to be repaid when the commodity was again available. Since most of these items depended on the harvest, annual repayment was the normal process. Likewise, since the lender expected to be repaid after no more than a year, simple annual interest was the usual transaction. When it became impossible to repay the loan after a year, the interest was calculated by multiplying the principal amount by the product of the interest rate and the number of periods (years):

$$I = Pni \tag{13.1}$$

where

I = interest accrued

P = principal amount

n = number of interest periods

i = interest rate per period (as a decimal, not as a percent)

Age	Annual Savings	Accumulation	Annual Savings	Accumulation
22	$1,200	$1,296		
23	$1,200	$2,696		
24	$1,200	$4,207		
25	$1,200	$5,840		
26	$1,200	$7,603		
27	$1,200	$9,507		
28	$1,200	$11,564		
29	$1,200	$13,785		
30	$1,200	$16,184		
31	$1,200	$18,775		
32		$20,277	$1,200	$1,296
33		$21,899	$1,200	$2,696
34		$23,651	$1,200	$4,207
35		$25,543	$1,200	$5,840
36		$27,586	$1,200	$7,603
37		$29,793	$1,200	$9,507
38		$32,176	$1,200	$11,564
39		$34,750	$1,200	$13,785
40		$37,530	$1,200	$16,184
41		$40,533	$1,200	$18,775
42		$43,776	$1,200	$21,573
43		$47,278	$1,200	$24,594
44		$51,060	$1,200	$27,858
45		$55,145	$1,200	$31,383
46		$59,556	$1,200	$35,189
47		$64,321	$1,200	$39,300
48		$69,466	$1,200	$43,740
49		$75,024	$1,200	$48,536
50		$81,025	$1,200	$53,714
51		$87,508	$1,200	$59,308
52		$94,508	$1,200	$65,348
53		$102,069	$1,200	$71,872
54		$110,234	$1,200	$78,918
55		$119,053	$1,200	$86,527
56		$128,577	$1,200	$94,745
57		$138,863	$1,200	$103,621
58		$149,973	$1,200	$113,207
59		$161,970	$1,200	$123,559
60		$174,928	$1,200	$134,740
61		$188,922	$1,200	$146,815
62		$204,036	$1,200	$159,856
63		$220,359	$1,200	$173,941
64		$237,988	$1,200	$189,152
65		$257,027	$1,200	$205,580

Figure 13.1

Therefore, if \$1 000 were to be loaned at 7 percent annual interest for 5 years, the interest would be

$$\begin{aligned} I &= Pni \\ &= (1\,000)(5)(0.07) \\ &= \$350 \end{aligned}$$

and the amount F to be repaid is

$$\begin{aligned} F &= P + I \qquad (13.2) \\ &= 1000 + 350 \\ &= \$1\,350 \end{aligned}$$

It can be seen that

$$\begin{aligned} F &= P + I = P + Pni \qquad (13.3) \\ &= P(1 + ni) \end{aligned}$$

13.3 Compound Interest

As time progressed and business developed, the practice of borrowing became more common, and the use of money replaced the barter system. It also became increasingly more common that money was loaned for longer periods of time. Simple interest was relegated to the single-interest period, and the practice of compounding developed. It can be shown by using Eq. (13.3), $n = 1$, that the amount owed at the end of one period is

$$P + Pi = P(1 + i)$$

The interest generated during the second period is then $(P + Pi)i$. It can be seen that interest is being calculated not only on the *principal* but on the previous interest as well. The sum F at the end of two periods becomes

P	principal amount
$+ Pi$	interest during first period
$+ Pi + Pi^2$	interest during second period
$P + 2Pi + Pi^2$	sum after two periods

This can be factored as follows:

$$P(1 + 2i + i^2) = P(1 + i)^2$$

The interest during the third period is

$$(P + 2Pi + Pi^2)i = Pi + 2Pi^2 + Pi^3$$

and the sum after three periods is

$P + 2Pi + Pi^2$	sum after second period
$+ Pi + 2Pi^2 + Pi^3$	interest during third period
$P + 3Pi + 3Pi^2 + Pi^3 = P(1 + i)^3$	sum after three periods

This procedure can be generalized to n periods of time and will result in

$$F = P(1 + i)^n \tag{13.4}$$

where F is the sum generated after n periods.

Consider the sum at the end of 5 years on a \$1 000 loan with 7 percent annual interest, compounded annually.

$$\begin{aligned} F &= P(1 + i)^n \\ &= (1\,000)(1.07)^5 \\ &= \$1\,402.55 \end{aligned}$$

Thus, the sum with annual compounding is \$1402.55, which compares with the previous sum of \$1 350 when simple interest was used.

The same compounded interest result can be obtained by the following arithmetic:

Principal amount	=	\$1 000.00
Interest during first year: (1 000)(0.07)	=	70.00
Sum after 1 year	=	1 070.00
Interest during second year: (1 070)(0.07)	=	74.90
Sum after 2 years	=	1 144.90
Interest during third year: (1 144.90)(0.07)	=	80.14
Sum after 3 years	=	1 225.04
Interest during fourth year: (1 225.04)(0.07)	=	85.75
Sum after 4 years		1 310.79
Interest during fifth year: (1 310.79) (0.07)	=	91.76
Sum after 5 years	=	\$1 402.55

Care must be exercised in using interest rates and payment periods to make sure that the interest rate used is the rate for the period selected.

Consider calculating the sum after 1 year. If the annual interest rate is 12 percent compounded annually, then $i = 0.12$ and $n = 1$. However, when the annual rate is 12 percent, but it is to be compounded every 6 months (semiannually), then $i = 0.12/2$ and $n = 2$. This idea can be extended to a monthly compounding period, with $i = 0.12/12$ and $n = 12$, or a daily compounding period, with $i = 0.12/365$ and $n = 365$.

Example problem 13.1 What lump sum must be paid at the end of 4 years if \$8 000 is borrowed from a bank at a 12 percent annual interest rate compounded (*a*) annually, (*b*) semiannually, (*c*) monthly, and (*d*) daily?

Solution

(a) Compounded annually:

$F = P(1 + i)^n$

where

$i = 0.12$

$n = 1 \times 4 = 4$ periods

$P = \$8\,000$

$F = 8\,000(1.12)^4$

$= \$12\,588.15$

Note: Always express the answer to the nearest penny.

(b) Compounded semiannually:

$F = P(1 + i)^n$

where

$i = 0.12/2$

$n = 2 \times 4 = 8$ periods

$P = \$8\,000$

$F = 8\,000(1.06)^8$

$= \$12\,750.78$

(c) Compounded monthly:

$F = P(1 + i)^n$

where

$i = 0.12/12$

$n = 12 \times 4 = 48$ periods

$P = \$8\,000$

$F = 8\,000(1.01)^{48}$

$= \$12\,897.81$

(d) Compounded daily:

$F = P(1 + i)^n$

where

$i = 0.12/365$

$n = 365 \times 4 = 1460$ periods

$P = \$8\,000$

$$F = 8\,000\left(1 + \frac{0.12}{365}\right)^{1460}$$

$$= \$12\,927.57$$

As you can see from the previous example, even though the stated interest is the same, 12 percent in this case, the change in the compounding period changes the sum. Thus, to compare different alternatives, we must know the *stated* or *nominal* annual interest rate and the compounding period. Or we can define an *effective annual rate,* often called *annual percentage rate* (APR), for comparison purposes. The annual percentage rate (APR) is then the interest rate that would have produced the final amount under annual (rather than semiannual, monthly, and so on) compounding.

Then, continuing with Example prob. 1, part (b), with a nominal interest rate of 12 percent and semiannual compounding, the APR can be found as follows:

$$F = \$12\,750.78 = 8\,000(1 + \text{APR})^4 = 8\,000\left(1 + \frac{0.12}{2}\right)^8$$

or

$$(1 + \text{APR})^4 = \left(1 + \frac{0.12}{2}\right)^8$$

then

$$\text{APR} = \left(1 + \frac{0.12}{2}\right)^2 - 1 = 0.1236 \qquad (12.36\%\ \text{APR})$$

Considering part (c) with 12 percent nominal and monthly compounding, the APR is found from

$$\$12\,897.81 = 8\,000(1 + \text{APR})^4 - 8\,000\left(1 + \frac{0.12}{12}\right)^{48}$$

$$\text{APR} = \left(1 + \frac{0.12}{12}\right)^{12} - 1 = 0.1268 \qquad (12.68\%\ \text{APR})$$

Financial institutions often give both nominal and APR values in their advertising. APR is always going to be larger than nominal interest if the compounding period is less than 1 year if the APR value is computed as previously defined. Beware, however: some financial institutions may state the nominal rate and simply call it APR if this makes the rate appear better in their advertising. Since you know how to compute APR, you can always check it out.

13.4 Cash-flow Diagram

The transaction described in Example prob. 13.1 for annual compounding is graphically illustrated in a *cash-flow diagram* (Fig. 13.2). Since a cash-flow diagram is very useful in the visualiza-

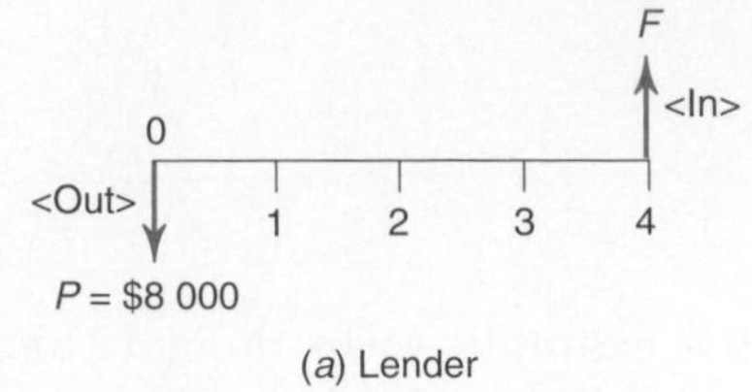

(a) Lender

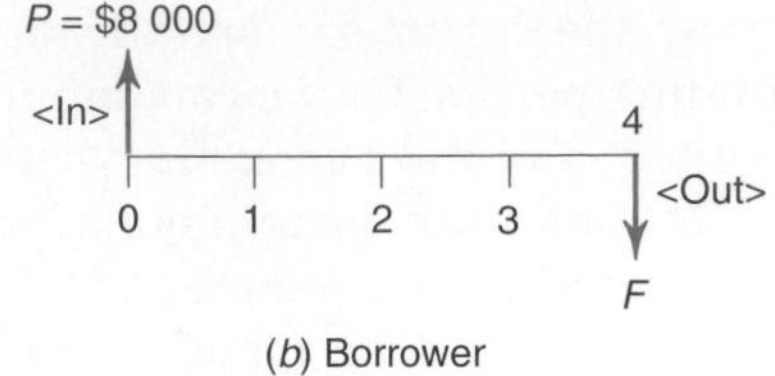

(b) Borrower

Figure 13.2 Cash-flow diagram: (a) As seen by the lender; (b) as seen by the borrower.

tion of any transaction, it will be used throughout this chapter. The following general rules apply:

1. The horizontal line is a time scale. Normally, years are given as the interval of time although sometimes the period number is more meaningful.
2. The arrows signify cash flow. A downward arrow means money out, and an upward arrow means money in.
3. The diagram is dependent on the point of view from which it is constructed, that is, on whether it is the lender's or the borrower's point of view (see Fig. 13.2).

13.5 Present Worth

It is important to keep in mind that the value of any transaction (loan, investment, and so on) changes with time because of interest. Thus, to express the value of a transaction, you must also give the point in time at which that value is computed. For example, the value of the loan described in Example prob. 13.1 (assume annual compounding) is $8 000 at year zero but is $12 588.15 4 years later. We will examine several methods of stating the value of a transaction; the first is present worth.

Present worth is a term that is used to describe the worth of a monetary transaction at the current time. It is the amount of money that must be invested now in order to produce a prescribed sum at another date.

In other words, if you were guaranteed an amount of money F 4 years from today, then P would be the present worth of F, where the interest is i and $n = 4$ (assuming annual compounding). Since this analysis is exactly the inverse of finding a future sum, we have

$$P = F(1 + i)^{-n} \tag{13.5}$$

As an example, if you can convince a lending institution that you will have a guaranteed amount of money available 4 years

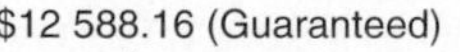

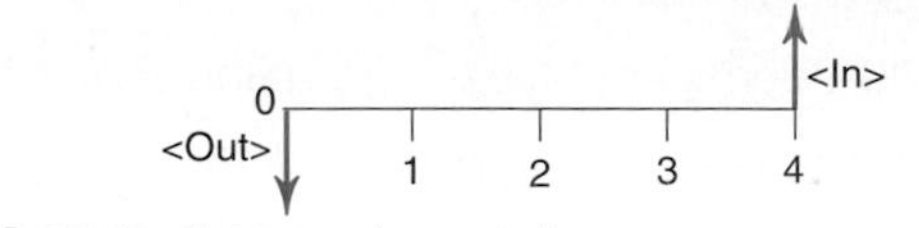

Figure 13.3
Banker's cash-flow diagram.

from today, it may be possible to borrow the present worth of that amount. If the guaranteed sum (4 years later) is equal to $12 588.15, the present worth at 12 percent annual interest (compounded annually) is $8 000. (See Fig. 13.3)

$$P = F(1 + i)^{-n}$$

where

$$F = \$12\,588.15$$

$$i = 0.12$$

$$n = 4$$

$$P = \frac{12\,588.15}{(1.12)^4}$$

$$= \$8\,000.00$$

In situations that involve economic decisions the following questions may arise:

1. Does it pay to make an investment now?

2. What is the current benefit of a payment that will be made at some other date?

3. If a series of payments is made over specific intervals throughout a designated time span, what is this worth now?

In such cases the answer can be examined by finding the present worth of the transaction.

Many businesses calculate their present worth each year since the change in their present worth is a measure of the growth of the company. The following example problem will help to demonstrate the concept of present worth.

Example problem 13.2 Listed below are five transactions. Determine their present worth if money is currently valued at 10 percent annual interest compounded annually. Determine the current net cash equivalent assuming no interest has been withdrawn or paid. Draw a cash-flow diagram for each.

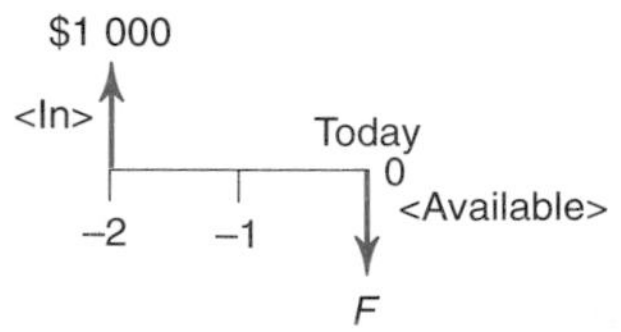

Solution

(a) $1 000 deposited 2 years ago. (See Fig. 13.4)

$1 000
<In>
Today
0
<Available>
–2
–1
F

Figure 13.4
Owner's cash-flow diagram.

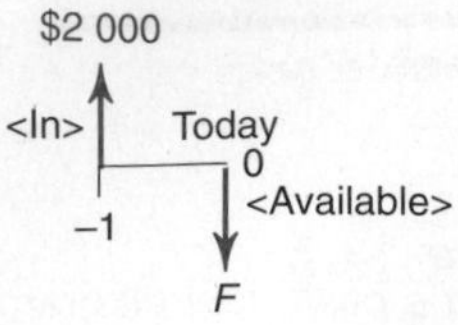

$2 000
<In> Today
0
<Available>
−1
F

Figure 13.5
Owner's cash-flow diagram.

$F = 1\,000(1.10)^2 = \$1\,210.00$

(b) $2 000 deposited 1 year ago. (See Fig. 13.5)

$F = 2\,000(1.10)^1 = \$2\,200.00$

(c) $3 000 to be received 1 year from now. (See Fig. 13.6)

$P = 3\,000(1.10)^{-1} = \$2\,727.27$

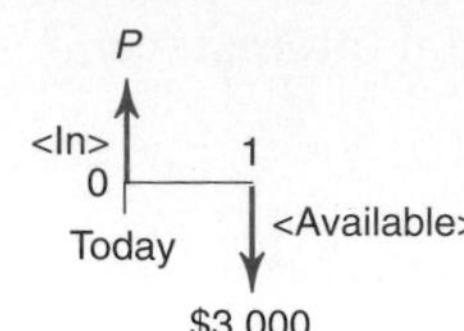

Figure 13.6
Owner's cash-flow diagram.

(d) $4 000 to be paid 2 years from now (treated as negative for the owner since it must be paid. (See Fig. 13.7)

$$P = -4\,000(1.10)^{-2}$$
$$= -\$3\,305.79$$

Figure 13.7
Owner's cash-flow diagram.

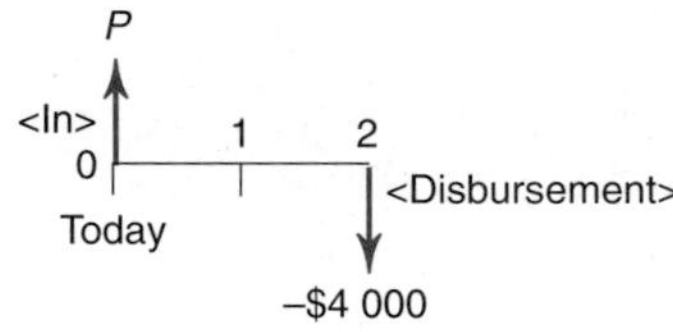

(e) $5 000 to be received 4 years from now. (See Fig. 13.8)

$P = 5\,000(1.10)^{-4} = \$3\,415.07$

Present worth of the five transactions: (*Note:* The values can be added or subtracted because each value has been computed on the same date.)

Figure 13.8
Owner's cash-flow diagram.

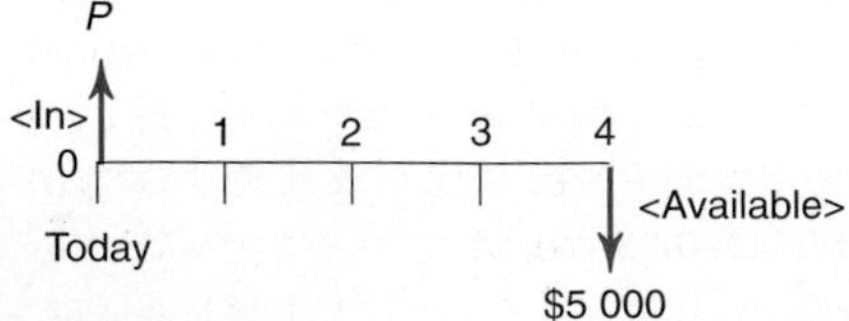

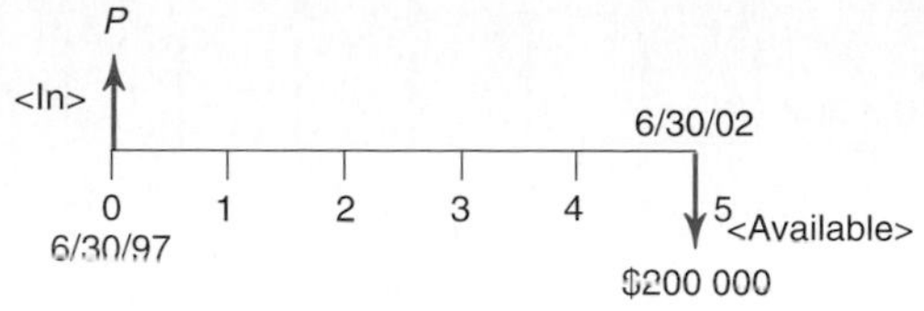

P
<In>
6/30/02
0 1 2 3 4 5 <Available>
6/30/97
$200 000

Figure 13.9
Owner's cash-flow diagram.

$$
\begin{array}{r}
\$1\,210.00 \\
2\,200.00 \\
2\,727.27 \\
-3\,305.79 \\
3\,415.07 \\
\hline
\text{Present worth} = \$6\,246.55
\end{array}
$$

Example problem 13.3 On June 30, 2002, there will be $200 000 available from the sale of a parcel of land. What value does this represent on June 30, 1997, if the annual interest is 12 percent compounded annually?

Solution (See Fig. 13.9)

$$P = F(1 + i)^{-n}$$
$$= 200\,000(1.12)^{-5}$$
$$= \$113\,485.37$$

13.6 Annuities

An *annuity* involves a series of equal payments at regular intervals. The value of such a series will be developed in the following sections from the idea of compound interest. Consideration of the point in time at which compounding begins will be of prime importance.

There are several forms of annuities, each with an assigned name, that will be discussed in this chapter.

13.6.1 Sinking Fund

A *sinking fund* is an annuity that is designed to produce an amount of money at some future time. It might be used to save for an expenditure that you know is going to occur—for instance, a Christmas gift fund or a new car fund. In business the fund may be used to provide cash needed to replace obsolete equipment or to upgrade software. The cash-flow diagram for the sinking fund is shown in Fig. 13.10.

If an amount A is deposited at the end of each period and interest is compounded each period at a rate of i, the sum F will be produced after n periods. Please note that the deposit period

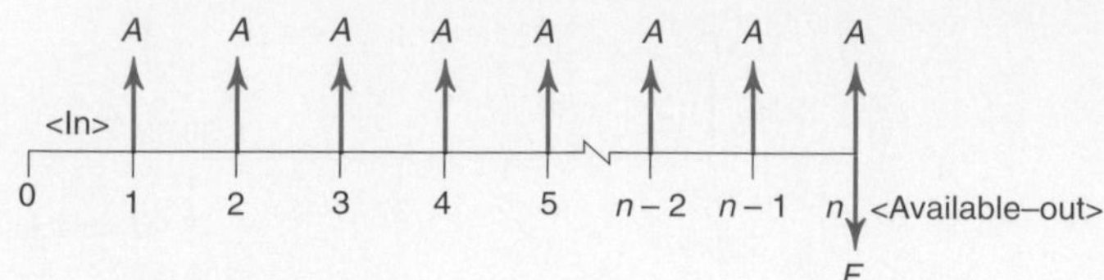

Figure 13.10 Saver's cash-flow diagram.

and the interest compounding period must be equal for the equations being developed to be valid.

It can be seen from Fig. 13.10 that the last payment will produce no interest, the payment at period $n - 1$ will produce interest equal to A times i, the payment at period $n - 2$ will produce a sum (interest and principal) of $A(1 + i)^2$, and so on. Hence, the sum produced will be as follows:

Deposit at end of period	Interest generated	Sum due to this payment
n	None	$A(1)$
$n-1$	$A(i)$	$A(1+i)$
$n-2$	$A(1+i)i$	$A(1+i)^2$
$n-3$	$A(1+i)^2 i$	$A(1+i)^3$

Thus, for four payments

$$F = A(4 + 6i + 4i^2 + i^3)$$

If we multiply and divide this expression by i, and then add and subtract 1 from the numerator, F becomes

$$F = A\left[\frac{(4i + 6i^2 + 4i^3 + i^4 + 1) - 1}{i}\right]$$

$$= A\left[\frac{(1 + i)^4 - 1}{i}\right]$$

It can be shown that the general term is

$$F = A\left[\frac{(1 + i)^n - 1}{i}\right] \qquad (13.6)$$

Therefore, if you want to accumulate F during n periods, an amount A must be deposited at the end of each period at i interest rate compounded at each period.

Example problem 13.4 How much money would be accumulated by a sinking fund if $90 is deposited at the end of each month for 3 years with an annual interest rate of 10 percent compounded monthly?

Solution (See Fig. 13.11)

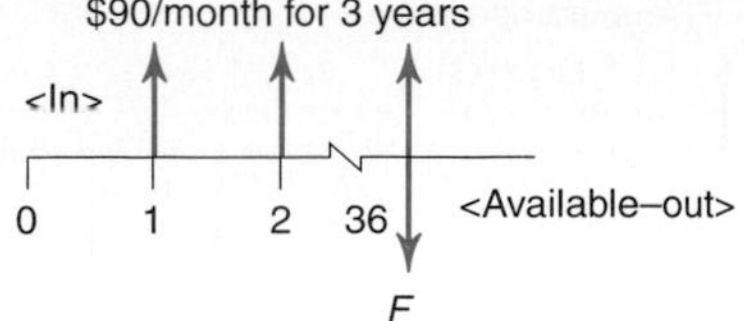

Figure 13.11
Saver's cash-flow diagram.

$$F = A\left[\frac{(1+i)^n - 1}{i}\right]$$

where

$A = \$90$

$i = 0.10/12$ (monthly compounding)

$n = 12 \times 3$

$$F = 90\left[\frac{(1+0.10/12)^{36} - 1}{0.10/12}\right]$$

$$= \$3\,760.36$$

Example problem 13.5 It is desired to accumulate \$10 000 to replace a piece of equipment after 8 years. How much money must be placed annually into a sinking fund that earns 7 percent interest? Assume the first payment is made today and the last one 8 years from today with interest compounded annually.

Solution (See Fig. 13.12)

$$F = A\left[\frac{(1+i)^n - 1}{i}\right]$$

$$A = \frac{F(i)}{}$$

$$= \frac{(10\,000)(0.07)}{(1.07)^9 - 1}$$

$$= \$834.86$$

Note that $n = 9$ in this example because the first payment must occur at the end of the first period so that the sinking-fund formula will be valid.

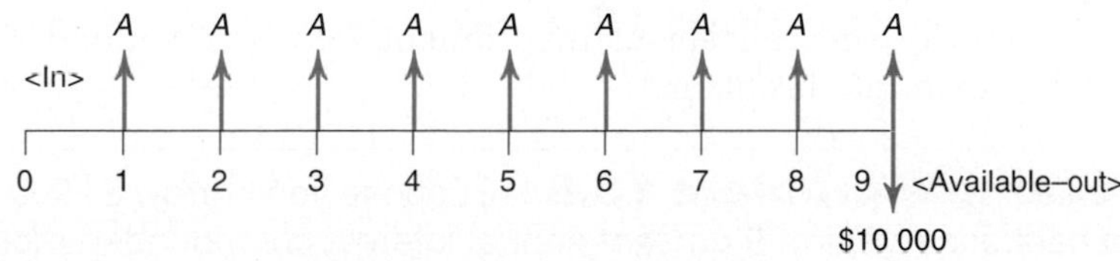

Figure 13.12
Company's cash-flow diagram.

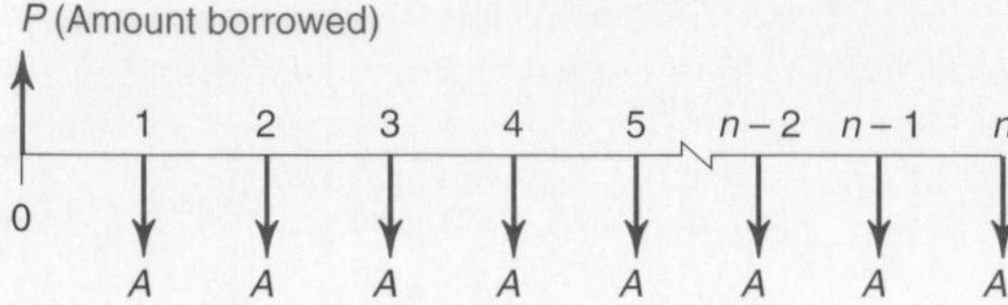

Figure 13.13
Buyer's cash-flow diagram.

13.6.2 Installment Loan

A second and very popular way that annuities are used to retire a debt is by making periodic payments instead of a single large payment at the end of a given time period. This time-payment plan used by most retail businesses and lending institutions is called an *installment loan.* It is also used to amortize (pay off with a sinking-fund approach) bond issues. A cash-flow diagram for this scheme is illustrated in Fig. 13.13.

In this case the principal amount P is the size of the debt and A is the amount of the periodic payment that must be made with interest compounded at the end of each period. It can be seen that if P were removed from the time line and F placed at the end of the nth period, it would be a sinking fund. Furthermore, it can be shown that F would also be the value of P placed at compound interest for n periods $[F = P(1 + i)^n]$. Likewise, P can be termed the present worth of the sinking fund that would be accumulated by the deposits. Therefore, since

$$F = P(1 + i)^n \qquad \text{and} \qquad F = A\left[\frac{(1 + i)^n - 1}{i}\right]$$

the present worth becomes

$$P = A\left[\frac{(1 + i)^n - 1}{i(1 + i)^n}\right] = A\left[\frac{1 - (1 + i)^{-n}}{i}\right] \tag{13.7}$$

The term within the brackets is known as the *present worth* of a sinking fund, or the *uniform annual payment present-worth factor.*

It follows that

$$A = P\left[\frac{i(1 + i)^n}{(1 + i)^n - 1}\right] \tag{13.8}$$

where the term in brackets is most commonly called the *capital recovery factor,* or the *uniform annual payment annuity factor,* and is the reciprocal of the uniform annual payment present-worth factor.

To clarify the concept of installment loans, consider the following example problem.

Example problem 13.6 Suppose you borrow $1000 from a bank for 1 year at 9 percent annual interest compounded monthly. Consider paying the loan back by two different methods:

1. You keep the \$1 000 for 1 year and pay back the bank at the end of the year in a lump sum. What would you owe? (See Fig. 13.14; note that the time line shown is in years but interest is compounded monthly.)

$$F = 1\,000(1 + 0.0075)^{12}$$

$$= \$1\,093.81$$

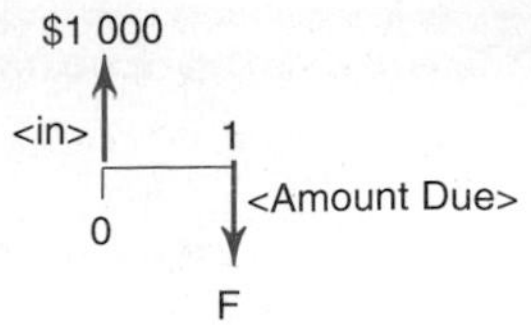

Figure 13.14 Borrower's cash-flow diagram.

2. The second method is the installment loan. You borrow \$1 000 from the bank and repay it in equal monthly payments. What is the amount of each payment? (See Fig. 13.15)

From Eq. (13.8)

$$A = P\left[\frac{i(1+i)^n}{}\right]$$

$$= 1\,000\left[\frac{0.0075(1+0.0075)^{12}}{(1+0.0075)^{12}-1}\right]$$

$$= \$87.45$$

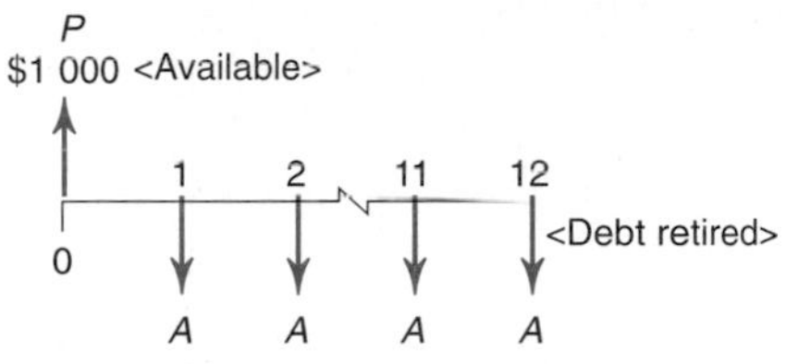

Figure 13.15 Borrower's cash-flow diagram.

Example problem 13.7 An automobile that has a total cost of \$15 500 is to be purchased by trading in an older vehicle for which \$6 250 is allowed. If the interest rate is 8.5 percent per year, compounded monthly, and the payments will be made monthly for 3 years, what are the monthly payments? The first payment is to be made at the end of the first month. (See Fig. 13.16)

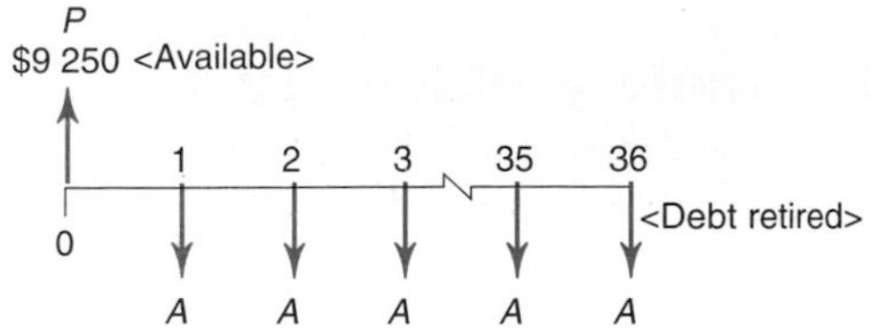

Figure 13.16 Purchaser's cash-flow diagram.

Solution From Eq. (13.8)

$$A = (15\,500 - 6\,250)\left[\frac{\left(\frac{0.085}{12}\right)\left(1+\frac{0.085}{12}\right)^{36}}{\left(1+\frac{0.085}{12}\right)^{36}-1}\right]$$

$$= \$292.00$$

Another way of expressing the relationship is by saying that \$9 250 (that is, \$15 500 − \$6 250) is the percent worth of 36 monthly payments of \$292.00, beginning in 1 month at 8.5 percent annual interest compounded monthly.

Example problem 13.8 Suppose that the auto purchase described in Example Prob. 13.7 is modified so that no payment is made until 6 months after the purchase and then a total of 36 monthly payments are made. Now, what is the amount of each monthly payment? (See Fig. 13.17)

Solution Balance due after trade-in = \$15 500 − \$6 250 = \$9 250. Balance due after 5 months = $\$9250(1 + 0.085/12)^5$ = \$9 582.28. Note that the unpaid balance is compounded for only 5 months because the first payment marks the end of the first period of the annuity. This is referred to as a *deferred annuity.*

The monthly payment is then

$$A = 9\,582.28\left[\frac{\left(\frac{0.085}{12}\right)\left(1+\frac{0.085}{12}\right)^{36}}{\left(1+\frac{0.085}{12}\right)^{36}-1}\right]$$

$$= \$302.49$$

Figure 13.17 Purchaser's cash-flow diagram.

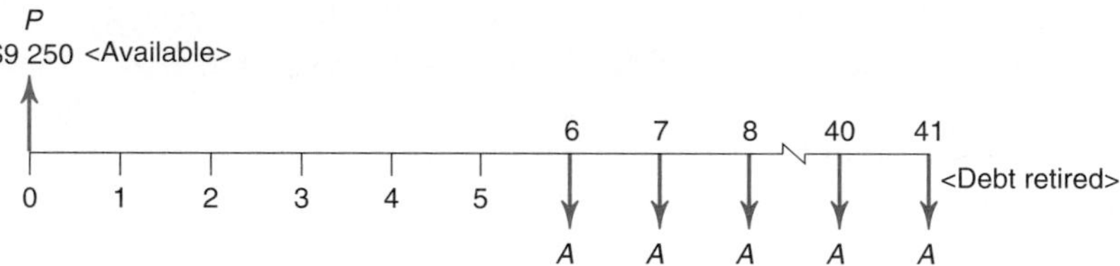

Example problem 13.9 Amy and Kevin are purchasing their first home and have arranged for a mortgage of \$65 000 at an annual interest rate of 8.75 percent compounded monthly for a period of 15 years.

(a) What will be their monthly payment?

(b) At the end of year 10, what amount will be necessary to pay off the loan?

(c) If at the end of the first year they pay an additional sum of \$25 000 on the principal, what is the remaining principal?

(d) If they continue with the monthly payments as found in part (a), how many payments will be necessary to retire the debt following the \$25 000 payment described in part (c)?

Solution

(a) This problem describes a standard installment loan so that Eq. (8) applies. (See Fig. 13.18)

$$A = P\left[\frac{i(1+i)^n}{(1+i)^n - 1}\right]$$

$$= 65\,000\left[\frac{\left(\frac{0.0875}{12}\right)\left(1+\frac{0.0875}{12}\right)^{180}}{\left(1+\frac{0.0875}{12}\right)^{180} - 1}\right]$$

$$= \$649.64$$

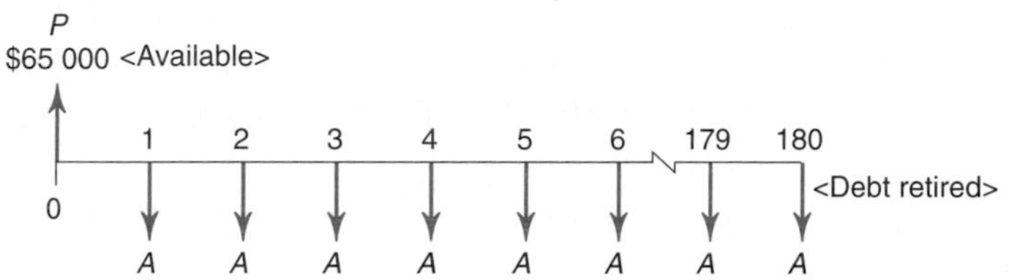

Figure 13.18
Home purchaser's cash-flow diagram.

(b) The amount necessary to pay off the loan at year 10 is the principal remaining. This can be viewed as the present worth of the monthly payments following year 10. (See Fig. 13.19) From Eq. (13.7)

$$P = A\left[\frac{(1+i)^n - 1}{i(1+i)^n}\right]$$

$$= 649.64\left[\frac{\left(1+\frac{0.0875}{12}\right)^{60} - 1}{\left(\frac{0.0875}{12}\right)\left(1+\frac{0.0875}{12}\right)^{60}}\right]$$

$$= \$31\,479.03$$

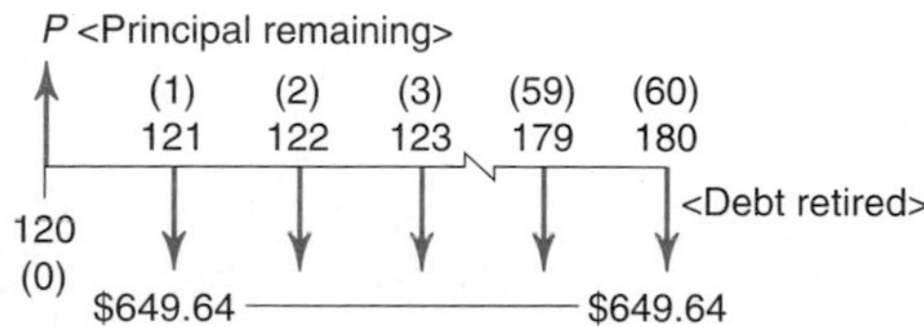

Figure 13.19
Purchaser's cash-flow diagram for last 60 months.

(c) The principal remaining at the end of the first year can be found following the procedure in part (b). (See Fig. 13.20)

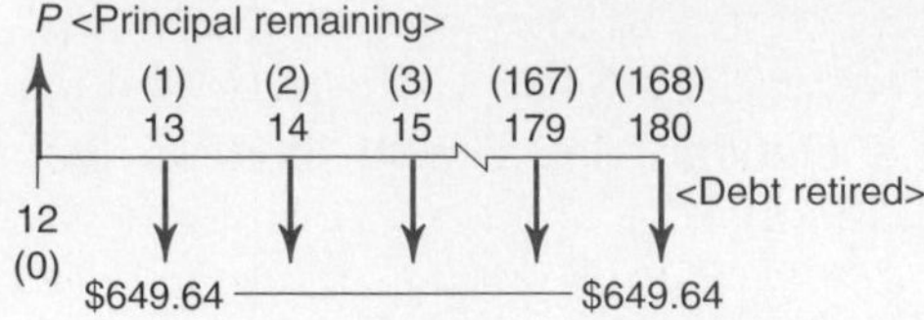

Figure 13.20 Purchaser's cash-flow diagram for last 14 years.

$$P = 649.64\left[\frac{\left(1+\frac{0.0875}{12}\right)^{168}-1}{\left(\frac{0.0875}{12}\right)\left(1+\frac{0.0875}{12}\right)^{168}}\right]$$

$$= \$62\,805.00$$

After the lump sum payment of \$25 000 the principal remaining is \$62 805.00 − \$25 000 = \$37 805.00.

(d) Equation (13.7) applies here with the present worth of \$37805.00, the monthly payment equal to \$649.64, and the number of payments unknown. (See Fig. 13.21)

$$37\,805.00 = 649.64\left[\frac{\left(1+\frac{0.0875}{12}\right)^{n}-1}{\left(\frac{0.0875}{12}\right)\left(1+\frac{0.0875}{12}\right)^{n}}\right]$$

thus

$$(0.42433)\left(1+\frac{0.0875}{12}\right)^{n} = \left(1+\frac{0.0875}{12}\right)^{n}-1$$

or

$$\left(1+\frac{0.0875}{12}\right)^{n} = 1.7371$$

Taking the logarithm of both sides, we have

$$n\log\left(1+\frac{0.0875}{12}\right) = \log 1.7371$$

then

$$n = \frac{\log 1.7371}{\log\left(1+\frac{0.0875}{12}\right)}$$

$$= 76.008$$

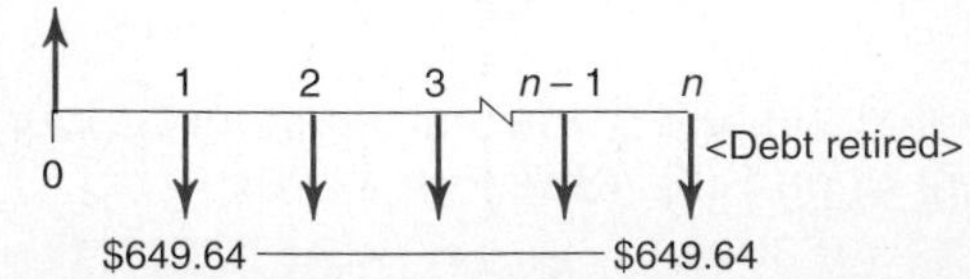

Figure 13.21 Purchaser's cash-flow diagram.

As is often the case, the computed number of periods is not an integer value, meaning that there will be 76 full payments of $649.64 with a small partial payment required to retire the mortgage.

Example problem 13.10 Suppose following graduation you decide to purchase a new vehicle and borrow $15 000 at 8.5 percent annual interest, compounded monthly, to finance the deal. The agreement requires monthly payments for a period of 2 years. Use a spreadsheet to document this and prepare it so that the amount borrowed and the annual interest rate can be changed with simple cell modifications. Include the amount of each payment, values for the amount of each payment going toward principal and toward interest, the principal remaining after each payment, and a running total of the interest paid.

Solution *Note:* The spreadsheet shown in this example is Microsoft Excel. Many other spreadsheets can be used with small coding changes. The general approach is similar, however.

The cash-flow diagram for you as the purchaser is shown in Fig. 13.22. As the problem is described, it fits the installment loan definition.

Figure 13.23 shows the cell contents (values and formulas) used to compute the results. Please note the following points:

1. Cell *C*3 contains the principal amount and can be changed.

2. Cell *C*4 contains the annual interest rate in percent and can be changed. (Coded as 8.5%, formula view shows as 0.085.)

3. Cell *C*5 is the loan term in years and can be changed but will not be in the example because this affects the total number of payments, and therefore the size of the spreadsheet.

4. A built-in function PMT is used in cell *C*6 to compute the payment based on the monthly interest rate, the number of payments, and the principal amount. Equation (13.7) could have been coded here rather than using the spreadsheet function.

5. Cells *B*12 through *B*35 all contain the value computed in *C*6.

6. The split of the payment into interest and principal is handled by first computing the interest for the first period (cell *D*12) and then subtracting it from cell *B*12 to obtain the portion of the payment applied to the principal. Note that cells *D*12 through *D*35 depend on

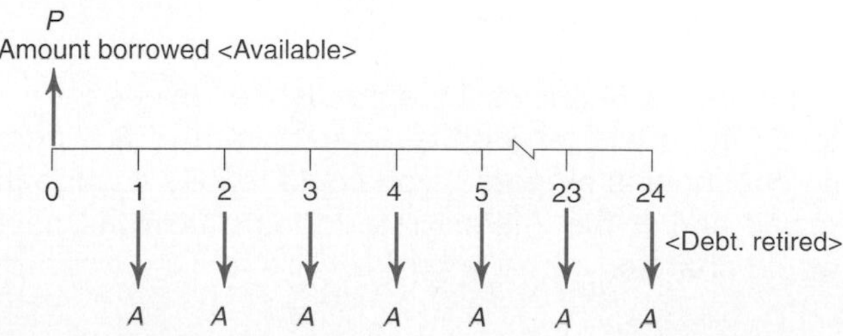

Figure 13.22
Purchaser's cash-flow diagram.

	A	B	C	D	E	F
1			AUTO FINANCE SCHEDULE			
2						
3	Amount borrowed		15000			
4	Annual interest rate(percent)		0.085			
5	Loan term(years)		2			
6	Monthly payment		=PMT(C4/(12),C5*12,-C3)			
7						
8			PAYMENT SCHEDULE			
9						
10	Number	Amount paid	Principal	Interest	Principal remaining	Total interest
11	0				=C3	
12	1	=$C$6	=B12-D12	=$C$4/(12)*E11	=E11-C12	=F11+D12
13	2	=$C$6	=B13-D13	=$C$4/(12)*E12	=E12-C13	=F12+D13
14	3	=$C$6	=B14-D14	=$C$4/(12)*E13	=E13-C14	=F13+D14
15	4	=$C$6	=B15-D15	=$C$4/(12)*E14	=E14-C15	=F14+D15
16	5	=$C$6	=B16-D16	=$C$4/(12)*E15	=E15-C16	=F15+D16
17	6	=$C$6	=B17-D17	=$C$4/(12)*E16	=E16-C17	=F16+D17
18	7	=$C$6	=B18-D18	=$C$4/(12)*E17	=E17-C18	=F17+D18
19	8	=$C$6	=B19-D19	=$C$4/(12)*E18	=E18-C19	=F18+D19
20	9	=$C$6	=B20-D20	=$C$4/(12)*E19	=E19-C20	=F19+D20
21	10	=$C$6	=B21-D21	=$C$4/(12)*E20	=E20-C21	=F20+D21
22	11	=$C$6	=B22-D22	=$C$4/(12)*E21	=E21-C22	=F21+D22
23	12	=$C$6	=B23-D23	=$C$4/(12)*E22	=E22-C23	=F22+D23
24	13	=$C$6	=B24-D24	=$C$4/(12)*E23	=E23-C24	=F23+D24
25	14	=$C$6	=B25-D25	=$C$4/(12)*E24	=E24-C25	=F24+D25
26	15	=$C$6	=B26-D26	=$C$4/(12)*E25	=E25-C26	=F25+D26
27	16	=$C$6	=B27-D27	=$C$4/(12)*E26	=E26-C27	=F26+D27
28	17	=$C$6	=B28-D28	=$C$4/(12)*E27	=E27-C28	=F27+D28
29	18	=$C$6	=B29-D29	=$C$4/(12)*E28	=E28-C29	=F28+D29
30	19	=$C$6	=B30-D30	=$C$4/(12)*E29	=E29-C30	=F29+D30
31	20	=$C$6	=B31-D31	=$C$4/(12)*E30	=E30-C31	=F30+D31
32	21	=$C$6	=B32-D32	=$C$4/(12)*E31	=E31-C32	=F31+D32
33	22	=$C$6	=B33-D33	=$C$4/(12)*E32	=E32-C33	=F32+D33
34	23	=$C$6	=B34-D34	=$C$4/(12)*E33	=E33-C34	=F33+D34
35	24	=$C$6	=B35-D35	=$C$4/(12)*E34	=E34-C35	=F34+D35

Figure 13.23
Spreadsheet cell contents (values and formulas)

the principal at the beginning of the period. The principal at the beginning of the first period is the original amount borrowed.

7. The principal remaining is found by subtracting the portion of the payment going to principal (column *C*) from the previous principal amount.

8. Finally, the total interest paid is simply a running sum.

9. You should note that this spreadsheet calculates values to greater precision than is displayed and thus there will be places where sums or differences will be off by a penny in the display. This can be prevented by using an option "precision as displayed" which actually changes the data at that point in the spreadsheet to the value displayed. The general result will be that the last payment will then be in error by a few cents.

Figure 13.24 shows the spreadsheet values for a loan of $15000 for 2 years with an annual interest rate of 8.5 percent. Once the spreadsheet is prepared, you could readily change the amount borrowed and/or the interest rate to learn how the payment schedule would change.

	A	B	C	D	E	F
1			AUTO FINANCE SCHEDULE			
2						
3	Amount borrowed		$15,000			
4	Annual interest rate(percent)		8.50%			
5	Loan term(years)		2			
6	Monthly payment		$681.84			
7						
8			PAYMENT SCHEDULE			
9						
10	Number	Amount paid	Principal	Interest	Principal remaining	Total interest
11	0				$15,000.00	
12	1	$681.84	$575.59	$106.25	$14,424.41	$106.25
13	2	$681.84	$579.66	$102.17	$13,844.75	$208.42
14	3	$681.84	$583.77	$98.07	$13,260.98	$306.49
15	4	$681.84	$587.90	$93.93	$12,673.08	$400.42
16	5	$681.84	$592.07	$89.77	$12,081.01	$490.19
17	6	$681.84	$596.26	$85.57	$11,484.75	$575.76
18	7	$681.84	$600.48	$81.35	$10,884.27	$657.11
19	8	$681.84	$604.74	$77.10	$10,279.53	$734.21
20	9	$681.84	$609.02	$72.81	$9,670.51	$807.02
21	10	$681.84	$613.34	$68.50	$9,057.17	$875.52
22	11	$681.84	$617.68	$64.15	$8,439.49	$939.68
23	12	$681.84	$622.06	$59.78	$7,817.44	$999.46
24	13	$681.84	$626.46	$55.37	$7,190.98	$1,054.83
25	14	$681.84	$630.90	$50.94	$6,560.08	$1,105.77
26	15	$681.84	$635.37	$46.47	$5,924.71	$1,152.23
27	16	$681.84	$639.87	$41.97	$5,284.84	$1,194.20
28	17	$681.84	$644.40	$37.43	$4,640.44	$1,231.64
29	18	$681.84	$648.97	$32.87	$3,991.47	$1,264.51
30	19	$681.84	$653.56	$28.27	$3,337.91	$1,292.78
31	20	$681.84	$658.19	$23.64	$2,679.72	$1,316.42
32	21	$681.84	$662.85	$18.98	$2,016.87	$1,335.40
33	22	$681.84	$667.55	$14.29	$1,349.32	$1,349.69
34	23	$681.84	$672.28	$9.56	$677.04	$1,359.25
35	24	$681.84	$677.04	$4.80	$0.00	$1,364.04

Figure 13.24
Spreadsheet values for Example Prob. 13.10.

Example problem 13.11 Modify the spreadsheet developed in Example prob. 13.10 so that the following series of payments (always greater than or equal to the minimum payment for a 2-year loan except for the final one) can be made:

1. $ 681.84 **3.** $ 700.00
2. $ 681.84 **4.** $ 681.84

5. $1 000.00
6. $ 681.84
7. $1 000.00
8. $1 000.00
9. $1 500.00
10. $ 900.00
11. $ 700.00
12. $ 800.00
13. $1 000.00
14. $ 681.84
15. $1 000.00
16. $1 500.00
17. $1 000.00
18. $ 569.25

Assume that any amount paid that is greater than $681.84 will be applied toward the principal and that no prepayment penalty is added.

Solution Figure 13.25 shows the cell contents for this case. Column *B* is no longer constant and the specific value must be placed in the cells since no pattern of payments is evident. Column *E* is modified to test whether there is principal remaining. If not, a value of zero is entered in the cell. Finally, Fig. 13.26 provides the numerical solution to the problem.

Figure 13.25
Spreadsheet cell contents for Example prob 13.11.

	A	B	C	D	E	F
1			AUTO FINANCE SCHEDULE			
2						
3	Amount borrowed		15000			
4	Annual interest rate(percent)		0.085			
5	Loan term(years)		2			
6	Monthly payment(minimum)		=PMT(C4/(12), C5*12,-C3)			
7						
8			PAYMENT SCHEDULE			
9						
10	Number	Amount paid	Principal	Interest	Principal remaining	Total interest
11	0				=C3	
12	1	681.84	=B12-D12	=$C$4/(12)*E11	=IF(B12>0, E11-C12, 0)	=F11+D12
13	2	681.84	=B13-D13	=$C$4/(12)*E12	=IF(B13>0, E14-C13, 0)	=F12+D13
14	3	700	=B14-D14	=$C$4/(12)*E13	=IF(B14>0, E15-C14, 0)	=F13+D14
15	4	681.84	=B15-D15	=$C$4/(12)*E14	=IF(B15>0, E16-C15, 0)	=F14+D15
16	5	1000	=B16-D16	=$C$4/(12)*E15	=IF(B16>0, E17-C16, 0)	=F15+D16
17	6	681.84	=B17-D17	=$C$4/(12)*E16	=IF(B17>0, E18-C17, 0)	=F16+D17
18	7	1000	=B18-D18	=$C$4/(12)*E17	=IF(B18>0, E19-C18, 0)	=F17+D18
19	8	1000	=B19-D19	=$C$4/(12)*E18	=IF(B19>0, E18-C19, 0)	=F18+D19
20	9	1500	=B20-D20	=$C$4/(12)*E19	=IF(B20>0, E19-C20, 0)	=F19+D20
21	10	900	=B21-D21	=$C$4/(12)*E20	=IF(B21>0, E22-C21, 0)	=F20+D21
22	11	700	=B22-D22	=$C$4/(12)*E21	=IF(B22>0, E23-C22, 0)	=F21+D22
23	12	800	=B23-D23	=$C$4/(12)*E22	=IF(B23>0, E24-C23, 0)	=F22+D23
24	13	1000	=B24-D24	=$C$4/(12)*E23	=IF(B24>0, E25-C24, 0)	=F23+D24
25	14	681.84	=B25-D25	=$C$4/(12)*E24	=IF(B25>0, E26-C25, 0)	=F24+D25
26	15	1000	=B26-D26	=$C$4/(12)*E25	=IF(B26>0, E27-C26, 0)	=F25+D26
27	16	1500	=B27-D27	=$C$4/(12)*E26	=IF(B27>0, E28-C27, 0)	=F26+D27
28	17	1000	=B28-D28	=$C$4/(12)*E27	=IF(B28>0, E29-C28, 0)	=F27+D28
29	18	569.25	=B29-D29	=$C$4/(12)*E28	=IF(B29>0, E28-C29, 0)	=F28+D29
30	19		=B30-D30	=$C$4/(12)*E29	=IF(B30>0, E29-C30, 0)	=F29+D30
31	20		=B31-D31	=$C$4/(12)*E30	=IF(B31>0, E30-C31, 0)	=F30+D31
32	21		=B32-D32	=$C$4/(12)*E31	=IF(B32>0, E31-C32, 0)	=F31+D32
33	22		=B33-D33	=$C$4/(12)*E32	=IF(B33>0, E32-C33, 0)	=F32+D33
34	23		=B34-D34	=$C$4/(12)*E33	=IF(B34>0, E33-C34, 0)	=F33+D34
35	24		=B35-D35	=$C$4/(12)*E34	=IF(B35>0, E34-C35, 0)	=F34+D35

	A	B	C	D	E	F
1			AUTO FINANCE SCHEDULE			
2						
3	Amount borrowed		$15,000			
4	Annual interest rate(percent)		8.50%			
5	Loan term(years)		2			
6	Monthly payment		$681.84			
7						
8			PAYMENT SCHEDULE			
9						
10	Number	Amount paid	Principal	Interest	Principal remaining	Total interest
11	0				$15,000.00	
12	1	$681.84	$575.59	$106.25	$14,424.41	$106.25
13	2	$681.84	$579.66	$102.17	$13,844.75	$208.42
14	3	$700.00	$601.93	$98.07	$13,242.81	$306.49
15	4	$681.84	$588.04	$93.80	$12,654.77	$400.29
16	5	$1000.00	$910.36	$89.64	$11,744.41	$489.93
17	6	$681.84	$598.65	$83.19	$11, 145.76	$573.12
18	7	$1000.00	$921.05	$78.95	$10,224.71	$652.07
19	8	$1000.00	$927.57	$72.43	$9,297.13	$724.49
20	9	$1500.00	$1,434.15	$65.85	$7,862.99	$790.35
21	10	$900.00	$844.30	$55.70	$7,018.69	$846.05
22	11	$700.00	$650.28	$49.72	$6, 368.40	$895.76
23	12	$800.00	$754.89	$45.11	$5,613.51	$940.87
24	13	$1000.00	$960.24	$39.76	$4,653.27	$980.63
25	14	$681.84	$648.88	$32.96	$4,004.39	$1,013.59
26	15	$1000.00	$971.64	$28.36	$3,032.76	$1,041.96
27	16	$1500.00	$1,478.52	$21.48	$1,554.24	$1,063.44
28	17	$1000.00	$988.99	$11.01	$565.25	$1,074.45
29	18	$569.25	$565.25	$4.00	$0.00	$1,078.45
30	19		$0.00	$0.00	$0.00	$1,078.45
31	20		$0.00	$0.00	$0.00	$1,078.45
32	21		$0.00	$0.00	$0.00	$1,078.45
33	22		$0.00	$0.00	$0.00	$1,078.45
34	23		$0.00	$0.00	$0.00	$1,078.45
35	24		$0.00	$0.00	$0.00	$1,078.45

Figure 13.26 Spreadsheet values for Example prob. 13.11.

13.6.3 Retirement Plan

A third way to consider annuities is the classic way, that is, the time when a sum of money is returned in monthly installments at retirement. The formula that applies is Eq. (13.7), and the cash-flow diagram is shown in Fig. 13.27.

The problem could be restated as follows: How much money P must be available at retirement so that A dollars can be received

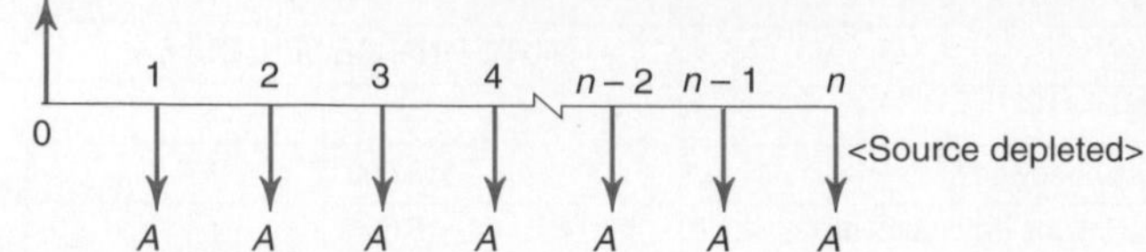

Figure 13.27
Series of monthly withdrawals.

for n periods, assuming i interest rate? Equation (13.7) can be solved for the amount of money P that must be accumulated by retirement if an amount A is to be withdrawn for n periods at a given interest rate.

To extend this idea, how much money P must be available if monthly amounts A are to be withdrawn forever? If we look at the equation,

$$P = A\left[\frac{(1 + i)^n - 1}{i(1 + i)^n}\right]$$

do we solve for $n = \infty$?

Perhaps we could restate the problem by asking what amount of money P must be available at retirement so that an amount A can be withdrawn each month and never affect the principal amount P?

If we use the compound interest formula, that is, $F = P(1 + i)^n$, and find the amount of interest generated for 1 month (here that is equal to A), then $F = P + A = P(1 + i)^1$. If we solve this equation for P, then

$$P = \frac{A}{(1 + i)^1 - 1} = \frac{A}{i}$$

and given an interest rate earned by the annuity (say, $i = 0.10/12$) and a fixed monthly income (say, $A = \$2000$), then $P = \$240000$.

If one can only earn 7 percent annual interest, the principal needed to have a perpetual monthly income of \$2 000 would be just over \$342 857.

13.7 Analysis of Alternatives

An engineer makes use of the material presented in this chapter in a very practical way. It is used to analyze a situation so that an intelligent decision can be made. Normally, several alternatives are available, each having some strong attributes. The task is to compare each alternative and to select the one that appears superior, all things considered.

Before examining some practical examples, a few definitions must be reviewed.

First Cost

This is the initial cost of purchase and includes items such as freight, sales tax, and installation.

Life

This refers to the number of years of service that the user expects from the item or property.

Salvage Value

This is the net sum to be realized from the disposal of an item or property after service. It normally includes removal costs, freight out, etc.

It is not always possible to determine these values with certainty, so the engineer must often work with data that are inexact. Even though the data are not perfect, an engineer can make better decisions with economic comparisons than without.

The most obvious method of comparing costs is to determine the total cost of each alternative. An immediate problem arises in that the various costs occur at different times, so the *total number of dollars spent is not a valid method of comparison*. You have seen that the present worth of an expenditure can be calculated. If this is done for all costs, the present worth of buying, operating, and maintaining two or more alternatives can then be compared. Simply stated, it is the sum of money necessary to buy, maintain, and operate a facility. The alternatives must obviously be compared for the same length of time, and replacements due to short-life expectancies must be considered.

A second method, preferred by those who work with annual budgets, is to calculate the *average annual cost* of each. The approach is similar to the present worth method, but the numerical value is in essence the annual contribution to a sinking fund that would produce a sum identical to the present worth placed at compound interest.

Many investors approach decisions on the basis of the profit that a venture will produce in terms of percent per year. The purchase of a piece of equipment, a parcel of land, a new product line, etc., is thus viewed favorably only if it appears that it will produce an annual profit greater than a certain percent. The amount of an acceptable percent return fluctuates with the money market. Since there is doubt about the amount of the profit, and certainly there is a chance of a loss, it would not be prudent to proceed if the prediction of return was not considerably above "safe" investments such as bonds.

The example that follows illustrates the use of these two methods (present worth and average annual cost) and includes a third technique called *future worth* that provides a check.

Each method compares money at the same point in time or over the same time period. Each method is different yet each yields the same conclusion.

Example problem 13.12 Consider the purchase of two computer-aided design (CAD) systems. Assume the annual interest rate is 12 percent.

	System 1	System 2
Initial cost	\$100 000	\$65 000
Maintenance and operating cost	4 000/year	8 000/year
Salvage	18 000 after 5 years	5 000 after 5 years

Using each of the three methods below, compare the two CAD systems and offer a recommendation:

1. Annual cost
2. Present worth
3. Future worth

Solution

1. Annual cost (See Fig. 13.28)

System 1	System 2
a. Initial cost	a. Initial cost
$A = P\left[\frac{i(1+i)^n}{(1+i)^n - 1}\right]$	$A = P\left[\frac{i(1+i)^n}{(1+i)^n - 1}\right]$
$P = 100\ 000$ $i = 0.12$ $n = 5$ $A = \$27\ 740.97$/year	$P = 65\ 000$ $i = 0.12$ $n = 5$ $A = \$18\ 031.63$/year
b. Maintenance and operating costs	b. Maintenance and operating costs
$MC = 4\ 000$/year	$MC = 8\ 000$/year
c. Salvage	c. Salvage
$F = A\left[\frac{(1+i)^n - 1}{i}\right]$	$F = A\left[\frac{(1+i)^n - 1}{i}\right]$
$A = \frac{F(i)}{[(1+i)^n - 1]}$	$A = \frac{F(i)}{[(1+i)^n - 1]}$
$F = 18\ 000$ $i = 0.12$ $n = 5$ $A = \$(-)2\ 833.38$/year	$F = 5\ 000$ $i = 0.12$ $n = 5$ $A = \$(-)787.05$/year

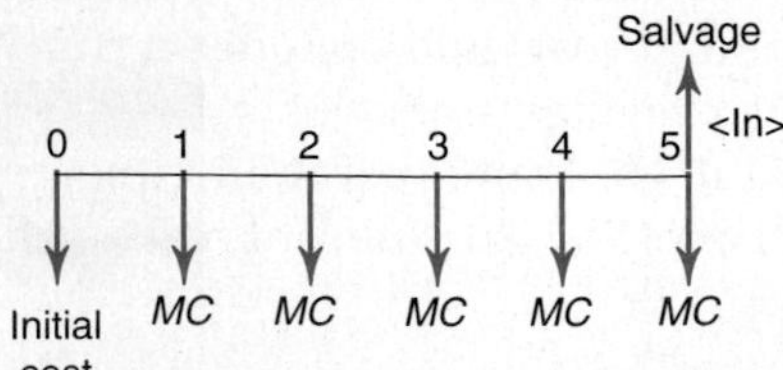

Figure 13.28
Company's cash-flow diagram (MC—maintenance cost).

System 1 (annual-cost analysis)	System 2 (annual-cost analysis)
+27 740.97 + 4 000.00 (−)2 833.38 $28 907.59	+18 031.63 + 8 000.00 (−) 787.05 $25 244.58

Conclusion: System 2 is less expensive.

2. Present worth

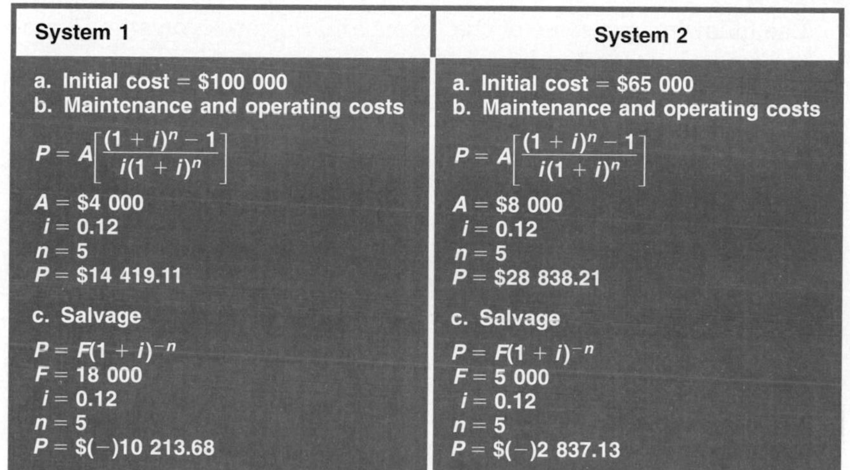

System 1	System 2
a. Initial cost = $100 000 b. Maintenance and operating costs $P = A\left[\frac{(1+i)^n - 1}{i(1+i)^n}\right]$ $A = \$4\ 000$ $i = 0.12$ $n = 5$ $P = \$14\ 419.11$ c. Salvage $P = F(1+i)^{-n}$ $F = 18\ 000$ $i = 0.12$ $n = 5$ $P = \$(-)10\ 213.68$	a. Initial cost = $65 000 b. Maintenance and operating costs $P = A\left[\frac{(1+i)^n - 1}{i(1+i)^n}\right]$ $A = \$8\ 000$ $i = 0.12$ $n = 5$ $P = \$28\ 838.21$ c. Salvage $P = F(1+i)^{-n}$ $F = 5\ 000$ $i = 0.12$ $n = 5$ $P = \$(-)2\ 837.13$

System 1 (present-worth analysis)	System 2 (present-worth analysis)
$100 000.00 14 419.11 (−)10 213.68 $104 205.43	$ 65 000.00 28 838.21 (−)2 837.13 $ 91 001.08

Conclusion: System 2 is less expensive.

3. Future worth

System 1	System 2
a. Initial cost = $100 000 $F = P(1+i)^n$ $P = \$100\ 000$ $i = 0.12$ $n = 5$ $F = \$176\ 234.17$ b. Maintenance and operating costs $F = A\left[\frac{(1+i)^n - 1}{\ }\right]$ $A = \$4\ 000$ $F = \$25\ 411.39$ c. Salvage = $(−)18 000	a. Initial cost = $ 65 000 $F = P(1+i)^n$ $P = \$65\ 000$ $i = 0.12$ $n = 5$ $F = \$114\ 552.21$ b. Maintenance and operating costs $F = A\left[\frac{(1+i)^n - 1}{\ }\right]$ $A = \$8\ 000$ $F = \$50\ 822.78$ c. Salvage = $(−)5 000

System 1 (future-cost analysis)	System 2 (future-cost analysis)
$176 234.17 25 411.39 (−)18 000.00 $183 645.56	$114,552.21 50 822.78 (−)5 000.00 $160 374.99

Conclusion: System 2 is less expensive.

13.8 Key Terms and Concepts

The following are some of the terms and concepts you should recognize and understand:

- Payback period
- Rate of return
- Simple interest
- Compound interest
- Principal
- Stated (nominal) interest
- Effective annual rate
- Annual percentage rate (APR)
- Cash-flow diagram
- Present worth
- Annuity
- Sinking fund
- Installment loan
- Uniform annual payment present-worth factor
- Capital recovery factor
- Uniform annual payment annuity factor
- Deferred annuity
- Retirement plan
- Analysis of alternatives
  - First cost
  - Life
  - Market value
  - Salvage value
  - Average annual cost
  - Future worth

Problems

13.1 What is the present worth of $10 000.00 to be paid to you in 4 years if the money is thought to be worth (*a*) 6 percent, (*b*) 12 percent, and (*c*) 24 percent? Assume annual compounding.

13.2 How much must be invested now to grow to $20 000 in 8 years if the annual interest rate is 7 percent compounded (*a*) annually, (*b*) semiannually, (*c*) monthly?

13.3 The taxes on a home are $1 800 per year. The mortgage stipulates that the owner must pay 1/12 of the annual taxes each month (in advance) to the bank so that the taxes can be paid on March 31 of the next year (taxes are due on April 1). Assuming that money costs 9.5 percent annual interest, compounded monthly, determine how much profit the bank makes each year on the owner's tax money.

13.4 Your hometown has been given $400 000 from the estate of a citizen. The gift stipulates that the money cannot be used for 3 full years but must be invested. If the money is invested at 8.8 percent annual interest, how much will be available in 3 years if it is compounded (*a*) annually, (*b*) semiannually, (*c*) daily?

13.5 If the interest rate is 9.8 percent per year, how long will it take for an investment to double in value with semiannual, monthly, and daily compounding? Is the time to double a linear function of the compounding period?

13.6 You have just made an investment that will repay $728.21 at the end of each month; the first payment is 1 month from today and the last one is 5 years from today.

(*a*) If the interest rate is 7.5 percent per year compounded monthly, what amount did you invest?

(*b*) If you can refuse all the payments and allow the money to earn at the same rate as stated, how much will you have at the end of 5 years?

13.7 You have made an investment that will yield $8 000 exactly 6 years from today. If the current interest rate is 8.3 percent compounded quarterly, what is your investment worth today?

13.8 You just borrowed $2 000 and have agreed to repay the bank $500 at the end of each of the next 5 years. What is the annual interest rate of the loan?

13.9 Today you have a savings account of $16 240. Based on an annual interest rate of 8 percent, what equal amount can you withdraw from the account at the end of each month for 4 years and leave $4 000 in the account? If you could earn 10 percent interest, what would be the value of your monthly withdrawals?

13.10 On January 1 this year you borrowed $20 000 toward component parts for a product you hope to have on the market in September of this year. You have agreed upon an annual interest rate of 11.6 percent compounded monthly. You also have agreed to begin repaying the debt on October 1, making equal monthly payments until the loan is repaid on February 1 next year. What are your monthly payments?

13.11 Referring to Prob. 13.10, suppose that you are unable to make the monthly payments and your creditor agrees to allow you to make a single payment on February 1 next year. How much will you owe at that time?

13.12 If sales at your company are doubling every 4 years, what is the annual rate of increase? What annual rate of increase would be necessary to double in 3 years? 2 years?

13.13 You have $50000 to invest, and you have decided to purchase bonds that will mature in 6 years. You have narrowed your choices to two types of bonds. The first class pays 9.5 percent annual interest. The second class pays 6.4 percent annual interest but is tax-free, both federal and state. Your income bracket is such that your highest income tax rate is 31 percent federal and the state income tax is 9 percent. Which is the better investment for you at this time? Assume that all conditions remain unchanged for the 6-year period.

13.14 On your 25th birthday you open an IRA (individual retirement account). At that time and on each succeeding birthday up to and including your 60th birthday, you deposit $2 000. During this period the interest rate on the account remains constant at 7.8 percent. No further payments are made, and beginning 1 month after your 65th birthday you begin withdrawing equal payments (the annual interest rate is the same as before). How much will you withdraw each month if the account is to be depleted with the last check on your 85th birthday? Assume annual compounding up to your 65th birthday and monthly compounding thereafter.

13.15 You wish to retire at age 60 and at the end of each month thereafter, for 25 years, to receive $3 000. Assume that you begin making monthly payments into an account at age 24. You continue these payments until age

60. If the interest rate is constant at 8.4 percent, how much must be deposited monthly between ages 24 and 60?

13.16 Two workstations are being considered to do a certain task. Workstation *A* costs \$22 000 new and \$2 600 to operate and maintain each year. Workstation *B* costs \$30 000 new and \$1 200 to operate and maintain each year. Assume both will be worthless after 6 years and that the interest rate is 11.0 percent. Determine by the annual-cost method which alternative is the better buy.

13.17 Assume you needed \$10 000 on April 1, 1995, and two options were available:

(*a*) Your banker would loan you the money at an annual interest rate of 12.0 percent, compounded monthly, to be repaid on September 1, 1995.

(*b*) You could cash in a certificate of deposit (CD) that was purchased earlier. The cost of the CD purchased September 1, 1994 was \$10 000. If left in the savings and loan company until September 1, 1995, the CD's annual interest is 7.8 percent compounded monthly. If the CD is cashed in before September 1, 1995, you lose all interest for the first 3 months and the interest rate is reduced to 3 percent, compounded monthly, after the first 3 months.

Which option is better and by how much? (Assume an annual rate of 7.5 percent, compounded monthly, for any funds for which an interest rate is not specified.)

13.18 Two machines are being considered for the same task. Machine *A* costs \$18 000 new and is estimated to last 6 years. The cost to replace machine *A* will be 4 percent more each year than it was the year before. It will cost \$1 200 per year to operate and maintain machine *A*. It will have a trade-in (salvage) value of \$1 500. Machine *B* costs \$38 000 to buy, will last 12 years, and will have a trade-in value of \$2 000. The cost of operation and maintenance is \$700 per year.

Compare the two machines and state the basis of your comparison. Include a cash-flow diagram for each alternative. Assume all interest rates at 8 percent per year unless otherwise stated.

13.19 Two systems are being considered for the same task. System *A* costs \$21 000 new and is estimated to last 4 years. It will then have a sale or trade-in value of \$1 500. The cost to replace system *A* will be 5 percent more each year than it was the year before. It will cost \$1 400 per year to operate and maintain system *A*, payable at the end of each year. System *B* costs \$40 000 to buy and will last 8 years. It will have a sale or trade-in value of \$2 000. The cost to operate and maintain system *B* will be \$800 per year, payable at the end of each year.

Assume the task will be performed for 8 years. Compare the two systems, state your basis of comparison, and include a cash-flow diagram. All interest rates are 8 percent per year unless otherwise stated.

13.20 You have borrowed as follows:

January 1, 1991	\$24 000
July 1, 1992	\$33 000
January 1, 1993	\$39 000

The agreed-upon annual interest rate was 7.75 percent compounded semiannually. How much did you owe on July 1, 1993?

You agreed to make the first of 15 monthly payments on October 1, 1993. Assume the interest rate was still 7.75 percent but was compounded monthly. How much was each of the 15 payments?

13.21 You have reached an agreement with an auto dealer regarding a new car. She has offered you a trade-in allowance of $5 500 on your old car for a new one that she has "reluctantly" reduced to only $16 995 (before trade-in). She has further agreed upon a contract that requires you to pay $315.22 each of the next 48 months, beginning 1 month from today. What is your interest rate, expressed as an annual percentage? Give your answer to the nearest 0.01 percent. With your instructor's approval, write a computer program or use a spreadsheet to solve the problem.

13.22 You have been assigned the task of estimating the annual cost of operating and maintaining a new assembly line in your plant. Your calculations indicate that during the first 4 years the cost will be $100 000 per year; the next 5 years will cost $130 000 per year; and the following 6 years will cost $155 000 per year. If the interest rate is constant at 13.8 percent over the next 15 years, what will be the average annual cost of operation and maintenance?

13.23 You are buying a new home for $185 000. You have an agreement with the savings and loan company to borrow the needed money if you pay 25 percent in cash and monthly payments for 30 years at an interest rate of 9.25 percent compounded monthly.

(*a*) What monthly payments will be required?
(*b*) How much principal reduction will occur in the first payment?
(*c*) Prepare a spreadsheet that will show each payment, how much of each will go to principal and how much to interest, the current balance, and the cumulative interest paid.

13.24 If you had started a savings account that paid 6.5 percent, compounded monthly, and your payments into the account were the same as in Prob. 13.23, how long would you have had to make payments in order to purchase the home for cash? (Assume the same down payment amount was available as in Prob. 13.23.)

13.25 Your bank pays 3.25 percent on Christmas Club accounts. How much must you put into an account weekly beginning on January 2 in order to accumulate $800 on December 4? Assume weekly compounding.

13.26 You can purchase a treasury note today for 94.2 percent of its face value of $10 000. Every 6 months you will receive an interest payment at the annual rate of 6.88 percent of face value. You can then invest your interest payments at the annual rate of 7.0 percent compounded semiannually. If the note matures 6 years from today, how much money will you receive from all the investments? Express this also as an annual rate of return.

13.27 What is the present worth of each of the following assets and liabilities and your net present worth?

(*a*) You deposited $3 000 exactly 3 years ago.
(*b*) You have a checking account with a current balance of $1 127.19.
(*c*) You must pay $4 150 exactly 4 years from now.
(*d*) Today you just made the 29th of 36 monthly payments of $129.52.
(*e*) You will receive $8 000 exactly 5 years from now.

Assume all annual interest rates are 7.5 percent. Assume monthly compounding in figuring part (*d*) and annual compounding for the rest.

13.28 Plant revisions are necessary to put a new product into production. These revisions could be done at one time or in several increments. It is estimated that a single project completed on July 1, 1998 would cost \$210 000. If completed over several years, the costs are estimated to be as follows:

1st phase	July 1, 1998	\$100 000
2nd phase	Jan. 1, 2001	\$ 80 000
3rd phase	July 1, 2005	\$ 60 000
4th phase	July 1, 2007	\$ 40 000
5th phase	July 1, 2010	\$ 50 000

If money can be borrowed at 11.2 percent annual interest compounded quarterly, which project is most economical? By how much? What other factors might affect your decision?

13.29 EJMN Engineering has estimated that the purchase of CAD workstations costing \$150 000 will reduce the firm's drafting expenses by \$12 500 per month during a 2-year period. If the workstations have zero salvage value in 2 years, what is the firm's expected annual rate of return on investment? (*Rate of return* is the equivalent interest rate that must be earned on the investment to produce the same income.) If approved by your instructor, write a computer program or use a spreadsheet to solve the problem.

13.30 Engineers at Specialty Manufacturing are writing a justification report to support the purchase of a DNC milling center (mill, controller, microcomputer, installation, etc.). They have learned that the total initial cost will be \$70 000. The labor savings and improved product quality will result in an estimated benefit to the company of \$2 100 each month over a 10-year time period. If the salvage value of the center is about \$10 000 in 10 years, what annual rate of return (annual percentage) on investment did the engineers calculate? Write a computer program or use a spreadsheet if assigned by your instructor.

13.31 Many new engineering graduates purchase and finance new cars. Automobiles are typically financed for 4 years with monthly payments made to the lending agency. Assume you will need to borrow \$16 000 with 48 monthly payments at 12.5 percent annual interest.

(*a*) Write a computer program or prepare a spreadsheet to produce the mortgage table below:

Payment number	Monthly payment	Amount to principal	Amount to interest	Cumulative interest	Current balance
1	\$xxx.xx	\$xxx.xx	\$xxx.xx	\$xxx.xx	\$xxx.xx
2	\$xxx.xx	\$xxx.xx	\$xxx.xx	\$xxx.xx	\$xxx.xx
3	.	.	.	.	.
.	.	.	.	.	.
.	.	.	.	.	.
.	.	.	.	.	.

(*b*) If you decided to pay the loan off at the end of 10 months, what amount is needed? At the end of 20 months? At the end of 40 months?

(*c*) What is the cumulative interest paid in the first 12 payments? Second 12? Third 12? Last 12?

(*d*) Repeat parts (*a*), (*b*), and (*c*) if the interest rate is 15 percent instead of 12.5. The amount borrowed remains the same.

(*c*) Repeat parts (*a*), (*b*), and (*c*) if you find it necessary to borrow \$19 000. The interest rate is still 12.5 percent.

(*f*) What is the result if you borrow \$19 000 and the interest rate is 15 percent?

13.32 Many of you will eventually purchase a house. Few will have the total cash on hand, so it will be necessary to borrow money from a home loan agency. Often you can borrow the money at a fixed annual interest for, say, 15 or 30 years. Monthly payments are made to the lending agency. Write a computer program or use a spreadsheet to prepare mortgage tables similar to the one described in Prob. 13.31 for the following situations:

(*a*) \$80 000 at 12 percent interest for 15 years

(*b*) \$80 000 at 9.75 percent interest for 15 years

(*c*) \$200 000 at 12 percent interest for 15 years

(*d*) \$200 000 at 12 percent interest for 30 years

(*e*) Other cases as may be assigned

(*f*) Critically examine the monthly payments and the cumulative interest amount produced by changing from a 30-year loan to a 15-year loan, all other parameters being constant.

14

Engineering Design—A Process

14.1 Introduction

Engineering design is a systematic process by which solutions to the needs of humankind are obtained. The process is applied to problems (needs) of varying complexity. For example, mechanical engineers will use the design process to find an effective, efficient method to convert reciprocating motion to circular motion for the drive train in an internal combustion engine; electrical engineers will use the process to design electrical generating systems using falling water as the power source; and materials engineers use the process to design high-temperature materials which enable astronauts to safely reenter the earth's atmosphere.

The vast majority of complex problems in today's high-technology society depend for solution not on a single engineering discipline but, rather, on teams of engineers, scientists, environmentalists, economists, sociologists, and legal personnel. Solutions are not only dependent on the appropriate applications of technology but also on public sentiment as executed through government regulations and political influence. As engineers we are empowered with the technical expertise to develop new and improved products and systems, but at the same time we must be increasingly aware of the impact of our actions on society and the environment in general and work conscientiously toward the best solution in view of all relevant factors.

Design is the culmination of the engineering educational process; it is the salient feature that distinguishes engineering from other professions.

A formal definition of *engineering design* is found in the curriculum guidelines of the Accreditation Board for Engineering and Technology (ABET). ABET accredits curricula in engineering schools and derives its membership from the various engineering professional societies. Each accredited curriculum has a well-defined design component which falls within the ABET guidelines. The ABET statement on design reads as follows:

> Engineering design is the process of devising a system, component, or process to meet desired needs. It is a decision-making process (of-

ten iterative), in which the basic sciences, mathematics, and engineering sciences are applied to convert resources optimally to meet a stated objective. Among the fundamental elements of the design process are the establishment of objectives and criteria, synthesis, analysis, construction, testing, and evaluation. The engineering design component of a curriculum must include most of the following features: development of student creativity, use of open-ended problems, development and use of modern design theory and methodology, formulation of design problem statements and specifications, consideration of alternative solutions, feasibility considerations, production processes, concurrent engineering design, and detailed system descriptions. Further, it is essential to include a variety of realistic constraints such as economic factors, safety, reliability, aesthetics, ethics, and social impact.

In order for you to gain the fundamental knowledge and experience needed to understand the design process, you must partake in meaningful design activities as a student. To assist you in your first activity, we will take you through an actual student design project as undertaken by a team of beginning engineering students. The project is guided by the 10-step design process listed in Fig. 14.1. As you study the design process and the description of how the student team accomplished each step, pay particular attention to the structure of the process rather than to the development of the particular solution arrived at by the team. By doing this you will understand how engineers approach a need, develop alternative solutions, select the best solution, and communicate the results.

Throughout this chapter you will see many examples of the utilization of computers in engineering design (see Fig. 14.2). The computer is a major tool for the engineer in the acquisition and analysis of data, and the definition of potential solutions. Throughout history engineers have used the best computational devices available at the time to obtain solutions to problems. The computer is the fastest and most powerful computational tool yet conceived; it produces numerical computations at heretofore unheard of rates. It also provides an insight to problems and solutions through its capability to simulate actual phenomena. This capability provides the engineer a tremendous advantage in developing new and improved products in a much shorter time frame than ever before. The rapidly developing computer envi-

Figure 14.1
A 10-step design process.

1. Identification of a need
2. Problem definition
3. Search
4. Constraints
5. Criteria
6. Alternative solutions
7. Analysis
8. Decision
9. Specification
10. Communication

Figure 14.2
A design team member views a three-dimensional model created on an engineering workstation.

ronment will enable you to be a better educated and more productive engineer than your predecessors.

14.1.1 The Design Process

A simple definition of *design* is "a structured problem-solving activity." A *process,* on the other hand, is a phenomenon identified

Figure 14.3
A student team reviews a part drawing for a design project with the instructor.

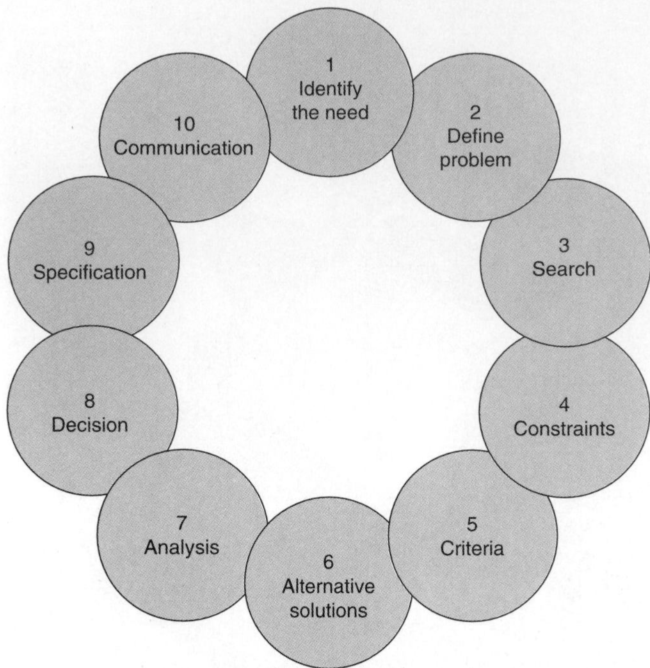

Figure 14.4
The design process is iterative in nature.

through step-by-step changes that lead toward a required result. Both these definitions suggest the idea of an orderly, systematic approach to a desired end. Figure 14.4 shows the design process approach as continuous and cyclic in nature. This idea has validity in that many problems arise during the design process that generate subsequent designs. You should not assume that each of your design experiences will necessarily follow the sequential steps without deviation. Experienced designers will agree that the steps as shown are quite logical; but on many occasions designers have had to repeat some steps or perhaps have been able to skip one or more steps.

Before beginning an overview of the entire design process, we must state that limits are always placed on the amount of time available. Normally, we establish a time frame or a series of deadlines for ourselves before we begin the process. It is almost impossible for us to tell you how much time should be allocated for each step because the problems are so varied. A sample of a time frame is shown in Fig. 14.5.

The whole process begins when a need is recognized: In essence, a human need has been identified. Often it is not the designers who are involved in step 1, but they usually assist in defining the problem (step 2) in terms that allow it to be scrutinized. Information is gathered in step 3, and then boundary conditions (constraints) are established (step 4). The criteria against which the alternative solutions are compared are chosen in step 5. At step 6 several possible solutions are entertained and the creative, innovative talents of the designer come into play. This is followed

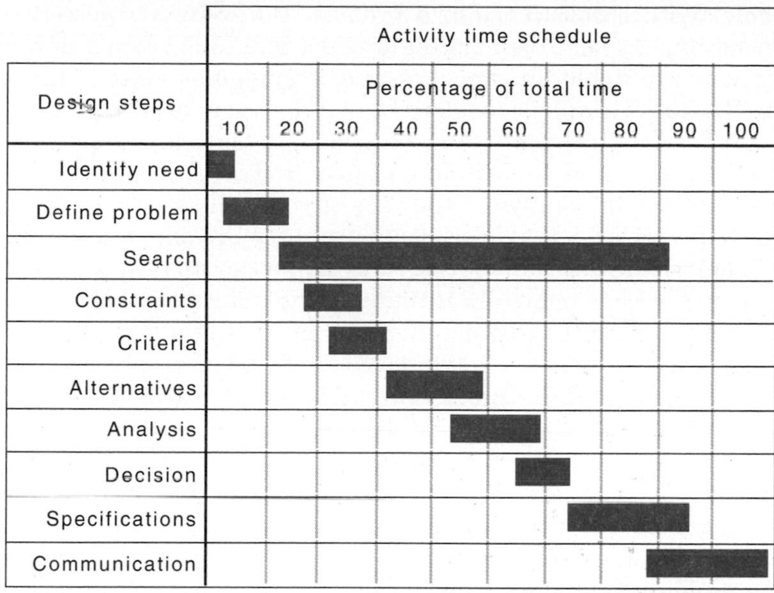

Figure 14.5
In order to control the design process, a time schedule must be developed early.

by detailed analysis of the alternatives (step 7), after which a decision is reached regarding which one should be completely developed (step 8). Specifications of the chosen concept are prepared (step 9), and its merits are explained to the proper people or agencies (step 10) so that implementation (construction, production, and so on) can be accomplished. A more detailed explanation of each step, as well as a reporting of the actions of the student design team, constitutes the remainder of this chapter.

14.2 Identification of a Need—Step 1

Before the process can begin, someone has to recognize that some constructive action needs to be initiated. This may sound vague, but understandably so because such is the way the process normally begins. Engineers do not have supervisors who tell them to come to work tomorrow and "identify a need." You might be asked to do so in the classroom because some professors may have you work on a project that you choose rather than one that is assigned. When most of us speak of a need, we generally refer to a lack or shortage of something we consider essential or highly desirable. Obviously, this is an extremely relative thing for what may be a necessity to some could be a luxury to others.

More often than not, then, someone other than the engineer decides that a need exists. In industry, or the private sector, it is essential that products sell for the company to survive. Most of the products have a life cycle that goes from the development stage, when the expenditures by the organization are high and sales are low, to the peak demand period, when profits are high, and eventually to the point where the product becomes obsolete. Even though a human need may still exist, the economic demand does not because a more attractive alternative has become available. With obsolescence of a product, the company perceives a need to phase out the existing product and to develop a new one that is profitable. Inasmuch as most companies exist to make a profit, profit can be considered to be the basic need.

A bias toward profit and economic advantage should not be viewed as a selfish position because products are purchased by people who feel that what they are buying will satisfy a need that they perceive as real. Society appreciates anyone who provides essential and desirable services, as well as goods that we use and enjoy. The consumers are ultimately the judges of whether there is truly a need. In like manner, the citizens of a community decide whether or not to have paved streets, parks, libraries, adequate police and fire protection, and scores of other things. City councils vote on the details of the programs. However, during the period when citizens and decision makers are formulating their plans, engineers are involved in supplying factual information to assist them. After the policy decisions have been made, engineers conduct studies, surveys, tests, and computations that allow them to prepare the detailed design plans, drawings, etc. that shape the final project.

14.2.1 The Chapter Example–Step 1 of the Design Process

Throughout the remainder of the chapter we will trace the steps that five beginning engineering students* followed to produce a design for their class project. As a starting point, a professor may assign students the task of identifying a need. It usually is easier to approach such an assignment by beginning with a very broad area of technology, such as energy.

The five students who were chosen were informed that all the student teams would be involved in some area dealing with the energy problem. Their professor began with construction of a decision tree, shown in Fig. 14.6. The class discussed sources of energy and jointly added the first level of subproblems: fossil, wind, geothermal, solar, nuclear, and organic. The class was then divided into groups; the groups began to subdivide further *one* of the energy sources listed above. There probably is no end to this procedure, but it does provide quickly a wide range of topics from which needs may be more easily recognized. Our student design team developed the organic section of Fig. 14.6, as shown, and thus began their discussion of the general topic of firewood for use in fireplaces. They recorded statements such as the following:

1. A large number of people have decided to install fireplaces in their homes and apartments.

---

*The five students were John W. Benike, Douglas L. Carper, Patrick J. Grablin, Rick Sessions, and David L. White.

2. Firewood is not as commercially available as it used to be.
3. The price of firewood has risen significantly.
4. People are now more willing to cut and split their own firewood than they were previously.
5. The small, inexpensive chain saw has made the cutting portion of the task more acceptable, but splitting the wood is still a major problem.

After a reasonable period of discussion involving these topics as well as others related to firewood, they agreed upon the following initial statement of need:

> There is a need for an inexpensive supply of firewood for use in the home.

(We will see later that this statement was changed by the students in much the same way that professional engineers refine and redefine problems during a design process.)

Figure 14.6
A decision tree pertaining to energy.

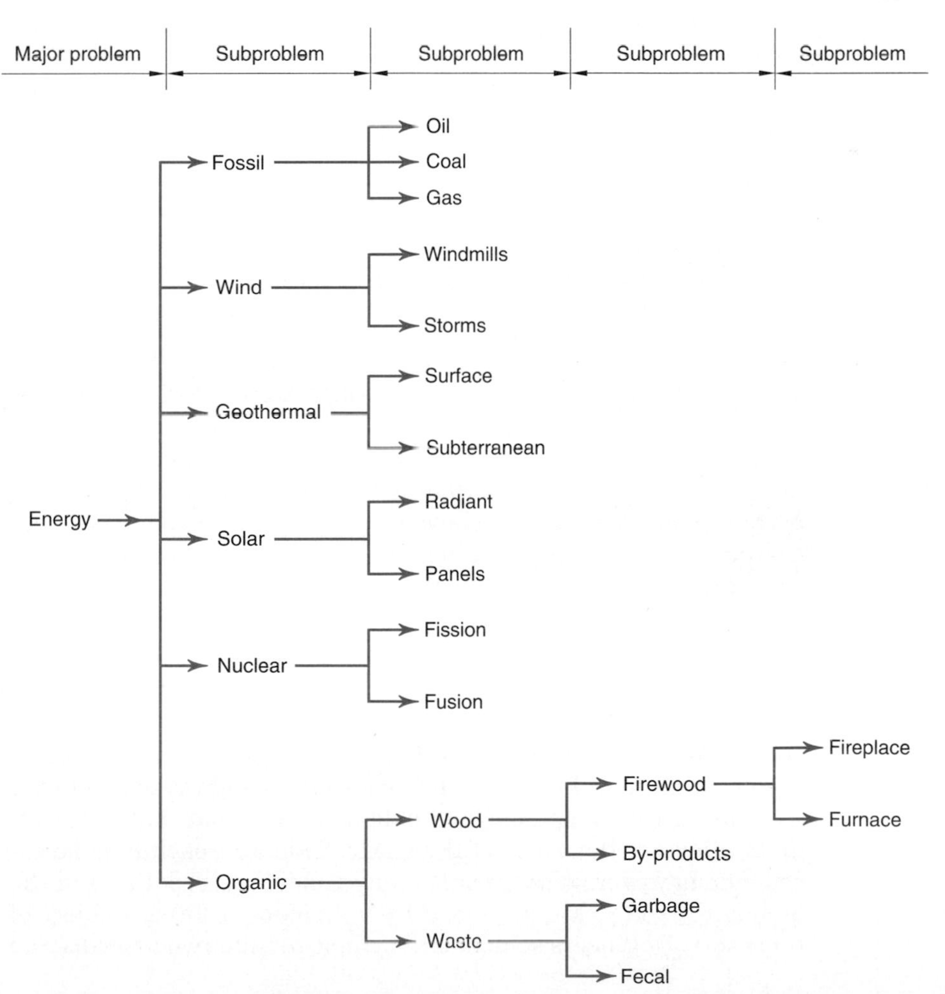

14.3 Problem Definition—Step 2

There is often a temptation to construct quickly a mental picture of a gadget that if properly designed and manufactured will satisfy a need. In the general case of a need for firewood we obviously know that we can call a supplier and have firewood delivered, but the cost factor is equally obvious. We could get a bit facetious and decide to burn our furniture. There have been emergency situations during extreme storms when this was the best solution available. If some friendly neighbor will supply us with firewood at no cost or effort, a problem does not exist. An important point to realize at this time is that had we allowed ourselves to focus on a specific piece of equipment or a single method of obtaining firewood in the very beginning, we would never have considered the statements mentioned earlier.

14.3.1 Broad Definition First

The need as previously stated does not point to any particular solution and thereby leaves us with the opportunity to consider a wide range of alternatives before we agree on a specific problem statement. Consider for a moment a partial array of possibilities that will satisfy the original statement that there is a need for an inexpensive supply of firewood for use in the home:

1. Purchase firewood from a supplier.
2. Purchase standing timber to be cut for firewood.
3. Purchase trees that have been cut or that have fallen.
4. Hire a portion of the work done (probably the splitting of logs).
5. Design improved equipment to facilitate a portion of the process.

This is not an exhaustive list. You may want to take a few minutes to add to it. The first item is the current solution for many people. The other items show some promise of being an improvement over simply purchasing the firewood. You may think that item 1 should not be listed because it offers no change. If so, you are mistaken because the status quo is the solution that is selected most often, at least as a temporary measure.

14.3.2 Symptom versus Cause

If you cough and do nothing but suck on a cough drop, you may be treating the symptom (the tickle in your throat) but doing little to alleviate the cause of the tickle. This approach may be expedient; however, it can result many times in a repetition of the problem if the tickle is caused by a virus or a foreign object of some sort. Engineers seldom tell a client to take two aspirins and call back tomorrow, but they can sometimes be guilty of failing to see the real or root problem.

For many years residential subdivisions were designed so that the rainfall would drain away quickly, and expensive storm sewer systems were constructed to accomplish this task. Not only were the sewers expensive, but they also resulted in transporting the water problem downstream for someone else to handle. In recent years perceptive engineers have designed land developments so that the rainfall is temporarily collected in "holding pools" and released gradually over a longer period of time. This approach employs smaller, less expensive sewers and reduces the likelihood of flooding downstream. The real problem was not how to get rid of the rainfall as rapidly as possible, but how to control the water.

14.3.3 Solving the Wrong Problem

In the 1970s the problem of increasing fatalities as a result of auto accidents was clearly recognized. It was shown that the fatality rate could be significantly reduced if the driver and front seat passenger used lap and shoulder belts. The solution technique that was implemented was to build in an interlock system that required the belts to be latched before the auto could be started. That solution certainly should have solved the problem but it did not. It attacked the problem of requiring that the belts be physically used, but it did nothing to solve the real problem—the driver's attitude. The driver and passenger still did not wish to use belts and did everything possible to avoid it, even by having the interlock system removed.

14.3.4 The Chapter Example—Step 2

The students whose progress we are following considered the possibilities outlined in Sec. 14.3.1. Their discussions covered a range of topics, and from their notebook we have listed a few of their pertinent recorded thoughts:

1. The range of possible solutions must be reduced before the problem can be solved.
2. People really like the smell of burning wood.
3. An adequate supply of wood already exists in most areas.
4. Chain saws are already well developed; hence, cutting the wood into proper lengths is not a pertinent problem.
5. Time is very short, so a direction must be selected that we (the students) can achieve.

Most assuredly these young people had other thoughts, many of which were not recorded. The result of their consideration was a slightly revised problem definition:

There needs to be available to the average household an inexpensive, efficient method of splitting a small quantity of firewood.

In practice, you and other engineers will face restrictions that will affect the quality of your solutions. Many times your solution will have to meet governmental regulations in order to qualify for grants of money; or perhaps safety requirements by some agency cannot be met if certain materials are used. In almost all of your projects, there will be cost and time constraints that force you to make decisions that are not what you really want to do. Such decisions, once made, then control many of your subsequent actions on that project.

This situation was faced by our student design team. By limiting the range of possible solutions and accepting the present method of cutting wood, they eliminated even the consideration of other burning materials. We are not being critical of their decision because we have experienced similar time and resource constraints.

14.4 Search—Step 3

Most of your productive professional time will be spent locating, applying, and transferring information—all sorts of information. This is not the popular opinion of what engineers do, but it is the way it will be for you. Engineers are problem solvers, skilled in applied mathematics and science, but they seldom, if ever, have enough information about a problem to begin solving it without first gathering more data. This search for information may reveal facts about the situation that result in redefinition of the problem (see Fig. 14.7).

14.4.1 Types of Information

The problem usually dictates what types of data are going to be required. The one who recognizes that something was needed (step 1) probably listed some things that are known and some things that need to be known. The one or ones who defined the problem had to have knowledge of the topic or they could not have done their part (step 2). Generally, there are several things that we look for in beginning to solve most problems. For example,

1. What has been written about it?
2. Is something already on the market that may solve the problem?
3. What is wrong with the way it is being done?
4. What is right with the way it is being done?
5. Who manufactures the current "solution"?
6. How much does it cost?
7. Will people pay for a better one if it costs more?
8. How much will they pay (or how bad is the problem)?

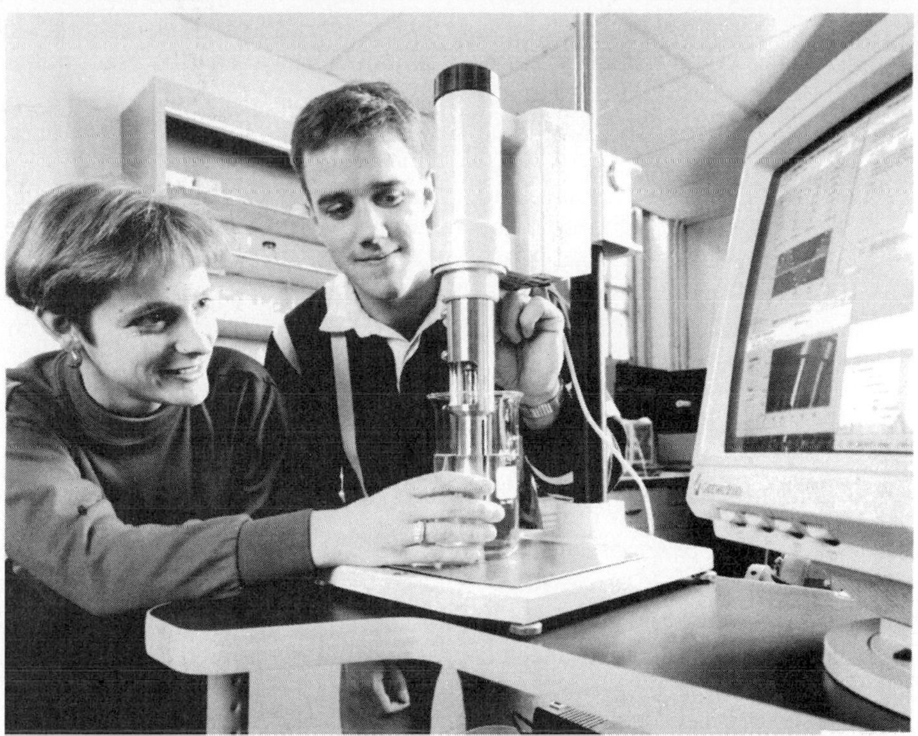

Figure 14.7
An analysis is under way in a materials science laboratory.

14.4.2 Sources of Information

If anything can be said about the last half of the twentieth century, it is that we have had an explosion of information. The amount of data that can be uncovered on most subjects is overwhelming. People in the upper levels of most organizations have assistants who condense most of the things that they must read, hear, or watch. When you begin a search for information, be prepared to scan many of your sources and document their location so that you can find them easily if the data subsequently appear to be important.

Some sources that are available include the following:

1. *Existing solutions.* Much can be learned from the current status of solutions to a specific need if actual products can be located, studied and, in some cases, purchased for detailed analysis. If a product can be acquired in the marketplace, a process called *reverse engineering* can be performed to determine answers to some questions posed in the section on Types of Information. An improved solution or an innovative new solution cannot be found unless the existing solutions are thoroughly understood. Reverse engineering is an excellent learning technique for students and engineers in industry who are beginning to apply the design process. You should look to local industries and retail outlets for existing solutions that

currently satisfy the need you have identified. In some instances you may be able to purchase the product or at least observe a demonstration of its capability.

2. *Internet.* The electronic Internet that today connects millions of computer users in nearly 100 nations to a global "information superhighway" provides rapid access to a wealth of knowledge. Industries, businesses, government organizations, and educational institutions are now connected to the Internet enabling nearly instantaneous access to research data, product information, special interest groups, and experts throughout the world. Although you may not yet have access to the Internet from your personal computer, your university has facilities from which you can gain access. The World Wide Web (WWW) is an Internet-based navigational system which provides the capability to find information and move to different locations (servers) with point and click ease. The WWW provides access to graphic images, photographs, audio, and full-motion video. Accessing the Internet for information should be a major component in any research effort.

3. *Your library.* Many universities have courses that teach you how to use your library. Such courses are easy when you compare them with those in chemistry and calculus, but their importance should not be underestimated. There are many sources in the library that can lead you to the information that you are seeking. You may find what you need in an index such as the *Engineering Index,* but do not overlook the possibility that a general index, such as *The Reader's Guide* or *Business Periodicals Index,* may also be useful. The *Thomas Register of American Manufacturers* may direct you to a company that makes a product that you need to know more about. *Sweets Catalog* is a compilation of manufacturer's information sheets and advertising material about a wide range of products. There are many other indexes that provide specialized information. The nature of your problem will direct which ones may be helpful to you. Do not hesitate to ask for assistance from the librarian. You should use to advantage the computer databases found in libraries and often available through CD-ROM technology.

4. *Government documents.* Many of these are housed in special sections of your library, but others are kept in centers of government—city halls, county court houses, state capitols, and Washington, D.C. The regulatory agencies of government, such as the Interstate Commerce Commission, the Environmental Protection Agency, and regional planning agencies, make rules and police them. The nature of the problem will dictate which of the myriad of agencies can fill your needs.

5. *Professional organizations.* The American Society of Civil Engineers is a technical society that will be of interest to students majoring in civil engineering. Each major in your college is associated with not one but often several such societies. The National Society of Professional Engineers is an organization that most engineering students will eventually join, as well as at least one technical society such as the American Society of Mechanical Engineers (ASME), the Institute of Electrical and Electronics Engineers (IEEE), or any one of dozens that serve the technical interests of the host of specialties with which professional practices seem most closely associated. Many engineers are members of several associations and societies. Other organizations, such as the American Medical Association and the American Bar Association, serve various professions, and all have publications and referral services.

6. *Trade journals.* They are published by the hundreds, usually specializing in certain classes of products and services.

7. *Vendor catalogs.* Perhaps your college subscribes to one of the several information services that gather and index journals and catalogs. These data banks may have tens of thousands of such items available to you on microfilm. You need only learn how to use them.

8. Individuals that you have reason to believe are experts in the field. Your college faculty has at least several such individuals, maybe many. There are, no doubt, some practicing engineers in your city.

14.4.3 Recording your Findings

The purpose of a bibliography is to direct you to more information than is included in the article you are reading. The form of the bibliography makes it easy to find the reference. So it seems reasonable for you to record your information sources in proper form so that if that reference is to be cited in your report, you are ready to do it properly. By so doing, you are ensuring that it can be found again quickly and easily. Few things are more discouraging than to be unable to locate an article that you found once and know will be helpful if you could locate it again.

It is usually a good procedure to record each reference on a card or sheet of paper or maintain a computer database. English teachers usually recommend the use of file cards, but engineers

TA 152.17 G273	Inganere, M.E. *Heat, Air, and Gas Power.* McGraw-Hill, New York, 1973.
	Has good tables in appendix and formulas on pages 52 - 55 covering heat transfer cases that may occur on project.

Figure 14.8
Documentation of research findings is essential if the findings are to be useful later. Recording pertinent information on a card or computer database will permit easy retrieval of the findings when needed.

seem to prefer information put in a bound notebook or the computer. Whatever your choice, Fig. 14.8 is recommended as a reasonable form of record.

As we are looking at something like a piece of equipment, we often have thoughts and ideas that should be recorded for future reference. At such times our ability to sketch is an invaluable tool because so many details can be graphically shown but are very difficult to describe in words.

14.4.4 The Chapter Example—Step 3

The team of engineering students whose project we are following realized the importance of the research phase but were also aware of the overall time constraints on the design process. They decided that a detailed research plan was needed so that specific assignments could be made to avoid conflict or overlap of effort. After considerable discussion specific research areas were assigned to each team member. They consulted home builders about the demand for fireplaces, suppliers of firewood, several manufacturers of chain saws, companies that sell chain saws, a landscape architect, city government, the library, a county extension service, an engineer, and a company that sells commercial log-splitters.

One of the team members was assigned the task of checking the library to determine what products were currently available. The librarian explained the various indexes, so the student selected the *Thomas Register.* After failing to find anything listed under "Logs," he tried "Splitters." Figure 14.9 is a reproduction of his notes. (The appendix of the student report includes copies of letters sent by the design team along with the responses received.)

With this information, the student went to the *Yellow Pages* of the local telephone book, wherein he learned that one of the products was sold locally; so a team member was assigned to visit the dealer. The dealer was temporarily out of advertising pamphlets, so the team member sketched the floor model and recorded pertinent data about it, for which the notes are shown as Fig. 14.10. Their procedure is to be commended and highly recommended in that they adequately documented the information in sufficient detail so that it could be used or the manufacturer could be contacted for additional data.

It must be noted here that this design project was performed prior to the availability of the WWW at this university. Today, of course, the team members would be searching for data on the Web.

The initial research stage has provided us with added information about the problem so that we are now ready to begin to describe the design in terms of things it must be or must have and what attributes are most important.

Source: Thomas Register of American Manufacturers
Thomas Register Catalog File

Listing: Splitters: Wood, Firewood, Kindling, etc.

Location: Engineering Library—Reference Tables

Manufacturers

1. Gordon Corporation
P.O. Box 244-TR
Farmington, CT
(Hydraulic)

2. H. L. Diehl Co.
South Windham, CT

3. Vermeer Mfg. Co.
3804 New Sharon Rd.
Pella, IA
(Powered, log, hydraulic, trailer)
(515) 628-3141

4. Tree King Mfg. & Engineering, Inc.
North St.
Showhegan, ME
(Hydraulic)

5. Lindig Mfg. Corp.
1831 West Country Rd.
St. Paul, MN
(612) 633-3072

6. Carthage Machine Co., Inc.
571 West 3rd Ave.
Carthage, NY
(Wood for pulp mills)

7. Equipment Design & Fabrication, Inc.
722 N. Smith St.
Charlotte, N.C.
(Log)

8. Pabco Fluid Power Co.
5752 Hillside Ave.
Cincinnati, OH
(Log)
(513) 941-6200

9. Piqua Engineering, Inc.
234-52 First Street
Piqua, OH
(513) 773-2464

10. Hencke Mfg. Co., Inc
433 W. Florida St.
Milwaukee, WI
(Log)

11. Didier Mfg. Co.
1652 Philips Ave.
P.O. Box 806
Racine, WI
(Hydraulic, log)

12. Murray Machinery, Inc.
104 Murray Road
Wausau, WI
(Hydraulic, paper roll)

Figure 14.9
A record of manufacturers who produce log-splitters.

14.5 Constraints—Step 4

Up to this point we have kept the problem definition as broad as possible, allowing for a maximum number of potential solutions. However, there are physical and practical limitations (called constraints) that will reduce the number of solutions for any problem. For example, in order to remain competitive in the marketplace, a retail cost of $50 for a particular solution cannot be exceeded. Another example is that any conceived solution must be able to operate using a standard household 120-V (volt) outlet. In many cases the physical size of the solution is limited by market competition or legal restrictions. The laptop computer is a good example of placing size (and weight) constraints on a solution in order to meet the competition in the marketplace. Whenever a *constraint* is applied, the solution possibilities are reduced; therefore, you should take care that the constraint is not "artificial," that is, not overly restrictive to the process of finding an innovative solution.

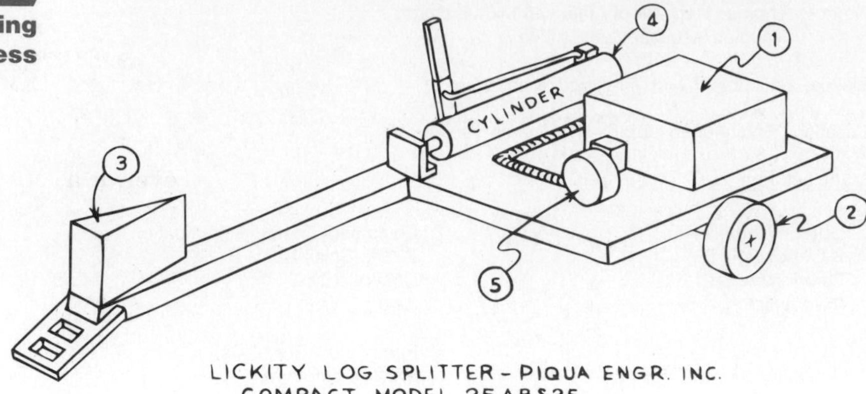

OVERALL SPEC'S
LENGTH 74.5"
WIDTH 32.5"
HEIGHT 23.0"(W/CONTROL LEVER FOLDED)

1. 5 H.P. BRIGGS & STRATTON
2. 8" SEMI-PNEUMATIC WHEELS
3. HEAT TREATED WEDGE
4. RAM FORCE – 10 TONS
5. HYDRAULIC PUMP

Figure 14.10
A sketch is an excellent method of recording certain types of information.

We face a similar situation in almost every decision we make, even though sometimes they are not really important. When you arose this morning, you had to choose some clothes to wear. You probably limited the choice to those hanging in your closet (or maybe in your roommate's closet). This was a constraint *you* placed on the situation, not one that really existed until you made it so. In most fields of engineering formulas have been developed and are used in designs of various kinds. Many, and probably most, of them are valid in a certain range of physical conditions. For instance, the hydraulic conditions of the flow of water are not valid below 0 °C or above 100 °C and are restricted to normal pressure ranges. We normally refer to these constraints or limits as *boundary conditions,* and they occur in many different ways.

14.5.1 The Chapter Example—Step 4

The student design team did not list any specific constraints, but we must assume that they are practical individuals who would assign a very low rating to a concept that exceeded some level of performance. For instance, if one of their ideas had an estimated cost of $500, it would no doubt be rated at near zero. They did not, however, tell us at what cost the zero rating begins; is it $200 or $100 or $75? In like manner, they did not say at what weight they would consider a concept to be too heavy to be portable. They did give us a little help by restricting the projected users to adults (excluding children).

14.6 Criteria—Step 5

Criteria are desirable characteristics of the solution which are established from experience, research, market studies, and customer preferences. In most instances the criteria are used to judge alternative solutions on a qualitative basis. However, if a performance function can be defined mathematically, then an optimum solution can be obtained numerically. The mathematical method of optimization is beyond the scope of this introductory text. Instead we concentrate on the selection and weighting of a set of criteria that will produce the best solution for the stated problem and solution constraints.

14.6.1 Design Criteria

Whereas each project or problem has a personality all its own, there are certain characteristics that occur in one form or another in a great many projects. We should ask ourselves, "What characteristics are most desirable and which are not applicable?" Typical design criteria are listed below:

1. Cost—almost always a heavily weighted factor
2. Reliability
3. Weight (either light or heavy)
4. Ease of operation and maintenance
5. Appearance
6. Compatibility
7. Safety features
8. Noise level
9. Effectiveness
10. Durability
11. Feasibility
12. Acceptance

There will be other criteria, and perhaps some of those given are of little or no importance in some projects, so a design team in industry or in the classroom must decide which *criteria* are important to the design effort (see Fig. 14.11). Since value judgments have to be made later, it probably makes little sense to include those criteria that will be given relatively low weights. There are often mild disagreements at this point, not about which criteria are valid but, rather, about how much weight should be assigned to each. It is frequently better if the team members make their assignments of weight independently and then compile all the results. This tends to dampen the effect of the more persuasive members at the same time that it forces all team members to contribute consciously. Usually, there are not many instances where one of the members strongly disagrees

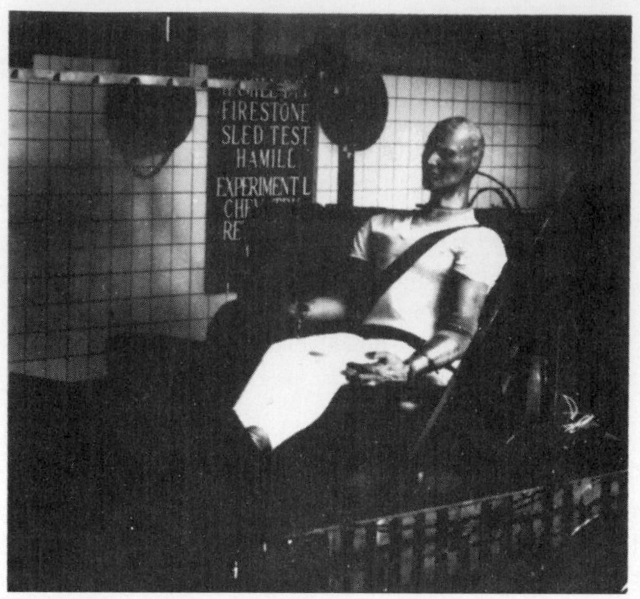

Figure 14.11
The simulation of accidents on a computer system assists engineers in the establishment of design criteria for safer vehicles. (Courtesy of the ISU Visualization Laboratory.)

with the mean value of the weight assigned to each criterion. Some negotiation may be required, but it is seldom a difficult situation to resolve.

14.6.2 The Chapter Example—Step 5

Most people feel comfortable when they talk in general terms about a great many things because as long as they do not have to get specific, an avenue of escape from their position is left open. The students whose progress we are following were not so fortunate for they were facing a real problem. A review of their time-line schedule indicated that it was time to make some decision. They therefore agreed upon the following criteria and assigned weights:

1. Cost: 30 percent
2. Portability: 20 percent
3. Ease of operation: 15 percent
4. Safety: 15 percent
5. Durability: 10 percent
6. Use of standard parts: 10 percent

What areas of agreement and disagreement do you see between our list of 12 criteria and their list of 6?

The most obvious difference is that they included the use of standard parts as an important criterion, but it was not listed at all in the section on Design Criteria. Just why this is important is not clear unless you try to place yourself in their position. If the team has plans to manufacture the log-splitter, then it will be much easier and less expensive to begin operations if many components can be purchased rather than manufactured in their own plant.

They agreed that cost, weight (portability), ease of operation, and safety were important. The others on our list of 12 were either not considered or considered to be of low importance (less than 5 percent).

It is our conclusion that at this point the students have decided that the solution to their problem will be a portable log-splitter that can be operated by a single adult. If this is true, then we can conclude that a loop has been installed in the design process. This is not unusual. Figure 14.4 might very well have a number of arrows to show that problem solvers do return to steps in the design process that have supposedly been completed. In this particular case we can assume that our students have redefined their problem even though they do not say so and do not report having undertaken any additional research. Again, this is not unusual.

14.7 Alternative Solutions—Step 6

Suppose that you are chief engineer for a manufacturing company and are faced with appointing someone to the position of director of product testing. This is an important position because all the company's products are given rigorous testing under this person's direction before they are approved. You must compare all the candidates with the job description (criteria) to see who would do the best job. This seems to be a ridiculously simple procedure, doesn't it? Well, we think it does too, but many times such a process is not followed and poor appointments are made. In the same way that the list of candidates for the position has to be made, so must we produce a list of possible answers to our problem before we can go about the job of selecting the best one.

14.7.1 The Nature of Invention

The word "invention" strikes fear into the minds of many people. They say, "Me, an inventor?" The answer is "Why not?" One reason that we do not fashion ourselves as inventors is that some of our earliest teaching directed us to be like other children. Since much of our learning was by watching others, we learned how to conform. We also learned that if we were like the other kids, no one laughed at us. We can recall in our early days, even preschool, that the worst thing that could happen would be if people laughed at us. We will bet that most of you have a similar feeling when you say something that is not too astute, and it is followed by smiles and polite laughs. Moreover, we do not like to experiment because many experiments fail. It is a very secure person who never has to try something that he or she has not already done well. Think about it: When you were in the first few grades at school, didn't you feel great when you were called on by the teacher and you knew the answer? Don't you, even today, try to avoid asking your professors a question because you do not want the professor or your classmates to know that you do not know the answer? Most of us like to be in the majority. Please do not assume that we are saying that the majority of the people are wrong or that it is bad to be like other people. However, if we dwell on such behavior, then we will never do anything new. A degree of inventiveness or creativity is essential if we are to arrive at solutions to problems that are better than the way things are being done now. If we can remove the blocks to creativity, then we have a good chance of being inventive.

The father of one of the authors had a motto above his desk that read as follows:

> Life's greatest art,
> Learned through its hardest knocks,
> Is to make
> Stepping stones of stumbling blocks.

He does not know whether or not this was original with his father, but he remembers it after not having seen it for over 20 years, and it surely applies to the process of developing ideas.

14.7.2 Building the List

There are a great number of techniques that can be used to assist us in developing a list of possible solutions. Three of the more effective methods will be briefly discussed.

Checkoff lists, designed to direct your thinking, have been developed by a number of people. Generally, the lists suggest possible ways that an existing solution to your problem might be changed and used. Can it be made a different color, a different shape, stronger or weaker, larger or smaller, longer or shorter, of a different material, reversed or combined with something else? It is suggested that you write your list down on paper and try to conceive of how the current solution to the problem might be if you changed it according to each of the words on your list. Ask yourself: Why is it like it is; will change make it better or worse; did the original designers have good reason for doing what they did or did they simply follow the lead of their predecessors?

Consider, as an application of checkoff lists, the log-splitter solution shown in Fig. 14.10. Use the checklist words "modify" and "rearrange" to guide or focus our efforts to obtain new solutions.

modify?	Use electric power instead of gas.
	Mount in the box of a pickup truck.
rearrange?	Vertical cylinder action (easier to handle logs).
	Wedge moved by cylinder (not fixed as shown).

Morphological listing gives a visual conception of the possible combinations that might be generated. These listings are usually shown as grids or diagrams. It is easy to visualize a rectangular prism as shown in Fig. 14.12. This example indicates that we are

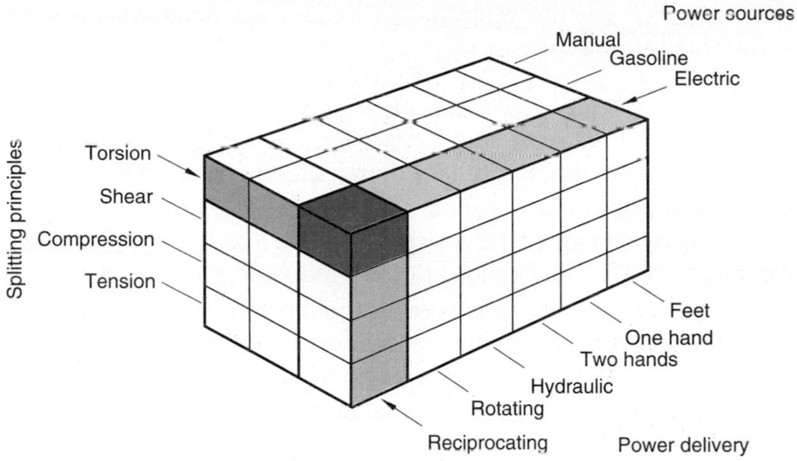

Figure 14.12
A morphological chart. There are 72 combinations of the three alternatives. The shaded segment, for example, would employ an electric reciprocating device that splits by torsion.

considering the log-splitting problem as composed of three subdivisions: power source, power delivery, and splitting principles. These are then subdivided, as shown in Fig. 14.12. The prism produces 72 different combinations such as the one indicated by the shaded volume. Here we would have the logs split by torsion applied by a reciprocating motion generated by an electric power source. Surely you can think of more than the three major subdivisions shown in Fig. 14.12 and can provide additional ideas for each of them.

Brainstorming is a technique that has received wide discussion and support. The mechanics of a brainstorming session are rather simple. The leader states the problem clearly and ideas about its solution are invited. The length of productive sessions varies, but it is usually in the half-hour range. Often it takes a few minutes for a group to rid itself of its natural reserved attitude. But brainstorming can be fun, so choose a problem area and try it with some friends. Be prepared for a surprise at the number of ideas that we will develop (see Fig. 14.13).

Figure 14.13
A student design team brainstorms for ideas to detect and resist motion in two directions, a requirement for their design of a stepper-climber exercise machine.

There are many descriptions of this process, most of which can be summed up as follows:

1. The size of the group is important. We have read of successful groups that range from 3 to 15; however, it is generally agreed that 6 to 8 is an optimum number for a brainstorming session.
2. Free expression is essential. That is what brainstorming is all about. Any evaluation of the exposed ideas is to be avoided. Nothing should be said to discourage a group member from speaking out.
3. The leader is a key figure, even though free expression is the hallmark. The leader sets the tone and tempo of the session and provides a stimulus when things begin to drag.
4. The members of the group should be equals. No one should feel any reason to impress or support any other member. If your supervisor is also a member, you must steer clear of concern for his or her feelings or support for his or her ideas.
5. Recorders are necessary. Everything that is said should be recorded, mechanically or manually. Evaluation comes later.

We have discussed a few techniques that are recommended to stimulate our thought processes. You may choose one of the free-wheeling techniques or perhaps a well-defined method. Regardless of your preferences, we think you will be pleased and even surprised at the large list of ideas that you can develop in a short period of time.

14.7.3 The Chapter Example—Step 6

Our team of engineering students approached their task of generating ideas by setting up a brainstorming session. Their minds were already tuned in on the problem; they had produced a list of candidates for the ultimate decision. They admit that they erred by giving preliminary evaluation of some of the ideas, which is a hard rule not to violate. The following is a list of ideas as it appeared in their written report. It is not, however, their total list:

1. Hydraulic cylinder (vertical or horizontal) used as a method to apply force.
2. Auto-jack or fence-tightener concept in order to apply pressure through a mechanical advantage (see Fig. 14.14).
3. Use of compressed air to force the wedge through the log.
4. Adaptations of conventional hand tools such as the axe, mall, or wedge.
5. Power or manual saws.
6. Heavy pile driver with a block and tackle for raising the weight.
7. High-voltage arc between electrodes; similar to a lightning bolt.

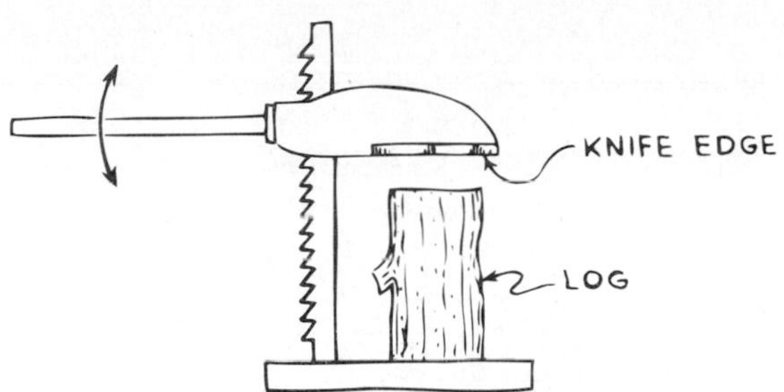

Figure 14.14
Thumbnail sketches are often helpful in describing ideas.

8. Spring-powered wedge using either compression or tension.

9. Sliding mass that drives the wedge into the wood.

10. Drop the wedge from the elevated position on the log.

11. Electronic sound that produces compression waves strong enough to split logs.

12. Wedge driven by explosive charge.

13. Spinning hammermill that breaks by shearing and concussion like a rock crusher.

14. Separate or split with intensive concentrated high energy such as a laser beam.

15. Force a conical wedge into a log and apply a torsional force.

16. Use a large mechanical vice with one jaw acting as a wedge.

17. Drill core (hole) in wood, fill with water, cap, and freeze.

18. Cut wood into slabs rather than across the grain.

19. Apply a force couple to the ends of a log, causing a shearing action.

20. Drop a log from an elevated position onto a fixed wedge.

14.8 Analysis—Step 7

At this point in the design process we have defined the problem, expanded our knowledge of the problem with a concentrated search for information, established constraints for a solution, selected criteria for comparing solutions, and generated alternative solutions. In order to determine the best solution in light of available knowledge and criteria, the alternative solutions must be analyzed to determine performance capability. Analysis thus becomes a pivotal point in the design process. Potential solutions which do not prove out during the analysis phase may be discarded or, under certain conditions, may be retained with a redefinition of the problem and a change in constraints or criteria. Thus, one may need to repeat segments of the design process (Fig. 14.4) after completing the analysis.

Analysis involves the use of mathematical and engineering principles to determine the performance of a solution. Consider a system, such as the cantilever beam in Fig. 14.15a, constrained by the laws of nature. When there is an input to the system (the applied load P), analysis will determine the performance of the

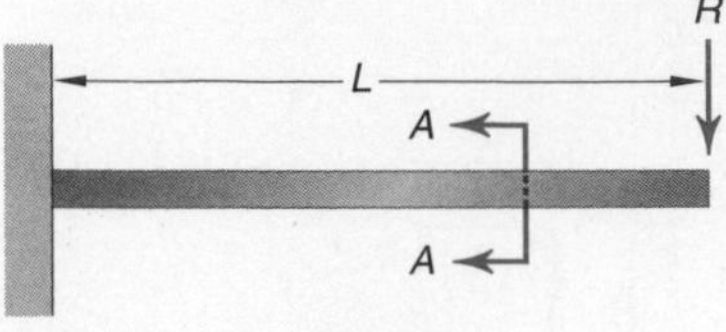

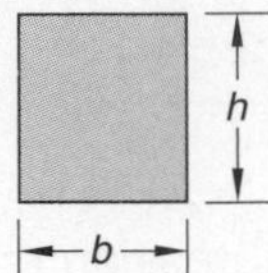

Figure 14.15 (a)

	P, N	L, m	h,m	b,m	E, Pa	I, m^4	d,m
	1.00E+05	4	0.1	0.2	2.07E+11	1.67E-05	0.618357
	1.00E+05	4	0.2	0.2	2.07E+11	0.000133	0.077295
	1.00E+05	4	0.3	0.2	2.07E+11	0.00045	0.022902
	1.00E+05	4	0.4	0.2	2.07E+11	0.001067	0.009662
	1.00E+05	4	0.5	0.2	2.07E+11	0.002083	0.004947
	1.00E+05	4	0.6	0.2	2.07E+11	0.0036	0.002863
	CANTILEVER BEAM DEFLECTION FOR RECTANGULAR SECTION						

Figure 14.15 (b)

system (beam)—that is, the deflection, stress buildup, etc. Keep in mind that the objective of design is to determine the best solution (system) to a need.

The following example should be studied for the process rather than the content. The goal is to obtain quantitative information for the decision step in the design process.

Example problem 14.1 Determine the deflection of the beam in Fig. 14.15 under the following conditions. Assume the beam is structural steel.

$L = 4.0$ m

$h = 0.40$ m

$b = 0.20$ m

$P = 1.0 \times 10^5$ N

Solution The deflection of the end of a cantilever beam for the configuration shown is given by

$$d = \frac{PL^3}{3EI} \qquad \text{(constraint equation)}$$

where

d = deflection, m

E = modulus of elasticity, a material constant, Pa

$= 2.07(10^{11})$ Pa for structural steel

I = moment of inertia, m^4

For a rectangular cross section

$$I = \frac{bh^3}{12}$$

$$= \frac{(0.2)(0.4)^3}{12}$$

$$= 1.067(10^{-3})m^4$$

Therefore,

$$d = \frac{(10^5)(4)^3}{3(2.07)(10^{11})(1.067)(10^{-3})}$$

$$= 9.66(10^{-3})m$$

$$= 9.7 \text{ mm}$$

This result would be forwarded to the designer for incorporation into the decision phase. The time required to produce an analysis is critical to the design process. If it takes longer to do an analysis than the schedule (Fig. 14.5) permits, then the results are somewhat meaningless. The engineer must exercise some judgment in selecting the method of analysis in order to assure results within the time limit. You can visualize the potential of computers in the analysis effort. Many alternatives can be investigated in a brief amount of time. It is a simple task to change any of the parameters—*b, h, P, L,* or *E*—and to see the effect on the deflection immediately. Fig. 14.15(b) shows a spreadsheet printout of cantilever beam deflections for a rectangular section for varying depths of the section while holding all other values constant. It would have been possible to produce plots of each parameter vs. deflection by using computer graphics. It is obvious that the more possibilities one can investigate, the better the problem is understood and the better the design will be.

We will discuss the beam of Example prob. 14.1 in more detail in Sec. 14.8 to illustrate further the role of analysis in engineering design.

Analysis performed by engineers in most design projects is based on the laws of nature, the laws of economics, and common sense.

14.8.1 The Laws of Nature

You have already come into contact with many of the laws of nature, and you will no doubt be exposed to many more. At this point in your education you may have been exposed to the conservation principles: the conservation of mass, of energy, of momentum, and of charge. From chemistry you are familiar with the laws of Charles, Boyle, and Gay-Lussac. In mechanics of materials Hooke's law is a statement of the relationship between load and deformation. Newton's three principles serve as the basis of analysis of forces and the resulting motion and reactions.

Many methods exist to test the validity of an idea against the laws of nature. We might test the validity of an idea by constructing a mathematical model, for example. A good model will allow us to vary one parameter many times and to examine the behavior of the other parameters. We may very well determine the limits within which we can work. Other times we will find that our boundary conditions have been violated, and therefore the idea must be modified or discarded.

Results of an analysis of a mathematical model are frequently presented as graphs. Often the slopes of tangents to curves, points of intersection of curves, areas under or over or between curves, or other characteristics provide us with data that can be used directly in our designs.

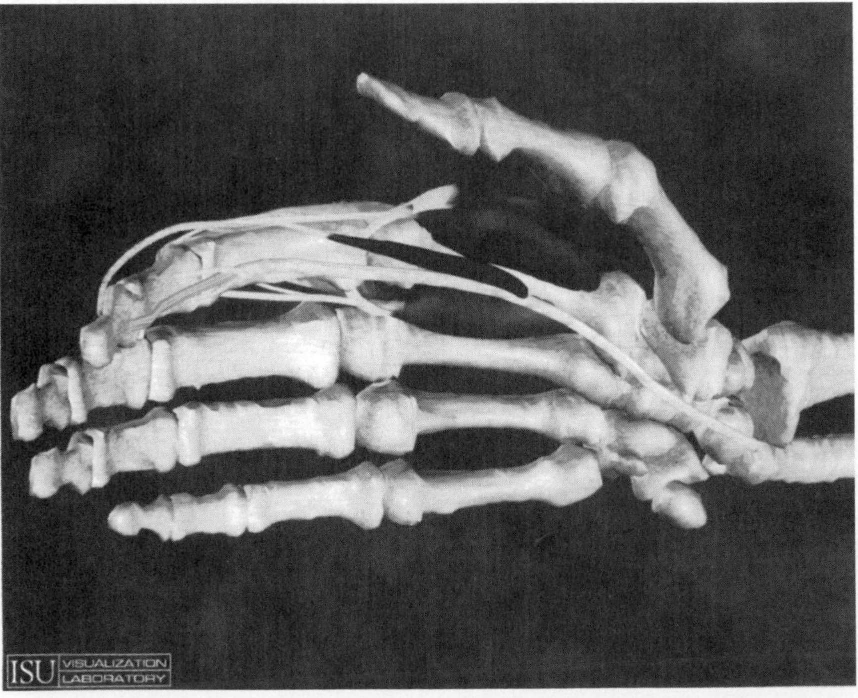

Figure 14.16 Advanced solid modeling software is capable of generating very complex models such as that of this human hand. Note the tendons and muscle fiber.

Figure 14.17
Students obtain a plot of data directly as the experiment progresses. The experiment can quickly be adjusted or redone if the plot does not follow theoretical predictions.

Computer graphics enables a mathematical model to be displayed on a screen. As parameters are varied, the changes in the model and its performance can also be quickly displayed to the engineer.

The preparation of scale models of proposed designs is often a necessary step. This can be a simple cardboard cutout or it can involve the expenditure of great sums of money to test the model under simulated conditions that will predict how the real thing will perform under actual use. A prototype or pilot plant is sometimes justified because the cost of a failure is too great to chance. Such a decision usually comes only after other less expensive alternatives have been shown to be inadequate.

You probably have surmised that the more time and money that you allot to your model, the more reliable are the data that you receive. This fact is often distressing because we want and need good data but have to balance our needs against the available time and money.

14.8.2 The Laws of Economics

Section 14.8.1 introduced the idea that money and economics are part of engineering design and decision making. We live in a society that is based on economics and competition. It is no doubt true that many good ideas never get tried because they are deemed to be economically infeasible. Most of us have been aware of this condition in our daily lives. We started with our parents explaining why we could not have some item that we wanted because it cost too much. Likewise, we will not put some very desirable component into our designs because the value gained will not return enough profit in relation to its cost.

Industry is continually looking for new products of all types. Some are desired because the current product is not competing well in the marketplace. Others are tried simply because it appears that people will buy them. How do manufacturers know that a new product will be popular? They seldom know with certainty. Statistics is an important consideration in market analysis. Most of you will find that probability and statistics are an integral part of your chosen engineering curriculum. The techniques of this area of mathematics allow us to make inferences about how large groups of people will react based on the reactions of a few. It is beyond our study at this time to discuss the techniques, but industry routinely employs such studies and invests millions of dollars based on the results.

14.8.3 Common Sense

We must never allow ourselves the luxury of failing to check our work. We must also judge the reasonableness of the result.

During the 1930s, the depression years, a national magazine conducted a survey of voters and predicted a Republican victory. The magazine was wrong, and the public lost confidence in it to the point that the magazine went out of business. The editors had sampled the population by taking all the telephone books in the United States and, by a system of random numbers, selecting people to be called. They then applied good statistical analysis and made their prediction. Why did they miss so badly? It is a bit hard for us today to imagine the depression years, but the facts are that large percentages of voters did not have telephones, so this economic class of people was not included in the analysis. But this group of people did vote, and they voted largely Democratic. Nothing was wrong with the analytical method, only the basic premise. The message is rather obvious: No matter how advanced our mathematical analysis, the results cannot be better than our basic assumptions. Likewise, we must always test our answers to see if they are reasonable.

14.8.4 The Chapter Example—Step 7

Our design team generated many ideas for splitting wood, 20 of which are listed in Sec. 14.7.3. In addition, the criteria had been previously determined (see Sec. 14.6.2) and value decisions made with regard to evaluating the importance of each criterion. At this point decisions had to be made to reduce the number of alternative solutions. The time available for completing the design project did not allow the team the luxury of thorough analysis of each of the ideas. Therefore, a decision was made to reduce the number of alternative solutions to five. These five were then investigated and developed in more detail. The following is the result of the team's analysis.

Analysis of Alternative Solutions

(Items marked with asterisks were kept by the team for further development.)

1. Hydraulic cylinder (vertical or horizontal)
   a. Extreme cost for materials and manufacturing
   b. High operational and maintenance costs
   c. Nonportable for one person
   d. Lack of standard parts

***2.** Auto-jack principle or fence tightener (force by creating a mechanical advantage)
   a. Reasonably portable
   b. Minimum manual labor required

3. Use of compressed air (pneumatic)
   a. Minimum portability
   b. Extensive material, manufacturing, and operational costs

4. Adaptations of conventional hand tools such as the axe, mall, or wedge
   a. Inefficient operation
   b. Is the current solution
   c. Unsafe for an inexperienced user

5. Power or manual saws
   a. High cost of materials and manufacturing
   b. Not a low-volume solution

6. Heavy pile driver with block and tackle used to raise weight
   a. Not portable
   b. Expensive

7. High-voltage arc between electrodes, similar to lightning bolt
   a. Inefficient
   b. Expensive
   c. Impractical to use

***8.** Spring-powered wedge using either compression or tension
   a. Relatively easy to use
   b. Portable
   c. Low manufacturing cost

***9.** Sliding mass that drives a wedge into wood
   a. Good for low-volume usage
   b. Portable
   c. Low initial cost and operational costs

*10. Drop a wedge from an elevated position onto the log
  a. Uses a mechanical advantage
  b. Simple construction
  c. Good for low-volume production

11. Electronically produced sound that produces compression waves strong enough to split logs
  a. Impractical because of other damage that could be done
  b. Dangerous for average person to use

*12. Wedge driven by explosive charge
  a. Minimum work
  b. Low cost
  c. Not easily portable

13. Hammermill that would chop wood much like a coal or rock crusher
  a. Expensive
  b. Much waste material (chips)
  c. Not easily portable

14. Separate or split with concentrated high energy from a laser beam
  a. Expensive
  b. Potentially dangerous

15. Force a conical wedge into log and apply a torsional force
  a. Complicated mechanical design for a single piece of equipment
  b. Expensive

16. Use a large vise with one jaw acting as a wedge
  a. Slow operating if powered by human
  b. Probably not easily portable

17. Drill core (hole) in wood, fill with water, cap, and freeze
  a. Time-consuming
  b. Inefficient

18. Cut wood into slabs rather than splitting
  a. Inefficient
  b. Not suitable for ordinary fireplace

19. Apply a force couple to the ends of a log, causing a shearing action
  a. Expensive
  b. Inefficient
  c. Destructive to wood

20. Drop a log from an elevated position onto a fixed wedge
  a. High amount of manual labor required
  b. Inefficient

You will note that the analysis is based solely on the stated criteria and at best is very general. Many of the comments made show that no computations involving the mechanics of wood splitting were made. No testing (or test data) was used for any of the potential solutions. [A professional engineer would almost surely have spent considerable time and money on models, simulations, prototypes, and tests to verify (or disprove) his or her ideas about the many aspects of the concepts before attempting a decision.] You should decide for yourself whether the team should be criticized at this point in their design effort. Remember, the team had only a few weeks together and limited experience in engineering analysis courses.

We will list at this time a few analyses that may have been made by beginning engineering students. You may agree or disagree with some of them and perhaps add other items.

1. Determine the force necessary to split a log of a given type of wood. This may be done experimentally or analytically. Some data on this exist in the literature.

2. Determine the stamina of a human being with regard to lifting a specified weight a number of times in a given time period. This would be valuable information for the manually operated splitters.

3. For impact-type splitters, make a preliminary estimation of impulse required to split logs. That is, ascertain what combinations of mass and velocity are needed to cleave the log.

4. Find out if there is a wedge angle that would be more efficient than others.

5. Decide what masses we are talking about with respect to the alternative solutions. Since portability is a major consideration, locate or generate some data to be able to make comparisons based on the total mass of the various devices.

We will illustrate the analysis necessary to estimate the sliding mass that would drive a wedge into wood (alternative solution 9 given above).

Team members who are just beginning their study of engineering may find the following analysis difficult to understand completely. A better understanding will come after engineering physics, statics, and dynamics. Another method to obtain a reasonable value for the mass of the wedge is to conduct an experiment using the concept shown. Still another method is to consult an instructor or upper-level engineering student to guide the team through the analysis.

Example problem 14.2 Estimate the mass required for the sliding-mass wood-splitter.

Solution We assume that the peak force is the critical parameter in causing the wood fibers to separate. We know from experience that logs can be split with a wedge driven by a sledge hammer (see Fig. 14.18a). We will attempt to generate the same impact

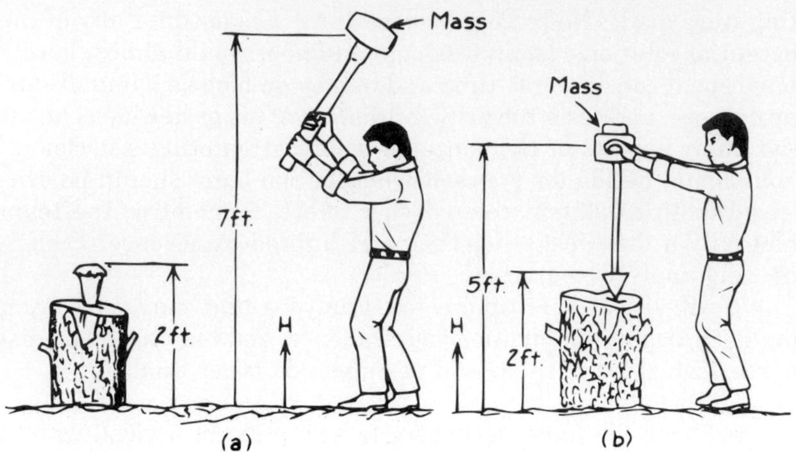

Figure 14.18

momentum with the sliding mass as with the sledge. The configuration for the sliding mass is shown in Fig. 14.18b. The problem will be illustrated using British units since that is what the design team used for their work.

From physics

Change in momentum = impulse

$$\Delta mV = F\,\Delta t \text{ for a constant force}$$

Energy change = work done

$$\Delta KE + \Delta PE = F\,\Delta H \text{ for a constant force}$$

where m = mass, slugs

V = velocity, ft/s

F = force, lbf

t = time, s

KE = kinetic energy, lbm · ft^2/s^2

$= \frac{1}{2}mV^2$

PE = potential energy, lbm · ft^2/s^2

$= mgH$

g = acceleration of gravity, ft/s^2

H = height, ft

Assumptions (refer to Fig. 14.18a):

Mass of sledge = 8/32.2 slug

= 0.2484 slug

$$\text{Peak height} = 7 \text{ ft}$$

$$\text{Average wedge height} = 2 \text{ ft}$$

$$\text{Initial velocity} = 0$$

$$\text{Downward force generated} = 10 \text{ lbf}$$

Using the subscripts i for initial conditions and f for final conditions, we have for the sledge configuration

$$\Delta KE + \Delta PE = F \, \Delta H$$

$$\frac{1}{2}m(V_f - V_i)^2 + mg(H_f - H_i) = F(H_f - H_i)$$

$$\frac{1}{2}mV_f^2 = F(H_f - H_i) + mg(H_i - H_f)$$

$$\frac{1}{2}(0{\cdot}248\ 4)\ V_f^2 = (-10)\ (2 - 7) + (0.248\ 4)(32.2)(7 - 2)$$

$$V_f^2 = \frac{50 + 40}{0.1242}$$

$$= 724.64$$

$$V_f = 26.92 \text{ ft/s}$$

Thus, the change in momentum of the sledge is

$$\Delta mV = m(V_f - V_i)$$

$$= 0.248\ 4(26.92 - 0)$$

$$= 6.687 \text{ slug} \cdot \text{ft/s}$$

Now with the subscript s representing the sliding mass, we make the following assumptions (refer to Fig. 14.18b):

$$\text{Peak height} = 5 \text{ ft}$$

$$\text{Average wedge height} = 2 \text{ ft}$$

$$\text{Downward force generated} = 10 \text{ lbf}$$

$$V_{is} = 0$$

Therefore, for the sliding mass configuration

$$\Delta KE + \Delta PE = F\Delta H$$

$$\frac{1}{2}m_s(V_{fs} - V_{is})^2 + m_s g(H_{fs} - H_{is}) = F_s(H_{fs} - H_{is})$$

For the same impact momentum

$$m_s V_{fs} = 6.687 \text{ slug} \cdot \text{ft/s}$$

$$\frac{1}{2}m_s V_{fs}^2 = m_s(32.2)(5 - 2) + (-10)(2 - 5)$$

$$\frac{m_s}{2}\left(\frac{6.687}{m_s}\right)^2 = 96.6m_s + 30$$

$$22.36 = 96.6m_s^2 + 30m_s$$

$$96.6m_s^2 + 30m_s - 22.36 = 0$$

Solving for m_s and discarding the unrealistic answer, we obtain

$$m_s = 0.350 \text{ slug}$$
$$= 11.3 \text{ lbm}$$

14.9 Decision—Step 8

As we mentally review our own professional practices, we can honestly say that the most difficult times for us have not been when the analysis of a problem was difficult but, rather, when it required a "tough" decision. We have known many engineers who are technically knowledgeable but are unable to make a final decision. They may be happy to suggest several possible solutions and to outline the strong and weak points of each—indeed they may feel that their function is to do just that—but let someone else decide which course is to be followed. The truth of the matter is that most engineering assignments require both: providing information and making decisions.

What makes reaching a decision so difficult? The answer is *trade-offs*. If we can be certain about anything in the future, it is that with your decisions, the necessity to compromise will come. Review the criteria in Sec. 14.5.2 that our team selected. In order to sell the log-splitter that they are to design, it must be competitive in cost—so the lower the cost, the better. If the engineers wish to make it more durable, chances are that they will use materials that are expensive or use more material (heavier construction perhaps). If they use more material, they are adding weight, which limits portability.

Each time one criterion is optimized, another moves away from its optimum position. If the relationships are complicated, you may have to go through very complex processes to reach a decision. You can be certain that no one idea will be better in all respects than all the others; hence, you may have to choose a concept that you know is inferior to others in one or more of the decision criteria.

14.9.1 Organization for Decision

In order to decide among several alternatives, you need as much information as possible about each. In design you need information in order to evaluate each alternative against each of the criteria. Analysis can provide the answers, as described in Sec. 14.7. If time and money are available, a prototype may be constructed and tested. In most cases judgment must be made with much less information. Computer models and engineering drawings are used to describe the form and function of the design.

Whatever information is available, it should fairly and accurately represent all the alternatives so that an equitable decision can be made.

14.9.2 The Chapter Example—Step 8

The analysis process (see Sec. 14.7.4) reduced our student team's list of ideas to five, which are repeated here for reference.

1. Auto-jack principle—the pressure wedge (item 2)
2. Spring-powered wedge (item 8)
3. Sliding mass (item 9)
4. Wedge dropped from an elevation (item 10)
5. Wedge driven by explosive charge (item 12)

Each team member was assigned the difficult and creative task of taking one of the five very general ideas and beginning the process of shaping it into physical form. Again, the student design team was limited in time.

A solution or physical shape for any of the five ideas listed can assume an infinite number of different forms. Two of the five solutions are illustrated in this chapter: idea 1, the pressure wedge, and idea 3, the sliding mass. The other three areas were also developed but are not illustrated here.

Each student took one possible solution and prepared a series of idea sketches. An idea sketch can be a single view, a multiview, or a pictorial. The sketches include little detail but clearly depict the form and function of the idea. Figure 14.19 illustrates three of the better idea sketches created for the pressure wedge idea.

A second student took the sliding-mass idea and developed the three separate sliding configurations illustrated in Fig. 14.20.

Obviously, the end product will be a refinement of this effort, so as many idea sketches as possible should be developed in the available time. When no additional time remains for analysis, it is necessary to evaluate each idea sketch. This phase involves considerable detail and is often referred to as "concept development." The final concepts must be developed to the point that comparative judgments can be intelligently made in evaluating each concept in light of the criteria.

Concept sketches are shown for both the pressure wedge (Fig. 14.21) and the sliding mass (Fig. 14.22).

14.9.3 Criteria in Decision

The objective of the entire design process is to choose the best solution for a problem within the time allowed. The steps that pre-

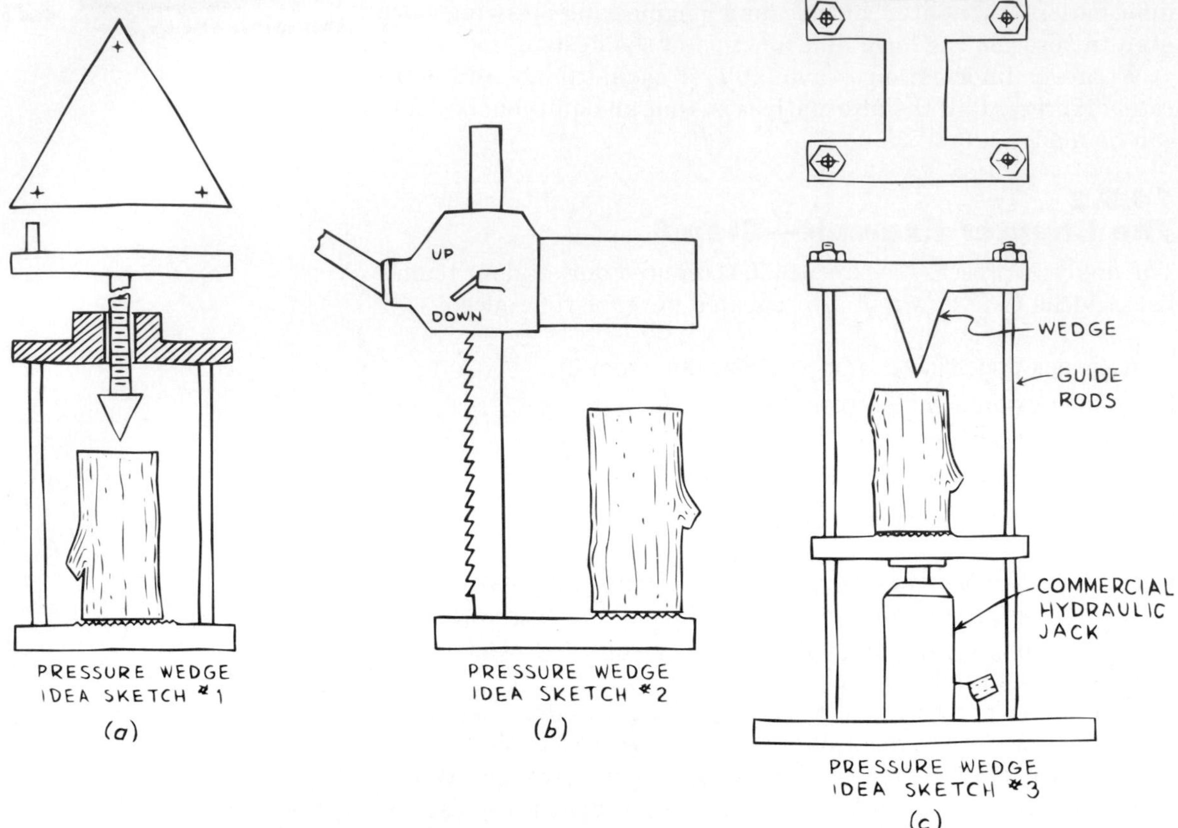

Figure 14.19
Three different idea sketches for the pressure wedge.

cede the decision phase are designed to give information that leads to the best decision. It should be quite obvious by now that poor research, a less-than-adequate list of alternatives, or inept analysis would reduce one's chances of selecting a good, much less the best, solution. Decision making, like engineering itself, is both an art and a science. There have been significant changes during the past few decades that have changed decision making from being primarily an art to what it is at the present, with probability, statistics, optimization, and utility theory all routinely used. It is not our purpose to explore these topics, but simply to note their influence and to consider for a moment our task of selecting the best of the proposed solutions to our problem. The term "optimization" is almost self-explanatory in that it emphasizes that what we seek is the best, or optimum, value in light of a criterion. As you study more mathematics, you will acquire more powerful tools through calculus and numerical methods for optimization.

In order to illustrate optimization, we will return to the beam problem illustrated in Fig. 14.15 and Example prob. 14.1. Our objective will be to determine the least mass of the beam for prescribed performance conditions. You will recall in our discussion

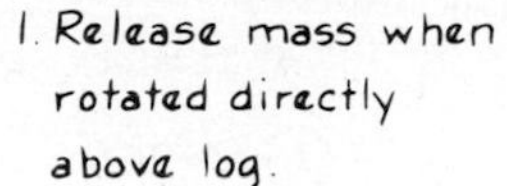

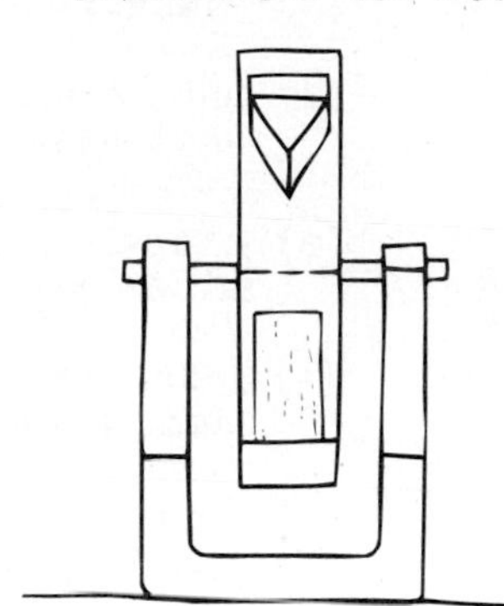

SLIDING MASS
IDEA SKETCH #1

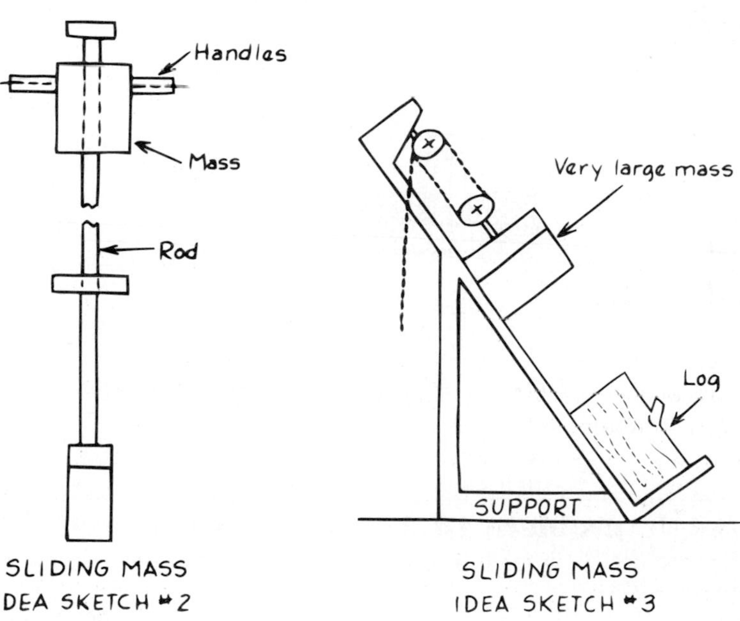

Figure 14.20
Three different idea sketches for the sliding mass.

of analysis that a system, the laws of nature and economics, an input to the system, and an output are involved. Analysis gives the output if the system, laws, and input are known.

If we consider the inverse problem—that is, if we were looking for a system, given the laws, input, and output—we would be using *synthesis* rather than analysis. Synthesis is not as straightforward as analysis since it is possible to have more than one system that will perform as desired. But if we specify a criterion for selecting the best solution, then a unique solution is possible.

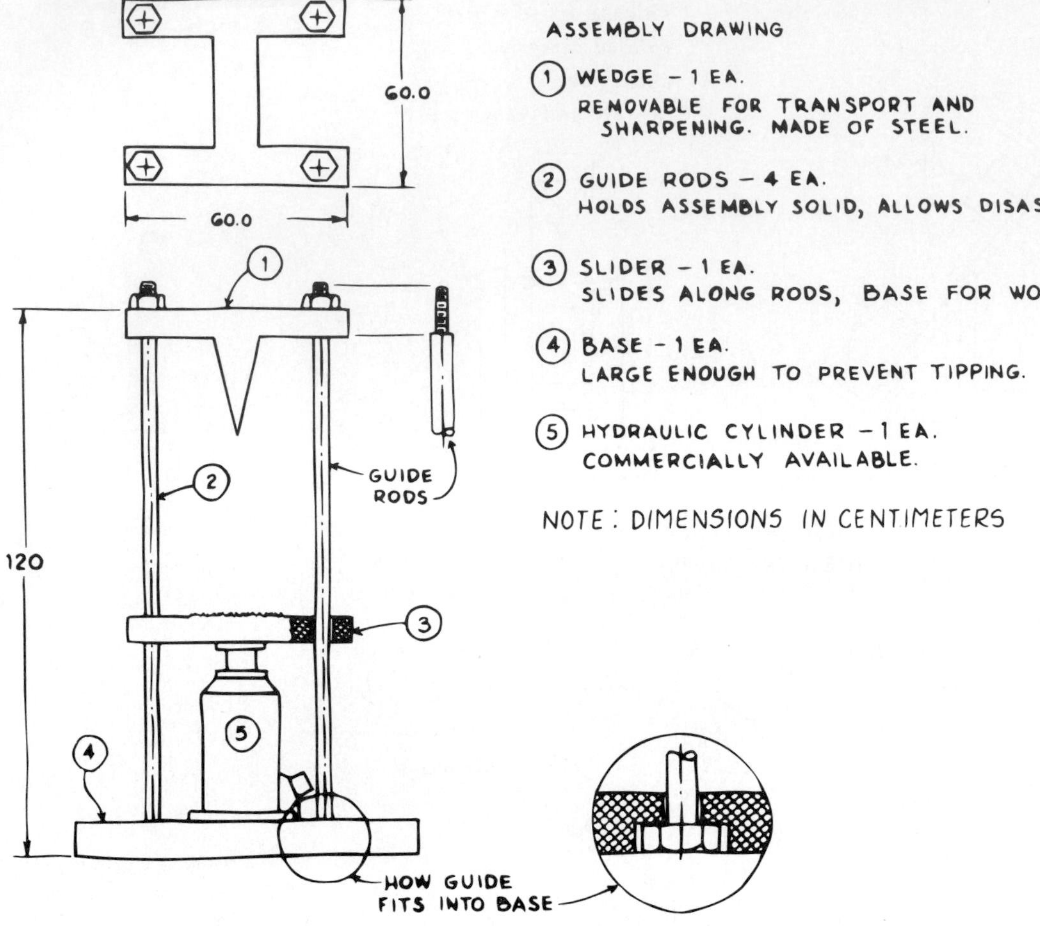

Figure 14.21
Concept development of the pressure-wedge idea, sketch no. 3 from Fig. 14.19.

The criterion used for selecting the best solution is often called a payoff function.

Example problem 14.3 (Refer to Fig. 14.15.) Determine the dimensions *b, h* for the least beam mass under the following conditions:

The deflection cannot exceed 40 mm.

The height *h* cannot be greater than three times the base *b*.

$E = 2.0 \times 10^{11}$ Pa

$L = 4.0$ m

$P = 1.0 \times 10^{5}$ N

If the beam has a constant cross section throughout, then the mass is a minimum when the cross-sectional area $A = bh$ is a minimum. Achieving minimum mass by finding minimum area (payoff function) will provide the best (optimum) solution.

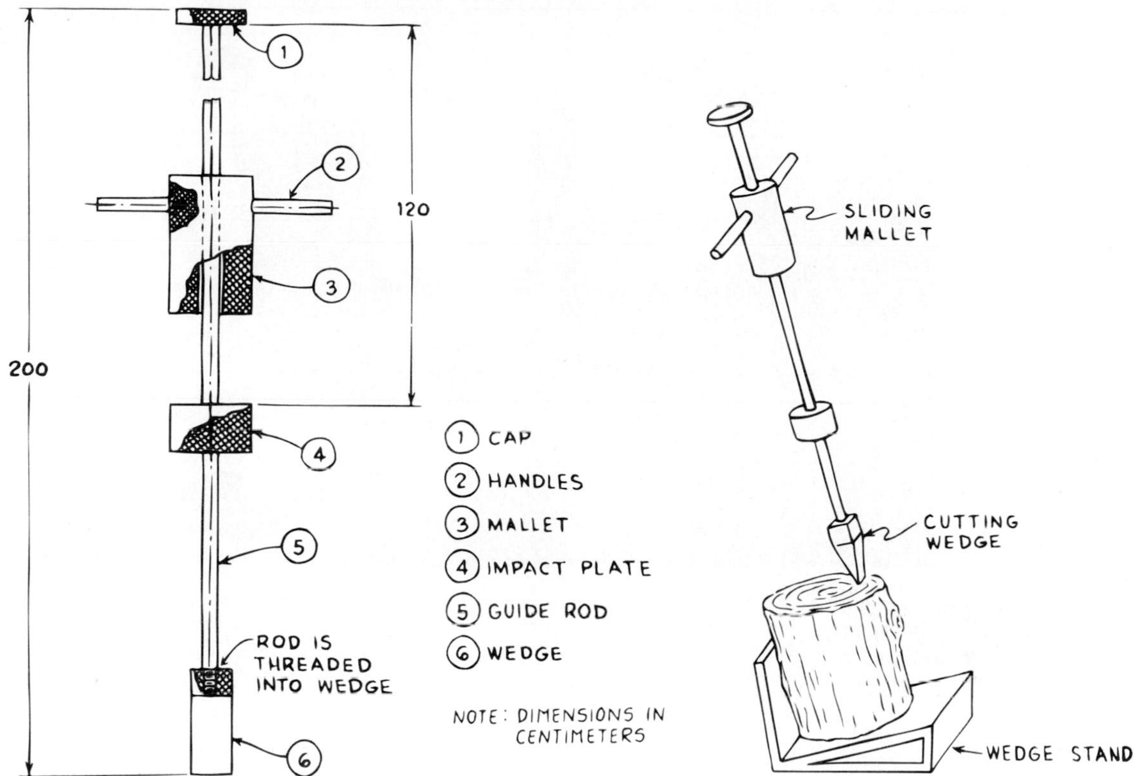

Figure 14.22
Concept development for the sliding-mass idea, sketch no. 2 from Fig. 14.20. This concept was selected as the best solution by the team.

Solution The system we are after is the beam shape $b \times h$ within the conditions specified above; the law is the deflection equation from Example prob. 14.1; the inputs are L, P, and E, and the output is the range of permissible deflection. The deflection equation becomes

$$d = \frac{PL^3}{3EI}$$

$$0.04 = \frac{10^5(4)^3}{3(2)(10^{11})I}$$

or

$$I = 2.667(10^{-4})\text{m}^4$$

Then

$$\frac{bh^3}{12} = 2.667(10^{-4})$$

Thus,

$$b = \frac{3.2(10^{-3})}{h^3}$$

Figure 14.23
Two members of an electrical engineering design team ponder results of a computer analysis.

This equation is a relationship for the beam under the condition that the deflection is a constant 40 mm. The expression is plotted in Fig. 14.24a. Note that values to the right of the curve represent beam dimensions for which the deflection would be less than 40 mm. Those to the left would cause the deflection to exceed 40 mm; thus, that portion of the *design space* for b and h is invalid.

Next we demonstrate the effect of the required relationship between b and h by plotting the line $b = h/3$, as shown in Fig. 14.24a. Points above this line represent valid geometric configurations; those below do not.

Now we have a better picture of the design space, or solution region for our problem. A point $h = 0.3$ m, $b = 0.3$ m, represents a satisfactory solution since it falls within all conditions *except* possibly minimum mass. Many designs stop at this point when a nominal solution has been found. These are the designs that may not survive in the marketplace because they are not optimum. In fact, to get a nominal solution, we could have guessed values for b and h and very quickly had an answer without going to all the effort we have up to this point.

We know the region in which the best solution (minimum area) lies. We again take advantage of the capability of the spreadsheet and have it search for the solution we desire. Figure 14.24(b) shows a table of values for h and the two constraints on the solution, namely a deflection of 40 mm or less and that h cannot be greater than three times b. We set up the spreadsheet to iterate on the payoff function, $A = bh$, within the constraints, and find a minimum. The solution, in this case, is the intersection of the two curves in Fig. 14.24a. Often, we find that optimum solutions lie on the boundary

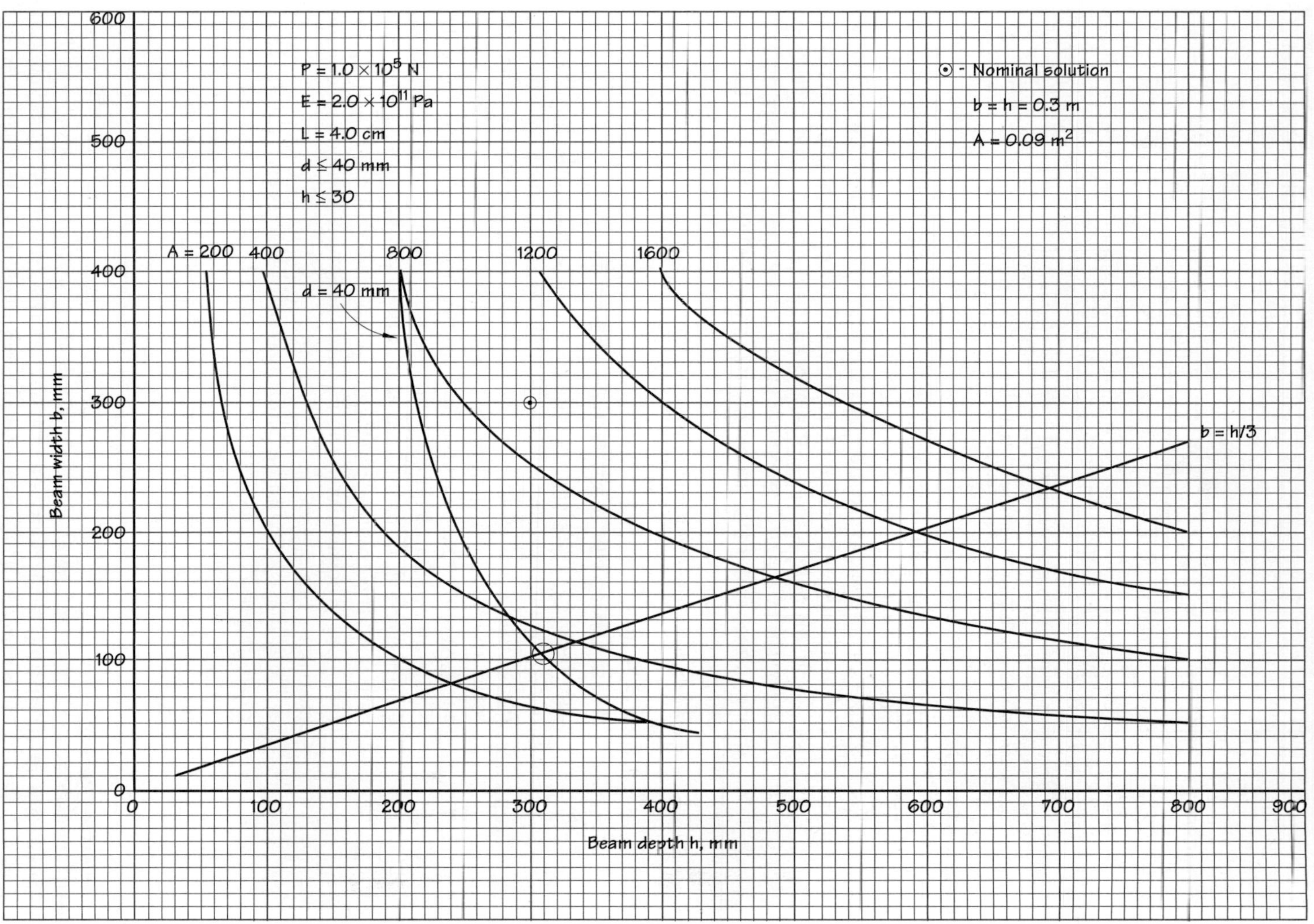

Figure 14.24 (a)

	h, m		0.0032/h^3		b=h/3		
	0.15		0.948148		0.05		
	0.2		0.4		0.066667		
	0.25		0.2048		0.083333		
	0.3		0.118519		0.1		
	0.35		0.074636		0.116667		
	0.4		0.05		0.133333		
	Minimum Area = 0.03266 m^2 when h = 0.313017 m and b = 0.104339 m						

Figure 14.24 (b)

of the design solution space. We also note that the function $A = bh$ does not have a minimum since it is asymptotic to the coordinate axes. However, in most design applications the payoff function will be of a complex nature and not have obvious shape characteristics.

You will note also that the spreadsheet analysis can quickly be extended to analyze cantilever beam designs for any range of loading, sizes, and materials simply by replacing the values in the governing equations. Now that you have seen this capability, you should apply it to appropriate analyses in your engineering courses.

14.9.4 The Chapter Example—Step 8

The method of decision employed by the student design team is one that has considerable merit, and one that you may find simple to use. As mentioned in Sec. 14.6.2, they established six criteria and assigned weights to them. They later examined each of the five surviving concepts and graded them on a 0 to 10 scale. The grade was multiplied by the weight in percent and the points were recorded. A total of 1 000 points would be perfect. The concept with the greatest number of points is considered the best alternative. The results of their evaluation are shown as a decision matrix in Fig. 14.25.

Note that the winning alternative did not receive the highest rating for safety, ease of operation, and durability, and it tied for highest in portability. So our team must report that this alternative has some shortcomings but that it is the best they could find.

14.10 Specification—Step 9

After progressing through the design process up to the point of reaching a decision, many feel that the "romance" has gone out of the object. The suspense and uncertainty of the solution are over, but much work still lies ahead. Even if the new idea is not a breakthrough in technology but simply an improvement in existing technology, it must be clearly defined to others. Many very

Decision matrix

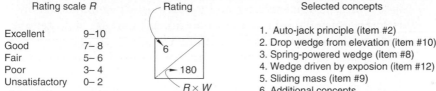

Criteria	Weight *W*, percent	Selected concepts (see below) 1	2	3	4	5	6
Cost	30	6 / 180	7 / 210	7 / 210	7 / 210	9 / 270	
Ease of operation	20	10 / 200	7 / 140	9 / 180	10 / 200	7 / 140	
Safety	15	9 / 135	7 / 105	6 / 90	5 / 75	8 / 120	
Portability	15	6 / 90	5 / 75	4 / 60	10 / 150	10 / 150	
Durability	10	8 / 80	9 / 90	10 / 100	9 / 90	9 / 90	
Use of standard parts	10	7 / 70	8 / 80	8 / 80	6 / 60	9 / 90	
Total	100	775	700	720	785	860	

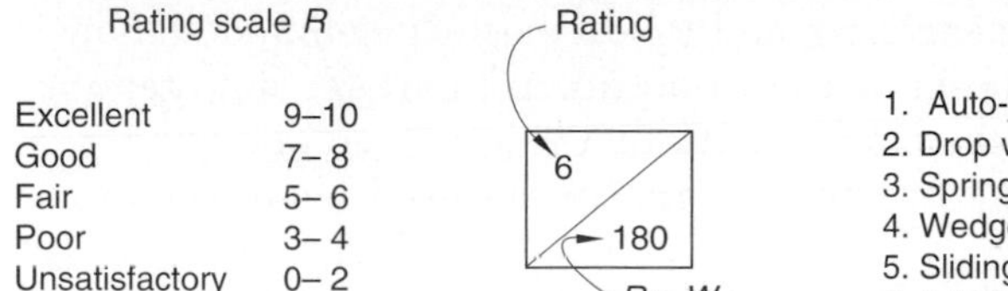

Selected concepts

1. Auto-jack principle (item #2)
2. Drop wedge from elevation (item #10)
3. Spring-powered wedge (item #8)
4. Wedge driven by exposion (item #12)
5. Sliding mass (item #9)
6. Additional concepts

Figure 14.25
Each concept was rated by the team on a scale of 0 to 10 for each criterion. The rating was multiplied by the criterion weight and then summed. Concept 5 was chosen as the optimum even though it did not receive the highest rating for three of the six criteria.

creative people are ill-equipped to convey to others just exactly what their proposed solution is. It is not the time to use vague generalities about the general scope and approximate size and shape of the chosen concept. One must be extremely specific about all the details regardless of the apparent minor role that each may play in the finished product.

14.10.1 Specification by Words

One medium of communication that the successful engineer must master is that of language, written and spoken. Your problems in professional practice would be considerably less complicated if you were required to defend and explain your ideas only to other engineers. Few engineers have such luxury; most must be able to write and speak clearly and concisely to people who do not have comparable technical competence and experience. They may be officials of government who are bound so tightly by budgets and the need for public acceptance that only the best explanation is good enough to pierce their protective armor. They may be people in business who know that capital is limited and that they cannot defend another poor report to stockholders. Without appropriate documentation, they will not be nearly as certain as you are that your idea is a good one. Therefore, this phase of specification—*communication*—is so important that we have assigned it as the final step in the design process. Before discussing it, however, we will discuss another means of communication.

14.10.2 Graphical Specifications

You will probably have many occasions in which to work closely with technicians and drafters as they prepare the countless drawings that are essential to the manufacture of your design. You will not be able to do your job properly if you cannot sketch well enough to portray your idea or to read drawings well enough to know whether the plans that you must approve will actually result in your idea being constructed as you desire.

A lathe operator in the shop, an electronics technician, a contractor, or someone else must produce your design. How is the person to know what the finished product is to look like, what materials are to be used, what thicknesses are required, how it is to work, what clearances and tolerances are demanded, how it is to be assembled, how it is to be taken apart for maintenance and repair, what fasteners are to be used, and so on?

Typical drawings that are normally required include:

1. A sufficient number of detail drawings describing the size and shape of each part
2. Layouts to delineate clearances and operational characteristics
3. Assembly and subassemblies to clarify the relationship of parts
4. Written notes, standards, specifications, etc., concerning quality and tolerances
5. A complete bill of materials

Included with the drawings are almost always written specifications, although certain classes of engineering work refer simply to documented standard specifications. For example, most cities have adopted one of several national building codes, so all structures constructed in those cities must conform to the code. It is quite common for engineers and architects to refer simply to the building code as part of the written specifications and to write detailed specifications for only items that are not covered in the code. This procedure saves time and money for all by providing uniformity in bidding procedures. Many groups have produced

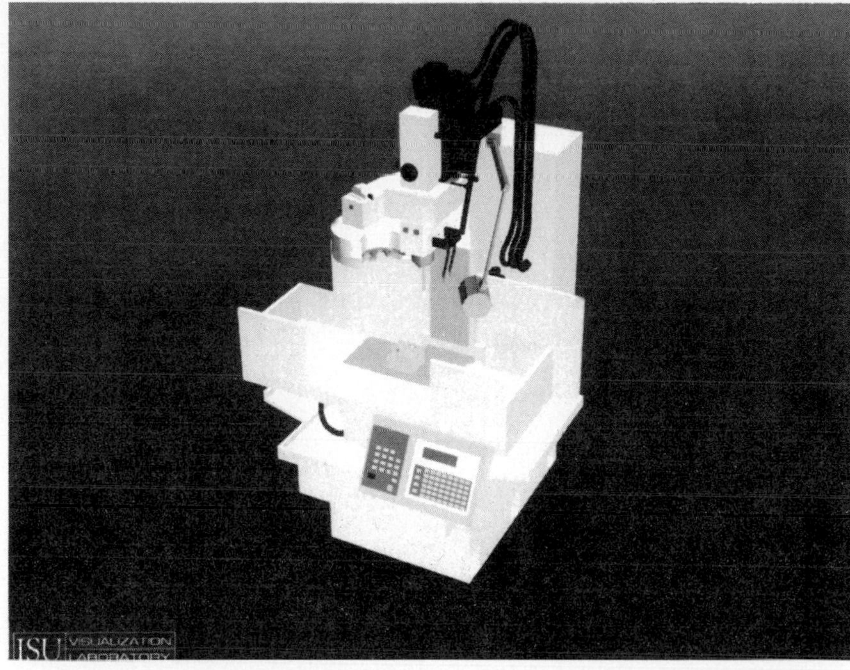

Figure 14.26 Manufacturing capability can be modeled and simulated on the computer as in this example of a milling machine.

standards that are widely recognized. For instance, there are standards for welds and fasteners for the obvious reasons—ease of specification and economy of manufacture. Moreover, there are such standards for each discipline of engineering.

14.10.3 The Chapter Example—Step 9

Reproductions of two drawings prepared by the student team are given as Figs. 14.27 and 14.28. The following drawings were part of their completed report:

1. Detailed drawings of all eight parts of the log-splitter (one shown)
2. Exploded pictorial of the log-splitter
3. Detail drawing of the wedge stand (see Fig. 14.22 for pictorial)
4. Welding assembly of the log-splitter
5. Welding assembly of the wedge stand
6. Complete parts list

The drawings were accompanied by a cost analysis, weight summary, and description of the operating characteristics.

None of the team worked in a plant that produces such items, so no doubt an experienced detailer or drafter would find reason to be critical of their drawings, but we feel that this phase of the process was performed quite well and that there will be few misunderstandings as a result of omissions on their part.

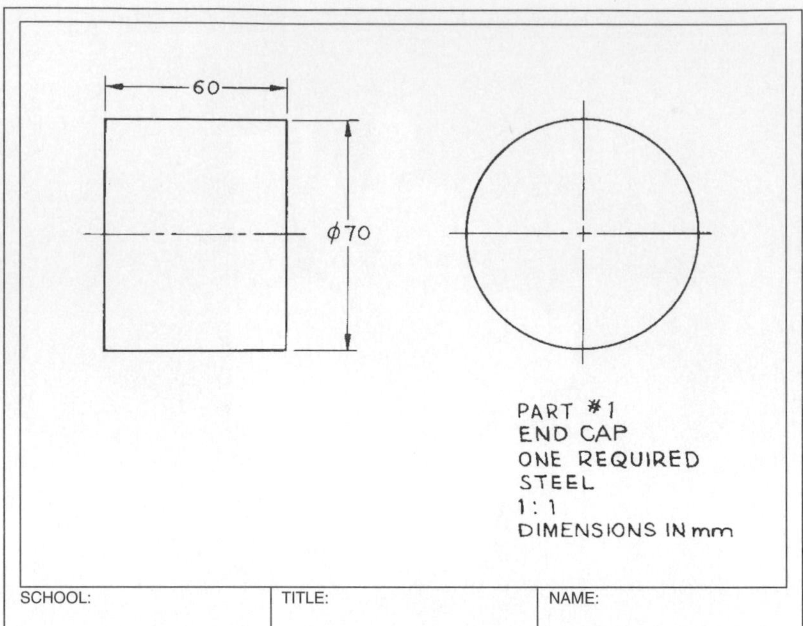

Figure 14.27
Detail drawing of part 1 (see Fig. 14.28).

14.11 Communication—Step 10

During the 1960s the word "communication" seemed to take on a very high priority at conferences and in the professional journals. The need for conveying information and ideas had not changed, but there was an awareness of too much incomplete and inaccurate rendering of information. At most of the professional conferences the authors have attended, one or more papers either discussed the need for engineers to develop greater skills in communication or demonstrated a technique for improving the skills. Students at most universities are required to complete freshman English courses and, in some colleges, a technical writing course in their junior or senior year; but many professors and employers feel that not enough emphasis is being placed on the application of communication skills.

For our purposes here, however, we will discuss only the salient points involved in design step 10.

14.11.1 Selling the Design

It is certainly the responsibility of any profession to inform people of findings and developments. Engineering is no exception in this regard. Our emphasis here will be on a second type of communication, however: selling, explaining, and persuading.

Selling takes place all the way through the design process. Individuals who are the most skillful at it will see many of their

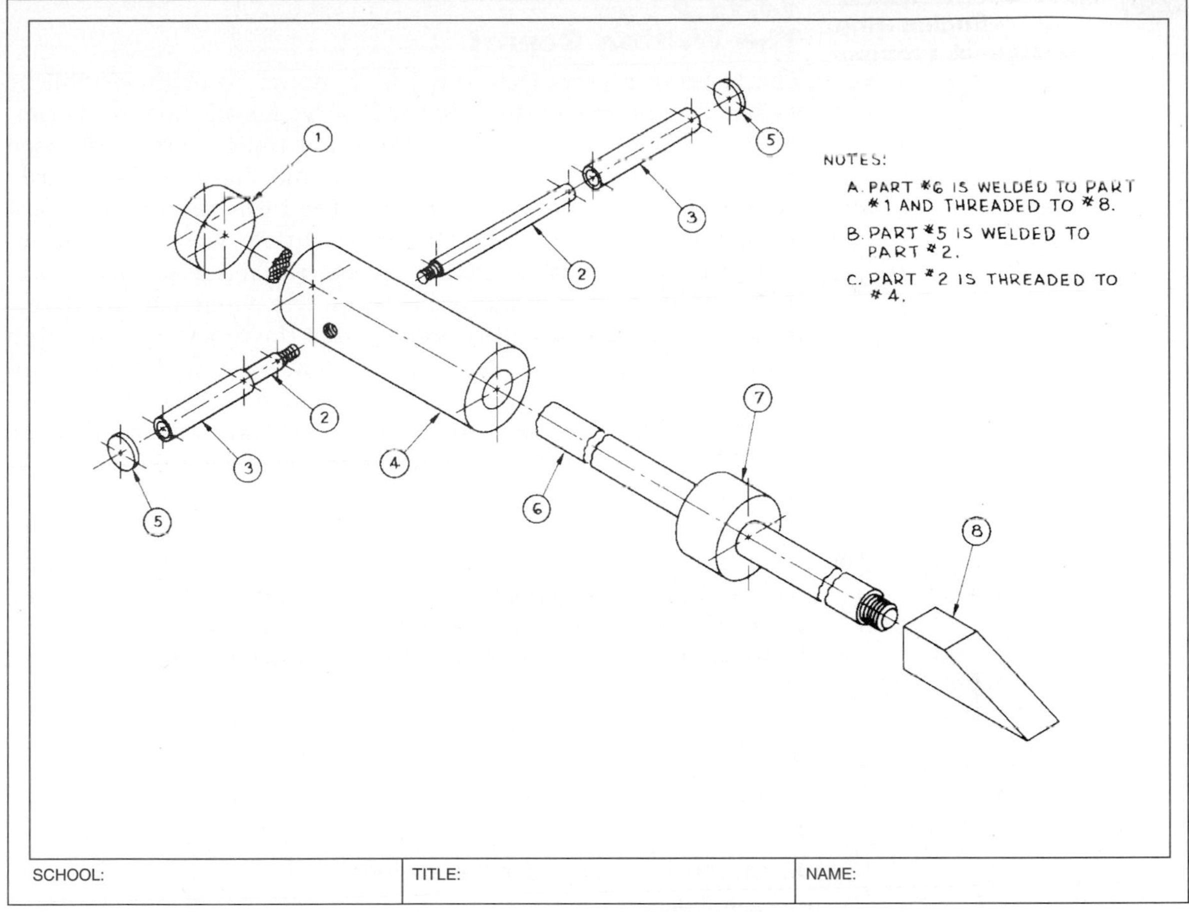
NOTES:
A. PART #6 IS WELDED TO PART #1 AND THREADED TO #8.
B. PART #5 IS WELDED TO PART #2.
C. PART #2 IS THREADED TO #4.
1
2
3
4
5
6
7
8
SCHOOL:
TITLE:
NAME:

Figure 14.28
Exploded pictorial drawing of the log-splitter.

ideas develop into realities. Those who are not so good at it will no doubt become frustrated with their supervisors for not exploring what they feel is a perfectly good idea in more depth.

If you are working as a design engineer for an industry, you cannot simply decide on your own that you will try to improve the product line. Industry is anxious for its engineers to initiate ideas, but will not necessarily approve all of them. As an engineer with a company, you must convince those who decide what assignments you get that the idea is worth the time and money required to develop it. Later, after the design has developed to the point where it can be produced and tested, you must again persuade management to place it into production.

It is a natural reaction to feel that your design has so many clear advantages that selling it should not be necessary. Such may be the case; but in actual situations, things seldom work so smoothly and simply. You will be selling or persuading or convincing others almost daily in a variety of ways. Among the many forms of communication are written and oral reports.

14.11.2 The Written Report

The types of reports that you will write as an engineer will be varied, so a precise outline that will serve for all the reports cannot be supplied. The two major types of reports are those used by individuals within the organization and those used primarily by clients or customers. Many times the in-house reports follow a strict form prescribed by the organization, whereas those intended for the client are usually designed for the particular situation. The nature of the project and the client usually determine the degree of formality employed in the report. Clients often state that they wish to use the report in some particular manner, which may direct you to the style of report to use. For instance, if you are a consultant for a city and have studied the needs for expansion of a power plant, the report may be very technical, brief, and full of equations, computations, and so on, if it is intended for the use of the city's engineer and public utilities director. However, if the report is to be presented to citizens in an effort to convince them to vote for a bond issue to finance the expansion, the report will take a different flavor.

Reports generally will have the following divisions or sections:

1. Appropriate cover page
2. Abstract
3. Table of contents
4. Body
5. Conclusions and recommendations
6. Appendixes

Abstract. A brief paragraph indicating the purpose and results of the effort being reported, the abstract is used primarily for archiving so that others can quickly decide if they want to obtain the complete report.

Body. This is the principle section of the report. It begins with an introduction to activities, including problem identification, background material, and the plan for attacking and solving the problem. If tests were conducted, research completed, and surveys undertaken, the results are recounted and their significance is underscored. In essence, the body of the report is the description of the individual or team effort on a project.

Conclusions and recommendations. This section tells why the study was done and explains the purpose of the report. Herein you explain what you now believe to be true as a result of the work discussed in the body and what you recommend should be done about it. You must lay the groundwork earlier in the report, and at this point you must sell your idea. If you have done the job carefully and fully, you may make a sale; but do not be discouraged if you do not. There will be other days and other projects.

Appendixes. Appendixes can be used to avoid interrupting your descriptions so that it can flow more smoothly. Those who do not want to know everything about your study can read it without digression. What is in the appendix completes the story by showing all that was done. But it should not contain information that is essential to one's understanding of the report.

It should be emphasized that all reports do not follow a specific format. For example, lengthy reports should have a summary section placed near the beginning. This one- or two-page section should include a brief statement of the problem, the proposed solution, the anticipated costs, and the benefits. The summary is for the use of higher-level management who in general do not have the time to read your entire report.

In many instances your instructor or supervisor will have specific requirements for a report. Each report is designed to accomplish a specific goal.

Student reports often must follow an instructor's directions regarding form and topics that must be included. If you are asked to write a report in an introductory design course, refer to Fig. 14.1 to make sure that all of the steps in the design process have been successfully completed. You might also study the bias of your instructor and make sure that you have done especially the steps that he or she considers most important. This may sound as though pleasing the instructor and getting a good grade is all that is important. But perhaps in this respect the academic situation is something like that in industry or private practice: Your report must take into account the audience—its biases and its expectations, whether professor in the classroom or supervisor in the business world.

14.11.3 The Oral Presentation

The objective of the oral presentation is the same as that of the written report—to furnish information and to convince the listener. However, the methods and techniques are quite different. The written report is designed to be glanced at, read, and then studied. The oral presentation is a one-shot deal that must be done quickly, so it must be simple. There is no time to go into detail, to show complicated graphs and tables of data or many of the things that are given in a written report. What can you do to make a good presentation?

First, you must be prepared. No audience listens to people who have not bothered to prepare themselves. So you should rehearse with a timer, a mirror, and a tape recorder.

Stand in such a way that you do not detract from what you are saying or showing.

Look at your audience and maintain eye contact. You will be receiving cues from those who are listening, so be prepared to react to these cues.

Project your voice by consciously speaking to the back row. The audience quickly loses interest if it has to struggle to hear.

Speak clearly. We all have problems with our voices—they are either too high, too low, or too accented and certain words or sounds are hard for us, but always be concerned for the listener.

Preparation obviously includes being thoroughly familiar with the material. It should also include determining the nature, size, and technical competence of the audience. You must know how much time will be allotted to your presentation and what else, if anything, is to be presented before or after your speech. It is essential that you know what the room is like because the physical conditions of the room—its size, lighting, acoustics, and seating arrangements—may very well control your use of slides, transparencies, video, and microphones.

The quality of your graphic displays can often influence the opinion of your audience. Again, consider to whom you are speaking carefully as you choose which and how many displays to use. Be certain that they can be read and understood or do not use them at all. Do not clutter your displays with so many details that the message is obscured. Do not try to make a single visual aid accomplish too many tasks: It is good to change the center of emphasis. By all means, test your visual aids before the meeting and never apologize for their quality. (If they are not good, do not use them.) Figure 14.29 shows an exploded pictorial of a small battery-powered grinding tool and provides a good overview of the tool's components.

Figure 14.30 narrows the focus to a subassembly of the tool. The quality of your visual aids can influence many people for you or against you before they hear all you have to say.

Have a good finish. Save something important for the last and make sure everyone knows when the end has come. By all means, do not end with "Well, I guess that's about all I have to say." You have much more to say, you just do not have the time to say it.

14.11.4 The Chapter Example—Step 10

The written and oral reports presented by the students were significant parts of their design experience. Both reports were regulated somewhat by their professor in much the same way that reports are in industry. They were told who would be reading the report and who would judge the oral presentation. They were given copies of the written report grading sheet and the oral pre-

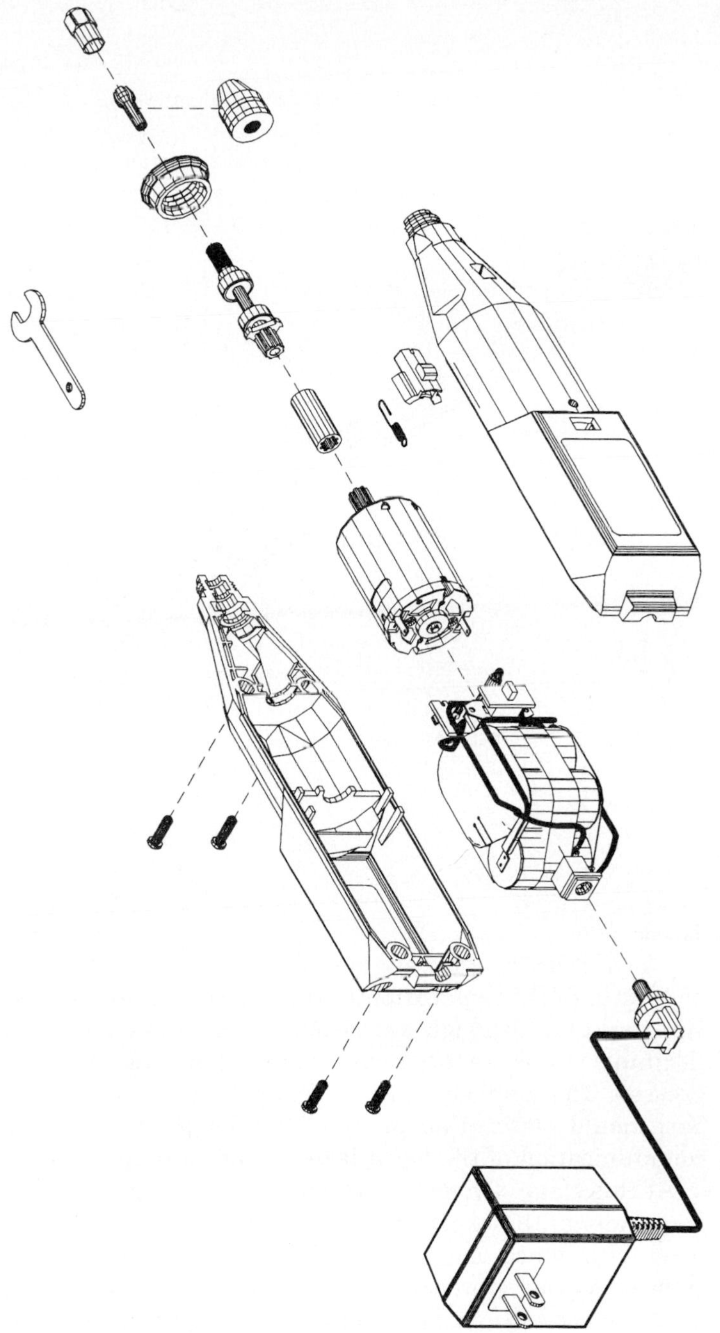

Figure 14.29
Exploded pictorial shows all parts of the grinding tool.

sentation judging card. They correctly accepted these constraints as real (not imagined), and they were given high ratings by their evaluators.

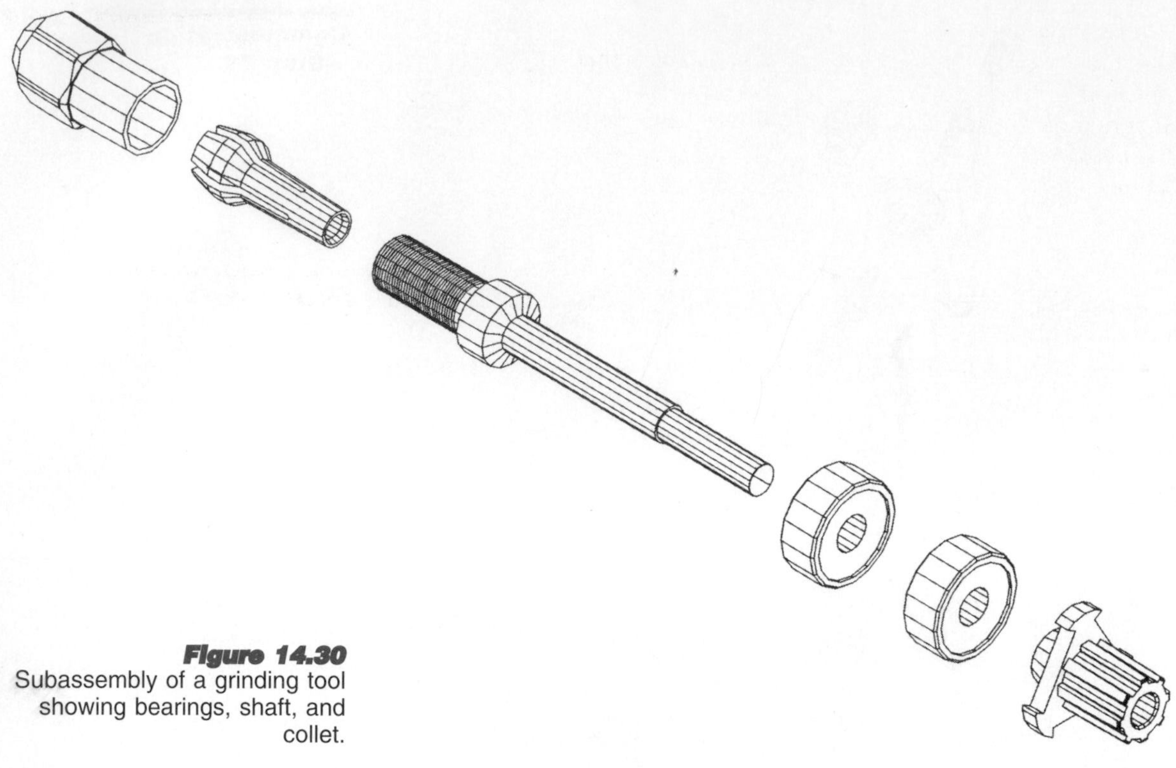

Figure 14.30
Subassembly of a grinding tool showing bearings, shaft, and collet.

14.12 Key Terms and Concepts

We have described the engineering design process with a series of 10 steps. The application of the process to an engineering design process, although structured, is an iterative process with flexibility to make necessary adjustments as the design progresses. The emphasis in this chapter is on conceptual design. You should realize that many of the details of specification and communication of the log-splitter design are not included.

At this stage of your engineering education it is important that you undergo the experience of applying the design process to a need with which you can identify based on your personal experiences. As you approach the baccalaureate degree, you will have acquired the technical capability to conduct the necessary analyses and to make the appropriate technical decisions required for complex products, systems, and processes.

Some of the terms and concepts you should recognize and understand are listed below:

Engineering design	Constraint
Design process	Criteria

Reverse engineering
Alternative solutions
Checkoff lists
Morphological listing
Brainstorming
Design space
Analysis
Decision matrix
Synthesis
Payoff function
Specification
Communication

Problems

Problems 14.1 and 14.2 involve synthesis that is similar to that illustrated in Example prob. 14.3.

14.1 For beam configuration of Fig. 14.15 determine the dimensions *b, h* for the least mass under the following conditions:

The deflection cannot exceed 50 mm.

The height *h* cannot be greater than *b*.

$E = 2.0 \times 10^{11}$ Pa

$L = 6.0$ m

$P = 1.0 \times 10^{5}$ N

Produce a brief report containing a computer plot that is similar to Fig. 14.14a and a discussion of the design space and how the solution was found.

14.2 A company transfers packages from point to point across the country. The limit on package size is that the girth plus the longest dimension (measured on the package) cannot exceed 60 in. Consider two kinds of packages, a rectangular-prism shape with square ends and a cylindrical shape where the cylinder height is greater than the diameter (girth = circumference). Determine for each shape the largest package volume that can be shipped and the package dimensions at the maximum volume. A *suggested procedure* follows:

(*a*) Write the constraint equation (60-in limit) and the payoff function (volume). Eliminate one of the unknowns in the payoff function by substituting the constraint equation.
(*b*) Plot the payoff function against the remaining variable.
(*c*) Determine the optimum values.
(*d*) Prepare a report of your findings.

14.3 Investigate current designs for one or more common items. If you do not have the items in your possession, purchase them or borrow from friends. Conduct the following "reverse engineering" procedures on each of the items:

(*a*) Write down the need that the design satisfies.
(*b*) Disassemble the item and list all the parts by name.
(*c*) Write down the function of each of the parts in the item.
(*d*) Reassemble the item.
(*e*) Write down answers to the following questions:

- Does the item satisfactorily solve the need you stated in part *a*?
- What are the strengths of the design?
- What are the weaknesses of the design?
- Can this design be easily modified to solve other needs? If so, what needs and what modifications need to be made?

- What other designs can solve the stated need?

The items for your study are the following:

- Mechanical pencil
- Safety razors from three vendors; include one disposable razor
- Flashlight
- Battery-powered slide viewer
- Battery-powered fabric shaver

14.4 The following list of potential projects can be approached in the manner used by the student design team featured in this chapter.

- Headlights that follow the wheels' direction
- A protective "garage" that can be stored in the car's trunk
- A device to prevent theft of helmets left on motorcycles
- A conversion kit for winter operation of motorcycles
- An improved rack for carrying packages or books on a motorcycle or bicycle
- A child's seat for a motorcycle or bicycle
- A tray for eating, writing, and playing games in the back seat of a car
- A system for improving traction on ice without studs or chains
- An inexpensive built-in jack for raising a car
- An auto-engine warmer
- A better way of informing motorists of speed limits, road conditions, hazards, etc.
- Theft- and vibration-proof wheel covers
- A better way to check the engine oil level
- A device to permit easier draining of the oil pan by weekend mechanics
- A heated steering wheel for cold weather
- A less expensive replacement for auto air-cleaner elements
- An overdrive system for a trail bike
- A sun shield for an automobile
- A well-engineered, efficient automobile instrument panel
- An SOS sign for cars stalled on freeways
- A remote car-starting system for warmup
- A car-door positioner for windy days
- A bicycle trailer
- Automatic rate-sensitive windshield wipers
- A corn detasseler
- An improved wall outlet
- A beverage holder for a card table
- A car wash for pickups
- A better rural mailbox
- A home safe
- An improved automobile traffic pattern on campus
- An alert for drowsy or sleeping drivers

- An improved automobile headlight
- An improved bicycle for recreation or racing
- Improved bicycle brakes
- A transit system for campus
- A pleasure boat with retractable trailer wheels
- Improved pedestrian crossings at busy intersections
- A transportation system within large airports
- An improved baggage-handling system at airports
- Improved parking facilities in and around campus
- A simple but effective device to assist in cleaning clogged drains
- A device to attach to a paint can for pouring
- An improved soap dispenser
- A better method of locking weights to a barbell shaft
- A shoestring fastener to replace the knot
- An automatic moisture-sensitive lawn waterer
- A better harness for seeing-eye dogs
- A better jar opener
- A system or device to improve efficiency of limited closet space
- A shoe transporter and storer
- A pen and pencil holder for college students
- An acceptable rack for mounting electric fans in dormitory windows
- A device to pit fruit without damage
- A riot-quelling device to subdue participants without injury
- An automatic device for selectively admitting and releasing an auxiliary door for pets
- A device to permit a person to open a door when loaded with packages
- A more efficient toothpaste tube
- A fingernail catcher for fingernail clippers
- A more effective alarm clock for reluctant students
- An alarm clock with a display to show it has been set to go off
- A device to help a parent monitor small children's presence and activity in and around the house
- A chair that can rotate, swivel, rock, or stay stationary
- A simple pocket alarm that is difficult to shut off, for discouraging muggers
- An improved storage system for luggage, books, etc. in dormitories
- A lampshade designed to permit one to study while his or her roommate is asleep
- A device that would permit blind people to vote in an otherwise conventional voting booth
- A one-cup coffee maker
- A solar greenhouse
- A quick-connect garden-hose coupling

- A device for recycling household water
- A silent wakeup alarm
- Home aids for the blind (or deaf)
- A safer, more efficient, and quieter air mover for room use
- A lock that can be opened by a secret method without a key
- A can crusher
- A rain-sensitive house window
- A better grass catcher for a riding lawnmower
- A winch for hunters of large game
- Gauges for water, transmission fluid, etc. in autos
- A built-in auto refrigerator
- A better camp cooler
- A dormitory-room cooler
- A device for raising and lowering TV racks in the classroom
- An impact hammer adapter for electric drills
- An improved method of detecting and controlling the level position of the bucket on a bucket loader
- Shields to prevent corn spillage where the drag line dumps into the sheller elevator (angle varies)
- An automatic tractor-trailer-hitch aligning device
- A jack designed expressly for motorcycle use (special problems involved)
- A motorbike using available (junk) materials
- Improved road signs for speed limits, curves, deer crossings, etc.
- More effective windshield wipers
- A windshield deicer
- Shock-absorbing bumpers for minor accidents
- A home fire-alarm device
- A means of evacuating buildings in case of fire
- Automatic light switches for rooms
- A carbon monoxide detector
- An indicator to report the need for an oil change
- A collector for dust (smoke) particles from stacks
- A means of disposing of or recycling soft-drink containers
- A way to stop dust storms, resultant soil loss, and air entrainment
- An attractive system for handling trash on campus
- A self-decaying disposable container
- A device for dealing with oil slicks
- A means of preventing heat loss from greenhouses
- A way of creating energy from waste
- A bookshelf with horizontally and vertically adjustable shelves and dividers
- A device that would make the working surface of graphics desks adjustable in height and slope, retaining the existing top and pedestal

- An egg container (light, strong, compact) for camping and canoeing
- Ramps or other facilities for handicapped students
- A multifunctional (suitcase/chair/bookshelf, etc.) packing device for students
- A self-sharpening pencil for drafting
- An adapter to provide tilt and elevation control on existing graphics tables
- A compact and inexpensive camp stove for backwoods hiking
- A road trailer operable from inside the car
- A hood lock for cars to prevent vandalism
- A system to prevent car thefts
- A keyless lock

Selected Topics from Algebra

A.1 Introduction

This appendix includes material on exponents and logarithms, simultaneous equations, and the solution of equations by approximation methods. The material can be used for reference or review. The reader should consult an algebra textbook for more detailed explanations of additional topics for study.

A.2 Exponents and Radicals

The basic laws of exponents are stated below along with an illustrative example.

Law	Example
$a^m a^n = a^{m+n}$	$x^5 x^{-2} = x^3$
$\frac{a^m}{\;} = a^{m-n} \quad a \neq 0$	$\frac{x^5}{\;} = x^2$
$(a^m)^n = a^{mn}$	$(x^{-2})^3 = x^{-6}$
$(ab)^m = a^m b^m$	$(xy)^2 = x^2 y^2$
$\left(\frac{a}{\;}\right)^m = \frac{a^m}{\;} \quad b \neq 0$	$\left(\frac{x}{\;}\right)^2 = \frac{x^2}{\;}$
$a^{-m} = \frac{1}{\;} \quad a \neq 0$	$x^{-3} = \frac{1}{\;}$
$a^0 = 1 \quad a \neq 0$	$2(3x^2)^0 = 2(1) = 2$
$a^1 = a$	$(3x^2)^1 = 3x^2$

These laws are valid for positive and negative integer exponents and for a zero exponent, and can be shown to be valid for rational exponents. Some examples of fractional exponents are illustrated here. Note the use of radical ($\sqrt{\;}$) notation as an alternative to fractional exponents.

Law	Example
$a^{m/n} = \sqrt[n]{a^m}$	$x^{2/3} = \sqrt[3]{x^2}$
$\frac{\sqrt[n]{a}}{\sqrt[n]{b}} = \sqrt[n]{\frac{a}{b}}$ $b \neq 0$	$\frac{\sqrt[3]{16}}{\sqrt[3]{2}} = \sqrt[3]{8} = 2$
$a^{1/2} = \sqrt[2]{a^1} = \sqrt{a}$ $a \geq 0$	$\sqrt{25} = 5$ (not ± 5)

A.3 Exponential and Power Functions

Functions involving exponents occur in two forms—power and exponential. The power function contains the base as the variable and the exponent is a rational number. An exponential function has a fixed base and variable exponent.

The simplest exponential function is of the form

$$y = b^x \quad b \geq 0$$

where b is a constant. Note that this function involves a power but is fundamentally different from the power function $y = x^b$.

The inverse of a function is an important concept for the development of logarithmic functions from exponential functions. Consider a function $y = f(x)$. If this function could be solved for x, the result would be expressed as $x = g(y)$. For example, the power function $y = x^2$ has as its inverse $x = \pm\sqrt{y}$. Note that in $y = x^2$, y is a single-valued function of x, whereas the inverse is a double-valued function. For $y = x^2$, x can take on any real value, whereas the inverse $x = \pm\sqrt{y}$ restricts y to only positive values or zero. This result is important in the study and application of logarithmic functions.

A.4 The Logarithmic Function

The definition of a logarithm may be stated as follows:

> A number L is said to be the logarithm of a positive real number N to the base b (where b is real, positive, and different from 1), if L is the exponent to which b must be raised to obtain N.

Symbolically, the logarithm function is expressed as

$$L = \log_b N$$

for which the inverse is

$$N = b^L$$

For instance,

$$\log_2 8 = 3 \qquad \text{since } 8 = 2^3$$

$$\log_{10} 0.01 = -2 \qquad \text{since } 0.01 = 10^{-2}$$

$$\log_5 5 = 1 \qquad \text{since } 5 = 5^1$$

$$\log_b 1 = 0 \qquad \text{since } 1 = b^0$$

Several properties of logarithms and exponential functions can be identified when plotted on a graph.

Example problem A.1 Plot graphs of $y = \log_2 x$ and $x = 2^y$ that are inverse functions.

Solution Since $y = \log_2 x$ and $x = 2^y$ are equivalent by definition, they will graph into the same line. Choosing values of y and computing x from $x = 2^y$ yields Fig. A.1.

Some properties of logarithms that can be generalized from Fig. A.1 are

1. $\log_b x$ is not defined for negative or zero values of x.
2. $\log_b 1 = 0$.
3. If $x > 1$, then $\log_b x > 0$.
4. If $0 < x < 1$, then $\log_b x < 0$.

Other properties of logarithms that can be proved as a direct consequence of the laws of exponents are, with P and Q being real and positive numbers,

1. $\log_b PQ = \log_b P + \log_b Q$.
2. $\log_b \dfrac{P}{Q} = \log_b P - \log_b Q$.
3. $\log_b (P)^m = m \log_b P$.
4. $\log_b \sqrt[n]{P} = \dfrac{1}{n} \log_b P$.

The base b, as stated in the definition of a logarithm, can be any real number greater than 0 but not equal to 1, since 1 to any

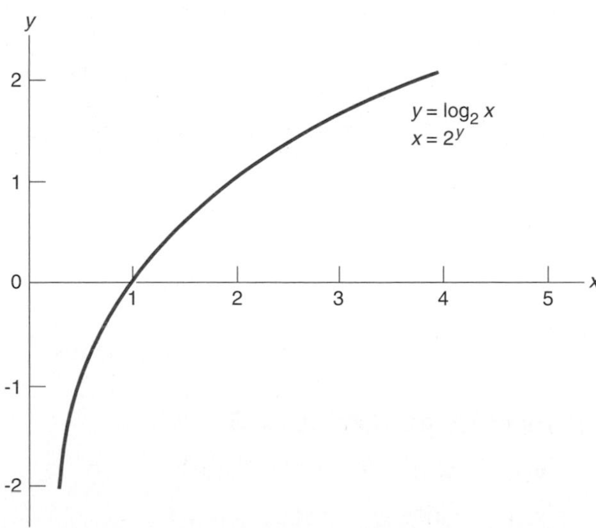

Figure A.1
The logarithmic function.

power remains 1. When using logarithmic notation, the base is always indicated, with the exception of base 10, in which case the base is frequently omitted. In the expression $y = \log x$, the base is understood to be 10. A somewhat different notation is used for the natural (Naperian) logarithms discussed in the Sec. A.5.

Sometimes it is desirable to change the base of logarithms. The procedure is shown by the following example.

Example problem A.2 Given that $y = \log_a N$, find $\log_b N$.

Solution

$$y = \log_a N$$

$$N = a^y \qquad \text{(inverse function)}$$

$$\log_b N = y \log_b a \qquad \text{(taking logs to base } b\text{)}$$

$$\log_b N = (\log_a N)(\log_b a) \qquad \text{(substitution for } y\text{)}$$

$$= \frac{\log_a N}{\log_a b} \qquad \left(\text{since } \log_b a = \frac{1}{\log_a b}\right)$$

A.5 Natural Logarithms and e

In advanced mathematics, the base e is usually chosen for logarithms to achieve simpler expressions. Logarithms to the base e are called natural, or Naperian, logarithms. The constant e is defined in the calculus as

$$e = \lim_{n \to 0} (1 + n)^{1/n} = 2.7182818284 \cdots$$

For purposes of calculating e to a desired accuracy, an infinite series is used.

$$e = \sum_{n=0}^{\infty} \frac{1}{n!}$$

The required accuracy is obtained by summing sufficient terms. For example,

$$\sum_{n=0}^{6} \frac{1}{n!} = 1 + 1 + \frac{1}{2} + \frac{1}{6} + \frac{1}{24} + \frac{1}{120} + \frac{1}{720}$$

$$= 2.718\ 055$$

which is accurate to four significant figures.

Natural logarithms are denoted by the symbol ln, and all the properties defined previously for logarithms apply to natural logarithms. The inverse of $y = \ln x$ is $x = e^y$. The following examples illustrate applications of natural logarithms.

Example problem A.3

$\ln 1 = 0$ since $e^0 = 1$

$\ln e = 1$ since $e^1 = e$

Example problem A.4 Solve for x:

$$2^x = 3^{x-1}$$

Specify answer to four significant figures.

Taking natural logarithms of both sides of the equation and using a calculator for evaluation of numerical quantities,

$$x \ln 2 = (x - 1)\ln 3$$

$$\frac{x}{x-1} = \frac{\ln 3}{\ln 2} = 1.585\ 0$$

$$x = 2.709 \qquad \text{(four significant figures)}$$

This problem could have been solved by choosing any base for taking logarithms. However, in general, base e or 10 should be chosen so that a scientific calculator can be used for numerical work.

A.6 Simultaneous Equations

Several techniques exist for finding the common solution to a set of n algebraic equations in n unknowns. A formal method for solution of a system of linear equations is known as Cramer's rule, which requires a knowledge of determinants.

A second-order determinant is defined and evaluated as

$$\begin{vmatrix} a_1 b_1 \\ a_2 b_2 \end{vmatrix} = a_1 b_2 - a_2 b_1$$

A third-order determinant is defined and evaluated as

$$\begin{vmatrix} a_1 b_1 c_1 \\ a_2 b_2 c_2 \\ a_3 b_3 c_3 \end{vmatrix} = a_1 \begin{vmatrix} b_2 c_2 \\ b_3 c_3 \end{vmatrix} - a_2 \begin{vmatrix} b_1 c_1 \\ b_3 c_3 \end{vmatrix} + a_3 \begin{vmatrix} b_1 c_1 \\ b_2 c_2 \end{vmatrix}$$

where the second-order determinants are evaluated as indicated above. The procedure may be extended to higher-order determinants.

Cramer's rule for a system of n equations in n unknowns can be stated as follows:

1. Arrange the equations to be solved so that the unknowns x, y, z, and so forth appear in the same order in each equation; if any unknown is missing from an equation, it is to be considered as having a coefficient of zero in that equation.

2. Place all terms that do not involve the unknowns in the right member of each equation.

3. Designate by D the determinant of the coefficients of the unknowns in the same order as they appear in the equations. Designate by D_i the determinant obtained by replacing the elements of the ith column of D by the terms in the right member of the equations.

4. Then, if $D \neq 0$ the values of the unknowns x y, z, and so forth, are given by

$$x = \frac{D_1}{} \qquad y = \frac{D_2}{} \qquad z = \frac{D_3}{} \ldots$$

Example problem A.5 Solve the following system of equations that have already been written in proper form for application of Cramer's rule.

$$3x + y - z = 2$$

$$x - 2y + z = 0$$

$$4x - y + z = 3$$

Solution

$$x = \frac{\begin{vmatrix} 2 & 1 & -1 \\ 0 & -2 & 1 \\ 3 & -1 & 1 \end{vmatrix}}{\begin{vmatrix} 3 & 1 & -1 \\ 1 & -2 & 1 \\ 4 & -1 & 1 \end{vmatrix}} = \frac{2\begin{vmatrix} -2 & 1 \\ -1 & 1 \end{vmatrix} - 1\begin{vmatrix} 0 & 1 \\ 3 & 1 \end{vmatrix} + (-1)\begin{vmatrix} 0 & -2 \\ 3 & -1 \end{vmatrix}}{3\begin{vmatrix} -2 & 1 \\ -1 & 1 \end{vmatrix} - 1\begin{vmatrix} 1 & 1 \\ 4 & 1 \end{vmatrix} + (-1)\begin{vmatrix} 1 & -2 \\ 4 & -1 \end{vmatrix}}$$

$$= \frac{2(-2 + 1) - 1(0 - 3) - 1(0 + 6)}{3(-2 + 1) - 1(1 - 4) - 1(-1 + 8)}$$

$$= \frac{2(-1) - 1(-3) - 1(6)}{3(-1) - 1(-3) - 1(7)}$$

$$= \frac{-5}{-7}$$

$$= \frac{5}{7}$$

The reader may verify the solutions $y = 6/7$ and $z = 1$.

There are several other methods of solution for systems of equations that are illustrated by the following examples.

Example problem A.6 Solve the system of equations:

$$9x^2 - 16y^2 = 144$$

$$x - 2y = 4$$

Solution The common solution represents the intersection of a hyperbola and straight line. The method used is substitution. Solving the linear equation for x yields

$$x = 2y + 4$$

Substitution into the second-order equation gives

$$9(2y + 4)^2 - 16y^2 = 144$$

which reduces to

$$20y^2 + 144y = 0$$

Factoring gives

$$4y(5y + 36) = 0$$

which yields

$$y = 0, \frac{-36}{5}$$

Substitution into the linear equation $x = 2y + 4$ gives the corresponding values of x:

$$x = 4, -\frac{52}{5}$$

The solutions are thus the coordinates of intersection of the line and the hyperbola:

$$(4.0), (-\frac{52}{5}, -\frac{36}{5})$$

which can be verified by graphical construction.

Example problem A.7 Solve the system of equations:

(a) $3x - y = 7$

(b) $x + z = 4$

(c) $t - z = -1$

Solution Systems of equations similar to these arise frequently in engineering applications. Obviously, they can be solved by Cramer's rule. However, a more rapid solution can be obtained directly by elimination.

From Eq. (*c*),

$$y = z - 1$$

From Eq. (*a*),

$$y = 7 - 3x$$

From Eq. (*b*),

$$x = 4 - z$$

Successive substitution yields

$$z - 1 = 7 - 3x$$

$$z - 1 = 7 - 3(4 - z)$$

$$-2z = -4$$

$$z = +2$$

Continued substitution gives

$$y = 1$$

$$x = 2$$

Every system of equations should first be carefully investigated before a method of solution is chosen so that the most direct method, requiring the minimum amount of time, is used.

A.7 Approximate Solutions

Many equations developed in engineering applications do not lend themselves to direct solution by standard methods. These equations must be solved by approximation methods to the accuracy dictated by the problem conditions. Experience is helpful in choosing the numerical technique for solution.

Example problem A.8 Find to three significant figures the solution to the equation

$$2 - x = \ln x$$

Solution One method of solution is graphical. If the equations $y = 2 - x$ and $y = \ln x$ are plotted, the common solution would be the intersection of the two lines. This would not likely give three-significant-figure accuracy, however. A more accurate method requires use of a scientific calculator or computer.

Inspection of the equation reveals that the desired solution must lie between 1 and 2. It is then a matter of setting up a routine that will continue to bracket the solution between two increasingly accurate numbers. Table A.1 shows the intermediate steps and indicates that the solution is $x = 1.56$ to three significant figures.

Table A.1 Solution of $2 - x = \ln x$

x	1	2	1.5	1.6	1.55	1.56	1.557
$2 - x$	1	0	0.500	0.400	0.450	0.440	0.443
$\ln x$	0	0.693	0.405	0.470	0.438	0.445	0.443

Computer spreadsheets and solvers or a programmable calculator could easily be used to determine a solution by the method just described. The time available and equipment on hand will always influence the numerical technique to be used.

Trigonometry

B.1 Introduction

This material is intended to be a brief review of concepts from plane trigonometry that are commonly used in engineering calculations. The section deals only with plane trigonometry and furnishes no information about spherical trigonometry. The reader is referred to standard texts in trigonometry for more detailed coverage and analysis.

B.2 Trigonometric Function Definitions

The trigonometric functions are defined for an angle contained within a right triangle, as shown in Fig. B.1.

$$\text{sine } \theta = \sin \theta = \frac{\text{opposite side}}{\text{hypotenuse}} = \frac{y}{r}$$

$$\text{cosine } \theta = \cos \theta = \frac{\text{adjacent side}}{\text{hypotenuse}} = \frac{x}{r}$$

$$\text{tangent } \theta = \tan \theta = \frac{\text{opposite side}}{\text{adjacent side}} = \frac{y}{x}$$

$$\text{cotangent } \theta = \cot \theta = \frac{\text{adjacent side}}{\text{opposite side}} = \frac{x}{y} = \frac{1}{\tan \theta}$$

$$\text{secant } \theta = \sec \theta = \frac{\text{hypotenuse}}{\text{adjacent side}} = \frac{r}{x} = \frac{1}{\cos \theta}$$

$$\text{cosecant } \theta = \csc \theta = \frac{\text{hypotenuse}}{\text{opposite side}} = \frac{r}{y} = \frac{1}{\sin \theta}$$

The angle θ is by convention measured positive in the counterclockwise direction from the positive x axis.

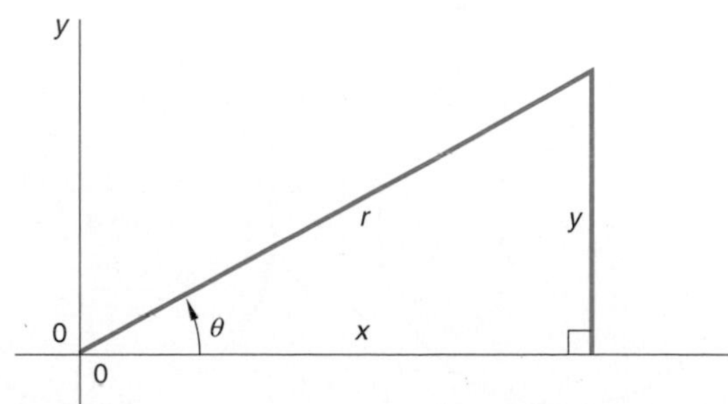

Figure B.1
Coordinate definition.

B.3 Signs of Trigonometric Functions by Quadrant

Table B.1

Quadrant 2	Quadrant 1
$x(-), y(+)$ sin and csc(+) cos and sec(−) tan and cot(−)	$x(+), y(+)$ sin and csc(+) cos and sec(+) tan and cot(+)
Quadrant 3	**Quadrant 4**
$x(-), y(-)$ sin and csc(−) cos and sec(−) tan and cot(+)	$x(+), y(-)$ sin and csc(−) cos and sec(+) tan and cot(−)

B.4 Radians and Degrees

Angles may be measured in either degrees or radians (see Fig. B.2). By definition,

$$1 \text{ degree } (^\circ) = \frac{1}{360} \text{ of the central angle of a circle}$$

1 radian (rad) = angle subtended at center 0 of a circle by an arc equal to the radius

The central angle of a circle is 2π rad or 360°. Therefore,

$$1^\circ = \frac{2\pi}{360^\circ} = \frac{\pi}{180^\circ} = 0.017\ 453\ 29 \cdots \text{rad}$$

and

$$1 \text{ rad} = \frac{360^\circ}{2\pi} = \frac{180^\circ}{\pi} = 57.295\ 78 \cdots {}^\circ$$

It follows that the conversion of θ in degrees to θ in radians is given by

$$\theta \text{ (rad)} = \theta\ (^\circ)\ \frac{\pi}{180^\circ}$$

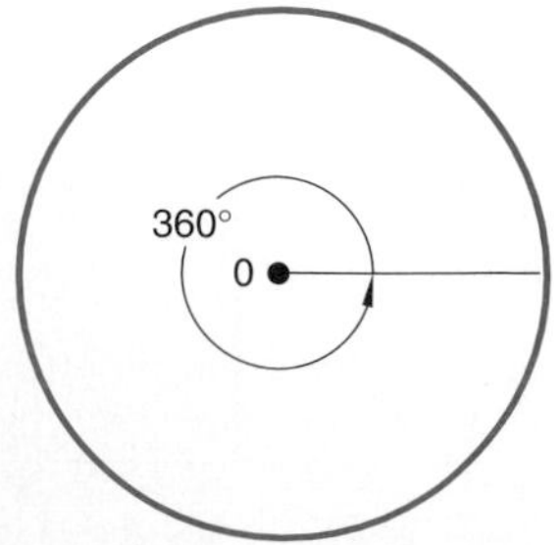

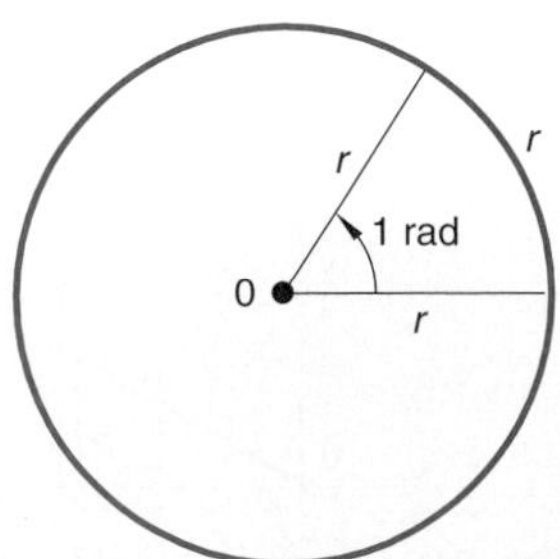

Figure B.2 Definition of degrees and radians.

and in like manner,

$$\theta\,(^\circ) = \theta\,(\text{rad})\,\frac{180^\circ}{\pi}$$

B.5 Plots of Trigonometric Functions

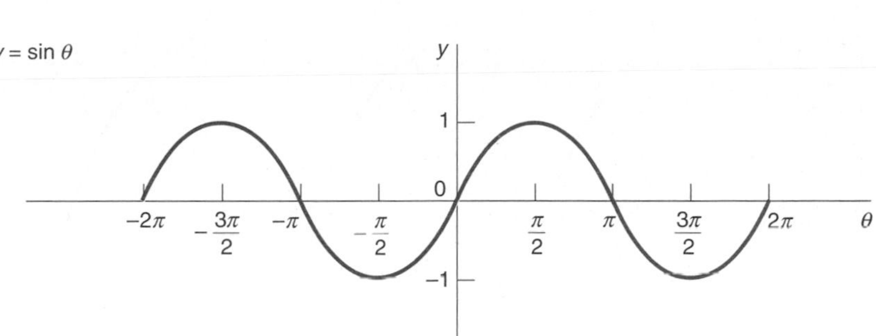

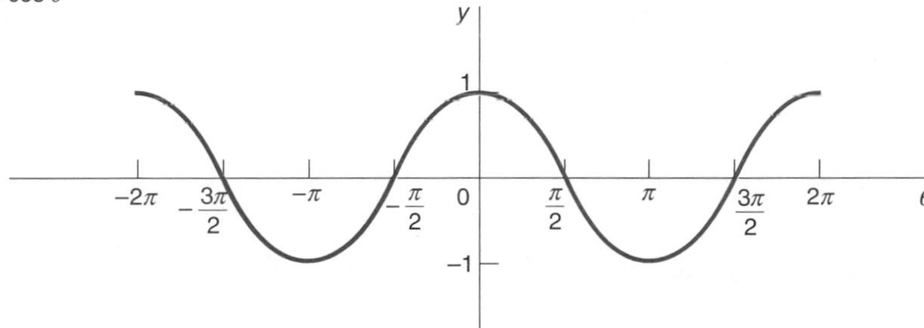

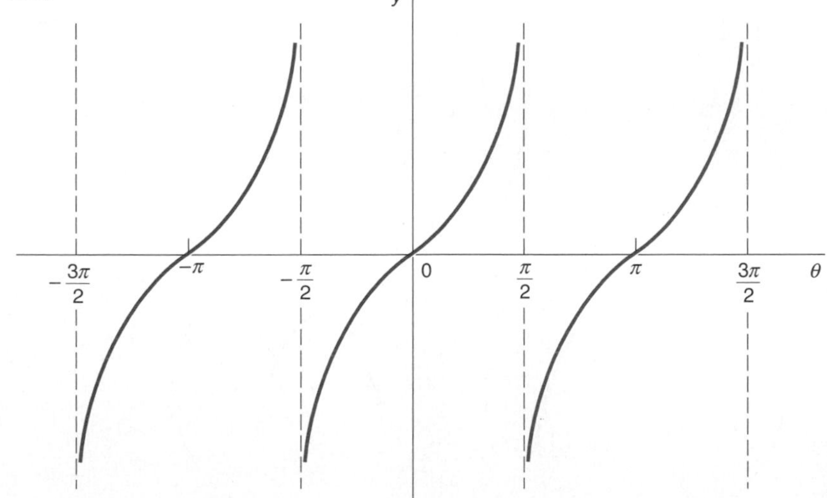

Figure B.3
Plots of trigonometric functions.

$y = \cot \theta$

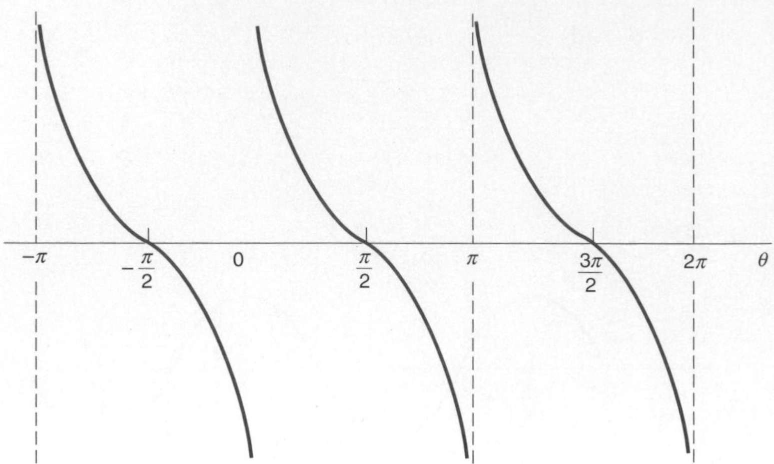

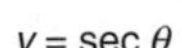

$y = \sec \theta$

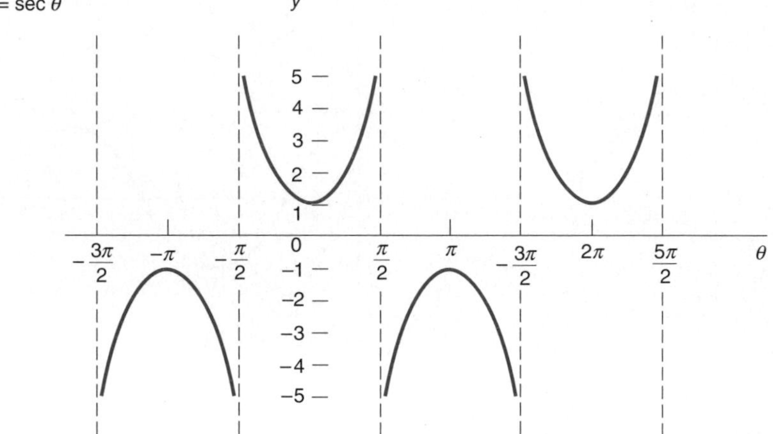

$y = \csc \theta$

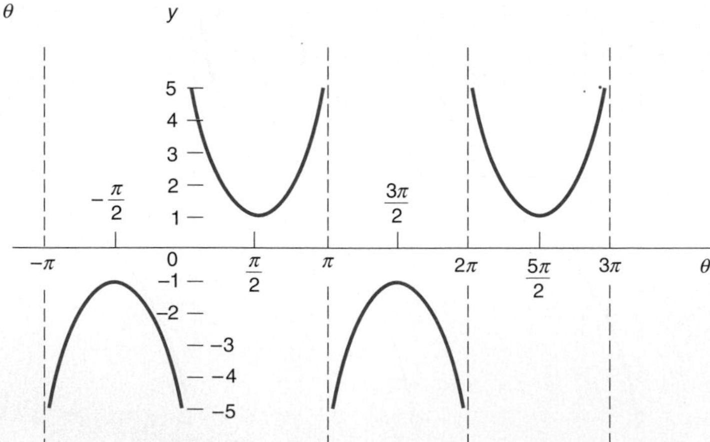

B.6 Standard Values of Often-Used Angles

From three basic triangles, it is possible to compute the values of the trigonometric functions for many standard angles such as 30°, 45°, 60°, 120°, 135°, etc. It is only necessary for us to recall that $\sin 30° = \cos 60° = \frac{1}{2}$, and $\tan 45° = 1$ to construct the necessary triangles from which values can be taken to obtain the other functions.

The functions for 0°, 90°, 180°, etc., can be found directly from the function definitions and a simple line sketch. See Table B.2.

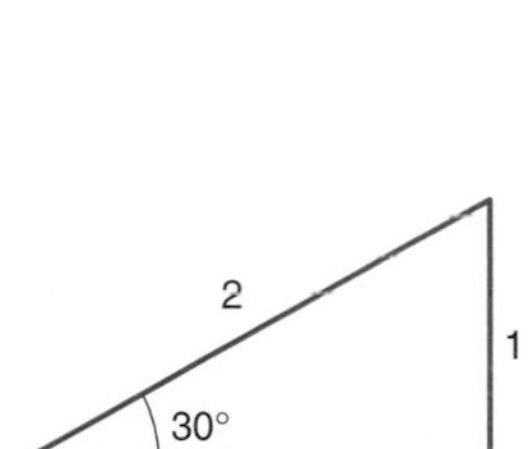

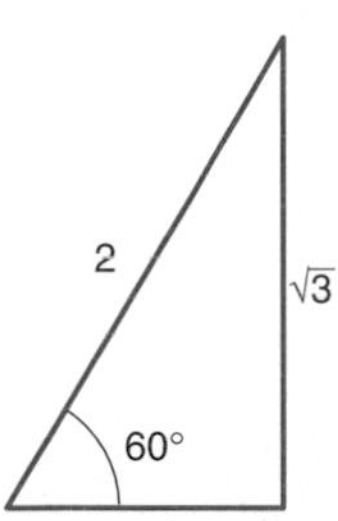

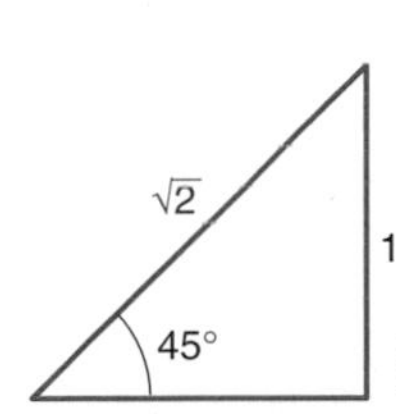

Figure B.4 Common triangles.

Table B.2 Functions of common angles

Function \ Angle	0°	30°	45°	60°	90°
sin	0	1/2	$\sqrt{2}/2$	$\sqrt{3}/2$	1
cos	1	$\sqrt{3}/2$	$\sqrt{2}/2$	1/2	0
tan	0	$\sqrt{3}/3$	1	$\sqrt{3}$	∞
cot	∞	$\sqrt{3}$	1	$\sqrt{3}/3$	0
sec	1	$2\sqrt{3}/3$	$\sqrt{2}$	2	∞
csc	∞	2	$\sqrt{2}$	$2\sqrt{3}/3$	1

B.7 Inverse Trigonometric Functions

Definition

If $y = \sin\theta$, *then* θ *is an angle whose sine is* y. The symbols ordinarily used to denote an inverse function are

$$\theta = \arcsin y$$

or

$$\theta = \sin^{-1} y$$

Note:

$$\sin^{-1} y \neq \frac{1}{\sin y}$$

This is an exception to the conventional use of exponents.

Inverse functions $\cos^{-1} y$, $\tan^{-1} y$, $\cot^{-1} y$, $\sec^{-1} y$, and $\csc^{-1} y$ are similarly defined. Each of these is a many-valued function of y. The values are grouped into collections called *branches.* One of

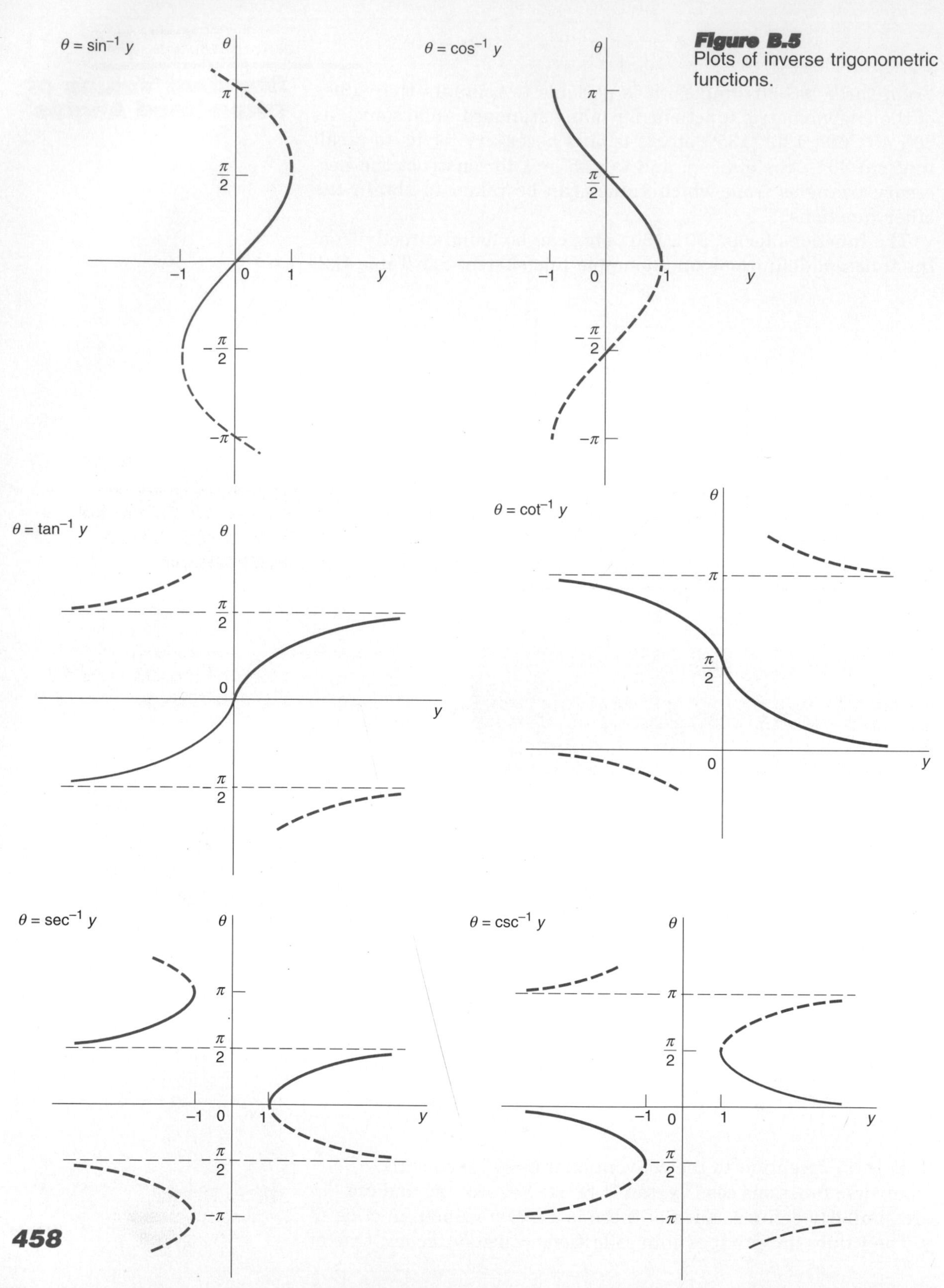

Figure B.5
Plots of inverse trigonometric functions.

these branches is defined to be the principal branch, and the values found there are the principal values.

The principal values are as follows:

$$-\frac{\pi}{2} \leq \sin^{-1} y \leq \frac{\pi}{2}$$

$$0 \leq \cos^{-1} y \leq \pi$$

$$-\frac{\pi}{2} < \tan^{-1} y < \frac{\pi}{2}$$

$$0 < \cot^{-1} y < \pi$$

$$0 \leq \sec^{-1} y \leq \pi \qquad \left(\sec^{-1} y \neq \frac{\pi}{2}\right)$$

$$-\frac{\pi}{2} \leq \csc^{-1} y \leq \frac{\pi}{2} \qquad (\csc^{-1} y \neq 0)$$

B.8 Plots of Inverse Trigonometric Functions

All angles are given in radians. Principal branches are shown as solid lines. See Fig. B.5.

B.9 Polar-Rectangular Coordinate

See Fig. B.6.

Conversion from polar to rectangular coordinates $(r,\theta) \rightarrow (x,y)$ is given by the following equations:

$$x = r \cos \theta$$

$$y = r \sin \theta$$

Conversion from rectangular to polar coordinates $(x,y) \rightarrow (r,\theta)$ requires the following equations:

$$r = [x^2 + y^2]^{1/2}$$

$$\theta = \tan^{-1}\left(\frac{y}{x}\right)$$

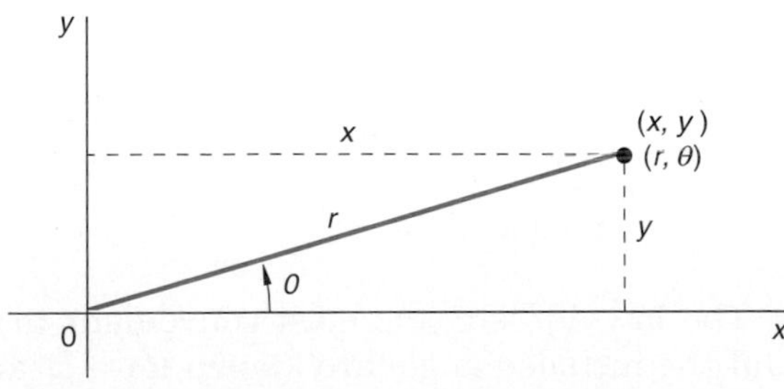

Figure B.6 Rectangular and polar coordinate definitions.

The conversion from polar to rectangular coordinates can also be thought of as the determination of the x and y components of a vector (r, θ). Likewise, conversion from rectangular to polar coordinates is the same as finding the resultant vector (r, θ) from its x and y components.

B.10 Laws of Sines and Cosines

The fundamental definitions of sine, cosine, etc., apply strictly to right triangles. Solutions needed for oblique triangles must then be accomplished by appropriate constructions that reduce the problem to a series of solutions to right triangles.

Two formulas have been derived for oblique triangles that are much more convenient to use than the construction technique. They are the law of sines and the law of cosines, which apply to any plane triangle. See Fig. B.7.

The *law of sines* states

$$\frac{\sin A}{a} = \frac{\sin B}{b} = \frac{\sin C}{c}$$

The *law of cosines* is

$$a^2 = b^2 + c^2 - 2bc \cos A$$

or

$$b^2 = a^2 + c^2 - 2ac \cos B$$

or

$$c^2 = a^2 + b^2 - 2ab \cos C$$

Application of the law of sines is most convenient in the case where two angles and one side are known and a second side is to be found.

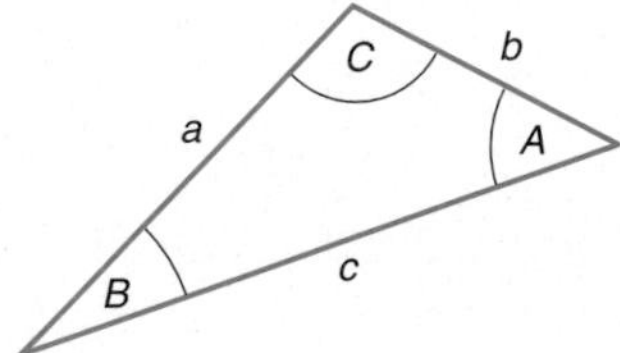

Figure B.7
Angle and side designations.

Example problem B.1 Determine the length of side x for the triangle with base 12.0 m as shown in Fig. B.8.

Solution The sum of the interior angles must be 180°; therefore,

$$\alpha = 180° - 43° - 59° = 78°.$$

Applying the law of sines,

$$\frac{\sin 78°}{12.0 \text{ m}} = \frac{\sin 43°}{x}$$

$$x = \frac{\sin 43°}{\sin 78°} 12.0 \text{ m}$$

$$= 8.37 \text{ m}$$

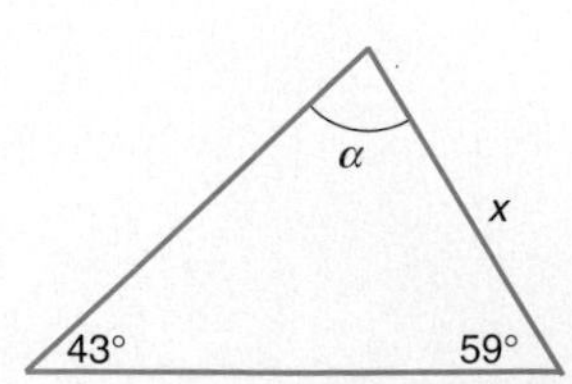

Figure B.8

The law of cosines is most convenient to use when two sides and the included angle are known for a triangle.

Example problem B.2 Calculate the length of side y of the triangle shown in Fig. B.9. Its base is 14.0 m.

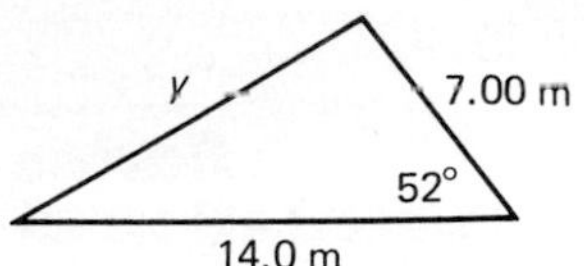

Figure B.9

Solution Substitute into the law of cosines:

$$y^2 = (14.0 \text{ m})^2 + (7.00 \text{ m})^2 - 2(14.0 \text{ m})(7.00 \text{ m})(\cos 52°)$$

$$= 124.33 \text{ m}^2$$

$$y = 11.2 \text{ m}$$

B.11 Area of a Triangle

Formulas for the area of a triangle (see Fig B.7) in terms of two sides and their included angle are

$$\text{Area} = \frac{1}{2} ab \sin C$$

$$\text{Area} = \frac{1}{2} ac \sin B$$

$$\text{Area} = \frac{1}{2} bc \sin A$$

Formulas written in terms of one side and three angles are

$$\text{Area} = \frac{1}{2} a^2 \frac{\sin B \sin C}{\sin A}$$

$$\text{Area} = \frac{1}{2} b^2 \frac{\sin A \sin C}{\sin B}$$

$$\text{Area} = \frac{1}{2} c^2 \frac{\sin A \sin B}{\sin C}$$

The formula for the area in terms of the sides is

$$\text{Area} = [s(s-a)(s-b)(s-c)]^{1/2}$$

where $s = \frac{1}{2}(a + b + c)$.

Example problem B.3 Determine the areas of the two triangles defined in Sec. B.10.

Solution For the 12.0 m base triangle,

$$\text{Area} = \frac{1}{2} (12.0 \text{ m})^2 \frac{\sin 59° \sin 43°}{\sin 78°}$$

$$= 43.0 \text{ m}^2$$

For the 14.0 m base triangle,

$$\text{Area} = \frac{1}{2} (7.00 \text{ m})(14.0 \text{ m}) \sin 52°$$

$$= 38.6 \text{ m}^2$$

B.12

Series Representation of Trigonometric

Infinite-series representations exist for each of the trigonometric functions. Those for sine, cosine, and tangent are

$$\sin\theta = \theta - \frac{\theta^3}{3!} + \frac{\theta^5}{5!} + \cdots + (-1)^{n-1}\frac{\theta^{2n-1}}{(2n-1)!} + \cdots$$

$$\theta \text{ in radians} \qquad (-\infty < \theta < +\infty)$$

$$\cos\theta = 1 - \frac{\theta^2}{2!} + \frac{\theta^4}{4!} + \cdots + (-1)^{n-1}\frac{\theta^{2n-2}}{(2n-2)!} + \cdots$$

$$\theta \text{ in radians} \qquad (-\infty < \theta < +\infty)$$

$$\tan\theta = \theta + \frac{\theta^3}{3} + \frac{2\theta^5}{15} + \cdots + \frac{2^{2n}(2^{2n}-1)\,B_n\theta^{2n-1}}{(2n)!} + \cdots$$

$$\theta \text{ in radians} \qquad \left(-\frac{\pi}{2} < \theta < \frac{\pi}{2}\right)$$

where B_n are the Bernoulli numbers

$$B_1 = \frac{1}{6}$$

$$B_2 = \frac{1}{30}$$

$$B_3 = \frac{1}{42}$$

$$B_4 = \frac{1}{30}$$

$$B_5 = \frac{5}{66}$$

$$B_n = \frac{(2n)!}{2^{2n-1}(\pi)^{2n}}\left(1 + \frac{1}{2^{2n}} + \frac{1}{3^{2n}} + \cdots\right)$$

and where the factorial symbol (!) is defined as

$$n! = n(n-1)(n-2)\cdots(3)(2)(1)$$

Trigonometric functions to any accuracy can be calculated from the series if enough terms are used.

For small angles, on the order of 5° or less, it may be sufficient to use only the first term of each series.

$$\sin\theta \cong \theta$$

$$\cos\theta \cong 1$$

$$\tan\theta \cong \theta$$

Functional Relationships

$$\tan\,\theta = \frac{\sin\,\theta}{\cos\,\theta}$$

$$\sin^2\,\theta + \cos^2\,\theta = 1$$

$$\sec^2\,\theta - \tan^2\,\theta = 1$$

$$\csc^2\,\theta - \cot^2\,\theta = 1$$

$$\sin\,\theta = \cos\,(90^\circ - \theta) = \sin(180^\circ - \theta)$$

$$\cos\,\theta = \sin\,(90^\circ - \theta) = -\cos\,(180^\circ - \theta)$$

$$\tan\,\theta = \cot\,(90^\circ - \theta) = -\tan\,(180^\circ - \theta)$$

$$\sin\,(-\theta) = -\sin\,\theta$$

$$\cos\,(-\theta) = \cos\,\theta$$

$$\tan\,(-\theta) = -\tan\,\theta$$

Sum of Angles Formulas

$$\sin\,(\theta \pm \alpha) = \sin\,\theta\,\cos\,\alpha \pm \cos\,\theta\,\sin\,\alpha$$

$$\cos\,(\theta \pm \alpha) = \cos\,\theta\,\cos\,\alpha \mp \sin\,\theta\,\sin\,\alpha$$

$$\tan\,(\theta \pm \alpha) = \frac{\tan\,\theta \pm \tan\,\alpha}{1 \mp \tan\,\theta\,\tan\,\alpha}$$

Multiple-Angle Formulas

$$\sin\,2\theta = 2\,\sin\,\theta\,\cos\,\theta$$

$$\cos\,2\theta = \cos^2\,\theta - \sin^2\,\theta = 2\,\cos^2\,\theta - 1 = 1 - 2\,\sin^2\,\theta$$

$$\tan\,2\theta = \frac{2\,\tan\,\theta}{1 - \tan^2\,\theta}$$

$$\sin\,3\theta = 3\,\sin\,\theta - 4\,\sin^3\,\theta$$

$$\cos\,3\theta = 4\,\cos^3\,\theta - 3\,\cos\,\theta$$

$$\tan\,3\theta = \frac{3\,\tan\,\theta - \tan^3\,\theta}{1 - 3\,\tan^2\,\theta}$$

$$\sin\,\frac{\theta}{2} = \pm\sqrt{\frac{1 - \cos\,\theta}{2}} \qquad \left(\text{sign depends on quadrant of } \frac{\theta}{2}\right)$$

$$\cos\,\frac{\theta}{2} = \pm\sqrt{\frac{1 + \cos\,\theta}{2}} \qquad \left(\text{sign depends on quadrant of } \frac{\theta}{2}\right)$$

$$\tan \frac{\theta}{2} = \pm \sqrt{\frac{1 - \cos \theta}{1 + \cos \theta}} = \frac{1 - \cos \theta}{\sin \theta}$$

$$= \frac{\sin \theta}{1 + \cos \theta} = \csc \theta - \cot \theta$$

$$\left(\text{sign depends on quadrant of } \frac{\theta}{2}\right)$$

Sum, Difference, and Product Formulas

$$\sin \theta + \sin \alpha = 2 \sin \left(\frac{\theta + \alpha}{2}\right) \cos \left(\frac{\theta - \alpha}{2}\right)$$

$$\sin \theta - \sin \alpha = 2 \cos \left(\frac{\theta + \alpha}{2}\right) \sin \left(\frac{\theta - \alpha}{2}\right)$$

$$\cos \theta + \cos \alpha = 2 \cos \left(\frac{\theta + \alpha}{2}\right) \cos \left(\frac{\theta - \alpha}{2}\right)$$

$$\cos \theta - \cos \alpha = 2 \sin \left(\frac{\theta + \alpha}{2}\right) \sin \left(\frac{\alpha - \theta}{2}\right)$$

$$\sin \theta \sin \alpha = \frac{1}{2}\left[\cos(\theta - \alpha) - \cos(\theta + \alpha)\right]$$

$$\cos \theta \cos \alpha = \frac{1}{2}\left[\cos(\theta - \alpha) + \cos(\theta + \alpha)\right]$$

$$\sin \theta \cos \alpha = \frac{1}{2}\left[\sin(\theta - \alpha) + \sin(\theta + \alpha)\right]$$

Power Formulas

$$\sin^2 \theta = \frac{1}{2} - \frac{1}{2} \cos 2\theta$$

$$\cos^2 \theta = \frac{1}{2} + \frac{1}{2} \cos 2\theta$$

$$\sin^3 \theta = \frac{3}{4} \sin \theta - \frac{1}{4} \sin 3\theta$$

$$\cos^3 \theta = \frac{3}{4} \cos \theta + \frac{1}{4} \cos 3\theta$$

Graphics

C.1 Introduction

As an engineer you will find that the ability to communicate graphically is an essential part of your professional activity. The old proverb "A picture is worth a thousand words" demonstrates graphics to be a valuable resource in the solution of engineering problems.

Graphics is the foundation of design and as such it is an essential method of communication from conceptualization to manufacture. The topics presented in App. C are intended to supplement Chap. 14. Each subject area is briefly and concisely outlined with illustrations to demonstrate fundamental concepts.

C.2 Engineering Lettering

Engineers in their daily professional activities find it necessary to maintain legible permanent records. Much of the work that they do must be communicated to others. This includes not only drawings but also notes, calculations, etc. It is important that en-

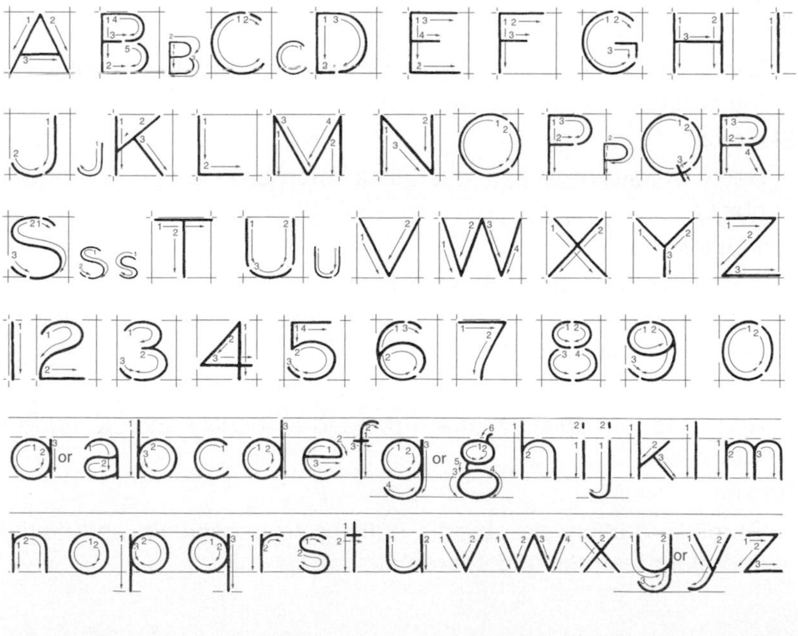

Figure C.1 Gothic letters.

gineers develop a lettering style and an ability to render that is simple, fast, and functional.

There are many types of letters, but perhaps the most universally accepted is the Gothic type illustrated in Fig. C. 1. Either uppercase or lowercase, vertical or slant letters are satisfactory. Perhaps it is wise at the start of an engineering education to select a lettering style and use it consistently, since practice develops the skill.

Guidelines are very helpful when lettering, since they ensure uniform height of both letters and numbers. Guidelines can be constructed freehand for many types of work.

C.3 Freehand Drawing

Freehand drawing, often referred to as sketching, differs from instrument work primarily in the amount of time and accuracy required. The end result, however, should be a clear, concise illustration suitable for the intended purpose. The ultimate use of the sketch will dictate the construction time and degree of accuracy needed.

The equipment necessary to do freehand work is quite simple: a pencil, soft eraser, and paper. For presentations or other specific communication tasks, for example, transparencies, flip charts, etc., the equipment needed may vary, but it is usually not very complex.

C.3.1 Key Elements in Sketching

Independent of the type of sketch or its eventual purpose there are a number of steps that must be completed. Each step may be emphasized differently depending on the purpose of the drawing, but all steps are to be considered any time a sketch is constructed.

Plan

Conceive the sketch with respect to size, orientation, location on the paper and degree of accuracy.

Skeleton

Begin by making a light, overall skeleton box construction of the object.

Proportions

Carefully check all geometric features as they are lightly constructed. Necessary changes should be made at each stage of the construction process.

Details

Add important details that are needed to identify and distinguish the object.

Darken

Carefully darken the object's outlines to accentuate and clarify important features and characteristics.

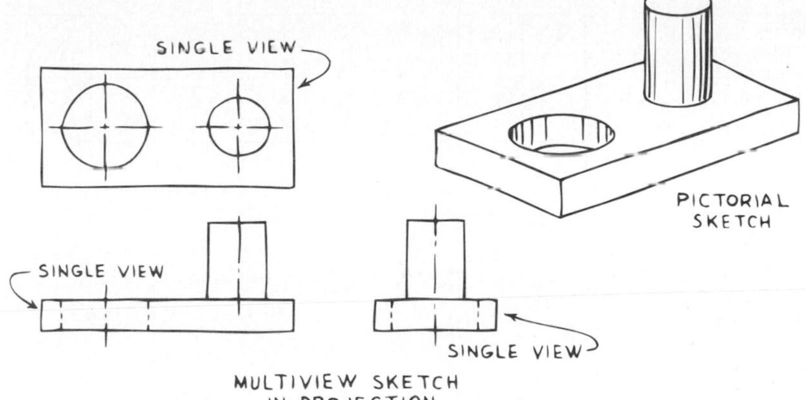

Figure C.2 Freehand drawings.

Label

Letter titles and notes in a neat, legible fashion with standard engineering numbers and letters.

These steps are general and apply to all types of freehand construction.

C.3.2 Types of Freehand Drawing

Three types of freehand drawings will be considered, as illustrated in Fig. C.2.

1. Single view: Objects with primarily two dimensions: that is, maps, charts, diagrams, graphs, etc. Included in this definition are single orthographic views or one view of multiview drawings.

2. Pictorial: Objects with three dimensions illustrated: that is, length, width, and height. It is an attempt to show in a single drawing what the eye would see.

3. Multiview: Separate, single orthographic views oriented in adjacent related positions to describe an object completely. Multiview drawings are discussed in Sec. C.4.

C.3.3 Construction of Single Views

The construction of single views primarily requires the ability to draw parallel lines and circles.

Circles should be sketched without construction lines only if the sketch is very rough or if the circles are small. See Fig. C.3.

A helpful construction technique is illustrated in Fig. C.4. The center of the circle is at point 0, so an arc can be drawn from *A* through *E* to *B* as follows. Construct a diagonal from *A* to *B* and then divide the line *DC* again. This will locate point *E*, which is very close to the exact location for the circular arc.

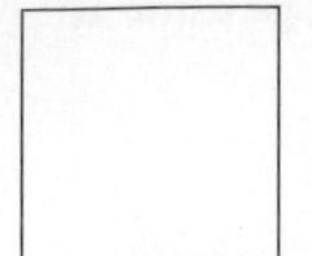

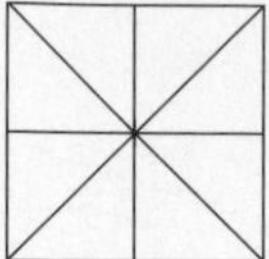
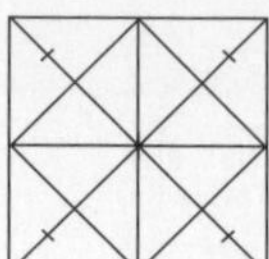
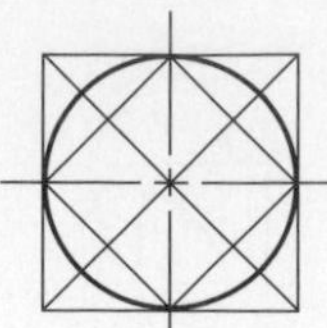

Figure C.3
Construction of a circle.

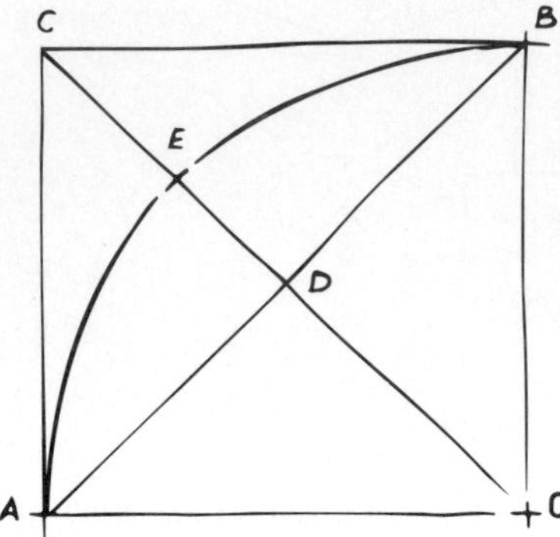

Figure C.4
Circle construction technique.

C.3.4 Construction of Pictorials

The correct selection of the axes is critical when constructing a pictorial. If all faces require equal emphasis, an isometric selection would be appropriate, as seen in Fig. C.5*a*, whereas oblique would better illustrate major features in one or parallel planes (Fig. C.5*b*).

A key element in the construction of pictorial sketches is the ellipse. Circles in the major plane of an oblique drawing are circles, but in the receding planes of most pictorials they are elliptical. The construction of an ellipse is illustrated in Fig. C.6.

Note the construction of step 4 in Fig. C.6. This procedure is identical to the construction technique developed in Fig. C.4.

Figure C.5
Axes selections.

Figure C.6
Ellipse construction.

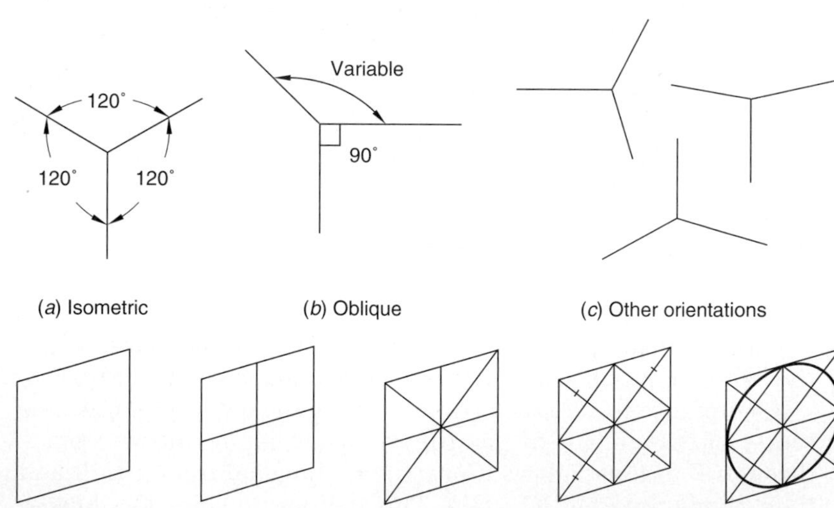

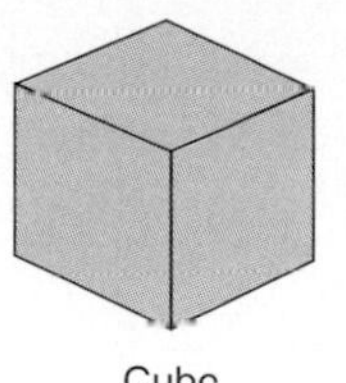

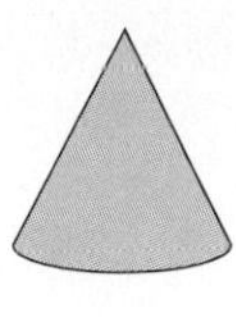

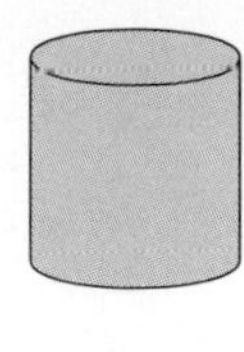

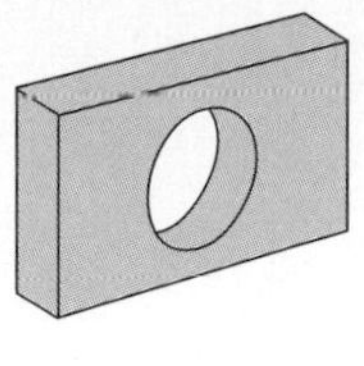

Figure C.7
Basic shapes.

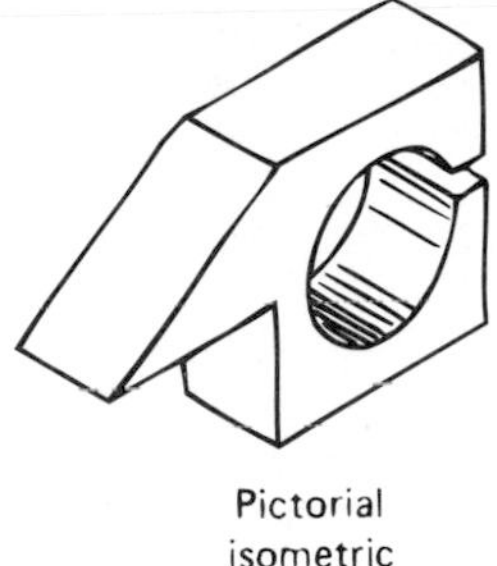

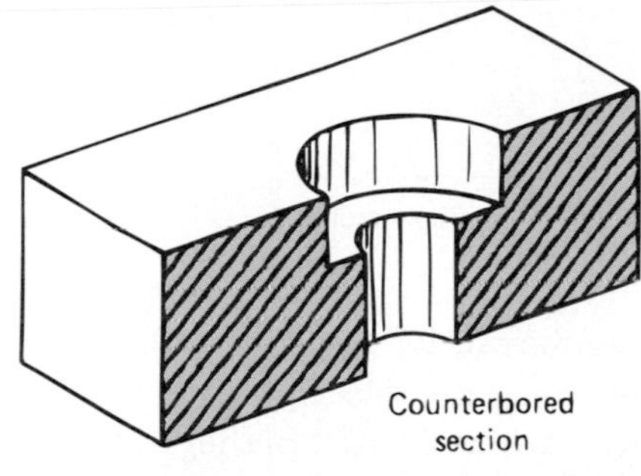

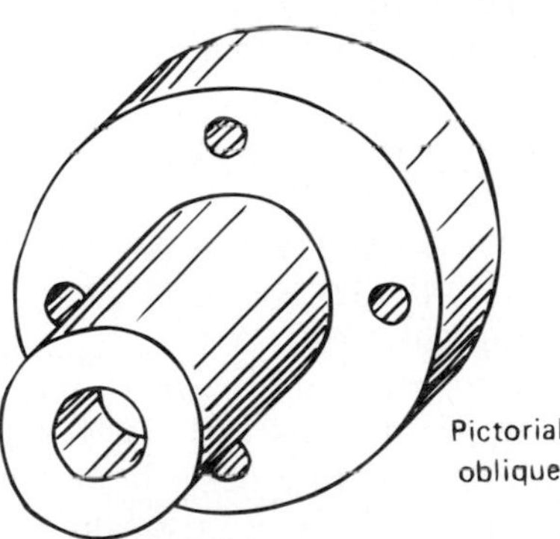

Figure C.8
Pictorials.

Although it is not precisely correct mathematically, it is helpful when sketching an ellipse.

The construction of many pictorials is a simple adaptation of the four basic shapes illustrated in Fig. C.7.

Practical application of the foregoing construction principles and six steps of sketching are demonstrated in the examples of Fig. C.8.

C.4 Multiview Drawings

The precise definition and delineation of objects can best be represented by a series of carefully selected single views. Although pictorials are an excellent method of conveying a visual image of the object, they do not provide the detail needed for manufacturing.

Multiview drawings define an object by placing the correct number of properly constructed single views in correct orthographic alignment. Figure C.9 illustrates a simple object show-

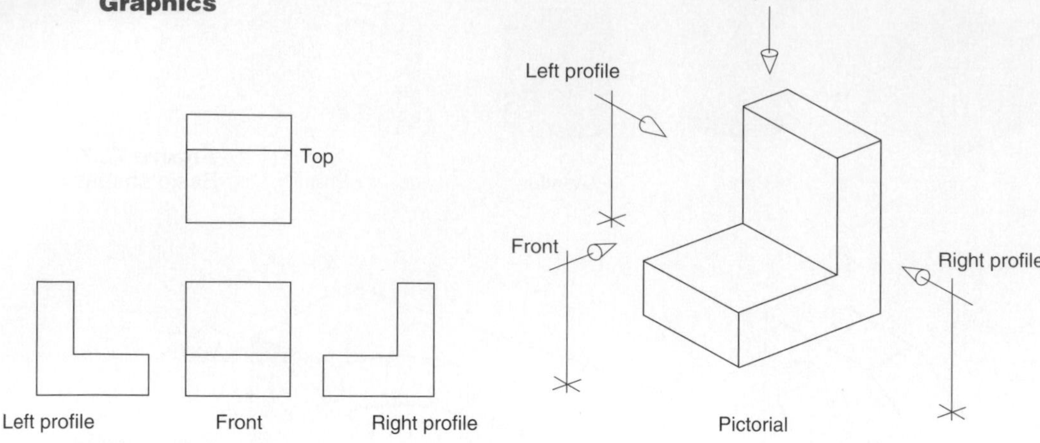

Figure C.9
Multiveiw drawings.

ing four of its six principal views, two of which are not essential for complete description of the object but are shown only to clarify various correct orthographic positions.

The number of views necessary to describe any object obviously varies with the complexity of the object, but constructing views that are not needed is a waste of time and money.

Objects with more detail may require more careful study because all contours must be represented in the correct construction of multiviews. Figure C.10*a* illustrates the use of centerlines; and Fig. C.10*b* shows proper use of both hidden lines and centerlines. Figure C.11 illustrates object lines, hidden lines, and centerlines on a single object.

The correct or necessary number of views that adequately describe an object is illustrated in Fig. C.12. The top and front views, in this case, do not completely describe a single object, but the five different right profile views, each taken independently with the given front view, should adequately describe five different objects.

An optimum number of properly constructed views is the way to communicate objects graphically without confusion and misunderstanding.

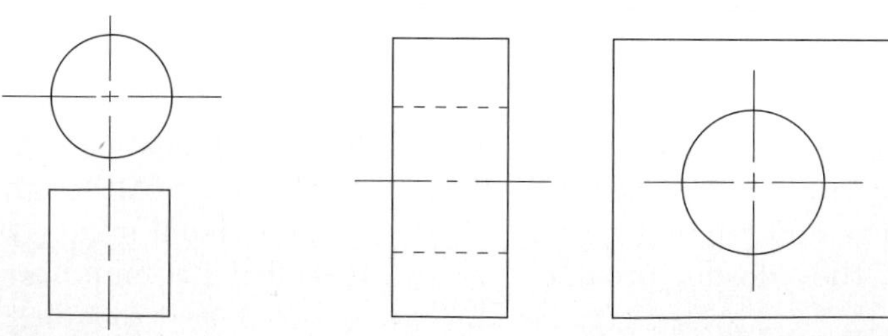

Figure C.10
Centerlines and hidden lines.

(*a*) Cylinder (*b*) Negative cylinder

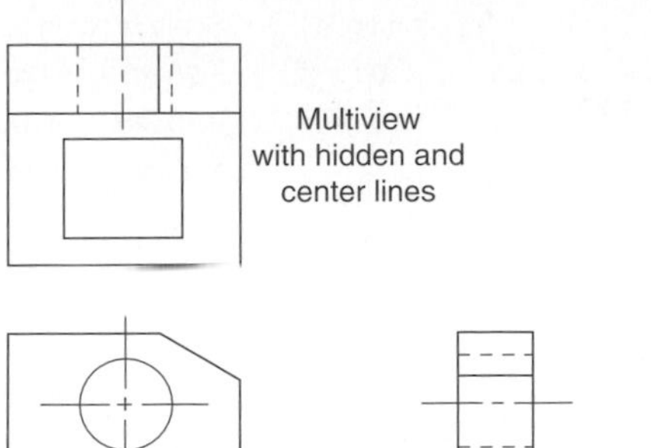

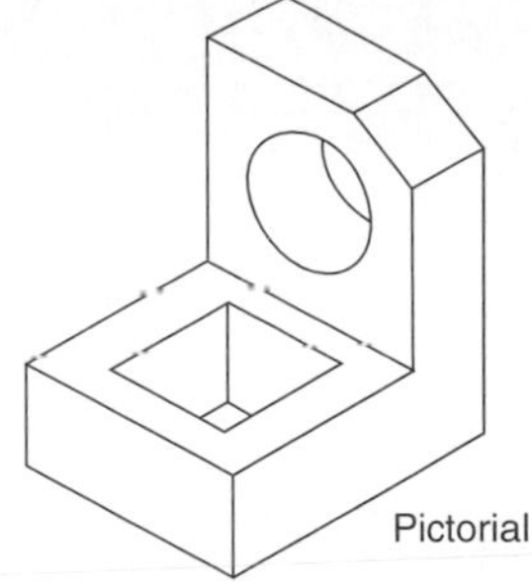

Figure C.11
Multiview drawing.

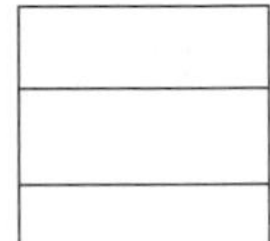

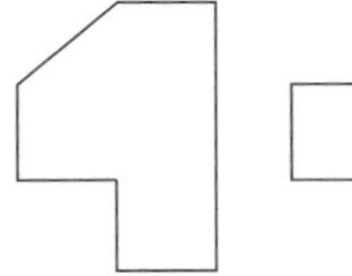

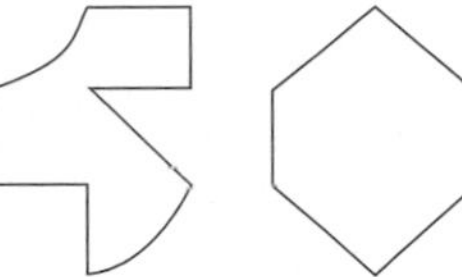

Figure C.12
Necessary views.

C.5 Scales

Graduations on metric scales are identified as a ratio: 1:100, 1:20, etc. This ratio signifies the drawing reduction from actual size; for example, 1:100 indicates that 1 unit on the drawing represents 100 units on the object.

A metric scale labeled with a ratio of 1:100 signifies that the distance from 0 to the number 1 is to represent a meter, as illustrated in Fig. C.13.

If you change the ratio from 1:100 to 1:10 but use the same scale, the distance from the 0 to the 1 now represents 0.1 m. Table C.1 demonstrates this functional characteristic of metric scales.

Table C.1 can be constructed for any metric scale or ratio. Figure C.14 illustrates how the scale varies with different ratios.

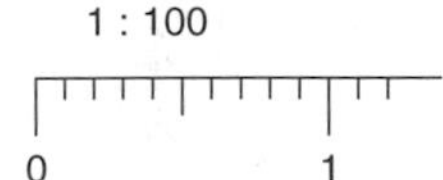

Figure C.13
Scale ratio.

Table C.1

Ratio	Distance from 0 to 1.0 is equal to
1:100	1 m = 10 dm = 100 cm = 1 000 mm
1:10	0.1 m = 1 dm = 10 cm = 100 mm
1:1	0.01 m = 0.1 dm = 1 cm = 10 mm
1:0.1	0.001 m = 0.01 dm = 0.1 cm = 1 mm

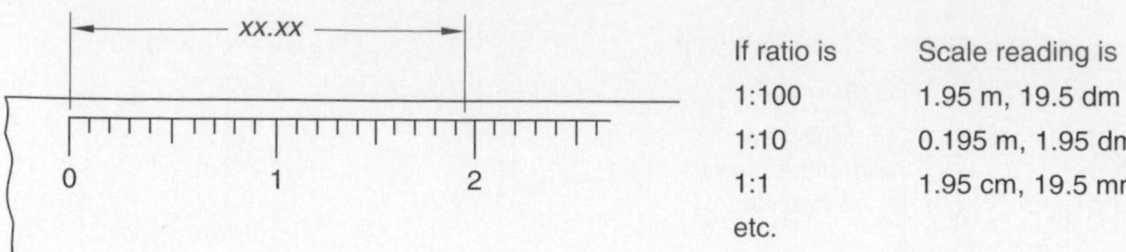

Figure C.14 Relation between ratio and scale reading.

C.6 Dimensions

In order that objects might be perceived in terms of precise physical measurement, dimensions must be correctly indicated. The size of each geometric shape, together with necessary location dimensions, has to be clearly understood.

Many standards that provide considerable uniformity to dimensioning practice have been recognized and adopted. One of the most widely used is that issued by the American National Standards Institute.

C.6.1 Definition of Terms

Dimensions are the numbers expressed in consistent units and used to indicate the physical lengths.

Dimension lines and arrowheads indicate the extent of the measurement. Arrowheads should be closed and consistently uniform in size and shape. Dimension lines are placed about 10 mm from the object and 6 mm from each other, as illustrated in Fig. C.15.

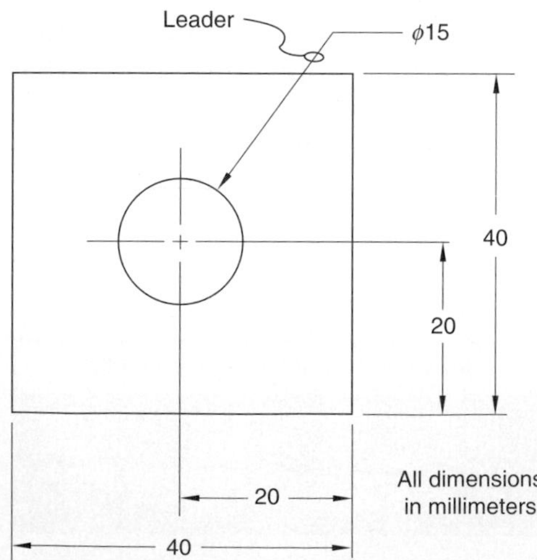

Figure C.15 Proper spacing.

Extension lines are used to locate the extension of the surface, and leaders are used to dimension circles or to direct notes to a specific place.

Geometric Shapes

A prism, cylinder, cone, and right pyramid are correctly dimensioned in Fig. C.16.

Holes, Radii, and Arcs

The illustrations in Fig. C.17 demonstrate correct dimensioning procedures for *through* holes and *blind* holes.

Circular arcs are dimensioned as shown in Fig. C.18. For castings with a large number of fillets and rounds, it is customary to

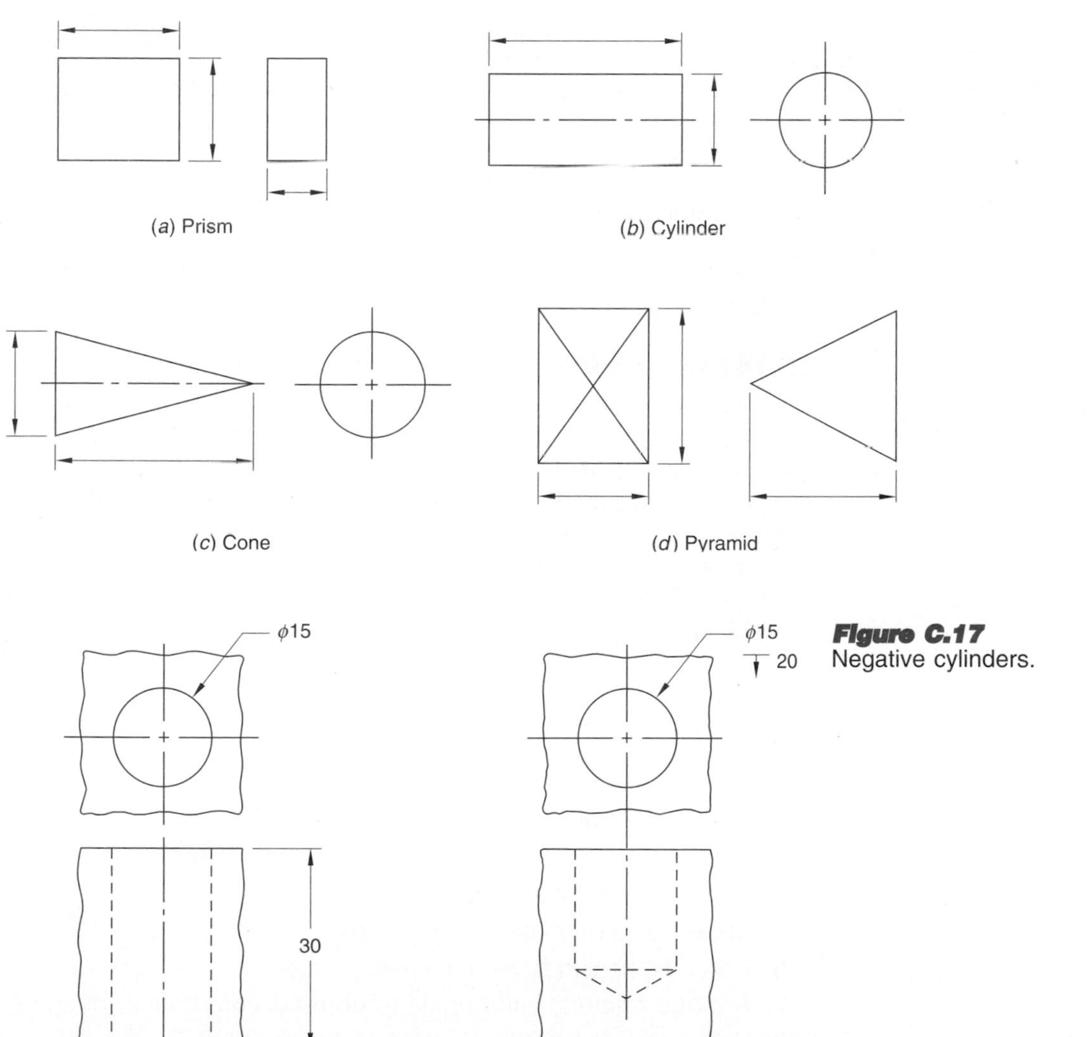

Figure C.16
Dimensioning basic shapes.

Figure C.17
Negative cylinders.

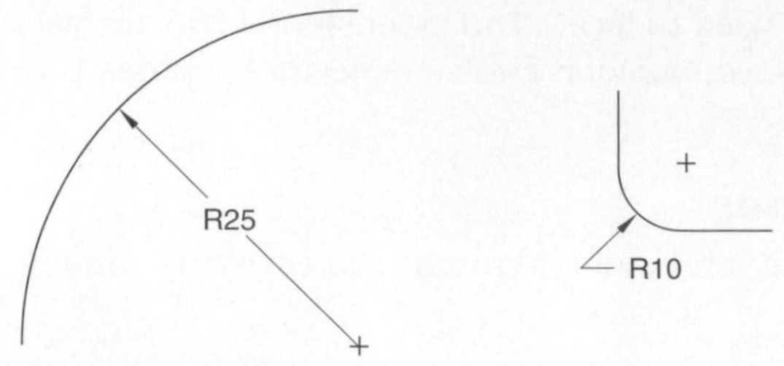

Figure C.18
Radii.

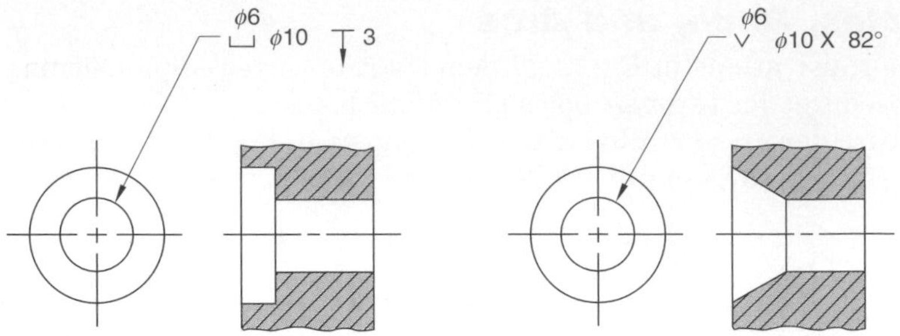

Figure C.19
Counterbore and countersink.

All dimensions in millimeters

indicate all radii by use of a general note: for example, "all radii 8 mm unless otherwise specified."

Counterbored and Countersunk Holes (see Fig. C.19)

Contour Rule

One of the most important guides to follow when dimensioning is *always* to *place the dimension in the view that shows the most characteristic feature.* Select a scale for the drawing, and use a minimum but sufficient number of views to completely describe and accurately dimension the object. Generally, three overall dimensions are included on all parts.

Figure C.20 illustrates three correctly dimensioned views of an object.

C.6.2 Recommended Dimensioning Practices

1. Never duplicate dimensions.
2. Don't crowd dimensions.
3. Place dimensions *between* views, not *on* views.
4. Avoid crossing dimension lines.
5. Break dimension lines at numerals.
6. Don't dimension to hidden lines.
7. Arrange numerals and notes to be read from bottom of sheet.
8. Use standard height for letters and numerals.
9. Provide dimensions so that calculations are not necessary.

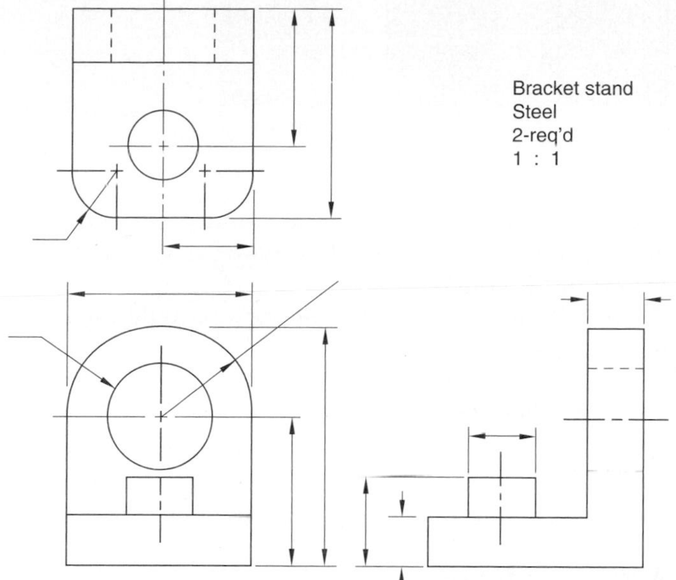

Figure C.20
Detail drawing with complete title.

10. Measure lengths in metric units.
11. Don't allow extension lines to touch object.
12. Use centerlines as extension lines when appropriate.

C.7 Sections

At times, visualization of an object that has considerable interior complexity can be enhanced by taking an appropriate section view. An imaginary plane, as illustrated in Fig. C.21, is cut through the object and the near portion is taken away, thereby exposing the interior (Fig. C.21*b*).

Figure C.21
Cutting plane.

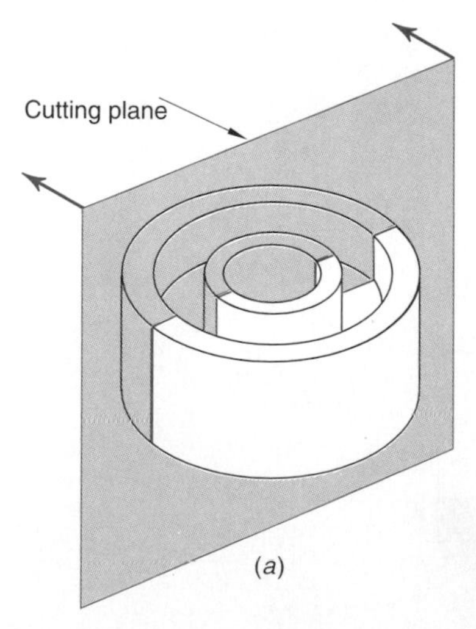

(*a*)

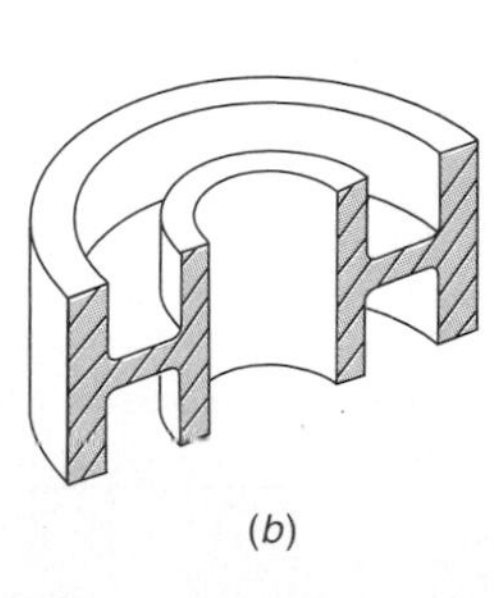

(*b*)

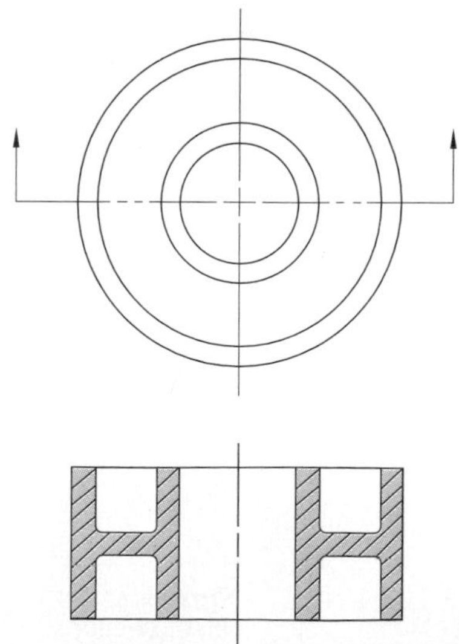

(*c*) Full section

Cast iron, or general

Steel

Wood

Figure C.22 Section lining.

Solids that are cut by the imaginary plane are cross-hatched according to preestablished standards. Three of these are illustrated in Fig. C.22.

Six different types of section views are defined and illustrated here.

Full Section

A full section results when the cutting plane passes completely through the object, as illustrated in Fig. C.21*c*. Hidden lines behind the cutting plane are normally omitted unless essential.

Figure C.23 Half-section.

Half-section

Half Sections

Half sections can be used on symmetrical objects when it is desirable to show an internal detail as well as a view of the exterior. Note that separation of the inside and outside is by a centerline, as seen in Fig. C.23.

Offset Section

An offset plane goes completely through the object but is staggered rather than straight. A convention normally practiced allows the lines of demarcation (change in direction of cutting plane) to be omitted in the sectioned view. The offset section is illustrated in Fig. C.24.

Revolved Section

Revolved sections are used to show the shape and contour of ribs, spokes, etc. The plane is passed through the object and the cut

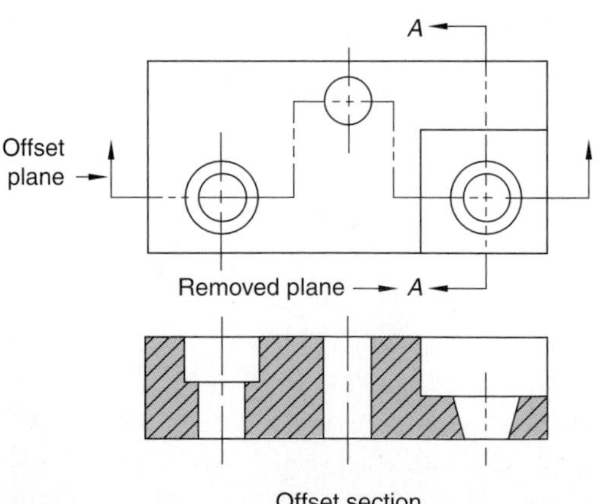

Figure C.24 Offset section.

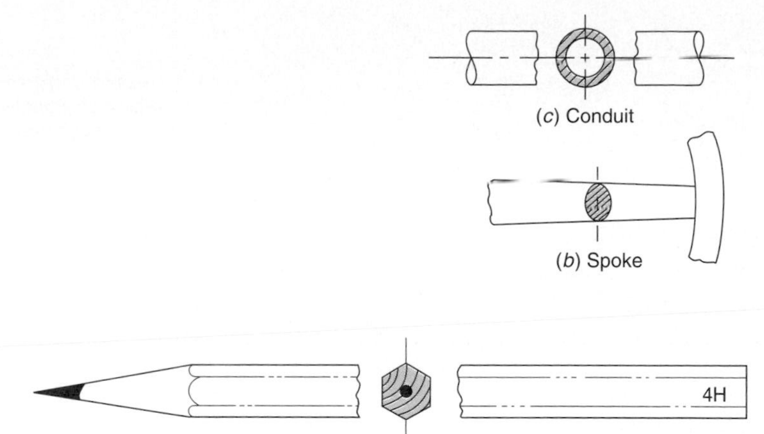

Figure C.25
Revolved section.

area is revolved 90° and cross-hatched. Examples of this section can be seen in Fig. C.25*a*, *b*, and *c*.

Removed Section

There are two significant advantages to a removed section. First, it can be moved to a separate location. A second advantage is that the scale can be changed, e.g., drawn at a larger scale. The cutting plane for the removed section illustrated in Fig. C.26 is found in Fig. C.24.

Figure C.26
Removed section.

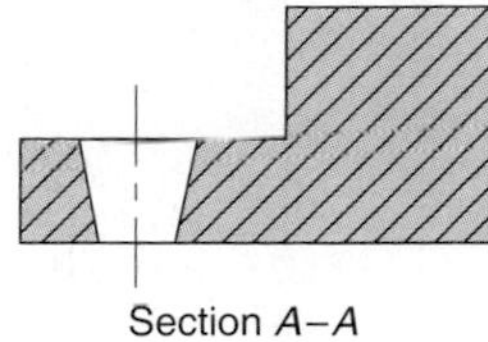

Partial Section

A partial section is used when a portion of an orthographic view or pictorial is broken away to expose the interior (see Fig. C.27). The cross-hatching indicates the type of material from which the object is constructed.

C.8 Design Drawings

Although each of the previous sections can be used at different times in the design process, they must be tied together to describe a total system more completely. It is the purpose of this section to outline some of the graphics necessary for a design to be delineated.

C.8.1 Layout Drawings

A layout is a very accurate, scaled instrument drawing used to determine operational characteristics, such as clearances, and the relation of one part to another.

Figure C.29 is a pictorial of an assembly, whereas Fig. C.28 is a layout to determine critical clearances at points 1 and 2. Because of the drawing time involved, the layout drawing shows a minimum of information. Centerlines and key features are used to verify operation.

Figure C.27
Partial section.

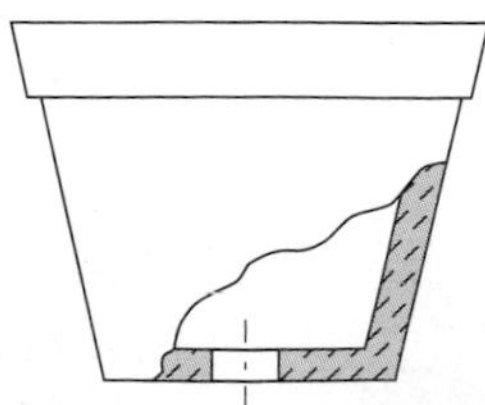

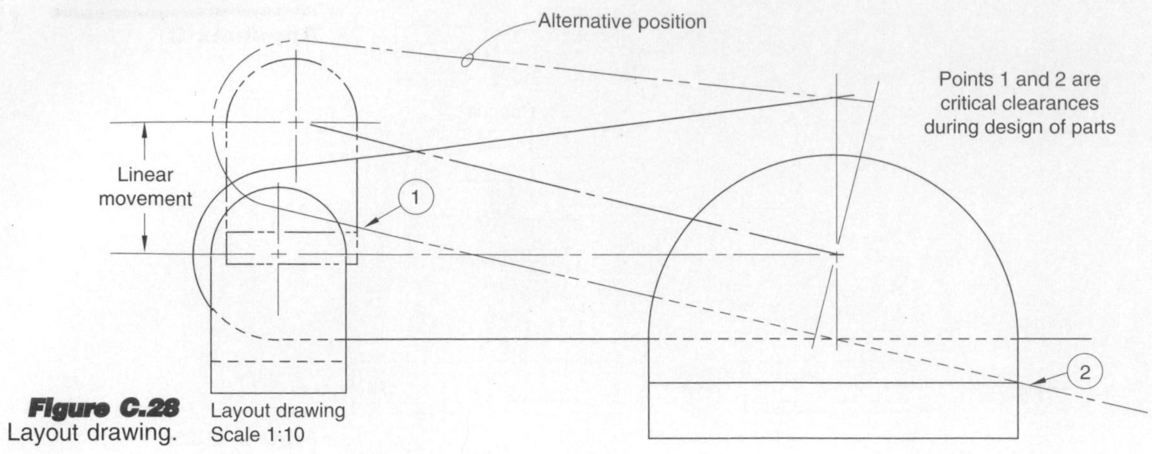

Figure C.28
Layout drawing.

C.8.2 Assembly Drawings

Assembly drawings illustrate, either in pictorial or orthographic form, the individual parts assembled. Figures C.29 and C.30 are both examples of assembly drawings. This type of drawing has several important functions. It demonstrates how the entire collection of individual items fit together. Critical dimensions, centerline dimensions, and overall dimensions are normally included to specify clearances, and so on. The drawing provides an opportunity to investigate the relationship of individual parts as they move or rotate to alternate positions.

Figure C.29
Pictorial assembly.

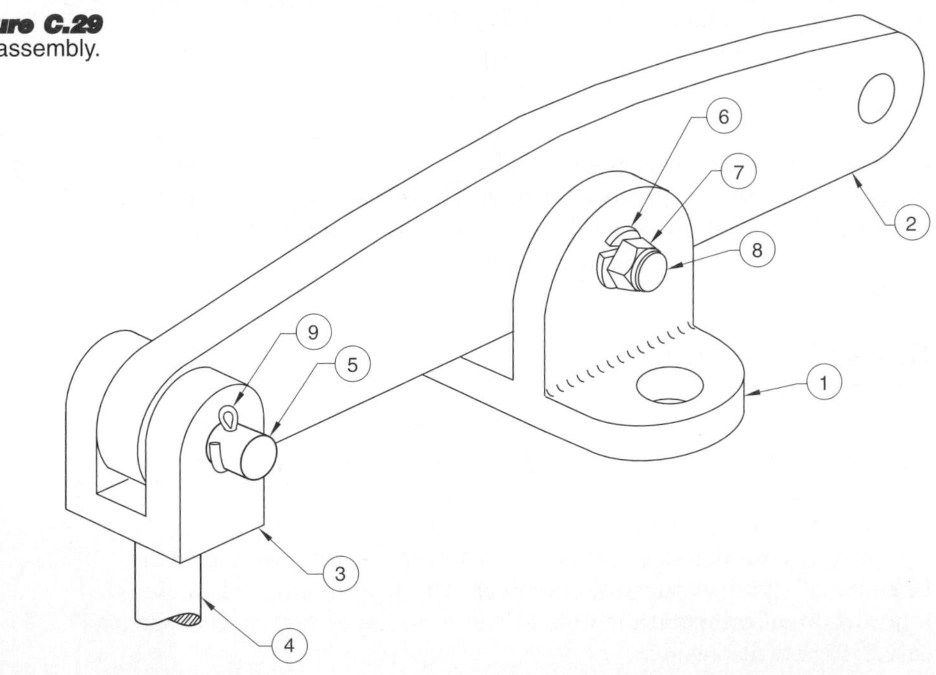

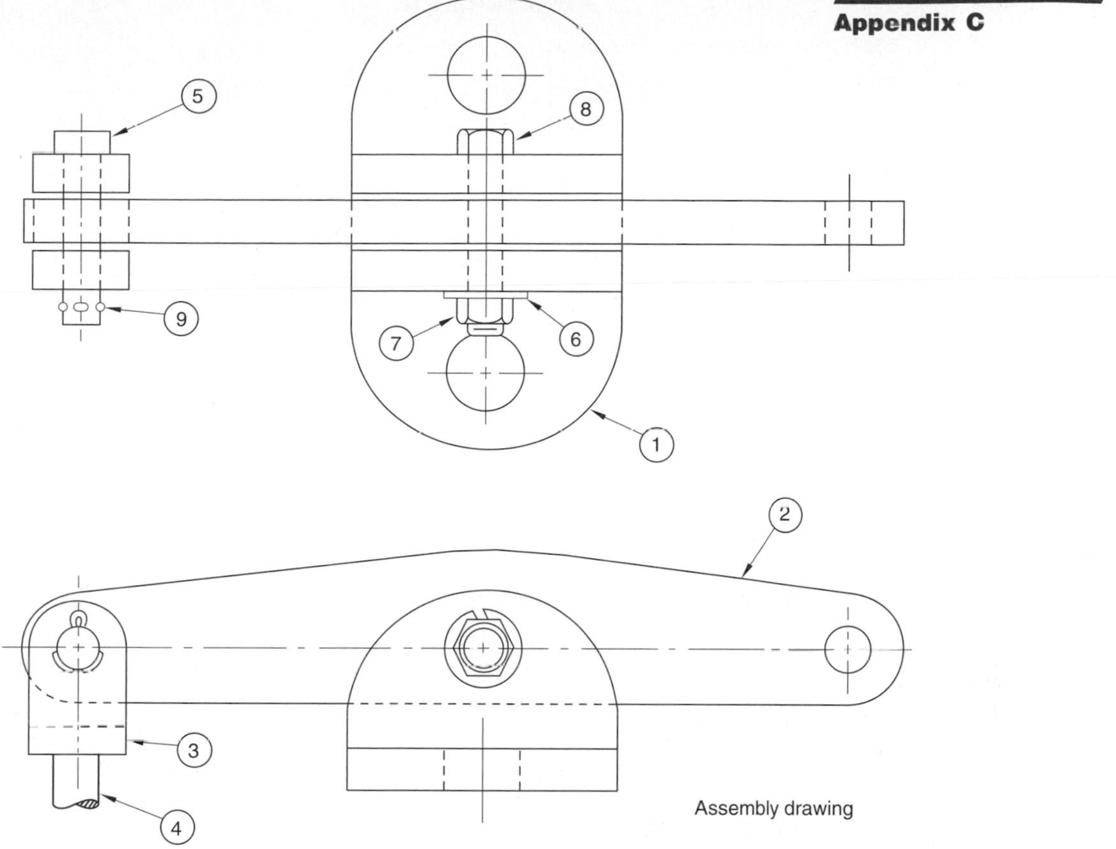

Figure C.30
Orthographic assembly.

Balloons (numbers within circles) and leaders pointing to the individual parts establish an identification system for the assembly of parts. Detail drawings (Sec. C.8.3) are keyed to the assembly by use of these identification numbers. The bill of material (Sec. C.8.4) also uses the same identification system.

C.8.3 Detail Drawings

A detail drawing, as illustrated in Fig. C.31, is a drawing of a *single* part completely specified. By working with the assembly and layout drawings the detail can be completed so that the individual part can be made. Standard parts, that is, nuts, bolts, etc., are not customarily detailed.

A detail drawing will consist of sufficient views, completely dimensioned, with appropriate section views, and a title that includes the identification number, name, number required in assembly, material, scale, and necessary notes.

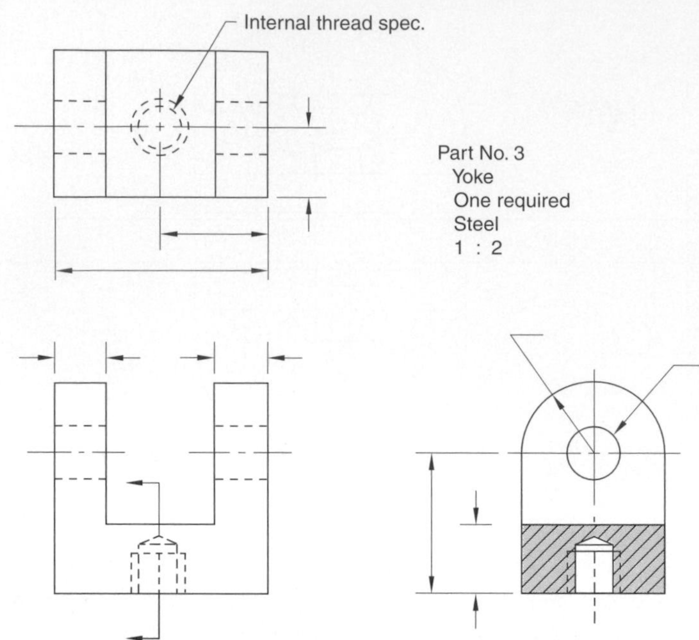

Figure C.31 Detail drawing.

C.8.4 Bill of Material

Every item in the assembly should appear in the bill of materials (see Tab. C.2). Materials, sizes, notes, etc., are added as the assembly is formalized.

C.9 Presentation Drawings

It is often necessary for the engineer to present data to people who are not familiar with technical graphs. A few of the many methods available to do this are included below.

Pie Diagrams

Pie diagrams are most popular when representing items that total 100 percent. All lettering and percentages should be placed on the sector or immediately adjacent. The circle should not con-

Table C.2

NO	NAME	# REQ'D	MATERIAL/SIZE AND NOTES
8			
7	HEX NUT		
6	WASHER	2	
5	PIN	1	
4	ROD	1	RD BAR STOCK
3	YOKE	1	CAST
2	ARM	1	STEEL
1	BASE	1	CAST

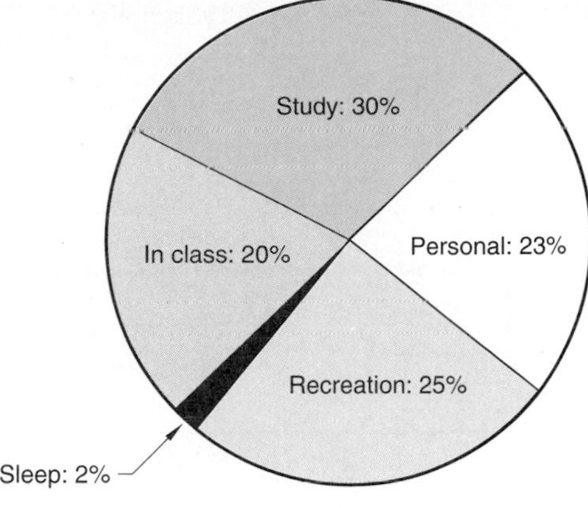

Figure C.32
Pie diagram. (Data collected at Iowa State University.)

tain more than five or six categories and each should be cross-hatched or marked differently. A title with pertinent information concerning source, and so on, should always be included. See Fig. C.32.

Column Charts

One of the most common nontechnical methods of representing information is the column chart, illustrated in Fig. C.33.

The quantity to be graphed is illustrated by bars, each of whose length is proportional to the value represented.

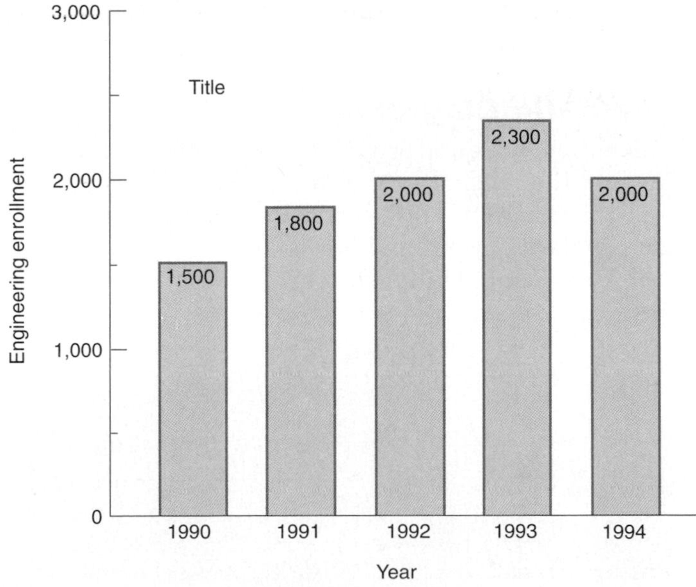

Figure C.33
Column chart.

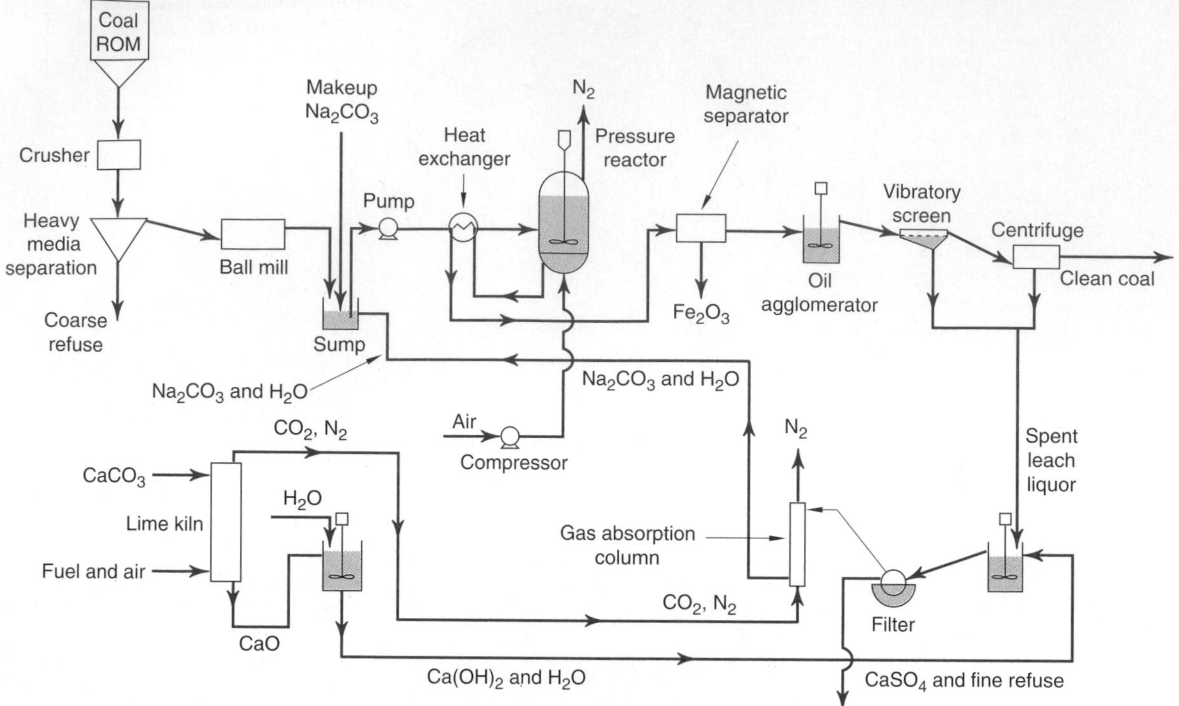

Figure C.34
Block diagram.

Block Diagrams

A block diagram, as represented in Fig. C.34, is an excellent technique to show in a simple fashion the overall process. Different symbols denote certain processes with size and balance considerations important for ease of understanding.

Schematic Diagrams

Combination block and schematic diagrams, as illustrated in Fig. C.35, and the schematic diagram in Fig. C.36 convey the operational characteristics of electrical as well as other complex systems. Symbols, layout, and labels provide a graphical expression of the functional relationship or interconnection of component parts.

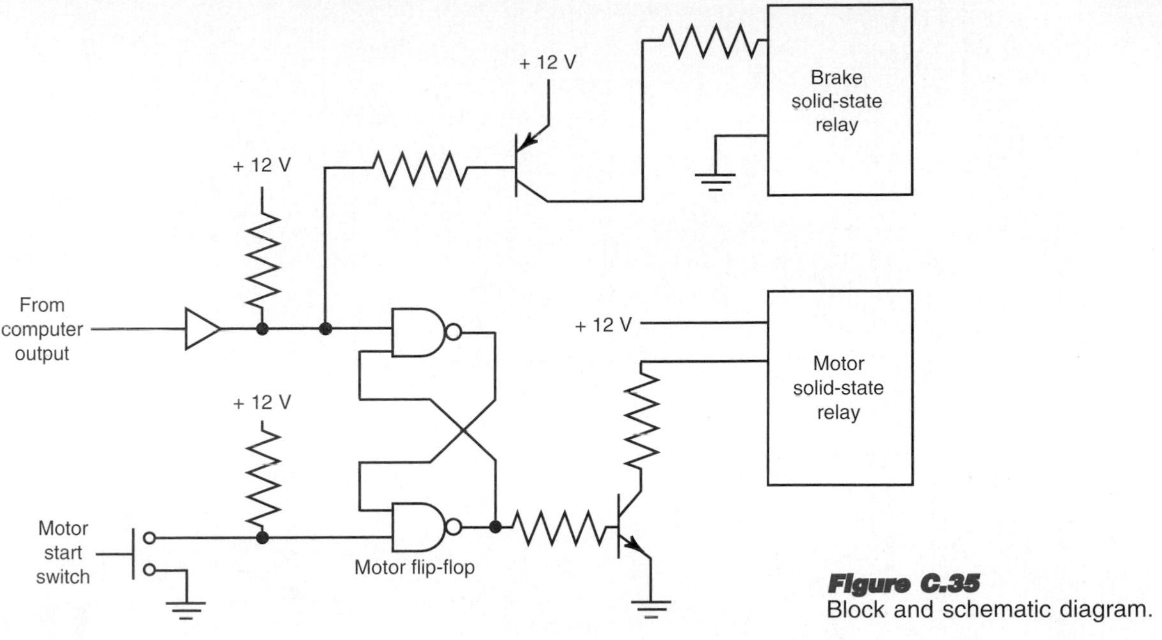

Figure C.35
Block and schematic diagram.

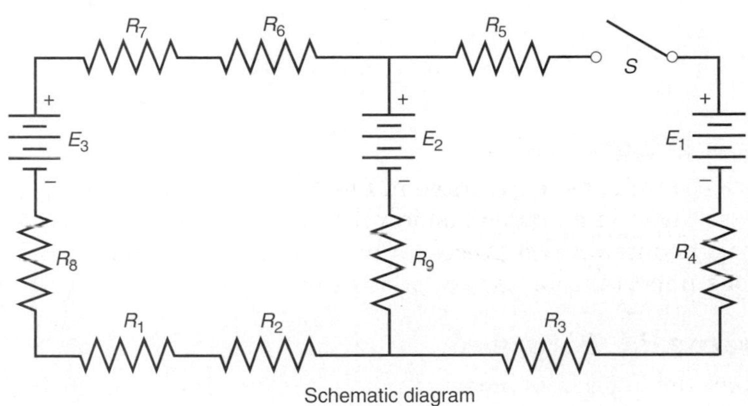

Figure C.36
Schematic diagram.

General

Approximate Specific Gravities and Densities

Material	Specific gravity	Average density lbm/ft^3	Average density kg/m^3
Gases (0°C and 1 atm)			
Air		0.080 18	1.284
Ammonia		0.048 13	0.771 0
Carbon dioxide		0.123 4	1.977
Carbon monoxide		0.078 06	1.251
Ethane		0.084 69	1.357
Helium		0.011 14	0.178 4
Hydrogen		0.005 611	0.089 88
Methane		0.044 80	0.717 6
Nitrogen		0.078 07	1.251
Oxygen		0.089 21	1.429
Sulfur dioxide		0.182 7	2.927
Liquids (20°C)			
Alcohol, ethyl	0.79	49	790
Alcohol, methyl	0.80	50	800
Benzene	0.88	55	880
Gasoline	0.67	42	670
Heptane	0.68	42	680
Hexane	0.66	41	660
Octane	0.71	44	710
Oil	0.88	55	880
Toluene	0.87	54	870
Water	1.00	62.4	1 000
Metals (20°C)			
Aluminum	2.55–2.80	165	2 640
Brass, cast	8.4–8.7	535	8 570
Bronze	7.4–8.7	510	8 170
Copper, cast	8.9	555	8 900
Gold, cast	19.3	1 210	19 300
Iron, cast	7.04–7.12	440	7 050
Iron, wrought	7.6–7.9	485	7 770
Iron ore	5.2	325	5 210
Lead	11.3	705	11 300
Manganese	7.4	462	7 400
Mercury	13.6	849	13 600
Nickel	8.9	556	8 900
Silver	10.4–10.6	655	10 500
Steel, cold drawn	7.83	489	7 830
Steel, machine	7.80	487	7 800
Steel, tool	7.70	481	7 700
Tin, cast	7.30	456	7 300
Titanium	4.5	281	4 500
Uranium	18.7	1 170	18 700
Zinc, cast	6.9–7.2	440	7 050

Nonmetallic Solids (20°C)

Material	Specific gravity	Average density	
		lbm/ft^3	kg/m^3
Brick, common	1.80	112	1 800
Cedar	0.35	22	350
Clay, damp	1.8–2.6	137	2 200
Coal, bituminous	1.2–1.5	84	1 350
Concrete	2.30	144	2 300
Douglas fir	0.50	31	500
Earth, loose	1.2	75	1 200
Glass, common	2.5–2.8	165	2 650
Gravel, loose	1.4–1.7	97	1 550
Gypsum	2.31	144	2 310
Limestone	2.0–2.9	153	2 450
Mahogany	0.54	34	540
Marble	2.6–2.9	172	2 750
Oak	0.64–0.87	47	750
Paper	0.7–1.2	58	925
Rubber	0.92–0.96	59	940
Salt	0.8–1.2	62	1 000
Sand, loose	1.4–1.7	97	1 550
Sugar	1.61	101	1 610
Sulfur	2.1	131	2 100

Unit Prefixes

Multiple and submultiple	Prefix	Symbol
1 000 000 000 000 = 10^{12}	tera	T
1 000 000 000 = 10^{9}	giga	G
1 000 000 = 10^{6}	mega	M
1 000 = 10^{3}	kilo	k
100 = 10^{2}	hecto	h
10 = 10	deka	da
0.1 = 10^{-1}	deci	d
0.01 = 10^{-2}	centi	c
0.001 = 10^{-3}	milli	m
0.000 001 = 10^{-6}	micro	μ
0.000 000 001 = 10^{-9}	nano	n
0.000 000 000 001 = 10^{-12}	pico	p
0.000 000 000 000 001 = 10^{-15}	femto	f
0.000 000 000 000 000 001 = 10^{-18}	atto	a

Element	Symbol	Atomic No.	Atomic Weight
Actinium	Ac	89	
Aluminum	Al	13	26.981 5
Americium	Am	95	
Antimony	Sb	51	121.750
Argon	Ar	18	39.948
Arsenic	As	33	74.921 6
Astatine	At	85	
Barium	Ba	56	137.34
Berkelium	Bk	97	
Beryllium	Be	4	9.012 2
Bismuth	Bi	83	208.980
Boron	B	5	10.811
Bromine	Br	35	79.904
Cadmium	Cd	48	112.40
Calcium	Ca	20	40.08
Californium	Cf	98	
Carbon	C	6	12.011 15
Cerium	Ce	58	140.12
Cesium	Cs	55	132.905
Chlorine	Cl	17	35.453
Chromium	Cr	24	51.996
Cobalt	Co	27	58.933 2
Columbium (see Niobium)			
Copper	Cu	29	63.546
Curium	Cm	96	
Dysprosium	Dy	66	162.50
Einsteinium	Es	99	
Erbium	Er	68	167.26
Europium	Eu	63	151.96
Fermium	Fm	100	
Fluorine	F	9	18.998 4
Francium	Fr	87	
Gadolinium	Gd	64	157.25
Gallium	Ga	31	69.72
Germanium	Ge	32	72.59
Gold	Au	79	196.967
Hafnium	Hf	72	178.49
Helium	He	2	4.002 6
Holmium	Ho	67	164.930
Hydrogen	H	1	1.007 97
Indium	In	49	114.82
Iodine	I	53	126.904 4
Iridium	Ir	77	192.2
Iron	Fe	26	55.847
Krypton	Kr	36	83.80
Lanthanum	La	57	138.91
Lead	Pb	82	207.19
Lithium	Li	3	6.939
Lutetium	Lu	71	174.97
Magnesium	Mg	12	24.312
Manganese	Mn	25	54.938 0
Mendelevium	Md	101	
Mercury	Hg	80	200.59
Molybdenum	Mo	42	95.94

Element	Symbol	Atomic No.	Atomic Weight
Neodymium	Nd	60	144.24
Neon	Ne	10	20.183
Neptunium	Np	93	
Nickel	Ni	28	58.71
Niobium	Nb	41	92.906
Nitrogen	N	7	14.006 7
Nobelium	No	102	
Osmium	Os	76	109.2
Oxygen	O	8	15.999 4
Palladium	Pd	46	106.4
Phosphorus	P	15	30.973 8
Platinum	Pt	78	195.09
Plutonium	Pu	94	
Polonium	Po	84	
Potassium	K	19	39.102
Praseodymium	Pr	59	140.907
Promethium	Pm	61	
Protactinium	Pa	91	
Radium	Ra	88	
Radon	Rn	86	
Rhenium	Re	75	186.2
Rhodium	Rh	45	102.905
Rubidium	Rb	37	85.47
Ruthenium	Ru	44	101.07
Samarium	Sm	62	150.35
Scandium	Sc	21	44.956
Selenium	Se	34	78.96
Silicon	Si	14	28.086
Silver	Ag	47	107.868
Sodium	Na	11	22.989 8
Strontium	Sr	38	87.62
Sulphur	S	16	32.064
Tantalum	Ta	73	180.948
Technetium	Tc	43	
Tellurium	Te	52	127.60
Terbium	Tb	65	158.924
Thallium	Tl	81	204.37
Thorium	Th	90	232.038
Thulium	Tm	69	168.934
Tin	Sn	50	118.69
Titanium	Ti	22	47.90
Tungsten	W	74	183.85
Uranium	U	92	238.03
Vanadium	V	23	50.942
Xenon	Xe	54	131.30
Ytterbium	Yb	70	173.04
Yttrium	Y	39	88.905
Zinc	Zn	30	65.37
Zirconium	Zr	40	91.22

Greek Alphabet

Alpha	A	α
Beta	B	β
Gamma	Γ	γ
Delta	Δ	δ
Epsilon	E	ϵ
Zeta	Z	ζ
Eta	H	η
Theta	Θ	θ
Iota	I	ι
Kappa	K	κ
Lambda	Λ	λ
Mu	M	μ
Nu	N	ν
Xi	Ξ	ξ
Omicron	O	o
Pi	Π	π
Rho	P	ρ
Sigma	Σ	σ
Tau	T	τ
Upsilon	Υ	υ
Phi	Φ	ϕ
Chi	X	χ
Psi	Ψ	ψ
Omega	Ω	ω

Physical Constants

Avogadro's number = 6.022 57 $\times$ 10^{23}/mol

Density of dry air at 0°C, 1 atm = 1.293 kg/m^3

Density of water at 3.98°C = 9.999 973 $\times$ 10^2 kg/m^3

Equatorial radius of the earth = 6 378.39 km = 3 963.34 mi

Gravitational acceleration (standard) at sea level = 9.806 65 m/s^2 = 32.174 ft/s^2

Gravitational constant = 6.672 $\times$ 10^{-11} N $\cdot$ m^2/kg^2

Heat of fusion of water, 0°C = 3.337 5 $\times$ 10^5 J/kg = 143.48 Btu/lbm

Heat of vaporization of water, 100°C − 2.259 1 $\times$ 10^6 J/kg = 971.19 Btu/lbm

Mass of hydrogen atom = 1.673 39 $\times$ 10^{-27} kg

Mean density of the earth = 5.522 $\times$ 10^3 kg/m^3 = 344.7 lbm/ft^3

Molar gas constant = 8.314 4 J/(mol $\cdot$ K)

Planck's constant = 6.625 54 $\times$ 10^{-34} J/Hz

Polar radius of the earth = 6 356.91 km = 3 949.99 mi

Velocity of light in a vacuum = 2.997 9 $\times$ 10^8 m/s

Velocity of sound in dry air at 0°C = 331.36 m/s = 1 087.1 ft/s

Areas Under the Standard Normal Curve from 0 to z

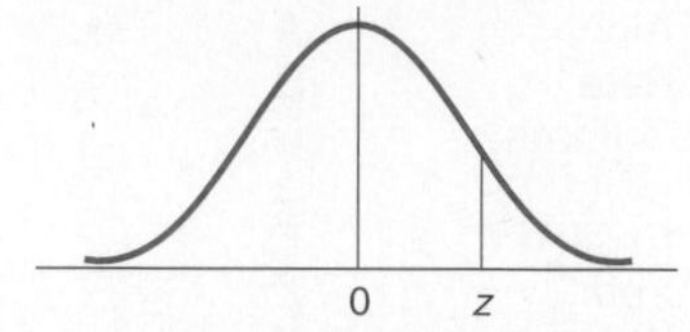

z	00	.01	.02	.03	.04	.05	.06	.07	.08	.09
0.0	.000 0	.004 0	.008.0	.012 0	.016 0	.019 9	.023 9	.027 9	.031 9	.035 9
0.1	.039 8	.043 3	.047 8	.051 7	.055 7	.059 6	.063 6	.067 5	.071 4	.075 4
0.2	.079 3	.083 2	.087 1	.091 0	.094 8	.098 7	102 6	.106 4	.110 3	.114 1
0.3	.117 9	.121 7	.125 5	.129 3	.133 1	.136 8	140 6	.144 3	.148 0	.151 7
0.4	.155 4	.159 1	.162 8	.166 4	.170 0	.173 6	177 2	.180 8	.184 1	.187 9
0.5	.191 5	.195 0	.198 5	.201 9	.205 4	.208 8	.212 3	.215 7	.219 0	.222 4
0.6	.225 8	.229 1	.232 4	.235 7	.238 9	.242 2	.245 4	.248 6	.251 8	.254 9
0.7	.258 0	.261 2	.264 2	.267 3	.270 4	.273 4	.276 4	.279 4	.282 3	.285 2
0.8	.288 1	.291 0	.293 9	.296 7	.299 6	.302 3	.305 1	.307 8	.310 6	.313 3
0.9	.315 9	.318 6	.321 2	.323 8	.326 4	.328 9	.331 5	.334 0	.336 5	.338 9
1.0	.341 3	.343 8	.346 1	.348 5	.350 8	.353 1	.355 4	.357 7	.359 9	.362 1
1.1	.364 3	.366 5	.368 6	.370 8	.372 9	.374 9	.377 0	.379 0	.381 0	.383 0
1.2	.384 9	.386 9	.388 8	.390 7	.392 5	.394 4	.396 2	.398 0	.399 7	.401 5
1.3	.403 2	.404 9	.406 6	.408 2	.409 9	.411 5	.413 1	.414 7	.416 2	.417 7
1.4	.419 2	.420 7	.422 2	.423 6	.425 1	.426 5	.427 9	.429 2	.430 6	.431 9
1.5	.433 2	.434 5	.435 7	.437 0	.438 2	.439 4	.440 6	.441 8	.442 9	.444 1
1.6	.445 2	.446 3	.447 4	.448 4	.449 5	.450 5	.451 5	.452 5	.453 5	.454 5
1.7	.455 5	.456 4	.457 3	.458 2	.459 1	.459 9	.460 8	.461 6	.462 5	.463 3
1.8	.464 1	.464 9	.465 6	.466 4	.467 1	.467 8	.468 6	.469 3	.469 9	.470 6
1.9	.471 3	.471 9	.472 6	.473 2	.473 8	.474 4	.475 0	.475 6	.476 1	.476 7
2.0	.477 2	.477 8	.478 3	.478 8	.479 3	.479 8	.480 3	.480 8	.481 2	.481 7
2.1	.482 1	.482 6	.483 0	.483 4	.483 8	.484 2	.484 6	.485 0	.485.4	.485 7
2.2	.486 1	.486 4	.486 8	.487 1	.487 5	.487 8	.488 1	.488 4	.488 7	.489 0
2.3	.489 3	.489 6	.489 8	.490 1	.490 4	.490 6	.490 9	.491 1	.491 3	.491 6
2.4	.491 8	.492 0	.492 2	.492 5	.492 7	.492 9	.493 1	.493 2	.493 4	.493 6
2.5	.493 8	.494 0	.494 1	.494 3	.494 5	.494 6	.494 8	.494 9	.495 1	.495 2
2.6	.495 3	.495 5	.495 6	.495 7	.495 9	.496 0	.496 1	.496 2	.496 3	.496 4
2.7	.496 5	.496 6	.496 7	.496 8	.496 9	.497 0	.497 1	.497 2	.497 3	.497 4
2.8	.497 4	.497 5	.497 6	.497 7	.497 7	.497 8	.497 9	.497 9	.498 0	.498 1
2.9	.498 1	.498 2	.498 2	.498 3	.498 4	.498 4	.498 5	.498 5	.498 6	.498 6
3.0	.498 7	.498 7	.498 7	.498 8	.498 8	.498 9	.498 9	.498 9	.499 0	.499 0
3.1	.499 0	.499 1	.499 1	.499 1	.499 2	.499 2	.499 2	.499 2	.499 3	.499 3
3.2	.499 3	.499 3	.499 4	.499 4	.499 4	.499 4	.499 4	.499 5	.499 5	.499 5
3.3	.499 5	.499 5	.499 5	.499 6	.499 6	.499 6	.499 6	.499 6	.499 6	.499 7
3.4	.499 7	.499 7	.499 7	.499 7	.499 7	.499 7	.499 7	.499 7	.499 7	.499 8
3.5	.499 8	.499 8	.499 8	.499 8	.499 8	.499 8	.499 8	.499 8	.499 8	.499 8
3.6	.499 8	.499 8	.499 9	.499 9	.499 9	.499 9	.499 9	.499 9	.499 9	.499 9
3.7	.499 9	.499 9	.499 9	.499 9	.499 9	.499 9	.499 9	.499 9	.499 9	.499 9
3.8	.499 9	.499 9	.499 9	.499 9	.499 9	.499 9	.499 9	.499 9	.499 9	.499 9
3.9	.500 0	.500 0	.500 0	.500 0	.500 0	.500 0	.500 0	.500 0	.500 0	.500 0

National Society of Professional Engineers: Code of Ethics for Engineers

Preamble

Engineering is an important and learned profession. The members of the profession recognize that their work has a direct and vital impact on the quality of life for all people. Accordingly, the services provided by engineers require honesty, impartiality, fairness and equity, and must be dedicated to the protection of the public health, safety and welfare. In the practice of their profession, engineers must perform under a standard of professional behavior which requires adherence to the highest principles of ethical conduct on behalf of the public, clients, employers and the profession.

I. Fundamental Canons

Engineers, in the fulfillment of their professional duties, shall:

1. Hold paramount the safety, health and welfare of the public in the performance of their professional duties.

2. Perform services only in areas of their competence.

3. Issue public statements only in an objective and truthful manner.

4. Act in professional matters for each employer or client as faithful agents or trustees.

5. Avoid deceptive acts in the solicitation of professional employment.

II. Rules of Practice

1. Engineers shall hold paramount the safety, health and welfare of the public in the performance of their professional duties.

a. Engineers shall at all times recognize that their primary obligation is to protect the safety, health, property and welfare of

---

Note: In regard to the question of application of the Code to corporations vis-a-vis real persons, business form or type should not negate nor influence conformance of individuals to the Code. The Code deals with professional services, which services must be performed by real persons. Real persons in turn establish and implement policies within business structures. The Code is clearly written to apply to the Engineer and it is incumbent on a member of NSPE to endeavor to live up to its provisions. This applies to all pertinent sections of the Code.

the public. If their professional judgment is overruled under circumstances where the safety, health, property or welfare of the public are endangered, they shall notify their employer or client and such other authority as may be appropriate.

b. Engineers shall approve only those engineering documents which are safe for public health, property and welfare in conformity with accepted standards.

c. Engineers shall not reveal facts, data or information obtained in a professional capacity without the prior consent of the client or employer except as authorized or required by law or this Code.

d. Engineers shall not permit the use of their name or firm name nor associate in business ventures with any person or firm which they have reason to believe is engaging in fraudulent or dishonest business or professional practices.

e. Engineers having knowledge of any alleged violation of this Code shall cooperate with the proper authorities in furnishing such information or assistance as may be required.

2. Engineers shall perform services only in the areas of their competence.

a. Engineers shall undertake assignments only when qualified by education or experience in the specific technical fields involved.

b. Engineers shall not affix their signatures to any plans or documents dealing with subject matter in which they lack competence, nor to any plan or document not prepared under their direction and control.

c. Engineers may accept assignments and assume responsibility for coordination of an entire project and sign and seal the engineering documents for the entire project, provided that each technical segment is signed and sealed only by the qualified engineers who prepared the segment.

3. Engineers shall issue public statements only in an objective and truthful manner.

a. Engineers shall be objective and truthful in professional reports, statements or testimony. They shall include all relevant and pertinent information in such reports, statements or testimony.

b. Engineers may express publicly a professional opinion on technical subjects only when that opinion is founded upon adequate knowledge of the facts and competence in the subject matter.

c. Engineers shall issue no statements, criticisms or arguments on technical matters which are inspired or paid for by interested parties, unless they have prefaced their comments by explicitly identifying the interested parties on whose behalf they are speaking, and by revealing the existence of any interest the engineers may have in the matters.

4. Engineers shall act in professional matters for each employer or client as faithful agents or trustees.

a. Engineers shall disclose all known or potential conflicts of interest to their employers or clients by promptly informing them of any business association, interest, or other circumstances which could influence or appear to influence their judgment or the quality of their services.

b. Engineers shall not accept compensation, financial or otherwise, from more than one party for services on the same project, or for services pertaining to the same project, unless the circumstances are fully disclosed to, and agreed to by, all interested parties.

c. Engineers shall not solicit or accept financial or other valuable consideration directly or indirectly, from contractors, their agents, or other parties in connection with work for employers or clients for which they are responsible.

d. Engineers in public service as members, advisors or employees of a governmental or quasi-governmental body or department shall not participate in decisions with respect to professional services solicited or provided by them or their organizations in private or public engineering practice.

e. Engineers shall not solicit or accept a professional contract from a governmental body on which a principal or officer of their organization serves as a member.

5. Engineers shall avoid deceptive acts in the solicitation of professional employment.

a. Engineers shall not falsify or permit misrepresentation of their, or their associates', academic or professional qualifications. They shall not misrepresent or exaggerate their degree of responsibility in or for the subject matter of prior assignments. Brochures or other presentations incident to the solicitation of employment shall not misrepresent pertinent facts concerning employers, employees, associates, joint venturers or past accomplishments with the intent and purpose of enhancing their qualifications and their work.

b. Engineers shall not offer, give, solicit or receive, either directly or indirectly, any political contribution in an amount intended to influence the award of a contract by public authority, or which may be reasonably construed by the public of having the effect or intent to influence the award of a contract. They shall not offer any gift, or other valuable consideration in order to secure work. They shall not pay a commission, percentage or brokerage fee in order to secure work except to a bona fide employee or bona fide established commercial or marketing agencies retained by them.

III. Professional Obligations

1. Engineers shall be guided in all their professional relations by the highest standards of integrity.

a. Engineers shall admit and accept their own errors when proven wrong and refrain from distorting or altering the facts in an attempt to justify their decisions.

b. Engineers shall advise their clients or employers when they believe a project will not be successful.
c. Engineers shall not accept outside employment to the detriment of their regular work or interest. Before accepting any outside employment they will notify their employers.
d. Engineers shall not attempt to attract an engineer from another employer by false or misleading pretenses.
e. Engineers shall not actively participate in strikes, picket lines, or other collective coercive action.
f. Engineers shall avoid any act tending to promote their own interest at the expense of the dignity and integrity of the profession.

2. Engineers shall at all times strive to serve the public interest.
a. Engineers shall seek opportunities to be of constructive service in civic affairs and work for the advancement of the safety, health and well-being of their community.
b. Engineers shall not complete, sign or seal plans and/or specifications that are not of a design safe to the public health and welfare and in conformity with accepted engineering standards. If the client or employer insists on such unprofessional conduct, they shall notify the proper authorities and withdraw from further service on the project.
c. Engineers shall endeavor to extend public knowledge and appreciation of engineering and its achievements and to protect the engineering profession from misrepresentation and misunderstanding.

3. Engineers shall avoid all conduct or practice which is likely to discredit the profession or deceive the public.
a. Engineers shall avoid the use of statements containing a material misrepresentation of fact or omitting a material fact necessary to keep statements from being misleading or intended or likely to create an unjustified expectation, or statements containing prediction of future success.
b. Consistent with the foregoing, Engineers may advertise for recruitment of personnel.
c. Consistent with the foregoing, Engineers may prepare articles for the lay or technical press, but such articles shall not imply credit to the author for work performed by others.

4. Engineers shall not disclose confidential information concerning the business affairs or technical processes of any present or former client or employer without his consent.
a. Engineers in the employ of others shall not without the consent of all interested parties enter promotional efforts or negotiations for work or make arrangements for other employment as a principal or to practice in connection with a specific project for which the Engineer has gained particular and specialized knowledge.

b. Engineers shall not, without the consent of all interested parties, participate in or represent an adversary interest in connection with a specific project or proceeding in which the Engineer has gained particular specialized knowledge on behalf of a former client or employer.

5. Engineers shall not be influenced in their professional duties by conflicting interests.

a. Engineers shall not accept financial or other considerations, including free engineering designs, from material or equipment suppliers for specifying their product.

b. Engineers shall not accept commissions or allowances, directly or indirectly, from contractors or other parties dealing with clients or employers of the Engineer in connection with work for which the Engineer is responsible.

6. Engineers shall uphold the principle of appropriate and adequate compensation for those engaged in engineering work.

a. Engineers shall not accept remuneration from either an employee or employment agency for giving employment.

b. Engineers, when employing other engineers, shall offer a salary according to professional qualifications.

7. Engineers shall not attempt to obtain employment or advancement or professional engagements by untruthfully criticizing other engineers, or by other improper or questionable methods.

a. Engineers shall not request, propose, or accept a professional commission on a contingent basis under circumstances in which their professional judgment may be compromised.

b. Engineers in salaried positions shall accept part-time engineering work only to the extent consistent with policies of the employer and in accordance with ethical considerations.

c. Engineers shall not use equipment, supplies, laboratory, or office facilities of an employer to carry on outside private practice without consent.

8. Engineers shall not attempt to injure, maliciously or falsely, directly or indirectly, the professional reputation, prospects, practice or employment of other engineers, nor untruthfully criticize other engineers' work. Engineers who believe others are guilty of unethical or illegal practice shall present such information to the proper authority for action.

a. Engineers in private practice shall not review the work of another engineer for the same client, except with the knowledge of such engineer, or unless the connection of such engineer with the work has been terminated.

b. Engineers in governmental, industrial or educational employ are entitled to review and evaluate the work of other engineers when so required by their employment duties.

c. Engineers in sales or industrial employ are entitled to make engineering comparisons of represented products with products of other suppliers.

9. Engineers shall accept personal responsibility for their professional activities; provided, however, that Engineers may seek indemnification for professional services arising out of their practice for other than gross negligence, where the Engineer's interests cannot otherwise be protected.

a. Engineers shall conform with state registration laws in the practice of engineering.
b. Engineers shall not use association with a nonengineer, a corporation, or partnership as a "cloak" for unethical acts, but must accept personal responsibility for all professional acts.

10. Engineers shall give credit for engineering work to those to whom credit is due, and will recognize the proprietary interests of others.

a. Engineers shall, whenever possible, name the person or persons who may be individually responsible for designs, inventions, writings, or other accomplishments.
b. Engineers using designs supplied by a client recognize that the designs remain the property of the client and may not be duplicated by the Engineer for others without express permission.
c. Engineers, before undertaking work for others in connection with which the Engineer may make improvements, plans, designs, inventions, or other records which may justify copyrights or patents, should enter into a positive agreement regarding ownership.
d. Engineers' designs, data, records, and notes referring exclusively to an employer's work are the employer's property.

11. Engineers shall cooperate in extending the effectiveness of the profession by interchanging information and experience with other engineers and students, and will endeavor to provide opportunity for the professional development and advancement of engineers under their supervision.

a. Engineers shall encourage engineering employees' efforts to improve their education.
b. Engineers shall encourage engineering employees to attend and present papers at professional and technical society meetings.
c. Engineers shall urge engineering employees to become registered at the earliest possible date.
d. Engineers shall assign a professional engineer duties of a nature to utilize full training and experience, insofar as possible, and delegate lesser functions to subprofessionals or to technicians.
e. Engineers shall provide a prospective engineering employee with complete information on working conditions and proposed status of employment, and after employment will keep employees informed of any changes.

"By order of the United States District Court for the District of Columbia, former Section 11(c) of the NSPE Code of Ethics prohibiting competitive bidding, and all policy statements, opinions, rulings or other guidelines interpreting its scope, have been rescinded as unlawfully interfering with the legal right of engineers, protected under the antitrust laws, to provide price information to prospective clients; accordingly, nothing contained in the NSPE Code of Ethics, policy statements, opinions, rulings or other guidelines prohibits the submission of price quotations or competitive bids for engineering services at any time or in any amount."

Statement by NSPE Executive Committee

In order to correct misunderstandings which have been indicated in some instances since the issuance of the Supreme Court decision and the entry of the Final Judgment, it is noted that in its decision of April 25, 1978, the Supreme Court of the United States declared: "The Sherman Act does not require competitive bidding."

It is further noted that as made clear in the Supreme Court decision:

1. Engineers and firms may individually refuse to bid for engineering services.
2. Clients are not required to seek bids for engineering services.
3. Federal, state, and local laws governing procedures to procure engineering services are not affected, and remain in full force and effect.
4. State societies and local chapters are free to actively and aggressively seek legislation for professional selection and negotiation procedures by public agencies.
5. State registration board rules of professional conduct, including rules prohibiting competitive bidding for engineering services, are not affected and remain in full force and effect. State registration boards with authority to adopt rules of professional conduct may adopt rules governing procedures to obtain engineering services.
6. As noted by the Supreme Court, "nothing in the judgment prevents NSPE and its members from attempting to influence governmental action . . ."

(Courtesy National Society of Professional Engineers)

Unit Conversions

Multiply:	by:	To obtain:
acres	4.356×10^{4}	ft^2
acres	$4.046\,9 \times 10^{-1}$	ha
acres	$4.046\,9 \times 10^{3}$	m^2
amperes	1	C/s
ampere hours	3.6×10^{3}	C
angstroms	1×10^{-8}	cm
angstroms	$3.937\,0 \times 10^{-9}$	in
atmospheres	1.013 3	bars
atmospheres	$2.992\,1 \times 10^{1}$	in of Hg
atmospheres	$1.469\,6 \times 10^{1}$	lbf/in^2
atmospheres	7.6×10^{2}	mm of Hg
atmospheres	$1.013\,3 \times 10^{5}$	Pa
barrels (petroleum, US)	4.2×10^{1}	gal (US liquid)
bars	$9.869\,2 \times 10^{-1}$	atm
bars	$2.953\,0 \times 10^{1}$	in of Hg
bars	$1.450\,4 \times 10^{1}$	lbf/in^2
bars	1×10^{5}	Pa
Btu	$7.776\,5 \times 10^{2}$	ft · lbf
Btu	$3.927\,5 \times 10^{-4}$	hp · h
Btu	$1.055\,1 \times 10^{3}$	J
Btu	$2.928\,8 \times 10^{-4}$	kWh
Btu per hour	$2.160\,1 \times 10^{-1}$	ft · lbf/s
Btu per hour	$3.927\,5 \times 10^{-4}$	hp
Btu per hour	$2.928\,8 \times 10^{-1}$	W
Btu per minute	$7.776\,5 \times 10^{2}$	ft · lbf/min
Btu per minute	$2.356\,5 \times 10^{-2}$	hp
Btu per minute	$1.757\,3 \times 10^{-2}$	kW
bushels (US)	1.244 5	ft^3
bushels (US)	$3.523\,9 \times 10^{1}$	L
bushels (US)	$3.523\,9 \times 10^{-2}$	m^3
candelas	1	lm/sr
candelas per square foot	$3.381\,6 \times 10^{-3}$	lamberts
centimeters	1×10^{8}	Å
centimeters	$3.280\,8 \times 10^{-2}$	ft
centimeters	$3.937\,0 \times 10^{-1}$	in
centipoises	1×10^{-2}	g/(cm · s)
circular mils	$5.067\,1 \times 10^{-6}$	cm^2
circular mils	$7.854\,0 \times 10^{-7}$	in^2
coulombs	1	A · s
cubic centimeters	$6.102\,4 \times 10^{-2}$	in^3
cubic centimeters	$3.531\,5 \times 10^{-5}$	ft^3
cubic centimeters	$2.641\,7 \times 10^{-4}$	gal (US liquid)
cubic centimeters	1×10^{-3}	L
cubic centimeters	$3.381\,4 \times 10^{-2}$	oz (US fluid)
cubic centimeters per gram	$1.601\,8 \times 10^{-2}$	ft^3/lbm
cubic centimeters per second	$2.118\,9 \times 10^{-3}$	ft^3/min
cubic centimeters per second	$1.585\,0 \times 10^{-2}$	gal (US liquid)/min
cubic feet	$2.295\,7 \times 10^{-5}$	acre · ft

Multiply:	by:	To obtain:
cubic feet	$8.035\,6 \times 10^{-1}$	bushels (US)
cubic feet	7.480 5	gal (US liquid)
cubic feet	1.728×10^{3}	in^3
cubic feet	$2.831\,7 \times 10^{1}$	L
cubic feet	$2.831\,7 \times 10^{-2}$	m^3
cubic feet per minute	7.480 5	gal (US liquid)/min
cubic feet per minute	$4.719\,5 \times 10^{-1}$	L/s
cubic feet per pound-mass	$6.242\,8 \times 10^{1}$	cm^3/g
cubic feet per second	$4.488\,3 \times 10^{2}$	gal (US liquid)/min
cubic feet per second	$2.831\,7 \times 10^{1}$	L/s
cubic inches	$4.650\,3 \times 10^{-4}$	bushels (US)
cubic inches	$1.638\,7 \times 10^{1}$	cm^3
cubic inches	$4.329\,0 \times 10^{-3}$	gal (US liquid)
cubic inches	$1.638\,7 \times 10^{-2}$	L
cubic inches	$1.638\,7 \times 10^{-5}$	m^3
cubic inches	$5.541\,1 \times 10^{-1}$	oz (US fluid)
cubic meters	$8.107\,1 \times 10^{-4}$	acre · ft
cubic meters	$2.837\,8 \times 10^{1}$	bushels (US)
cubic meters	$3.531\,5 \times 10^{1}$	ft^3
cubic meters	$2.641\,7 \times 10^{2}$	gal (US liquid)
cubic meters	1×10^{3}	L
cubic yards	$2.169\,6 \times 10^{1}$	bushels (US)
cubic yards	$2.019\,7 \times 10^{2}$	gal (US liquid)
cubic yards	$7.645\,5 \times 10^{2}$	L
cubic yards	$7.645\,5 \times 10^{-1}$	m^3
dynes	1×10^{-5}	N
dynes per square centimeter	$9.869\,2 \times 10^{-7}$	atm
dynes per square centimeter	1×10^{-6}	bars
dynes per square centimeter	$1.450\,4 \times 10^{-5}$	lbf/in^2
dyne centimeters	$7.375\,6 \times 10^{-8}$	ft · lbf
dyne centimeters	1×10^{-7}	N · m
ergs	1	dyne · cm
fathoms	6	ft
feet	3.048×10^{1}	cm
feet	1.2×10^{1}	in
feet	3.048×10^{-4}	km
feet	3.048×10^{-1}	m
feet	$1.893\,9 \times 10^{-4}$	mi
feet	$6.060\,6 \times 10^{-2}$	rods
feet per second	1.097 3	km/h
feet per second	$1.828\,8 \times 10^{1}$	m/min
feet per second	$6.818\,2 \times 10^{-1}$	mi/h
feet per second squared	3.048×10^{-1}	m/s^2
foot-candles	1	lm/ft^2
foot-candles	$1.076\,4 \times 10^{1}$	lux
foot pounds-force	$1.285\,9 \times 10^{-3}$	Btu
foot pounds-force	$1.355\,8 \times 10^{7}$	dyne · cm
foot pounds-force	$5.050\,5 \times 10^{-7}$	hp · h
foot pounds-force	1.355 8	J
foot pounds-force	$3.766\,2 \times 10^{-7}$	kWh
foot pounds-force	1.355 8	N · m
foot pounds-force per hour	$2.143\,2 \times 10^{-5}$	Btu/min
foot pounds-force per hour	$2.259\,7 \times 10^{5}$	ergs/min
foot pounds-force per hour	$5.050\,5 \times 10^{-7}$	hp
foot pounds-force per hour	$3.766\,2 \times 10^{-7}$	kW
furlongs	6.6×10^{2}	ft
furlongs	$2.011\,7 \times 10^{2}$	m
gallons (US liquid)	$1.336\,8 \times 10^{-1}$	ft^3

Multiply:	by:	To obtain:
gallons (US liquid)	2.31×10^{2}	in^3
gallons (US liquid)	3.785 4	L
gallons (US liquid)	$3.785\ 4 \times 10^{-3}$	m^3
gallons (US liquid)	1.28×10^{2}	oz (US fluid)
gallons (US liquid)	8	pt (US liquid)
gallons (US liquid)	4	qt (US liquid)
grams	$2.204\ 6 \times 10^{-3}$	lbm
grams per centimeter second	1	poises
grams per cubic centimeter	$6.242\ 8 \times 10^{1}$	lbm/ft^3
hectares	2.471 1	acres
hectares	1×10^{2}	ares
hectares	$1.076\ 4 \times 10^{5}$	ft^2
hectares	1×10^{4}	m^2
horsepower	$2.546\ 1 \times 10^{3}$	Btu/h
horsepower	5.5×10^{2}	ft · lbf/s
horsepower	$7.457\ 0 \times 10^{-1}$	kW
horsepower	$7.457\ 0 \times 10^{2}$	W
horsepower hours	$2.546\ 1 \times 10^{3}$	Btu
horsepower hours	1.98×10^{6}	ft · lbf
horsepower hours	$2.684\ 5 \times 10^{6}$	J
horsepower hours	$7.457\ 0 \times 10^{-1}$	kWh
hours	6×10^{1}	min
hours	3.6×10^{3}	s
inches	2.54×10^{8}	Å
inches	2.54	cm
inches	$8.333\ 3 \times 10^{-2}$	ft
inches	1×10^{3}	mils
inches	$2.777\ 8 \times 10^{-2}$	yd
joules	$9.478\ 2 \times 10^{-4}$	Btu
joules	$7.375\ 6 \times 10^{-1}$	ft · lbf
joules	$3.725\ 1 \times 10^{-7}$	hp · h
joules	$2.777\ 8 \times 10^{-7}$	kWh
joules	1	W · s
joules per second	$5.690\ 7 \times 10^{-2}$	Btu/min
joules per second	1×10^{7}	ergs/s
joules per second	$7.375\ 6 \times 10^{-1}$	ft · lbf/s
joules per second	$1.341\ 0 \times 10^{-3}$	hp
joules per second	1	W
kilograms	2.204 6	lbm
kilograms	$6.852\ 2 \times 10^{-2}$	slugs
kilograms	1×10^{-3}	t
kilometers	$3.280\ 8 \times 10^{3}$	ft
kilometers	$6.213\ 7 \times 10^{-1}$	mi
kilometers	$5.399\ 6 \times 10^{-1}$	nmi (nautical mile)
kilometers per hour	$5.468\ 1 \times 10^{1}$	ft/min
kilometers per hour	$9.113\ 4 \times 10^{-1}$	ft/s
kilometers per hour	$5.399\ 6 \times 10^{-1}$	knots
kilometers per hour	$2.777\ 8 \times 10^{-1}$	m/s
kilometers per hour	$6.213\ 7 \times 10^{-1}$	mi/h
kilowatts	$3.414\ 4 \times 10^{3}$	Btu/h
kilowatts	1×10^{10}	ergs/s
kilowatts	$7.375\ 6 \times 10^{2}$	ft · lbf/s
kilowatts	1.341 0	hp
kilowatts	1×10^{3}	J/s
kilowatt hours	$3.414\ 4 \times 10^{3}$	Btu
kilowatt hours	$2.655\ 2 \times 10^{6}$	ft · lbf
kilowatt hours	1.341 0	hp · h
kilowatt hours	3.6×10^{6}	J

Multiply:	by:	To obtain:
knots	1.687 8	ft/s
knots	1.150 8	mi/h
liters	$2.837\ 8 \times 10^{-2}$	bushels (US)
liters	$3.531\ 5 \times 10^{-2}$	ft^3
liters	$2.641\ 7 \times 10^{-1}$	gal (US liquid)
liters	$6.102\ 4 \times 10^{1}$	in^3
liters per second	2.118 9	ft^3/min
liters per second	$1.585\ 0 \times 10^{1}$	gal (US liquid)/min
lumens	$7.957\ 7 \times 10^{-2}$	candle power
lumens per square foot	1	foot-candles
lumens per square meter	$9.290\ 3 \times 10^{-2}$	foot-candles
lux	1	lm/m^2
meters	1×10^{10}	Å
meters	3.280 8	ft
meters	$3.937\ 0 \times 10^{1}$	in
meters	$6.213\ 7 \times 10^{-4}$	mi
meters per minute	1.666 7	cm/s
meters per minute	$5.468\ 1 \times 10^{-2}$	ft/s
meters per minute	6×10^{-2}	km/h
meters per minute	$3.239\ 7 \times 10^{-2}$	knots
meters per minute	$3.728\ 2 \times 10^{-2}$	mi/h
microns	1×10^{4}	A
microns	$3.280\ 8 \times 10^{-6}$	ft
microns	1×10^{-6}	m
miles	5.28×10^{3}	ft
miles	8	furlongs
miles	1.609 3	km
miles	$8.689\ 8 \times 10^{-1}$	nmi (nautical mile)
miles per hour	$4.470\ 4 \times 10^{1}$	cm/s
miles per hour	8.8×10^{1}	ft/min
miles per hour	1.466 7	ft/s
miles per hour	1.609 3	km/h
miles per hour	$8.689\ 8 \times 10^{-1}$	knots
miles per hour	$2.682\ 2 \times 10^{1}$	m/min
nautical miles	1.150 8	mi
newtons	1×10^{5}	dynes
newtons	$2.248\ 1 \times 10^{-1}$	lbf
newton meters	1×10^{7}	dyne · cm
newton meters	$7.375\ 6 \times 10^{-1}$	ft · lbf
ounces (US fluid)	$2.957\ 4 \times 10^{1}$	cm^3
ounces (US fluid)	$7.812\ 5 \times 10^{-3}$	gal (US liquid)
ounces (US fluid)	1.804 7	in^3
ounces (US fluid)	$2.957\ 4 \times 10^{-2}$	L
pascals	$9.869\ 2 \times 10^{-6}$	atm
pascals	$2.088\ 5 \times 10^{-2}$	lbf/ft^2
pascals	$1.450\ 4 \times 10^{-4}$	lbf/in^2
poises	1	g/(cm · s)
pounds-force	4.448 2	N
pounds-mass	$4.535\ 9 \times 10^{2}$	g
pounds-mass	$4.535\ 9 \times 10^{-1}$	kg
pounds-mass	$3.108\ 1 \times 10^{-2}$	slugs
pounds-mass	$4.535\ 9 \times 10^{-4}$	t
pounds-mass	5×10^{-4}	tons (short)
pounds-force per square foot	$4.725\ 4 \times 10^{-3}$	atm
pounds-force per square foot	$4.788\ 0 \times 10^{1}$	Pa
pounds-force per square inch	$6.804\ 6 \times 10^{-2}$	atm
pounds-force per square inch	$6.894\ 8 \times 10^{-2}$	bars
pounds-force per square inch	2.036 0	in of Hg

Multiply:	by:	To obtain:
pounds-force per square inch	$5.171\ 5 \times 10^{1}$	mm of Hg
pounds-force per square inch	$6.894\ 8 \times 10^{3}$	Pa
pounds-mass per cubic foot	$1.601\ 8 \times 10^{-2}$	g/cm^3
pounds-mass per cubic foot	$1.601\ 8 \times 10^{1}$	kg/m^3
radians	$5.729\ 6 \times 10^{1}$	°
radians	$1.591\ 5 \times 10^{-1}$	r (revolutions)
radians per second	9.549 3	r/min
slugs	$1.459\ 4 \times 10^{1}$	kg
slugs	$3.217\ 4 \times 10^{1}$	lbm
square centimeters	$1.076\ 4 \times 10^{-3}$	ft^2
square centimeters	$1.550\ 0 \times 10^{-1}$	in^2
square feet	$2.295\ 7 \times 10^{-5}$	acre
square feet	$9.290\ 3 \times 10^{2}$	cm^2
square feet	$9.290\ 3 \times 10^{-6}$	ha
square feet	$9.290\ 3 \times 10^{-2}$	m^2
square meters	$1.076\ 4 \times 10^{1}$	ft^2
square meters	$1.550\ 0 \times 10^{3}$	in^2
square miles	6.4×10^{2}	acres
square miles	$2.787\ 8 \times 10^{7}$	ft^2
square miles	$2.590\ 0 \times 10^{2}$	ha
square miles	2.590 0	km^2
square millimeters	$1.076\ 4 \times 10^{-5}$	ft^2
square millimeters	$1.550\ 0 \times 10^{-3}$	in^2
stokes	1	cm^2/s
stokes	$1.550\ 0 \times 10^{-1}$	in^2/s
tons (long)	2.24×10^{3}	lbm
tons (long)	1.016 0	t
tons (long)	1.12	tons (short)
tons (metric)	$9.017\ 2 \times 10^{-1}$	tons (short)
tons (short)	2×10^{3}	lbm
watts	3.414 4	Btu/h
watts	1×10^{7}	ergs/s
watts	$4.425\ 4 \times 10^{1}$	ft · lbf/min
watts	$1.341\ 0 \times 10^{-3}$	hp
watts	1	J/s
watt hours	3.414 4	Btu
watt hours	$2.655\ 2 \times 10^{3}$	ft · lbf
watt hours	$1.341\ 0 \times 10^{-3}$	hp · h

Plane Surfaces

PLANE SURFACES

Rectangle

Area = (base)(height) = ba

Diagonal (c) = $\sqrt{b^2 + a^2}$

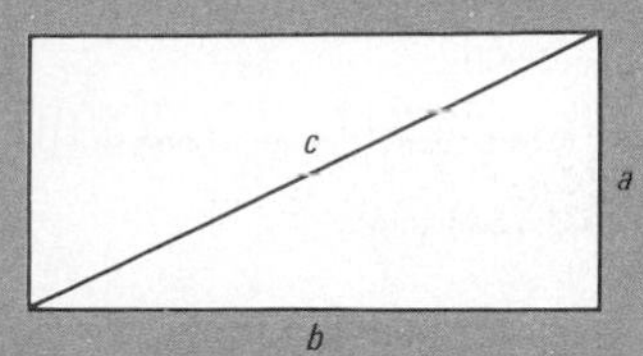

Right triangle

Area = $\frac{1}{2}$(base)(height) = $\frac{1}{2}ba$

$A + B + C = 180°$

Hypotenuse (c) = $\sqrt{b^2 + a^2}$

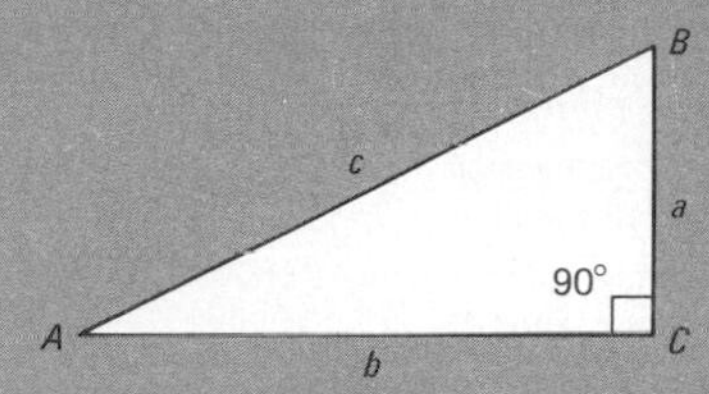

Any triangle

Area = $\frac{1}{2}$(base)(height) = $\frac{1}{2}bh$

Height (h) is perpendicular to base

$A + B + C = 180°$

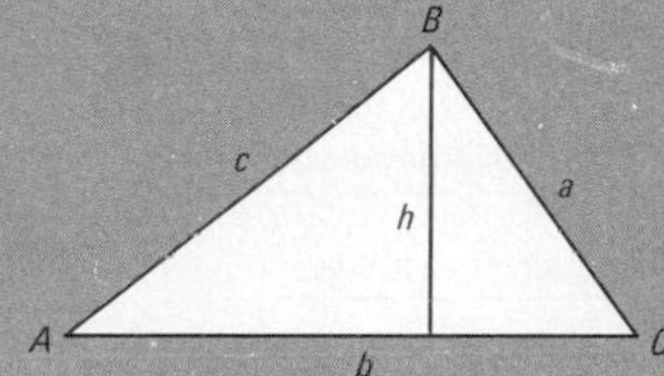

Equilateral triangle

Area = $\frac{1}{2}$(base)(height) = $\frac{1}{2}bh$

$A = B = C = 60°$

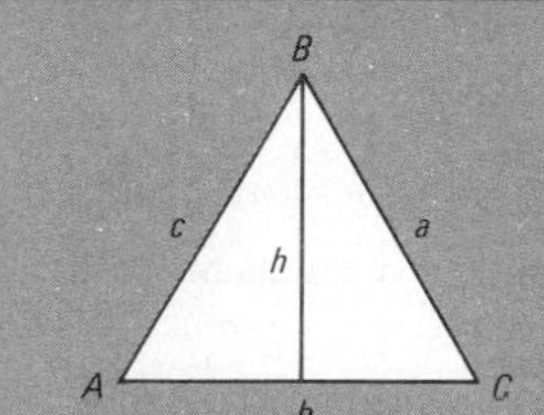

Isosceles triangle

Area = $\frac{1}{2}$(base)(height) = $\frac{1}{2}bh$

$A = C$

$A + B + C = 180°$

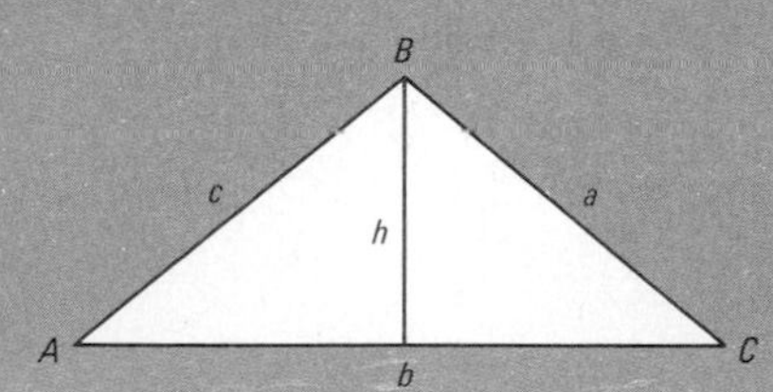

Parallelogram

Area = (base)(height)

Height (h) is perpendicular to base

$A + B + C + D = 360°$

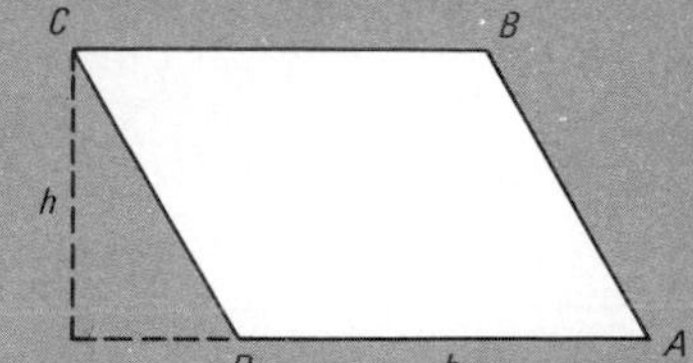

Trapezoid

Area = $\frac{1}{2}$(sum of parallel sides)(height)

Height (h) is perpendicular to parallel sides a and b

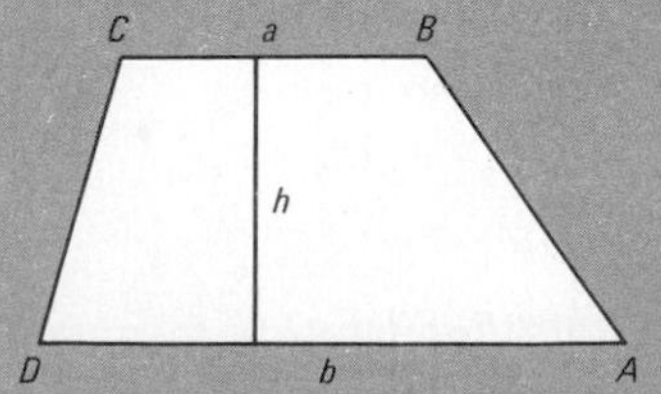

PLANE SURFACES

Regular polygon

Area $= \frac{1}{2}$(distance ST)(length one side)(number of sides)

Equal sides and angles

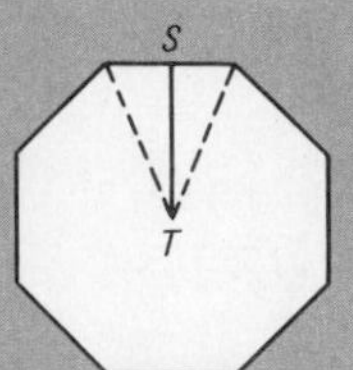

Circle

$$\text{Area} = \pi(CD)^2 = \frac{\pi(AB)^2}{4}$$

$$\pi = \frac{\text{circumference}}{\text{diameter}} = 3.14159...$$

$$\text{Circumference} = \pi(AB) = 2\pi(CD)$$

$$1 \text{ radian} = \frac{180^\circ}{\pi} = 57.2958^\circ$$

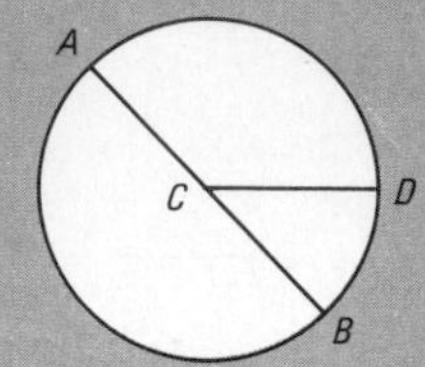

Sector of a circle

$$\text{Area} = \frac{(\text{arc length})(\text{radius})}{2}$$

$$= \frac{\pi(AB)^2(\text{angle } BAC)}{360^\circ}$$

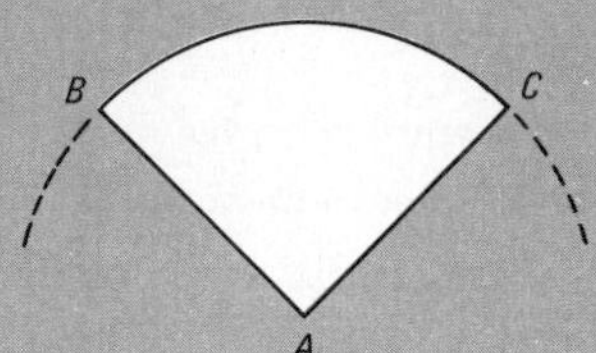

Segment of a circle

Area = area of sector (ABC) − area of triangle ABC

$$= \frac{(\text{radius})^2}{2}\left[\frac{\pi(\text{angle } BAC^\circ)}{180^\circ} - \sin BAC^\circ\right]$$

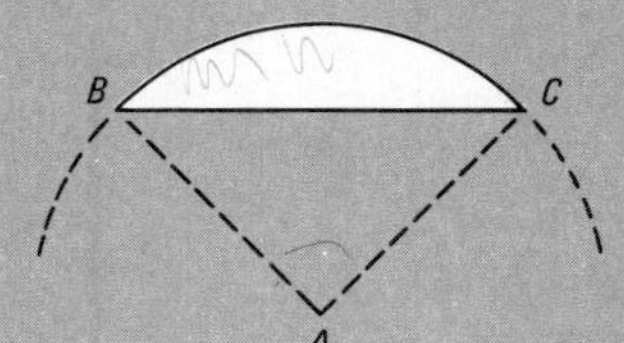

Ellipse

$$\text{Area} = \pi(AE)(ED)$$

$$= \frac{\pi}{4}(AB)(CD)$$

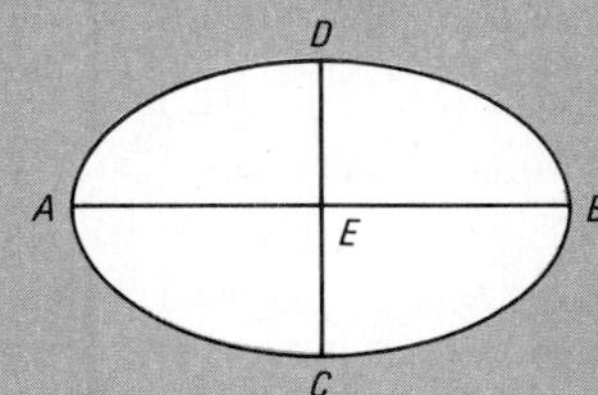

Rhombus

$$\text{Area} = \frac{1}{2}(c)(d)$$

$$4a^2 = c^2 + d^2$$

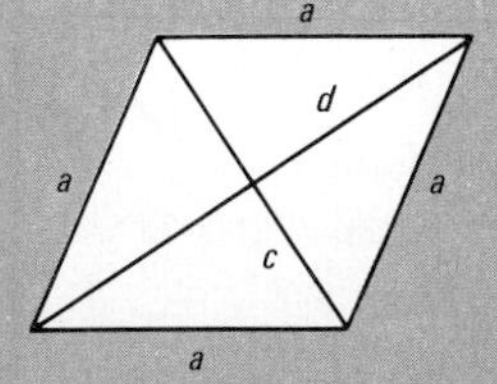

Sector of an annulus

$$\text{Area} = \frac{1}{2}\theta(R_1 + R_2)(R_1 - R_2)$$

$$= \frac{1}{2}h(P_1 + P_2)$$

$$= \frac{1}{2}h\theta(R_1 + R_2)$$

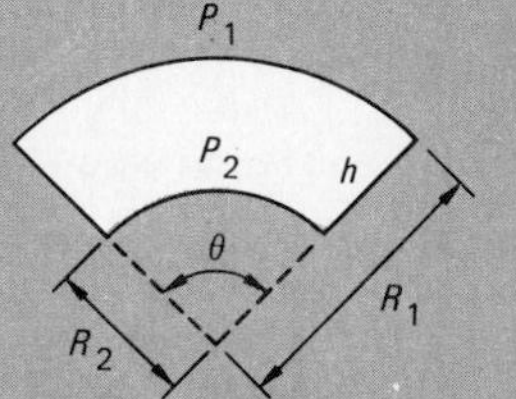

Rectangular prism

Volume = $(W)(B)(H)$

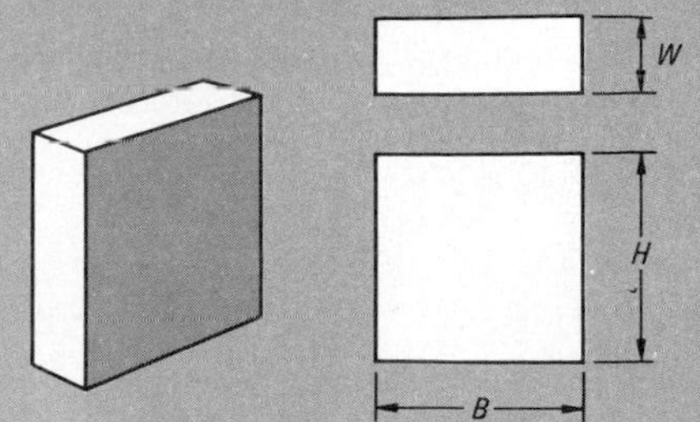

Pyramid

Applicable to base of any shape

Volume = $\frac{1}{3}$(area of base)(H)

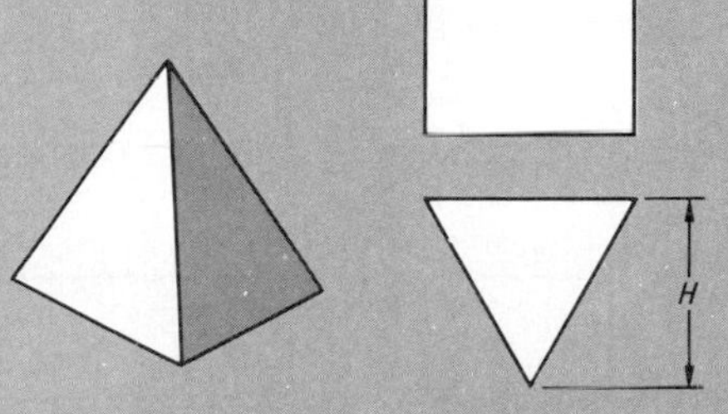

Cone

$$\text{Volume} = \frac{\pi}{12} D^2 H$$

$$= \frac{\pi}{3} R^2 H$$

$$\text{Surface area} = \frac{1}{2}(2\pi R)(S)$$

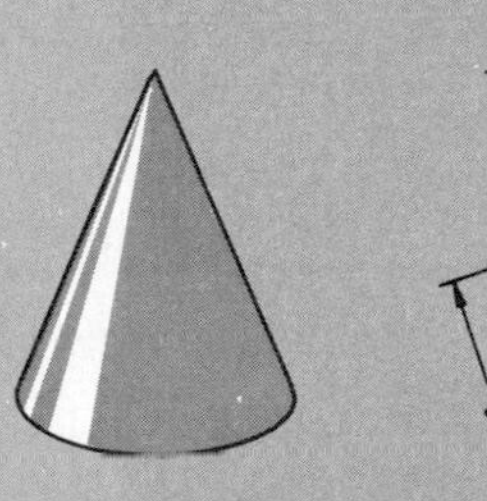

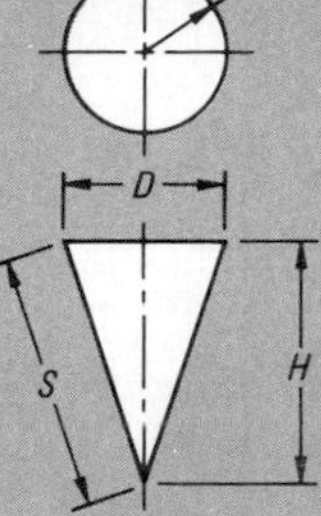

Cylinder

$$\text{Volume} = \pi R^2 H = \frac{\pi D^2}{4} H$$

Surface area = $2\pi RH$

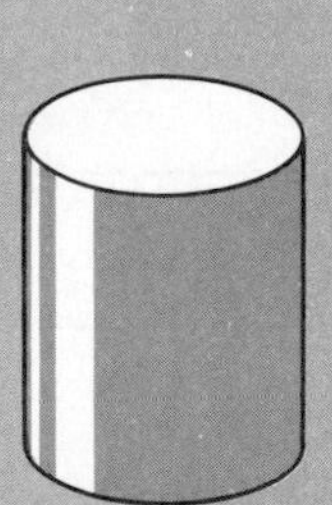

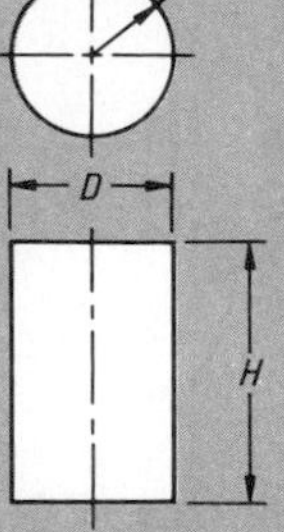

Hollow cylinder

$$\text{Volume} = \frac{\pi H}{4}(D_2^2 - D_1^2)$$

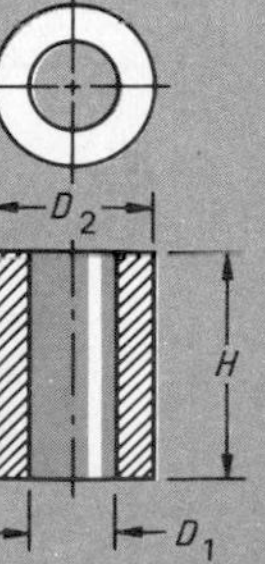

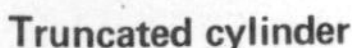

Truncated cylinder

Volume = $\frac{\pi}{8}D^2H$

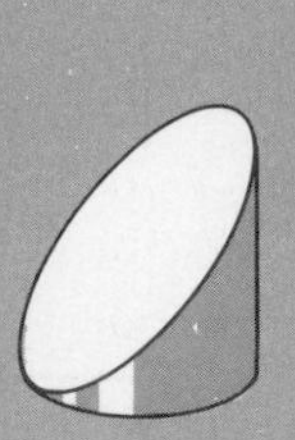
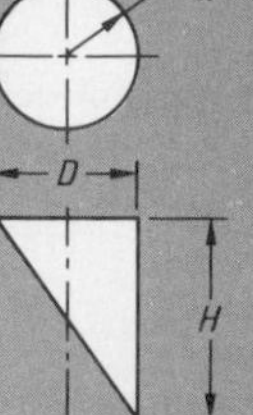

Sphere

Volume = $\frac{4}{3}\pi R^3$

$= \frac{\pi D^3}{6}$

Surface area = $4\pi R^2 = \pi D^2$

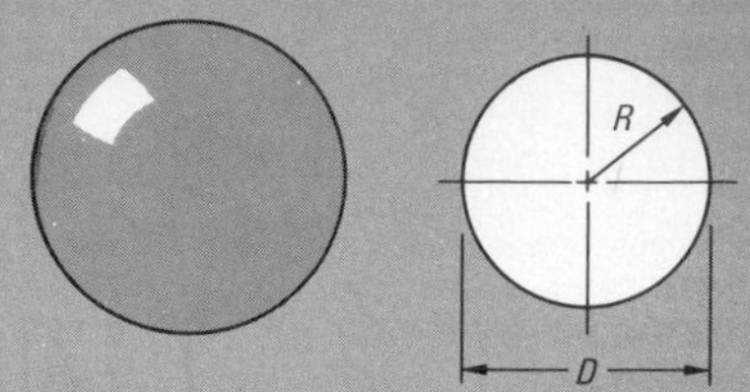

Spherical segment

Volume = $\pi H^2 (R - \frac{H}{3})$

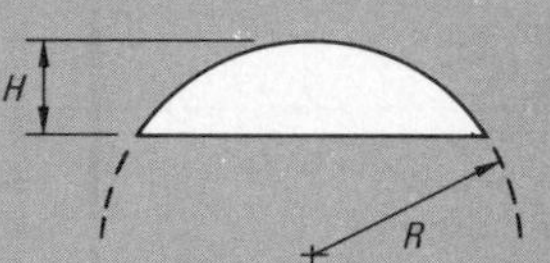

Hollow sphere

Volume = $\frac{4\pi}{3}(R_1^3 - R_2^3)$

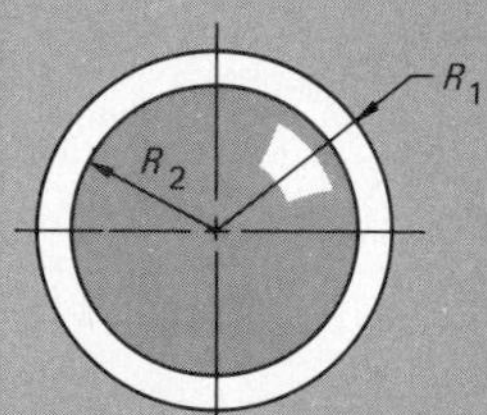

Torus

Volume = $\frac{1}{4}\pi^2 D_2^2 D_1$

Surface area = $\pi^2 D_1 D_2$

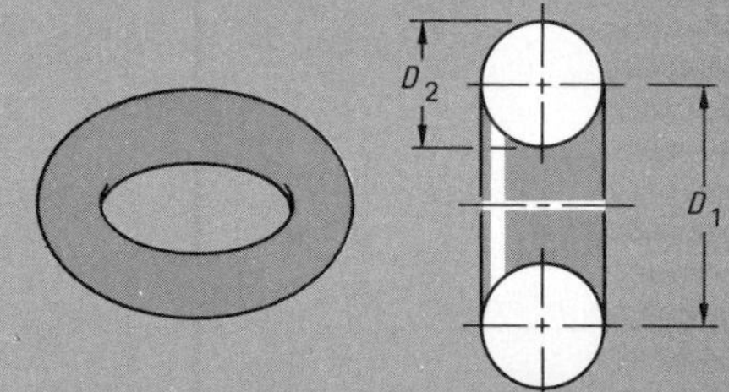

Hollow torus

Volume = $\frac{1}{4}\pi^2 D_1 (D_2^2 - D_3^2)$

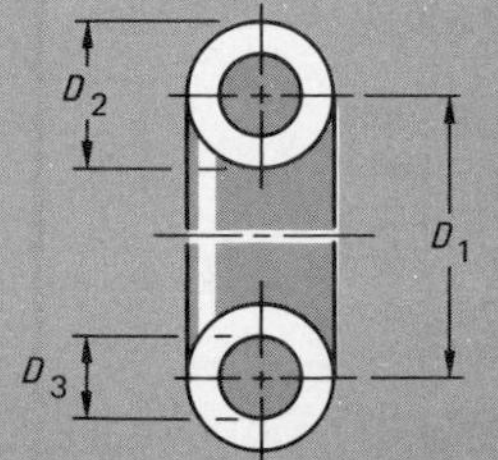

Answers to Selected Problems

2.2 $\measuredangle A = 105.5°$; $\measuredangle B = 34.5°$; $C = 50.1$ in

2.6 $\overline{V}_3 = 44.8$ mph $\angle 45°$

2.12 $\omega = 728$ rpm

2.17 Horizontal distance = 272 m
Vertical distance = 157 m

2.22 12.1 m

2.25 2.3 in

3.3 (*b*) emf $= 0.053t + 1.0$

3.8 (*b*) $R = 12.0A^{-0.964}$

3.13 (*b*) $P = 23.4e^{0.248(\text{month})}$

3.18 (*b*) $C = 5600e^{-0.075W}$
(*c*) $C = 124$ counts/second

4.2 (*c*) 3; (*f*) 2; (*i*) 4

4.5 (*a*) $Q = 46$; (*c*) $Y = -0.017$

4.6 (*b*) 80.4°

5.3 (*a*) 217 cm; (*b*) 17.8 ft^3; (*c*) 174°F; (*d*) 3788×10^5 cm^3;
(*e*) 1920 kW

5.7 (*a*) 1.22×10^3 N; (*b*) 1.22×10^3 N; (*c*) 950 N

5.11 $V_s = 1.073$ m^3; $m = 1.849 \times 10^6$ kg

5.18 $h_L = 1.28 \times 10^3 \dfrac{\text{ft-lbf}}{\text{lbm}} = 3.83 \times 10^3 \dfrac{\text{J}}{\text{kg}}$

5.23 (*a*) 2.12×10^5 mm^2; (*b*) 1.70 kg; (*c*) the ball will float

7.2 (*a*)

Weights	Frequency
206–212	2
213–219	1
220–226	6
227–233	15
234–240	4
241–247	2

(*c*) Median = 228.5 kg, Mode = 228 kg, Mean = 228.4 kg
(*d*) s = 8.27 kg

7.8 (*a*) 2%; (*b*) 2%; (*c*) 84%; (*d*) 84%

7.12 (*b*) $V = 3.72t + 1.69$; (*d*) $r^2 = 0.994$

7.16 (*b*) emf $= 0.0264t + 0.331$; (*d*) $r^2 = 0.997$

7.22 (*b*) $I = 4.01e^{-1.01t}$; (*d*) $r = -1$, $r^2 = 1$; (*e*) 0.532 A

9.2 $\bar{F}_y = 850$ N↑; $\bar{F}_x = 850$ N←

9.10 384 lb ↖101°

9.17 $M_A = 24\,600$ Nm ↺; $M_B = 15\,800$ Nm ↻

9.29 $\bar{R}_A = 1430$ lb→; $\bar{R}_B = 1650$ lb 30°↖

9.36 $\bar{R} = 290\hat{i} + 460\hat{j} + 180\hat{k}$, $\Theta_x = 60°$, $\Theta_y = 36°$, $\Theta_z = 72°$

9.43 (*a*) T = 13.2 kN; (*b*) d = 13.3 mm

10.2 16.1%

10.6 250 kg

10.11 23.1%

10.16 267 kg

10.21 Initial acid = 16.9 kg, Sulfuric acid = 56.2 kg, Nitric acid = 26.9 kg

11.2 1.40×10^5 C

11.6 180 W, 7.4 J/impact

11.11 (*b*) 9.6 Ω; (*c*) 1.2 A; (*d*) 0.77 A, 8.9 W; (*e*) 0.51W

11.15 (*a*) 71.3 V; (*b*) 1130 W; (*c*) 56.3 $\frac{J}{s}$; (*d*) 1.07 kW

12.4 20.7×10^6 J

12.8 55.2 mi/h

12.13 (*a*) 125 000 J; (*b*) 0.18 hp

12.19 6.18×10^4 Btu

12.25 330k = 59°C

12.29 (*a*) 660 W; (*b*) 1.61×10^4 gal

13.1 (*a*) \$7920.94; (*b*) \$6355.18; (*c*) \$4229.74

13.6 (*a*) \$36 341.54; (*b*) \$52 814.96

13.12 18.9% (4 year), 26.0% (3 year), 41.4% (2 year)

13.18 Using a present-worth comparison, A = \$39 854.95, B = \$42 481.02; conclusion: A is cheaper

13.23 (*a*) \$1141.46; (*b*) \$71.93

13.26 \$15 023.07, 8.09%

Selected Bibliography

Allen, Myron S., *Morphological Creativity,* Prentice-Hall, Englewood Cliffs, N.J., 1962.

Bassin, Milton G., Stanley M. Brodsky, and Harold Woloff, *Statics and Strengths of Materials,* McGraw-Hill, New York, 1969.

Beakley, George C., and Robert E. Lovell, *Computation, Calculators, and Computers,* Macmillan, New York, 1983.

Bullinger, Clarence E., *Engineering Economy,* McGraw-Hill, New York, 1958.

Burghardt M. David, *Introduction to the Engineering Profession,* Harper Collins, New York, 1991.

Chapra, Steven C., and Raymond P. Canale, *Numerical Methods for Engineers,* 2nd ed., McGraw-Hill, New York, 1988.

Craver, W. Lionel, Jr., Darrell C. Schroder, and Anthony J. Tarquin, *Introduction to Engineering and Technology,* Holt, Rinehart and Winston, New York, 1988.

Crosby, Philip B., *Quality is Free,* McGraw-Hill, New York, 1979.

DeJong, Paul S., James S. Rising, and Maurice W. Almfeldt, *Engineering Graphics,* 6th ed., Kendall/Hunt, Dubuque, Iowa, 1983.

Deming, W. Edwards, *Out of the Crisis,* MIT Center for Advanced Engineering Study, Cambridge, Mass., 1986.

Eide, Arvid R., Roland D. Jenison, Lane H. Mashaw, Larry L. Northup, and C. Gordon Sanders, *Engineering Graphics Fundamentals,* 2nd ed., McGraw-Hill, New York, 1995.

Erickson, William H., and Nelson H. Bryant, *Electrical Engineering, Theory and Practice,* 2nd ed., Wiley, New York, 1959.

Freund, John, *Modern Elementary Statistics,* 3rd ed., Prentice-Hall, Englewood Cliffs, N.J., 1967.

Gajda, Walter J., and William E. Biles, *Engineering: Modeling and Computation,* Houghton Mifflin, Boston, 1978.

Gordon, William J. J., *Synectics,* Harper, New York, 1961.

Grant, Eugene L., W. Grant Ireson, and Richard S. Leavenworth, *Principles of Engineering Economy,* 6th ed., Ronald, New York, 1976.

Harrisberger, Lee, *Engineersmanship,* 2nd ed., Brooks/Cole, Monterey, Calif., 1982.

Hogan, R. C., and D. W. Champagne, *Personal Style Inventory,* Annual Handbook for Group Facilitators, 1980.

Jayaraman, Sundaresan, *Computer-Aided Problem Solving for Scientists and Engineers,* McGraw-Hill, New York, 1991.

Kearnes, David T., and David A. Nadler, *Prophets in the Dark: How Xerox Reinvented Itself and Beat Back the Japanese,* Harper Business, New York, 1992.

Osborn, Alex F., *Applied Imagination,* Scribner, New York, 1963.

"A Report of the Total Quality Leadership Steering Committee and Working Councils," November, 1992.

Roberts, H. V., and B. F. Sergesketter, *Quality Is Personal,* Macmillan, New York, 1993.

Scarl, Donald, *How to Solve Problems,* 2nd ed., Dosoris Press, Glen Cove, N.Y., 1990.

Scholtes, P. R., *The Team Handbook,* Joiner Associates, Inc., 1988.

Smith, Gerald W., *Engineering Economy: Analysis of Capital Expenditures,* Iowa State University Press, Ames, Iowa, 1973.

"Standard for Metric Practice," American National Standards Institute E 388-76 268-1976.

Theusen, H. G., W. J. Fabrycky, and G. L. Thuesen, *Engineering Economy,* Prentice-Hall, Englewood Cliffs, N.J., 1977.

White, John A., *Principles of Engineering Economic Analysis,* 3rd ed., Wiley, New York, 1989.

Index